Essentials of Human Histology

MW01609224

Essentials of Human Histology

Second Edition

William J. Krause, Ph.D.

Professor, Department of Pathology and Anatomical Sciences,
University of Missouri School of Medicine, Columbia, Missouri

Little, Brown and Company
Boston   New York   Toronto   London

Copyright © 1996 by Little, Brown and Company (Inc.)

Second Edition
Previous edition copyright titled *Essentials of Histology,*
© 1994 by Little, Brown and Company (Inc.)

All rights reserved. No part of this book may be reproduced in
any form or by any electronic or mechanical means, including
information storage and retrieval systems, without permission in
writing from the publisher, except by a reviewer who may quote
brief passages in a review.

Library of Congress Cataloging-in-Publication Data
Krause, William J.
    Essentials of human histology / William J. Krause. — 2nd ed.
      p. cm.
    Rev. ed. of: Essentials of histology / William J. Krause, J. Harry
Cutts. 1st ed. c1994.
      Includes index.
      ISBN 0-316-50336-3
      1. Histology.   2. Histology—Examinations, questions, etc.
I. Krause, William J. Essentials of histology.   II. Title.
    [DNLM: 1. Histology. QS 504 K913e 1996]
QM551.K893 1996
611'.018—dc20
DNLM/DLC
for Library of Congress   95-45641
                                CIP

Printed in the United States of America
MV-NY

Editorial: Evan R. Schnittman, Suzanne Jeans
Production Services: Textbook Writers Associates, Inc.
Cover Designer: Hannus Design Associates
Cover Artist: Peg Gerrity

Contents

Preface

The format of *Essentials of Human Histology*, Second Edition, departs considerably from that of usual histology textbooks. In general, this text follows the traditional and logical sequence of cells to tissues to organs, but within this sequence the chapters on the eye and ear have been placed immediately after the coverage of nerve tissue. This seems a more logical arrangement because these two structures essentially are specialized nerve receptors. Similarly, it is more appropriate to consider mitosis with the cell and meiosis with the reproductive organs.

Use of the Text

To understand human histology, it is essential to learn a specialized vocabulary and to assimilate a large body of facts. Learning, as distinct from memorization, depends to a great degree on repetition and reinforcement and is made easier if the material to be learned can be presented in discrete, manageable segments. The format of *Essentials of Human Histology* is designed specifically to meet these requirements and, if used properly, will enable the student to master the discipline quickly and efficiently.

The subject matter is broken down into small learning units, each of which is introduced by a list of **key words** appropriate to that unit. The key words introduce the main features of the subject to be discussed and provide the basic vocabulary for that unit. As each segment is read, note the key words (identified by bold print) and how they contribute to the discussion. After completing the text segment, return to the key words, using them as **prompts** to recall the details of the material just read. The key words serve as a summary of the topic and provide a means for rapid review. If a key word fails to prompt a response, it and its associated text can be found quickly from the **bold** type in the appropriate segment.

A **review table** follows the text of each chapter and presents key structural features of identification of the tissue or organ presented, offering the briefest possible review.

Functional reviews briefly outline the structural/functional relationships and serve to draw the information together and to provide an additional review of the topic.

Each chapter concludes with a series of questions and answers to assist student understanding of selected topics. As a final short review, the **appendix** provides a few short tables that outline the basic differences of several structures that are frequently difficult to recognize.

During preparation of this text, I kept in mind three major considerations: (1) most curricula place considerable time constraints on the student; (2) function and structure are inextricably related; and (3) the learning process essentially is a matter of repetition and reinforcement. This text presents the vast material of histology in a concise and logical manner, without sacrificing the detail that is necessary for an understanding of the microscopic structure of the body. *Essentials of Human Histology*, Second Edition, is written to meet students' needs both within a classic classroom and a problem-based learning setting. In addition, this text is especially designed for reviewing for **USMLE** examinations. A series of discussion and objective questions is provided for each chapter to aid the student in this regard.

W.J.K.

Credits

The author and publisher gratefully acknowledge the permission of S.T.E.M. Laboratories, Inc. for the use of several figures in the text, developed previously by W.J. Krause as 2 x 2 color projection slides in conjunction with S.T.E.M. Laboratories, Inc., Kansas City, Missouri.

The following publishers provided permission to use previously published illustrative materials (listed by author):

Andrews, P., Georgetown University, Washington, DC
Figure 16-4, from *American Journal of Anatomy* 140:81, 1974, Wistar Press.
Figures 16-6, 16-8, 16-9 and 16-16, from *Laboratory Investigation* 32:610, 1975, © U.S.-Canadian Division of the International Academy of Pathology.
Berdan, R.C., University of Calgary, Calgary, Canada
Figure 15-28, from *Journal of Ultrastructural Research* 90:55, 1985, Academic Press.
Breipohl, W., Univ des Gesamthochschule, Essen, West Germany
Figure 14-6, from *Cell and Tissue Research* 183:105, 1977, Springer-Verlag.
Brink, P.R., State University of New York, Stony Brook, New York
Figure 2-11, from *American Journal of Anatomy* 154:1, 1979, Wistar Press.
Fujita, T., Niigata University, Asahi-Machi, Niigata, Japan
Figures 12-6, 12-7, and 16-11, from *Archivum Histologicum Japonicum* vol 37, 1974, Japan Society of Histological Documentation.
Haggis, G.H., Canada Department of Agriculture, Ottawa, Canada
Figure 15-42, from *Laboratory Investigation* 29:60, 1973, © U.S.-Canadian Division of the International Academy of Pathology.
Hirakow, R., Saitama Medical School, Iruma-Gun, Saitama, Japan
Figures 7-17 through 7-19, from *Electron Microscopy Technique,* Ishiyaku Publishers, Inc., Tokyo, Japan.
Huxley, H.E., University Medical School, Cambridge, England

Figures 7-4 and 7-8, from *Scientific American* vol. 199, 1958, Scientific American.
Köling, A., Uppsala University, Uppsala, Sweden
Figure 1-19, from "Freeze fracture electron microscopy of the nuclear envelope of the human odontoblast." *Archives of Oral Biology* 30:691, Copyright © 1985, Pergamon Journals Ltd.
Kuwabara, T., National Eye Institute, NIH, Bethesda, Maryland
Figure 9-8, from *Histology* by Weiss and Greep. Copyright © 1977 McGraw-Hill. Used with permission of McGraw-Hill Book Company.
Lenoir, M., University of Montpellier, Montpellier Cedex, France
Figures 10-14 and 10-15, from *Anatomy and Embryology* 175:477, 1987, Springer-Verlag.
Magney, J.E., University of Minnesota, Minneapolis, Minnesota
Figures 15-30 and 15-37, from *American Journal of Anatomy* 177:47, 1986, Alan R. Liss, Inc.
Murakami, T., Okayama University, Medical School, Okayama, Japan
Figure 16-3, from *Archivum Histologicum Japonicum* vol. 34, 1972, Japan Society of Histological Documentation.
Stranock, S.D., Merchiston Castle School, Edinburgh, Scotland
Figure 1-13, from *Journal of Anatomy* 129:885, 1979, Cambridge University Press.
Trotter, J.A., University of New Mexico, Albuquerque, New Mexico
Figures 7-15 and 7-16, from *Anatomical Record* 213:16 and 213:26, 1985, Alan R. Liss, Inc.
Tyson, G.E., Mississippi State University, Mississippi State, Mississippi
Figure 16-7, from *Vichows Archiv. B. Cell Pathology* 25:105, 1977, Springer-Verlag.

The following publishers provided permission to use illustrative materials that were previously published in their journals:

British Medical Association
    Figure 15-19, from *British Medical Journal* 2:1052, 1977.
Cambridge University Press—all from *Journal of Anatomy*
    Figure 2-9, 104:467, 1969;
    Figure 2-10, 125:85, 1978;
    Figure 2-21, 122:449, 1976.
The Wistar Press
    Figure 14-15, from *American Journal of Anatomy* 146:181, 1976.

Thanks to the following individual contributors for the illustrative materials they provided:

Allison, O.L., S.T.E.M. Laboratories, Inc., Kansas City, Missouri
    Figures 4-4, 4-15, 5-2, 5-3, 5-4, 5-9, 15-3

Dow, P., University of British Columbia, Vancouver, Canada
    Figure 7-14
Engstrom, H.H., Uppsala University, Uppsala, Sweden
    Figure 10-7
Feeney-Burns, L., University of Missouri, Columbia, Missouri
    Figures 9-9, 9-11, 9-13
King, J.S., Ohio State University, Columbus, Ohio
    Figures 8-3, 8-15
Lenoir, M., University of Montpellier, Montpellier Cedex, France
    Figure 10-8
Stachura, J., Institute of Pathology, Copernicus Medical Academy, Krakow, Poland
    Figures 5-7, 11-6

Essentials of Human Histology

1 The Cell

Multicellular organisms are made up of two distinct structural elements: cells and those products of cells, that form the intercellular substances. Cells are the fundamental units of living material and show a variety of functional specializations that are essential for the survival of the organism. Each cell is a distinct entity; each contains all the machinery necessary for independent existence and is separated from its surroundings by an individual plasma membrane.

Protoplasm

KEY WORDS: protein, carbohydrate, lipid, nucleic acid, inorganic material, water

Protoplasm, the living substance of a cell, consists of protein, carbohydrate, lipid, nucleic acids, and inorganic materials dispersed in water to form a complex, semifluid gel. The consistency of protoplasm differs in different cells and may change from a viscous to a more fluid state.

Protein, alone or combined with lipid or carbohydrate, forms the major structural component of the cell and the intercellular substances. Enzymes and many hormones are proteins. The major **carbohydrates** are glucose, which is the chief source of energy in human cells, and glycogen, the storage form of glucose. Complexes of carbohydrates and proteins form the main constituents of intercellular substances that bind cells together. Other carbohydrate-protein complexes form some enzymes and antibodies. **Lipids** serve as an energy source; they also have important structural functions and are major components of the membrane systems of cells.

Nucleic acids are divided into two classes. Deoxyribonucleic acid (DNA) represents the genetic material and is found primarily but not exclusively in the nucleus. Ribonucleic acid (RNA) is present in the cytoplasm and nucleus and carries information from the nucleus to the cytoplasm. It also serves as a template for synthesis of proteins by the cell.

Inorganic materials are as much an integral component of protoplasm as proteins, carbohydrates, and lipids; without them physiologic processes are impossible. Among the inorganic constituents of protoplasm are calcium, potassium, sodium, and magnesium present as carbonates, chlorides, phosphates, and sulfates; small quantities of iron, copper, and iodine; and trace elements such as cobalt, manganese, zinc, and other metals. The inorganic materials have many functions, including maintenance of intracellular and extracellular osmotic pressures, transmission of nerve impulses, contraction of muscle, adhesiveness of cells, activation of enzymes, transport of oxygen, and maintenance of the rigidity of tissues such as bone.

Water makes up about 75 percent of protoplasm. Part of the water is free and available as a solvent for various metabolic processes, and part is bound to protein.

Properties of Protoplasm

KEY WORDS: irritability, conductivity, contractility, absorption, metabolism, secretion, excretion, growth, reproduction

Protoplasm is characterized by several physiologic properties that distinguish it from inanimate material. All living cells show these properties, but in some cells a particular property may be emphasized.

Irritability is a fundamental property of all living cells and refers to the ability to respond to a stimulus.

Conductivity refers to the ability of a cell to transmit a stimulus from the point of origin to another point on the cell surface or to other cells. Conductivity also is a property of all cells but, like irritability, is most highly developed in nerve tissue.

Contractility is the ability of a cell to change shape in response to a stimulus and generally is indicated by a shortening of the cell in some direction. This property is most prominent in muscle.

Absorption involves transfer of materials across the cell membrane into the interior of the cell, where it might be used in some manner. All cells show the ability to absorb material, some very selectively.

Metabolism refers to the ability of a cell to break down absorbed material to produce energy.

Secretion is the process by which cells elaborate and release materials for use elsewhere. **Excretion,** on the other hand, is the elimination from cells of metabolic waste products.

Growth of an organism can occur by increasing the amount of cytoplasm in existing cells or by increasing the number of cells. There are limits to the size a cell can attain without sacrificing the efficiency with which nutrients and oxygen reach the interior of the cell. Beyond a maximum size, increase in the amount of protoplasm occurs through **reproduction** (division) of cells.

Organization of Protoplasm

KEY WORDS: nucleus, nuclear envelope, cytoplasm, plasmalemma (cell membrane), organelle, inclusion, cytoskeleton, cytoplasmic matrix (cytosol), karyolymph

Although cells differ in size, shape, and function, the protoplasm of each cell consists of two major components: nucleus and cytoplasm. The **nucleus** contains the hereditary or genetic material and is completely surrounded by cytoplasm, from which it is separated by a **nuclear envelope.** The **cytoplasm** is limited by a **plasmalemma (cell membrane),** which separates the cell from the external environment. Within the cytoplasm are several structures representing organelles and inclusions. **Organelles** are highly organized, living structural units of the cytoplasm that perform specific functions in the cell. **Inclusions** generally represent inert cell products or metabolites that often are only temporary components. Most, but not all, of the organelles are membranous structures whose size and concentration vary with the type and activity of the cell. The different organelles tend to be localized in discrete areas of the cytoplasm so that they and their associated metabolic processes remain separate from other components of the cell. Cell shape is maintained by a three-dimensional **cytoskeleton** that provides structural support for the cell and serves in cell motility and intracellular transport. The cytoskeleton consists of various microfilaments and microtubules.

The organelles and inclusions are suspended in an amorphous medium called the **cytoplasmic matrix (cytosol).** Nuclear material also is suspended in a structureless ground substance called the **karyolymph.** Other than their differences in location, karyolymph and cytosol appear to be equivalent.

Cytoplasmic Organelles

Organelles are specialized units of the cell that perform specific functions and constitute part of the living substance of the cell. Organelles include structures such as the plasmalemma, granular and smooth forms of endoplasmic reticulum, ribosomes, Golgi complexes, mitochondria, lysosomes, peroxisomes, and centrioles. Many organelles are limited by membranes that are similar in structure to the boundary of the cell itself. These membranes, including the cell membrane, are metabolically active sheets that are essential to the life of the cell. In electron micrographs, the membranes exhibit a trilaminar structure consisting of inner and outer dense lines separated by a light zone. Because the trilaminar structure is representative of all biologic membranes of a cell, it has been called a *unit membrane.* The thickness of the unit membrane varies from organelle to organelle and generally is greatest where it forms the plasmalemma. While the membranes of different organelles appear to have only minor cytologic variations, they vary considerably in the chemical composition, enzymatic properties, and functions.

Plasmalemma (Cell Membrane)

KEY WORDS: unit membrane, phospholipid bilayer, integral membrane protein, peripheral membrane protein, transmembrane protein, glycocalyx, endocytosis, phagocytosis, pinocytosis, micropinocytosis, fluid-phase micropinocytosis, transcytosis, adsorptive micropinocytosis, coated pit, coated vesicle, clathrin, endosome, exocytosis, regulated exocytosis, constitutive exocytosis

Each cell is enclosed by a plasmalemma that is 8 to 10 nm thick and has the typical trilaminar structure of the **unit membrane.** It appears as two dense lines, each 3 nm wide, separated by a 2- to 4-nm clear zone. The plasmalemma is composed of proteins, lipids, glycolipids, and carbohydrates. Lipid forms about 50 percent of the mass of the plasmalemma and consists of three major classes of lipid: phospholipids, glycolipids, and cholesterol. Four major types of phospholipid have been identified in the plasmalemma. These are sphingomyelin, phosphatidylserine, phosphatidylcholine, and phosphatidylethanolamine. About one-half the lipid content of the plasmalemma is phospholipid. Each phospholipid molecule has a polar head region that is charged and an uncharged, nonpolar tail region that consists of two chains of fatty acids. The polar heads are hydrophilic and so lie on both the inner and outer surfaces of the plasmalemma. The nonpolar tail regions are hydrophobic and line up in the center of the plasmalemma. The result is two parallel layers of phospholipid molecules lying tail to tail and forming a **phospholipid bilayer** in which the hydrocarbon (fatty acid) chains are directed inward and the polar groups are directed outward (Fig. 1-1). The plasmalemma acts as a fluid due to the lateral move-

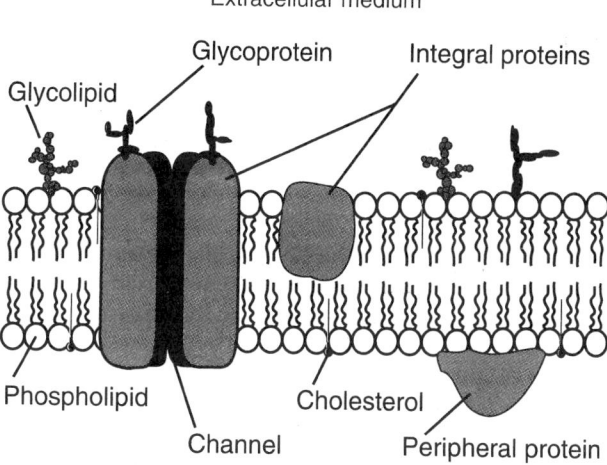

Fig. 1-1. Fluid mosaic model of the plasma membrane. Lipids are arranged in a bilayer. Integral proteins are embedded in the bilayer and often span it, forming a channel. Peripheral proteins do not penetrate the bilayer.

ment of adjacent phospholipid molecules. Therefore breaks and tears seal spontaneously due to the polar nature of the phospholipids. Cholesterol molecules within the plasmalemma tend to limit the lateral fluid movement of the phospholipids and contribute to the mechanical stability of the plasmalemma. Because of the polar lipid composition of the plasmalemma, it is highly permeable to substances such as oxygen, water, nitrogen, and small uncharged molecules. It is impermeable to charged ions such as potassium, sodium, chloride, calcium, larger uncharged molecules, and all charged large molecules. Entry of these substances depends on integral proteins of the plasmalemma. Glycolipids — lipids with attached sugar residues — are a more minor component of lipids comprising the plasmalemma. Sugar residues of this class of lipid usually face the external surface of the plasmalemma and contribute to the negative charge of the cell surface. Glycolipids are important in cell-to-cell and cell-to-interstitial matrix interactions. They also may play a role in immune reactions. Human glycolipids are produced primarily from ceramide and are called *glycosphingolipids*. Important glycolipids of the plasmalemma are gangliosides and galactocerebrosides, major components of nerve cell membranes and myelin, respectively.

Although membrane lipids form the foundation of the bilayered structure of the plasmalemma, membrane proteins are primarily responsible for its specialized func-

tions. The membrane proteins are able to move laterally in the lipid bilayer over the surface of the cell if they are not bound to filaments in the underlying cytoplasm. Membrane proteins function to transport molecules into or out of cells (membrane pump proteins, ion-channel proteins, carrier proteins), act as receptors for chemical signals between cells (hormone receptors) and generate messenger molecules that diffuse into the cytoplasm, attach elements of the cytoskeleton to the plasmalemma, attach cells to the extracellular matrix (cell adhesion molecules), or may even possess specific enzymatic activity when stimulated.

There are two basic types of membrane proteins: integral and peripheral proteins (see Fig. 1-1). **Integral membrane proteins** are firmly embedded in the lipid bilayer and cannot be removed. **Peripheral membrane proteins** are defined as those proteins which can be removed from the plasmalemma without disrupting the lipid bilayer. Various subtypes of integral and peripheral membrane proteins exist. Some integral proteins are **transmembrane proteins** that span the entire width of the plasmalemma and protrude from both surfaces. This type of integral protein has three parts: a region to the cell exterior, a region passing through the lipid bilayer, and a region to the interior of the cell. The amino acids of this type of integral protein form an alpha-helix, and the protein is referred to as a *tripartite single-pass integral membrane protein*. Not all transmembrane proteins are of this type. Transmembrane proteins that make multiple passes through the plasmalemma also occur, and all transporters and ion channels identified thus far are multipass transmembrane proteins. Specific transmembrane proteins occur in areas of the plasmalemma specialized for attachment to other cells or the extracellular matrix. Here they pass through the lipid bilayer and link cells together or anchor the cell to the extracellular matrix. Not all integral proteins are transmembrane proteins.

Peripheral membrane proteins are generally attached to the surface of the plasmalemma — usually the inner surface — and contribute to its stability. Peripheral membrane proteins can attach to the surface of the plasmalemma by ionic interactions with an integral protein, another peripheral membrane protein, or interaction with the polar head groups of the phospholipids. Examples of peripheral membrane proteins are spectrin and ankyrin, which are found on the cytoplasmic surface of the erythrocyte plasmalemma. Both function to anchor elements of the cytoskeleton to the cytoplasmic surface of the plasmalemma. Similar peripheral membrane proteins are present in most other cells. The membrane proteins include proteoglycans. The protein core of this molecule spans the lipid bilayer, and the portion of the long molecule bearing the carbohydrate side chains projects from the exterior surface of the plasmalemma. The sugar residues of the carbohydrate portion of these molecules, as well as glycoproteins and glycolipids, form the fuzzy coat observed by electron microscopy that is referred to as the **glycocalyx.** Such a coat is present on all cells, and the ionized carboxyl and sulfate groups of the polysaccharides units give it a strong negative charge. The glycocalyx plays an important role in determining the immunologic properties of the cell and its relationships and interactions with other cells. Carbohydrates offer far greater structural diversity for recognition than do proteins. The infinite variety of molecular configurations of the subunits of the large polysaccharides that extend from the plasmalemma forms the basis for cell recognition. Some integral membrane proteins consist of a polypeptide chain linked to a carbohydrate (sialic acid) and function as adhesion molecules; these are called *cell-adhesion molecules* (CAMs). Several CAMs have been identified: one for neurons, one for hepatocytes, one for muscle, one for the adhesion of glial cells to neurons, and several others.

Thus the plasmalemma is a selectively permeable membrane in which ions and small water-soluble molecules (amino acids, glucose) must be pumped through protein-lined channels that traverse the plasmalemma to gain access to the cell interior. The most common protein channels are voltage-gated ion channels that require a transmembrane potential to open, receptors that require binding of a specific molecule to open, and mechanically gated channels that sense movement in the plasmalemma. The channel proteins undergo an allosteric change that opens the channel when stimulated. Thus the movement of solutes across the plasmalemma depends on the activity of specific transmembrane transport proteins. Movement of a single solute (molecule) by transmembrane transport proteins is referred to as a *uniport mechanism*. The movement of two or more solutes across the plasmalemma in the same direction involves a *symport* or *cotransport mechanism.* Coupled transport involving the movement of two or more solutes, but in opposite directions across a cell membrane, is referred to as a *countertransport* or *antiport mechanism.*

The plasmalemma also plays an active role in bringing macromolecular materials into the cell (**endocytosis**), as well as discharging materials from the cell.

Phagocytosis is a form of endocytosis in which particulate matter is taken into a cell. During its attachment phase, particles bind to receptors on the plasmalemma, while in its ingestion phase, the cytoplasm forms pseudopods that flow around the particles, engulf them, and take them into the cell in membrane-bound vacuoles. In a similar way, fluid may be incorporated into the cell in small cytoplasmic vesicles in a process called **pinocytosis.** Phagocytosis and pinocytosis can be seen by light microscopy.

Some materials may enter the cell in even smaller vesicles formed by minute invaginations of the cell surface. This process, called **micropinocytosis,** is visible only with the electron microscope. Nonselective fluid-phase and selective adsorptive types of micropinocytosis have been recognized (Fig. 1-2). In most cells, **fluid-phase micropinocytosis** occurs with formation of small vesicles from the plasmalemma. The vesicles, which contain fluid and anything dispersed in the fluid, traverse the cytoplasm to the opposite side of the cell, where they discharge their contents (Figs. 1-2A and 1-3). In certain cells, such as endothelial cells, this process is termed **transcytosis** and is an important mechanism for moving material across a cellular barrier to the extracellular spaces.

Adsorptive micropinocytosis is the selective uptake of specific macromolecules at certain receptor-binding sites in the cell membrane. This receptor-mediated micropinocytosis is important in the ingestion of regulatory and nutritional proteins. Short, bristle-like projections may be present on the cytoplasmic surface at these sites and form **coated pits** from which **coated vesicles** arise (Fig. 1-2B). The cytoplasmic surfaces of the vesicles are coated with **clathrin,** a protein that appears as radiating spikes that give a fuzzy appearance to the vesicles. Clathrin may prevent fusion of coated vesicles with membranous organelles. Following the formation of the coated vesicle, the clathrin coat is lost and the vesicles fuse with preexisting vacuoles to form structures called **endosomes.** As the molecules being transported are released from these early endosomes into the cytosol, portions of the endosome membrane containing unoccupied receptors bud off as small vesicles, return to the plasmalemma, and fuse with it. Thus many receptors are recycled. Late endosomes fuse with lysosomes and are broken down. Then their products are released into the cytoplasm.

Materials such as secretory granules are released from the cell by **exocytosis,** a process in which the limiting membranes of the granules fuse with the plasmalemma before discharging their contents. In this way, a breach in the limiting plasmalemma is avoided. The excess membrane incorporated into the plasmalemma during exocytosis is removed by endocytosis of small vesicles. Two types of exocytosis are known to occur: regulated and constitutive. **Regulated exocytosis** is stimulus-dependent and occurs in cells specialized for the release of a large volume of secretory material. **Constitutive exocytosis** occurs in cells in which the secretory product is not concentrated in a number of large granules but is released continuously in small vesicles.

Fig. 1-2. A. Diagrammatic representation of fluid-phase micropinocytosis. B. Adsorptive micropinocytosis via coated vesicles.

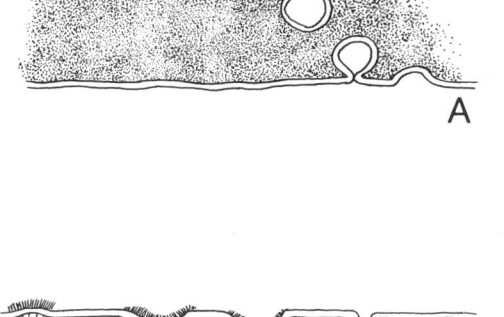

Ribosomes

KEY WORDS: **ribonucleoprotein, free ribosome, polyribosome (polysome), messenger RNA, protein synthesis, codon, transfer RNA**

Ribosomes are small, uniformly sized particles of **ribonucleoprotein** 12 to 15 nm in diameter and composed of large and small subunits. They may be attached to the membranes of the endoplasmic reticulum or be present as

Fig. 1-3. A portion of a vascular capillary illustrating the nucleus and cytoplasm of an endothelial cell (human connective tissue). Note the limiting basal lamina (TEM, × 6,000).

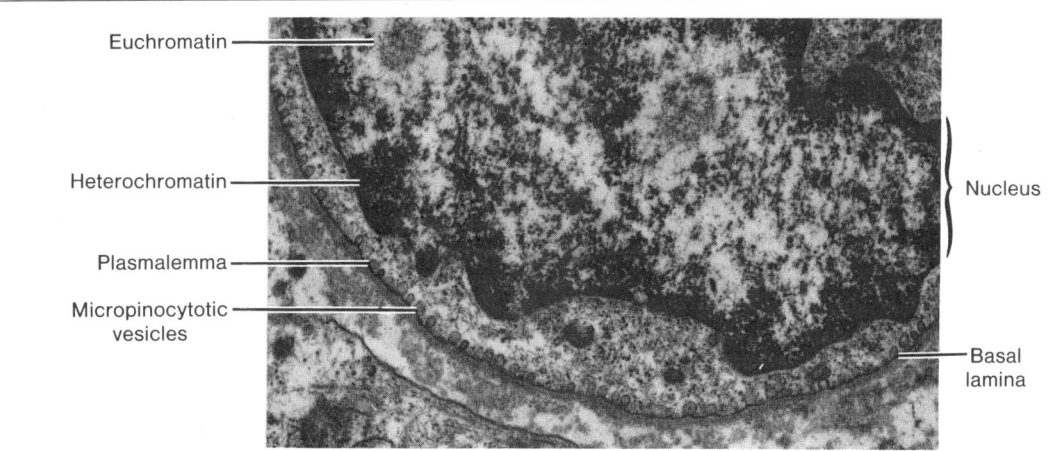

free ribosomes suspended in the cytoplasm, unassociated with membranes (Fig. 1-4). Free ribosomes often occur in clusters called **polyribosomes (polysomes),** in which individual ribosomes are united by a thread of ribonucleic acid (RNA) called **messenger RNA** (mRNA). Free ribosomes are sites of **protein synthesis,** the protein formed being used by the cell itself rather than secreted. Free ribosomes synthesize cytosolic proteins, peripheral membrane proteins, and proteins destined for use by mitochondria, peroxisomes, and the nucleus. Individual free ribosomes are not active; only when they are attached to mRNA to form polyribosomes do they become active in protein synthesis. Similarly, ribosomes on the endoplasmic reticulum must be associated with mRNA before they engage in the synthesis of protein.

Messenger RNA is formed in the nucleus on a template of uncoiled deoxyribonucleic acid (DNA). It contains a message that is encoded as successive sets of three nucleotides called **codons** that specify the sequence in which amino acids are to be incorporated into newly forming protein. During synthesis, mRNA enters the cytoplasm and attaches to ribosomes that move along the mRNA, translating the code and assembling amino acids in the proper order. On reaching the end of the mRNA, ribosomes detach and simultaneously release the newly synthesized protein molecule. Amino acids are brought to the ribosomes for incorporation into the protein by **transfer RNA** (tRNA), another form of ribonucleoprotein. There is a specific tRNA for each of the amino acids. Ribosomal and transfer RNA are formed by transcription of specific segments of the DNA molecule. Formation of ribosomal RNA is directed by a specific region of a chromosome.

Endoplasmic Reticulum

KEY WORDS: tubule, cisternae, granular endoplasmic reticulum, ribosome, smooth endoplasmic reticulum

The cytoplasm of almost all cells contains a continuous, irregular network of membrane-bound channels called the *endoplasmic reticulum.* Typically, this organelle appears as

Fig. 1-4. A region of cytoplasm (intestinal epithelial cell) illustrating parallel cisternae of granular endoplasmic reticulum, polyribosomes, a mitochondrion, and the plasmalemma at the lower left (TEM, × 20,000).

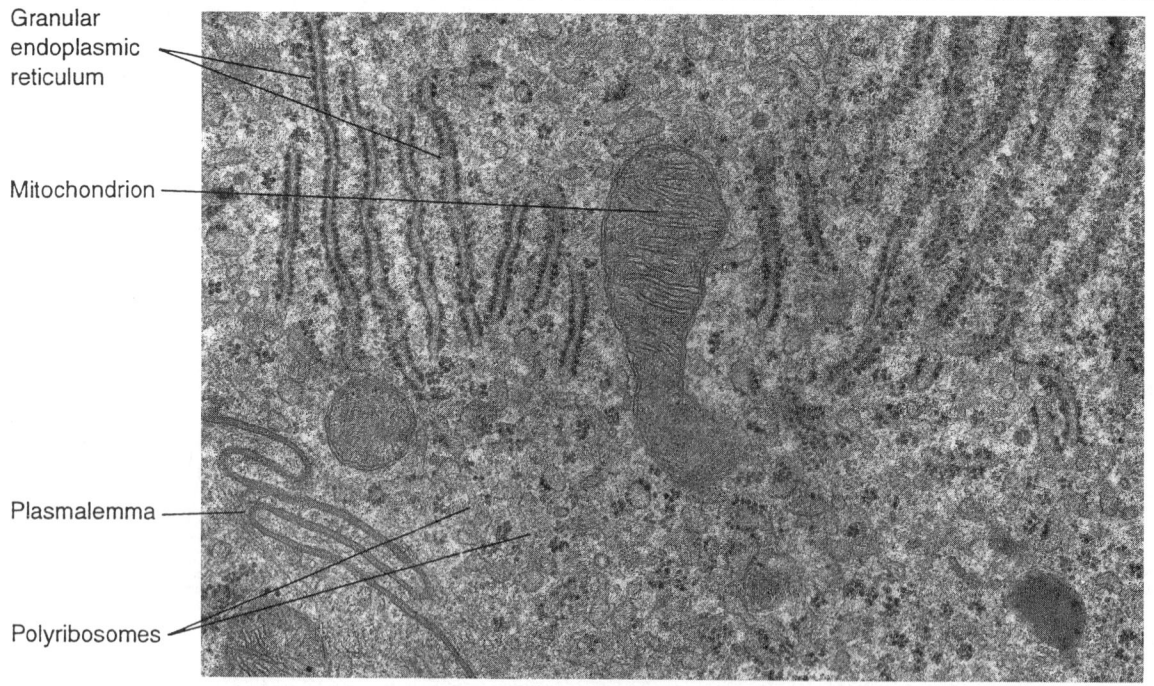

Granular endoplasmic reticulum

Mitochondrion

Plasmalemma

Polyribosomes

anastomosing **tubules,** but the membranes also form parallel, flattened saccules called **cisternae.** Small vesicles not attached to tubules or cisternae may be present also and are considered to be part of the endoplasmic reticulum.

Smooth and granular forms of endoplasmic reticulum can be distinguished. **Granular endoplasmic reticulum** (GER) usually consists of an array of flattened cisternae bounded by a membrane (Figs. 1-4 and 1-5). The outer surface is studded with numerous **ribosomes.**

The granular endoplasmic reticulum is the site of protein synthesis for secretory and some organelle proteins. The proteins are synthesized by ribosomes on the external surface of the GER, then enter the lumen of the reticulum, where they are isolated from the surrounding cytoplasm.

Proteins destined for secretion, organelles, and some membranes have an NH_2-terminal signal sequence (a leader sequence) that is recognized by a signal-recognition particle in the cytosol as it is being translated from mRNA. The signal-recognition particle (SRP) binds to both the signal peptide and the ribosome, an event that stops the translation process. The SRP together with the ribosome and the newly translated NH_2 terminus of the forming polypeptide become bound to an integral membrane-docking protein called a *signal-recognition receptor* (ribosome receptor protein) in the membrane of the endoplasmic reticulum. After being bound translation resumes, and the signal peptide together with the forming polypeptide enters a transmembrane channel (pore) to gain access to the cisternal lumen of the granular endoplasmic reticulum. Once inside the lumen of the endoplasmic reticulum the signal sequence of the polypeptide is clipped off by a signal peptidase located along the inner membrane surface of the granular endoplasmic reticulum. Following translation of the message and synthesis of the secretory polypeptide, the ribosome is released from the endoplasmic reticulum and the trans-

membrane channel is obliterated by lateral diffusion of molecules in the membrane of the endoplasmic reticulum. Once inside the cisternal lumen these polypeptides and forming proteins remain isolated from the remainder of the cytoplasm. Here these unfolded proteins become associated with a binding protein. The latter is thought to be involved in protein folding that occurs in the cisternal lumen. Most proteins synthesized by the granular endoplasmic reticulum also become glycosylated after entry into the cisternal lumen. **Smooth endoplasmic reticulum** (SER) lacks ribosomes and consists primarily of a system of interconnecting tubules without cisternae (Figs. 1-6 and 1-7).

In most cells, one form of endoplasmic reticulum usually predominates. Protein-secreting cells such as pancreatic acinar cells or plasma cells are characterized by an abundance of GER, whereas the smooth type predominates in cells that secrete steroid hormones. In still other cells, such as liver cells, both types of endoplasmic reticulum are present in nearly equal amounts and may be continuous. Granular endoplasmic reticulum is known to be involved in the synthesis of protein, but the exact role of SER remains obscure. Smooth endoplasmic reticulum has been implicated in different functions in different cells: synthesis of steroid hormones in certain endocrine cells, synthesis of lipoproteins and detoxification of lipid-soluble drugs by liver cells, synthesis of triglycerides, and metabolism of

Fig. 1-6. Diagrammatic representation of smooth endoplasmic reticulum.

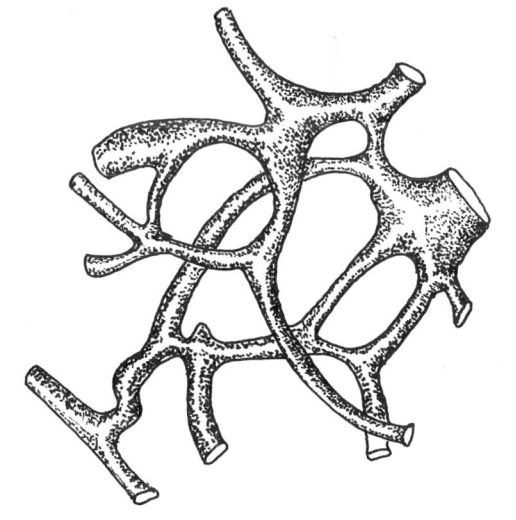

Fig. 1-5. Diagrammatic representation of granular endoplasmic reticulum.

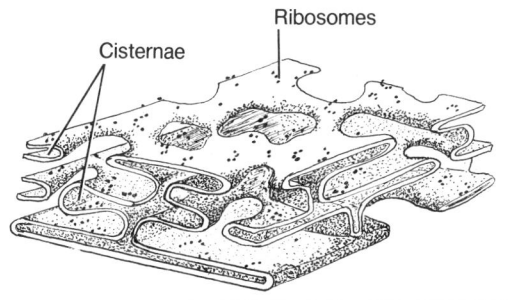

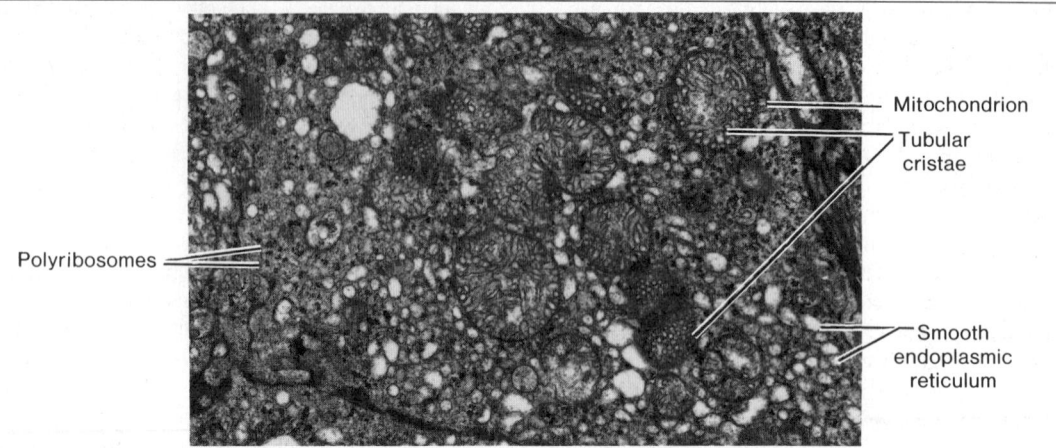

Fig. 1-7. A region of cytoplasm from a cell of the adrenal cortex that contains numerous profiles of smooth endoplasmic reticulum. Mitochondria contain tubular cristae. Both features indicate a cell type engaged in the synthesis of steroid (TEM, × 3,500).

lipid and cholesterol. The synthesis of membrane lipids (triglycerides, phospholipids, ceramide) usually occurs in the SER. These lipids are then moved to recipient organelles by phospholipid-exchange proteins or small transport vesicles. Smooth endoplasmic reticulum also has a role in the release and recapture of calcium ions during contraction and relaxation of striated muscle.

Golgi Complex

KEY WORDS: negative image, saccule, forming face, maturing face, transport vesicle, condensing vacuole, secretory granule, trans-Golgi network

The Golgi complex (apparatus) does not stain in ordinary histologic preparations, nor is it visible in living cells; it does sometime appear as a **negative image** — a nonstaining area of the cytoplasm usually close to the nucleus. The size and appearance of the Golgi complex vary with the type and activity of the cell and may be small and compact or large and netlike. Some cells contain multiple Golgi complexes.

In electron micrographs, the Golgi complex is seen to consist of several flattened **saccules** (cisternae), each limited by a smooth membrane (Fig. 1-8). The saccules are disc-shaped, slightly curved, and often appear to be compressed near the center and dilated at the edges. The saccules are arranged in stacks and are separated by spaces 20 to 30 nm wide. Because of the curvature of the saccules,

the Golgi complex has convex and concave faces. The convex face usually is directed toward the nucleus and is called the **forming face;** the concave or **maturing face** is oriented toward the cell membrane. The forming face is associated with numerous small vesicles, and at this face the outer saccule is perforated by many small openings. The saccules at the maturing face tend to be more dilated than those at the concave face (Fig. 1-9).

Secretory products are concentrated in the Golgi complex, whose size varies with the activity of the cell. In protein-secreting cells, peptides first accumulate within the lumen of the GER and then are transported to the Golgi complex in small **transport vesicles,** which are formed from ribosome-free areas of the GER adjacent to the Golgi complex. The vesicles carry small quantities of protein to the Golgi body, where they coalesce with and contribute membrane to the developing outer saccule at the forming face. Proteins accumulate within the cisternae of the Golgi membranes and are concentrated as they pass through the Golgi complex. At the maturing face, the saccules expand and bud off to form limiting membranes that enclose the protein in structures called **condensing vacuoles.** Addition of new membrane to the forming face balances loss of membrane from the maturing face. Secretory materials within the vacuoles become more concentrated, and the condensing vacuoles eventually mature into **secretory granules.**

It is now known that this simplified version of Golgi function is much more complex. The GER synthesizes a

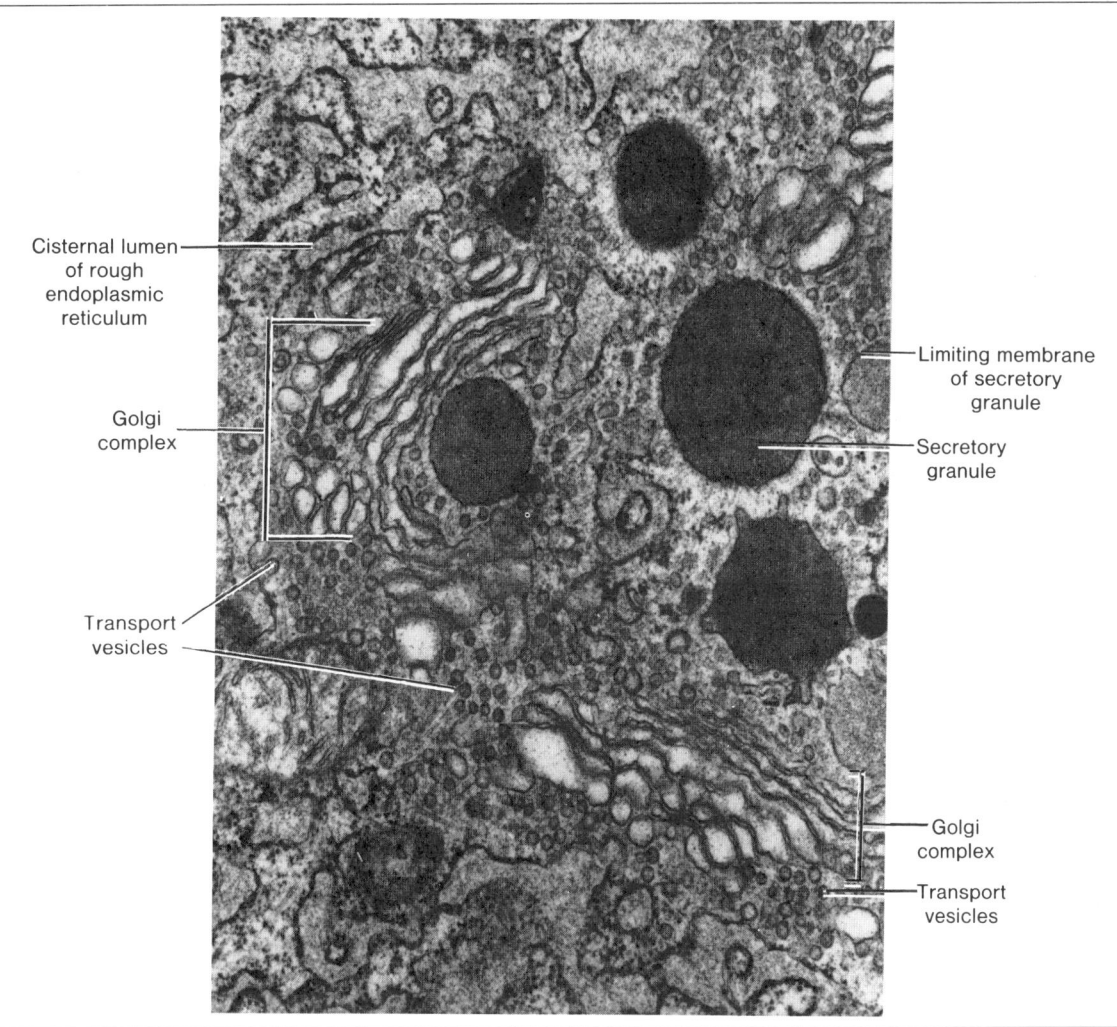

Cisternal lumen
of rough
endoplasmic
reticulum

Golgi
complex

Transport
vesicles

Limiting membrane
of secretory
granule

Secretory
granule

Golgi
complex

Transport
vesicles

Fig. 1-8. A portion of cytoplasm taken from a cell of the duodenal glands. This cell is actively involved in the synthesis and secretion of glycoprotein. Note the amorphous material (protein) within the cisternal lumen of the GER. The protein is transported to the Golgi complex, where it is complexed to carbohydrate. The formed glycoprotein is then "packaged" by surrounding Golgi membranes to form a secretory granule (TEM, × 37,500).

large number of proteins: some for export as secretory products and others destined to become incorporated into the structural components of the cell itself. The Golgi complex functions in the posttranslational modification and sorting of proteins and lipids synthesized by the endoplasmic reticulum. Essential enzymes involved in glycosylation and other functions are found on the luminal side of the endoplasmic reticulum and Golgi membrane cisternae.

As a result, the Golgi complex can be subdivided into four functional compartments depending on the enzymes present within its cisternae. The cis compartment of the forming face receives transport vesicles that have budded off from the transitional elements of the endoplasmic reticulum. This compartment is highly fenestrated and appears as a network of anastomosing tubules and vesicles. Following modification of the proteins and lipids received

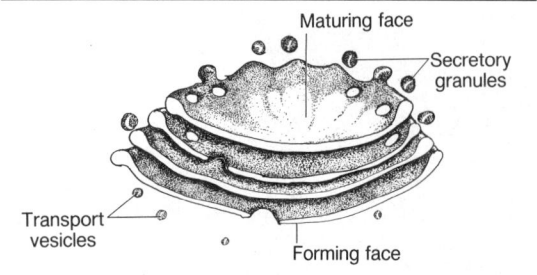

Fig. 1-9. Diagrammatic representation of the Golgi complex.

from the endoplasmic reticulum, vesicles bud off from cisternae of this compartment and fuse with cisternae of a medial compartment. After the transported lipids and proteins are acted on by enzymes in this compartment, vesicles form once again and transport these molecules to the cisternae of the trans compartment at the maturing face of the Golgi complex. Some terminal cisternae of this region are highly fenestrated and form a network of anastomosing tubules and vesicles. This specialized region of the Golgi complex is called the **trans-Golgi network.** It is in the trans-Golgi network that proteins, glycoproteins, and lipids are sorted into different transport vesicles. The specific chemical groups added to the proteins in the endoplasmic reticulum and later modified in the various compartments of the Golgi complex designate where specific proteins will go within the cell. Thus sorting takes place at the maturing face within the trans-Golgi network, where certain pro-

teins are designated to be packaged into larger secretory granules, some are enclosed in small, smooth-surfaced vesicles, and others are placed in clathrin-coated vesicles. The larger, membrane-limited secretory granules are usually destined for regulated exocytosis (stimulated secretion). Those products in small, smooth-surfaced vesicles are involved primarily in constitutive exocytosis as well as transport of membrane to other organelles. The small coated vesicles transport enzymes (acid hydrolases) that fuse specifically with developing lysosomes.

During release of secretory granules, the membranes of the secretory granules fuse with the plasmalemma and become incorporated into the cell membrane. The membranes of secretory granules must have some special properties because they fuse specifically with the plasmalemma and not with the membranes of other organelles.

There appears to be a continuous movement of membrane through the cell, from endoplasmic reticulum to transport vesicles, to Golgi complex, to secretory granules, and then to the plasmalemma. Internalization of plasmalemma occurs during phagocytosis, pinocytosis, and micropinocytosis.

Lysosomes

KEY WORDS: **acid hydrolase, autolysis, primary lysosome, secondary lysosome, phagocytosis, heterophagy, phagosome, heterophagic vacuole, residual body, autophagy, cytolysosome, multivesicular body**

Lysosomes are small, membrane-bound, dense bodies measuring 0.2 to 0.55 μm in diameter (Fig. 1-10). More

Fig. 1-10. A portion of a human gastric lining epithelial cell that contains five large secondary lysosomes (TEM, × 12,000).

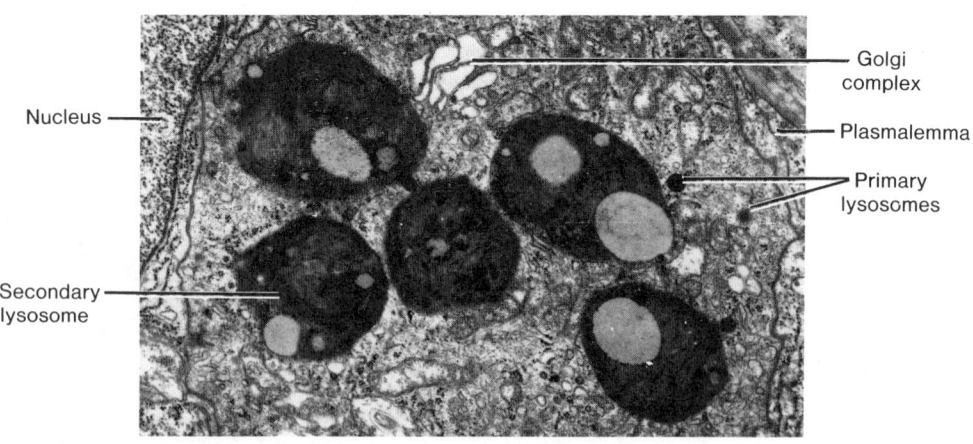

than 50 enzymes have been identified in lysosomes. Since they are active at an acid pH, lysosomal enzymes often are referred to as **acid hydrolases.** The limiting membrane of lysosomes protects the remainder of the cell from the effects of the contained enzymes which, if released into the cytoplasm, would digest or lyse the cell. Such an occurrence is called **autolysis** and is presumed to occur normally during such diverse events as regression of the mesonephros in kidney development and regression of mammary tissue after cessation of lactation. Increased lysosomal activity occurs during the regression of some tumors.

The appearance of lysosomes varies according to the phase and activity state in which they are observed and their association with cell structures or material brought into the cell. Because lysosomes exhibit such variable morphologic features, a histochemical test identifying at least two acid hydrolases must be done for positive identification. **Primary lysosomes** are those which have been newly released at the Golgi complex and have not engaged in digestive activities. The enzymes of the primary lysosomes are synthesized in the GER, transported to the Golgi complex, and released from the maturing face as membrane-bound, electron-dense granules.

Within the Golgi cisternae mannose components of enzymes destined for lysosomes are phosphorylated, and as they move through the Golgi complex, they are bound to a transmembrane glycoprotein (mannose-6-phosphate receptor) on the luminal side of membranes forming the trans-Golgi network. These receptor-enzyme complexes are then segregated to regions of the trans-Golgi network that form small vesicles coated with the protein clathrin. Coated vesicles containing receptor-bound acid hydrolases bud from the trans-Golgi network and soon after formation lose their clathrin coat. The small transport vesicles, filled with acid hydrolases, then fuse with prelysosomal vacuoles (endolysosomes). The origin of the structures with which the transport vesicles fuse to form primary lysosomes remains controversial. However, some evidence does suggest that these vacuoles are late endosomes formed by endocytosis of the plasmalemma. Primary lysosomes usually remain within the cell and are not secreted.

Secondary lysosomes are vacuolar structures that represent sites of past or current lysosomal activity and include heterophagic vacuoles, residual bodies, and cytolysosomes. The relationships of these structures are best understood from a description of the processes involved in **phagocytosis** (Fig. 1-11).

Some cells, such as macrophages and some granular leukocytes of the blood, have a special capacity to engulf extracellular materials and destroy them. The process by which substances are taken into the cell from the exterior environment and broken down by lysosomal activity is called **heterophagy.** The process involves invagination of the cell membrane and containment of the material in a membrane-bound vacuole. Thus the extracellular material taken into the cell is sequestered in a vacuole called a **phagosome** and remains isolated from the cytoplasm. As the phagosome moves through the cytoplasm of the cell, it encounters a primary lysosome. The membranes of the two structures fuse, and the enzymes of the lysosome are discharged into the phagosome. The combined primary lysosome and phagosome is now called a **heterophagic vacuole,** a type of secondary lysosome. The material within the heterophagic vacuole is digested by the lysosomal enzymes, and any useful materials are transferred into the cytoplasm for use by the cell. Nondegradable materials such as some dye particles, asbestos fibers, silica, or carbon may remain within the vacuole, now called a **residual body.** A residual body is another form of secondary lysosome that is thought by some to be eliminated from the cell by exocytosis. However, in many cells the residual bodies accumulate and persist for long periods of time.

Autophagy refers to the lysosomal breakdown of cytoplasmic organelles in normal, viable cells (see Fig. 1-11). The lysosomal system is involved in the destruction of worn or damaged organelles and the remodeling of the cytoplasm. During the process, a portion of the cytoplasm containing aged or damaged organelles becomes surrounded by a membrane to form an autophagic vacuole. The membrane is thought to be derived from SER. The vacuole fuses with a primary lysosome to form another type of secondary lysosome called a **cytolysosome.** The fate of the materials within the autophagic vacuoles (which may be the cell's own mitochondria, ribosomes, endoplasmic reticulum, and so forth) is the same as that of heterophagic vacuoles and again results in formation of residual bodies. In many cells, indigestible substances within autophagic vacuoles form a brownish material called *lipofuscin pigment*, the amount of which increases with age. In addition to these activities, lysosomes form an intracellular digestive system with the capacity of taking in and breaking down most naturally occurring molecules within cells. The absence of a specific lysosomal enzyme results in the accumulation of its normal substrate within the lysosome. Large accumulations of substrate seriously affect lysosome function and are the basis of numerous lysosomal storage disorders such as Tay-Sachs and Gaucher's diseases.

Another form of lysosome is the **multivesicular body,** a membrane-bound vacuolar structure, 0.5 to 0.8 μm in

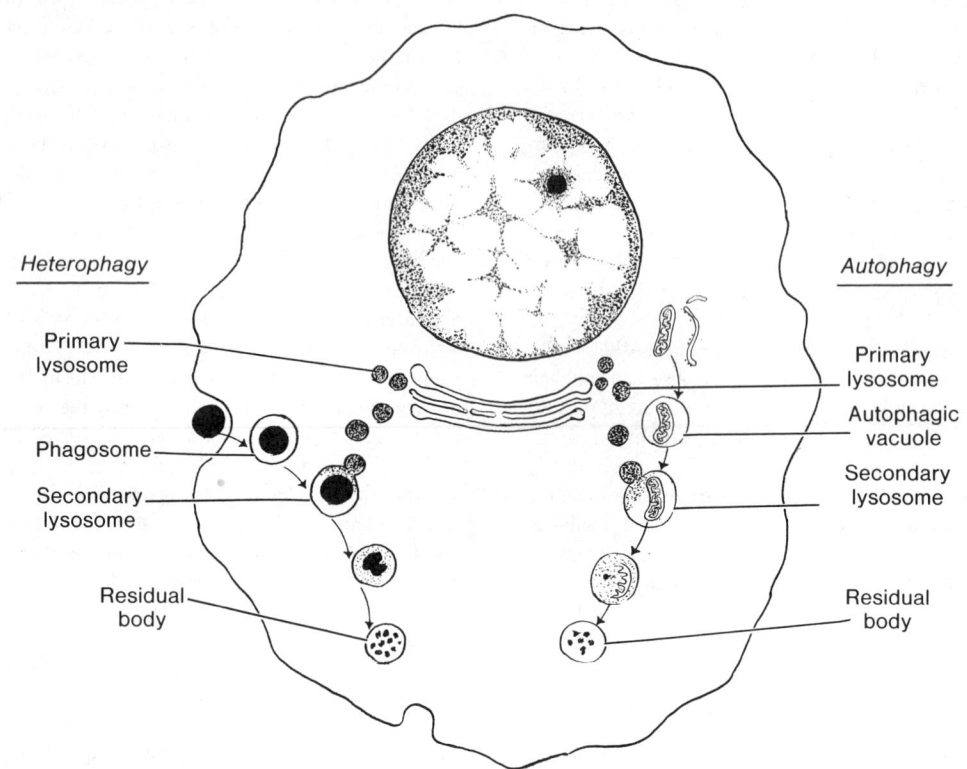

Fig. 1-11. Schematic representation of autophagy and heterophagy.

diameter, that contains several small, clear vesicles. Its origin, function, and exact relationship to other lysosomes remain obscure.

Peroxisomes

KEY WORDS: nucleoid, hydrogen peroxide, catalase

Peroxisomes, or microbodies, comprise another class of membrane-bound organelles. Usually larger than lysosomes, their internal structure varies and can be crystalline or dense. The crystalline structures are called **nucleoids.** Peroxisomes lack acid hydrolases but do contain several enzymes that can remove hydrogen atoms from organic substrates and produce **hydrogen peroxide,** which, although essential for many cellular functions and capable of destroying microorganisms, in excess is lethal to cells. Peroxisomes contain the enzyme **catalase,** which can make up as much as 40 percent of the total peroxisomal enzyme, and utilize the hydrogen peroxide formed to oxidize other substrates. Excess hydrogen peroxide is converted to water by catalase. More than 40 enzymes have been associated with peroxisomes. Peroxisomes have been implicated in the oxidation of substrates, particularly very long chain fatty acids, and also are abundant in cells involved in steroid synthesis and cholesterol metabolism.

Formation of peroxisomes does not appear to involve the Golgi complex. Peroxisomes are thought to form by the growth and fission of existing peroxisomes and their contents (proteins, phospholipids, membrane lipids) imported from the cytosol.

Mitochondria

KEY WORDS: outer mitochondrial membrane, inner mitochondrial membrane, cristae, membrane space, intracristal space, intercristal space, mitochondrial matrix, Krebs cycle, elementary particle, electron transfer, DNA, ribonucleoprotein, matrix granule

Mitochondria are membranous structures that play a vital role in the production of the energy required by cells. They are visible in living cells examined by phase contrast microscopy and can change shape as they slowly move about the cytoplasm. Mitochondria can be stained in appropriately fixed tissues, where they appear as rods or thin filaments. They usually are not visible in routine tissue sections because of the lipid solvents used during tissue preparation.

Ultrastructurally, mitochondria show a variety of shapes and sizes, but all are enclosed by two membranes, each of which has the typical trilaminar substructure (Fig. 1-12). The inner and outer mitochondrial membranes differ markedly in their chemical composition and physiologic properties. The **outer mitochondrial membrane** is a continuous, smooth structure that completely envelops the organelle. It contains specialized transmembrane transport proteins (porins) that allow permeability to certain molecules from the cytosol. An **inner mitochondrial membrane** runs parallel to the outer membrane but is thrown into numerous folds, the **cristae,** that extend into the interior of the mitochondrion.

The narrow space between inner and outer membranes, the **membrane space,** is continuous with the small **intracristal space** within each crista. The inner mitochondrial membrane surrounds the larger **intercristal space** that contains a slightly more electron-dense material called the **mitochondrial matrix.** Enzymes of the **Krebs cycle,** responsible for the final breakdown of fatty acids, monosugars, and some amino acids, reside within the mitochondrial matrix.

Cristae greatly increase the surface area of the inner

Fig. 1-12. Structure of a mitochondrion.

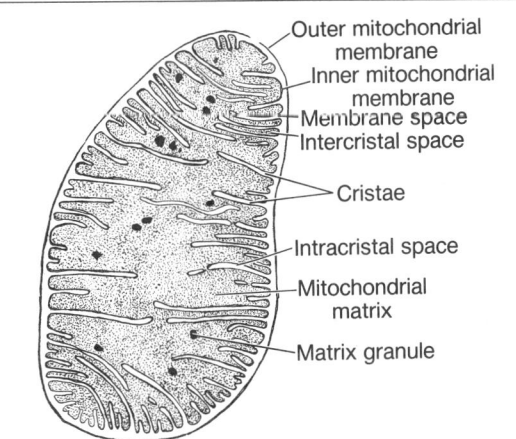

Outer mitochondrial membrane
Inner mitochondrial membrane
Membrane space
Intercristal space
Cristae
Intracristal space
Mitochondrial matrix
Matrix granule

mitochondrial membrane and may be either shelf-like or tubular in shape; the tubular form is seen most often in cells involved in steroid synthesis. The inner mitochondrial membrane is the site of many enzymatic reactions and is studded with club-shaped structures called **elementary particles.** Each particle consists of a spherical head, 9 to 10 nm in diameter, attached to the inner mitochondrial membrane by a narrow stalk 5 nm long. Phosphorylating enzymes and enzymes of the **electron transfer** system are located either on the elementary particles or within the inner mitochondrial membranes that form the cristae. The inner mitochondrial membrane is impermeable to small ions, a feature essential for development of electrochemical gradients and the formation of high-energy metabolites. Energy released by the enzymes operating in the Krebs cycle is accepted by the electron transfer system of cytochromes and then incorporated into high-energy phosphate compounds such as adenosine triphosphate (ATP), the primary source of energy for cell activities. Thus a primary function of mitochondria is synthesis of ATP, which is transported into the adjacent cytoplasm.

Mitochondria are unique among organelles because they contain their own complement of **DNA** and are capable of self-replication. Mitochondrial DNA differs from nuclear DNA in its lower molecular weight and is unusual in that it consists of branched filaments of variable thickness arranged in a circular manner. Mitochondrial DNA represents less than 1 percent of the total DNA in a cell. The human mitochondrial genome contains 16,569 nucleotides. These nucleotides can encode only 13 protein subunits involved in electron transport and oxidative phosphorylation. Therefore, mitochondria must import the majority of their proteins from the cytosol. The mitochondrial matrix also contains small particles of **ribonucleoprotein** that are 12 nm in diameter and are similar in structure and function to cytoplasmic ribosomes. Messenger RNA and transfer RNA also have been identified in mitochondria. Scattered throughout the mitochondrial matrix are the more conspicuous **matrix granules,** which measure 30 to 50 nm in diameter. These granules are thought to regulate the internal ionic composition of the mitochondrion. In addition to the enzymes of the Krebs cycle, enzymes involved in protein and lipid synthesis are present in the mitochondrial matrix of some cell types. Mitochondria also contain several enzymes associated with the synthesis of steroid hormones, and at several steps in the synthetic pathway, substrate enters the mitochondrion for processing and then returns to other cytoplasmic compartments for the completion of additional steps. More than 50 enzymes have been localized to mitochondria.

Annulate Lamellae

The annulate lamellae are membranous organelles consisting of parallel cisternae arranged in stacks. At regular intervals along their lengths, the cisternae show numerous small pores that appear to be closed by thin, electron-dense diaphragms (Fig. 1-13). Because they contain pores, they exhibit a morphologic similarity to the nuclear envelope. The cisternae are spaced uniformly throughout the stack, and the pores in successive cisternae may be aligned. Annulate lamellae often have a perinuclear location and might be continuous with elements of GER. They have been seen in germ cells, various somatic cells, and tumor cells but are relatively uncommon. The functional significance of the annulate lamellae is unknown.

Centrioles

KEY WORDS: centrosome, diplosome, microtubule, triplet, procentriole, centriolar satellite, basal body, ciliogenesis, procentriole organizer

Under the light microscope, centrioles appear as minute rods or granules usually located near the nucleus in a spe-

Fig. 1-13. An annulate lamella taken from a differentiating germ cell (TEM, × 15,000).

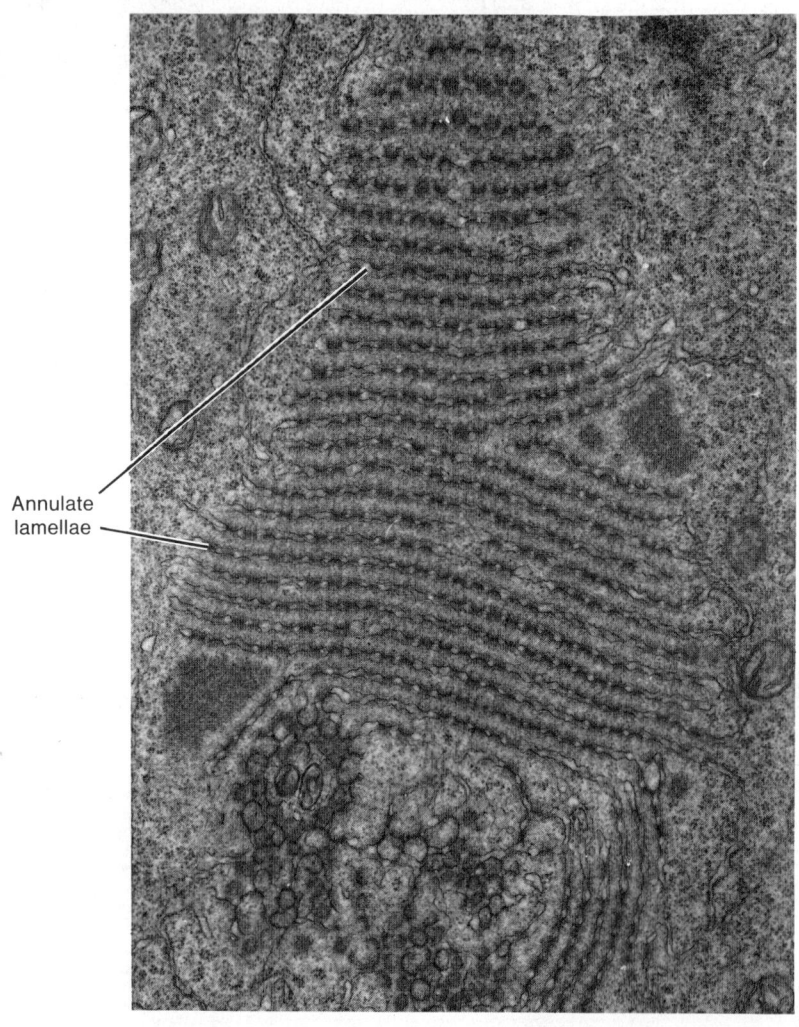

Annulate
lamellae

cialized region of the cytoplasm called the **centrosome.** In some cells centrioles are located between the nucleus and the free surface of the cell at some distance from the nucleus. Two centrioles usually are present in the nondividing cell and together form the **diplosome.** As seen in electron micrographs, the two centrioles that make up the diplosome are perpendicular to each other. The wall of each centriole consists of nine subunits, each of which is made up of three fused **microtubules;** the subunits are referred to as **triplets.** The nine sets of triplets are so arranged in the centriolar wall that they resemble a pinwheel when seen in cross section. The microtubules within each triplet are called the A, B, and C microtubules, the innermost being the A microtubule, the central tubule being the B, and the most peripheral tubule being the C (Figs. 1-14 and

1-15). The clear center of the centriole contains a thin filament that passes in a helix immediately adjacent to the inner surface of the centriolar wall. Multinucleated cells contain several centrioles.

Centrioles are self-replicating organelles that duplicate just before cell division. A new centriole, called a **procentriole,** forms at right angles to each of the parent centrioles. Initially, the wall of the procentriole consists of a ring of amorphous material with no microtubules. As the procentriole elongates by addition of material at its distal end, microtubules appear in the wall, and the new structure assumes the configuration of the parent centriole. Immediately after duplication, each parent centriole together with a newly formed daughter centriole migrates to the opposite poles of the cell, where they function in the development of the mitotic spindle. Small, dense bodies, the **centriolar satellites,** are associated with the centrioles and initiate the development and polymerization of microtubules during formation of the mitotic spindle.

In some cells, centrioles migrate close to the surface, where they form **basal bodies** from which arise cilia or flagella; these also are microtubule-containing structures. During the process of **ciliogenesis,** dense spherical bodies (**procentriole organizers**) appear in the cytoplasm, and numerous procentrioles form around each. Multiple newly formed centrioles migrate to sites immediately beneath the plasmalemma and become oriented perpendicular to it. The two innermost microtubules (A and B) of each centriolar triplet begin a rapid polymerization of tubulin and serve as templates for formation of the microtubules of the axoneme. Elongation of the cilium is complete within 1 hour.

Fig. 1-14. Diagrammatic representation of a centriole seen in cross section.

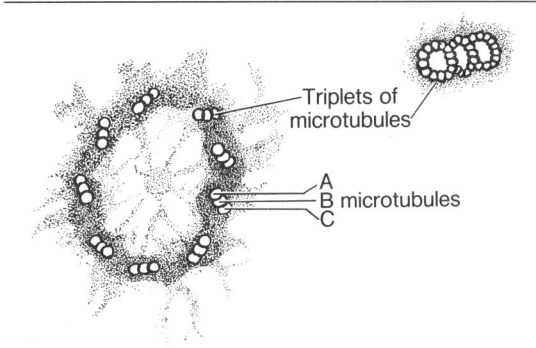

Triplets of microtubules

A
B microtubules
C

Fig. 1-15. Centrioles form a submandibular gland ductal cell (A) and a human fibroblast (B) are shown in both longitudinal and cross-sectional profiles, respectively (A. TEM, × 48,000; B. TEM, × 51,000).

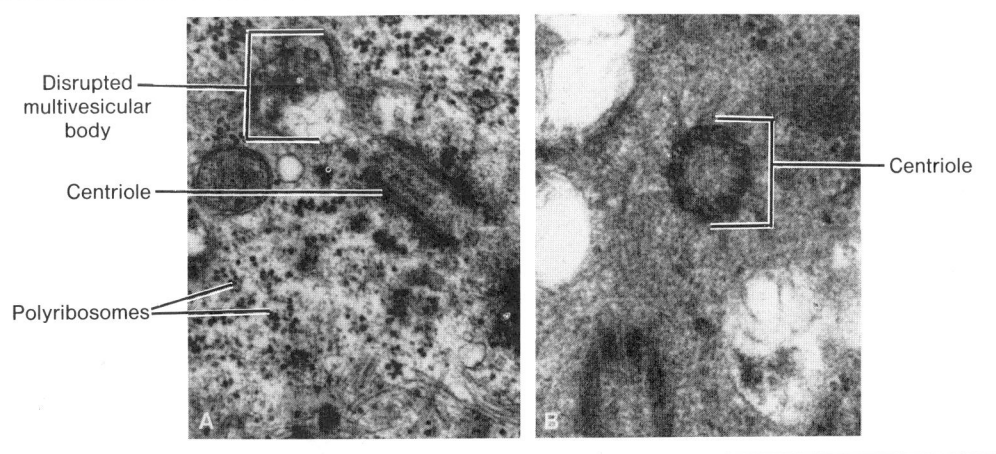

Disrupted multivesicular body

Centriole

Polyribosomes

Centriole

Cytoskeleton

The cytoskeleton gives structural support to the cytoplasm and consists of microfilaments, intermediate filaments, microtubules, and a microtrabecular lattice. Interaction between the cytoskeleton and the plasmalemma is essential for cell movement, intracellular transport, endocytosis, focal mobility of the plasmalemma, maintenance of cell shape, stabilization of cell junctions, and spatial orientation of enzymes and other molecules in the hyaloplasm.

Microfilaments

KEY WORDS: actin, myosin, actin-binding protein

Cytoplasmic filaments are responsible for contractility, a property shown in some degree by almost all cells. Filaments are best developed in muscle cells, where the proteins **actin** and **myosin** form two different types of filaments whose interactions are responsible for the contractile properties of muscle cells.

Microfilaments have a diameter of 7 nm and usually are located immediately beneath the plasmalemma. Actin shows the same dimensions as the microfilaments and has been identified chemically in microfilaments. Actin is an important component of the cytoskeleton and occurs in two molecular forms: G-actin, which consists of globular subunits, and F-actin, the polymerized form that consists of two identical strands helically coiled around one another. Actin filaments may be organized into parallel bundles or randomly distributed to form an extensive, interwoven network. **Actin-binding proteins** such as filamin link actin filaments together under certain conditions to form a stiff supportive framework for the plasmalemma. Actin also may be linked to transmembrane proteins in specialized regions of the plasmalemma at junctional complexes, contributing further to the cytoskeleton of individual cells. Under different conditions actin filaments participate in cell locomotion, ruffling of the cell membrane, invagination of the plasmalemma during endocytosis, and formation of the contraction ring of dividing cells. In this case, a thin layer of actin filaments located along the cytoplasmic surface of the cell membrane interacts with myosin. Myosin also is present in all cells and interacts with actin filaments to generate movement of the plasmalemma. Myosin is an actin-activated ATPase protein and may exist in monomeric forms. Actin-myosin interactions are involved in the movement of transport vesicles and certain organelles within the cell. A number of actin-binding proteins exist and interact with either G-actin monomers or F-actin filaments and influence the distribution and function of actin in the cytoplasm (Table 1-1).

Intermediate Filaments

KEY WORDS: keratin, vimentin, desmin, neurofilament, glial filament

The thicker intermediate filaments measure 9 to 12 nm in diameter and form a more diverse population of filaments; they may be present as individual strands or loose bundles. Intermediate filaments often are attached to the internal surface of the plasmalemma at cell junctions and contribute to the cytoskeleton of the cell, providing support. Microfilaments and intermediate filaments may occur in the same cell.

Table 1-1. Actin-Binding Proteins

Actin-Binding Protein	Action
Profilin	Binds to G-actin monomers preventing polymerization and thus providing a pool of monomer for future use
Capping protein	Binds to ends of actin filaments limiting a further increase in length
Fimbrin	Binds parallel actin filaments into bundles
Filamin	Cross-links actin filaments into a three-dimensional network
Gelsolin	Fragments actin filaments into shorter segments
Vinculin	Mediates binding of actin to plasmalemma
α-Actinin	Mediates binding of actin to plasmalemma
Spectrins I and II	Links actin protofilaments to plasmalemma and cytoplasmic vesicles
Tropomyosin	Stabilizes actin filaments
Myosin I	Interacts with actin filaments in locomotion of cells and movement of vesicles and organelles
Myosin II	Slides actin filaments during contraction of muscle cells

Although indistinguishable structurally, five types of intermediate filaments have been identified by immunocytochemical means. Most cells contain only a single type, but occasionally a cell type may contain two. **Keratin** filaments are present in epithelial cells and form bundles that end in cell attachments. They are most abundant in the stratified squamous epithelia, especially in the skin. **Vimentin** filaments occur in fibroblasts and other cells derived from mesenchyme; **desmin** filaments are found in skeletal, cardiac, and smooth muscle cells and provide for the mechanical integration of the contractile proteins actin and myosin; **neurofilaments** are found only in nerve cell bodies and their processes and often occur in bundles. **Glial filaments** are confined to the supporting (glial) cells of the central nervous system and consist of glial fibrillary acidic protein. Glial filaments and neurofilaments provide internal support for their respective cell types. A knowledge of this restricted distribution is often useful for the assessment of malignant tumors to determine their site of origin.

Microtubules

KEY WORDS: tubulin, microtubule-associated proteins (MAPs)

Microtubules are straight or slightly curved, nonbranching tubules measuring 21 to 25 nm in diameter and several micrometers in length. Their walls, about 6 nm thick, consist of 13 parallel protein filaments, each 4 to 5 nm in diameter, arranged in a helix. The central zone is electron-lucent, so the structure appears to be hollow. Microtubules are composed primarily of the protein **tubulin,** which occurs in alpha and beta forms. The number of microtubules in the cytoplasm varies from one cell type to another and at different times within the same cell. Microtubules can rapidly form or disappear depending on cell activity. Microtubules are in dynamic equilibrium with a large reserve of soluble tubulin in the cytoplasm. Once formed, microtubules elongate by adding subunits at one end and usually depolymerize at the distal end, recycling subunits to the cytoplasm. Small quantities of high-molecular-weight proteins called **microtubule-associated proteins (MAPs)** also have been isolated and are thought to correspond to the slender, lateral filamentous projections of the microtubules. Four different types of MAPs have been identified: kinesin, dynein, dynamin, and axonemal dynein. Kinesin links transport vesicles to microtubules and through cyclic interaction moves the vesicle toward the forming end of the microtubule. Dynein also links transport vesicles to microtubules, but those destined to move toward the end of the microtubule where depolymerization occurs, providing for two-way traffic within a cell. Dynamin links neighboring microtubules into bundles, and axonemal dynein is responsible for the sliding action of microtubules within cilia and flagella, resulting in their beating action.

Microtubules usually are scarce in resting cells but are present in large numbers in dividing cells, where they make up the mitotic spindle. After the cell has divided, most of the microtubules disappear. In interphase cells, microtubules form prominent components of centrioles, cilia, and flagella and, in some cells, contribute to the cytoskeleton. Within platelets microtubules form stiffening elements that help maintain their shape. In nerve cells, microtubules are important for transport of material from one region of the cytoplasm to another. A transient increase in the number of microtubules is seen in nondividing cells during changes in shape associated with cell movement and during differentiation. It has been suggested that a three-dimensional framework (a microtrabecular lattice) in the cytoplasmic matrix links the cell organelles and cytoskeleton into a coordinated functional unit during cell movement and other activities.

Cytoplasmic Inclusions

Inclusions are nonliving elements found in the cytoplasm and include such diverse materials as pigment granules, glycogen, lipid droplets, and crystals. They are not essential to the life or functioning of the cell and represent metabolic products, storage materials, or foreign substances taken into the cell from the environment.

Pigment Granules

KEY WORDS: melanin, melanosome, hemosiderin, lipofuscin

Naturally occurring pigments in human cells include melanin, hemosiderin, and lipofuscin, each of which has its own inherent color. **Melanin** is contained within **melanosomes,** which are membrane-bound granules formed and found in melanocytes and sometimes secondarily deposited in certain other cells, especially prominently at the skin's dermal-epidermal junction, contributing to the color of the skin. The pigment epithelium of the retina and iris and certain cells of the brain also contain melanin.

Hemosiderin is a golden brown pigment derived from breakdown of hemoglobin present in red blood cells. Phagocytic cells of the liver, bone marrow, and spleen normally contain this type of pigment.

Lipofuscin is found in many cells throughout the body, particularly in older persons. Sometimes called the "wear and tear" pigment, lipofuscin is light brown in color, increases with age, and represents an end product of lysosomal activity. Neurons normally contain this type of pigment.

Glycogen

KEY WORDS: glucose polymer, beta particle, alpha particle

Glycogen is a large **glucose polymer** and is the storage form of carbohydrate. It cannot be seen in usual tissue preparations unless selectively stained. Ultrastructurally, glycogen appears either as **beta particles,** which are irregular, small, dense particles 15 to 45 nm in diameter, or as **alpha particles,** which measure 90 to 95 nm in diameter and represent several smaller particles of glycogen clumped to form rosettes.

Lipid

KEY WORDS: fat cell, lipid droplet

Fat cells are the chief storage sites for lipid, but many other cell types store **lipid droplets** of various sizes. Lipid synthesized by a cell accumulates in cytoplasmic droplets that lack a limiting membrane. Intracellular lipid serves as a source of energy and as a supply of short-chain carbon compounds for the synthesis of membranes by the cell. During preparation of routine tissue sections, lipid usually is extracted, and sites of lipid storage appear as clear vacuoles. In electron micrographs, preserved lipid droplets appear as homogeneous spheres of different densities.

Crystals

Crystalline inclusions are normal constituents in several cell types and may be free in the cytoplasm or contained in secretory granules, mitochondria, endoplasmic reticulum, the Golgi complex, and even the nucleus. Many of the observed intracellular crystals are believed to represent a storage form of protein.

Nucleus

The nucleus is an essential organelle present in all complete cells. The only cytoplasmic structures in which nuclei are absent are mature mammalian erythrocytes and blood platelets; these should not be regarded as true cells. Generally, each cell has a single nucleus, but some, such as the parietal cells of the stomach, cardiac muscle cells, and liver cells, may possess two nuclei. Giant cells, such as the osteoclasts of bone, megakaryocytes of marrow, and skeletal muscle cells, may have several nuclei.

The shape of the nucleus varies and may be spherical, ovoid, or elongated corresponding to the cell shape, or it might be lobulated as in the granular leukocytes of the blood.

The nucleus contains all the information necessary to initiate and control the differentiation, maturation, and metabolic activities of each cell. The nondividing nucleus is enclosed in a nuclear envelope and contains the chromatin material and one or two nucleoli (Figs. 1-16 and 1-17). These are suspended in a nuclear ground substance called the *karyolymph* or *nuclear matrix.*

Chromatin and Chromosomes

KEY WORDS: karyosome, heterochromatin, histone, euchromatin, diploid, haploid, homologous chromosome, sex chromosome, autosome, karyotype, metacentric, acrocentric, submetacentric

Genetic information is stored in the molecules of DNA that make up the chromosomes. Each chromosome contains a single long molecule of DNA that consists of two linear polymers of nucleotide subunits each made up of a phosphate group, a pentose sugar (deoxyribose), and four organic bases (adenosine, cytosine, guanine, thymidine). The bases project toward and bind to complementary bases

Fig. 1-16. Diagrammatic representation of nuclear structures.

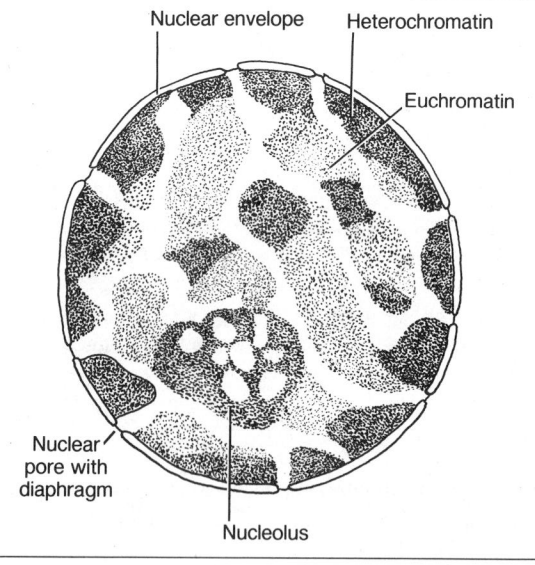

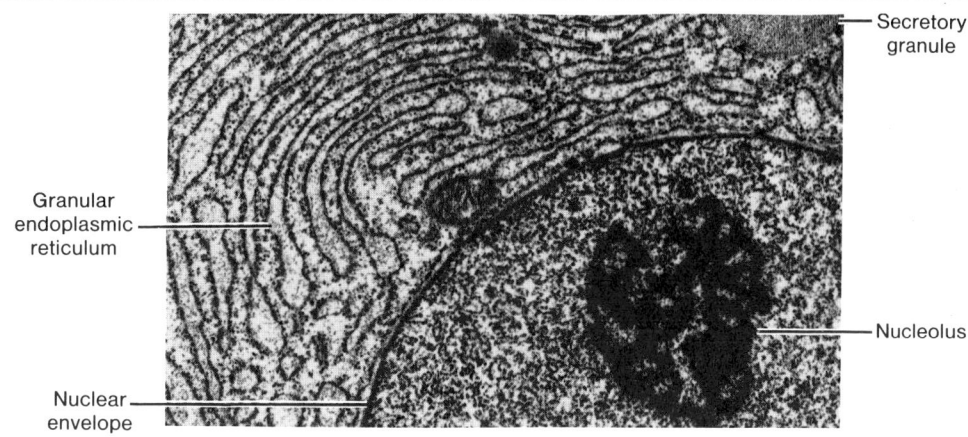

Secretory granule

Granular endoplasmic reticulum

Nucleolus

Nuclear envelope

Fig. 1-17. A portion of a chief cell taken from a human stomach illustrates regions of both the nucleus and cytoplasm. Note the abundance of GER and the nucleolus in this protein-secreting cell (TEM, × 11,000).

of the adjacent polymer. Together the two polynucleotide chains intertwine and form the antiparallel double helix of the DNA molecule. Genetic information of the DNA molecules is encoded in the sequence of the bases. A *gene* refers to that unit of heredity that involves a sequence of bases necessary for the synthesis of a protein or a nucleic acid. In nondividing nuclei, chromosomes are largely uncoiled and dispersed, but some regions of the chromosomes remain condensed, stain deeply, and are visible by the light microscope as chromatin. Nuclear DNA is associated with a variety of proteins that form chromatin. These proteins can be divided into two general categories: histones and nonhistone proteins. The latter include structural proteins, regulatory proteins, and the enzymes needed for nuclear function (DNA and RNA polymerases). Histones are the most abundant proteins in the nucleus and form the inner core of a DNA-protein complex called a *nucleosome*. The nucleosome is the basic unit of the chromatin fiber. Ultrastructurally, chromatin appears as fibers of nucleoprotein, 20 to 30 nm in diameter. Each fiber consists of a double helix of DNA wrapped about a core of histone. Individual masses of chromatin are called **karyosomes,** and although not entirely constant, the chromatin masses do tend to be characteristic in size, pattern, and quantity for any given cell type. Collectively, the karyosomes form the **heterochromatin** of the nucleus and represent coiled portions of the chromosomes. Heterochromatin is believed to be complexed to **histones** and to be nonactive. Histones are simple proteins that contain a high proportion of basic amino acids. The dispersed regions of the chromosomes

stain lightly and form **euchromatin,** which is active in controlling the metabolic processes of the cell (see Figs. 1-16 and 1-17). The distinction between heterochromatin and euchromatin disappears during cell division when all the chromatin condenses and becomes metabolically inert.

Chromosomes are permanent entities of the cell and are present at every stage of the cell cycle, but their appearance depends on the physiologic state of the cell. At interphase the chromosomes form delicate, tortuous threads, and it is only during cell division that they assume the appearance of discrete, solid, rodlike structures (Fig. 1-18). Analysis and study of chromosomes can be carried out most conveniently in dividing cells that have been arrested in metaphase. Alkaloids such as the *Vinca* drugs and colchicine interfere with spindle formation and permit intracellular accumulation of metaphase chromosomes for study.

The number of chromosomes typically is constant for each species but varies considerably between species. In humans, the chromosome number is 46. The figure given is for the **diploid** number in somatic cells. Germ cells (ova and sperm) contain half this number and are said to be **haploid.** The chromosomes present in somatic cells represent the inheritance of two sets of chromosomes, one from each male and female parent. In the male and female sets, chromosomes that are similar are called **homologous chromosomes.** In many diploid higher animals, a pair of **sex chromosomes** is specialized for and participates in the determination of sex; all other chromosomes are called **autosomes.** In humans there are 44 autosomes and a pair of

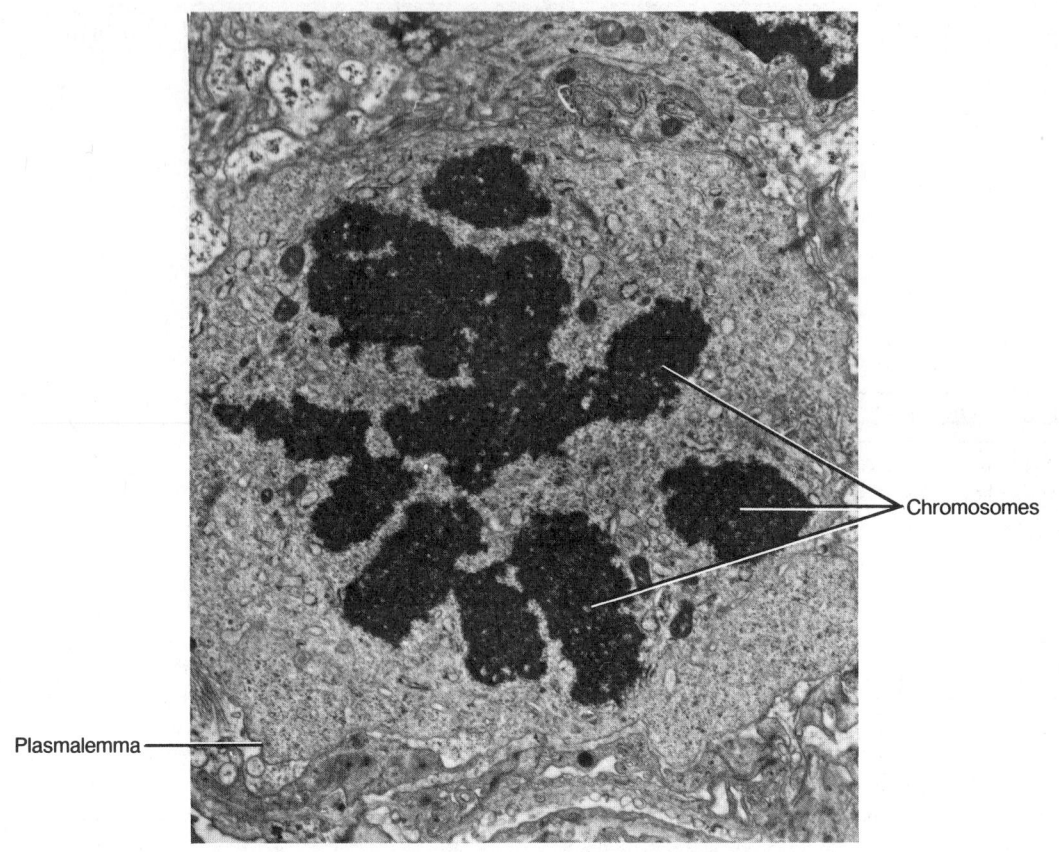

Fig. 1-18. A dividing cell taken from liver shows several distinct chromosomes (TEM, × 5,000).

sex chromosomes that are homologous (XX) in the female and heterologous (XY) in the male.

Homologus chromosomes can be recognized at metaphase and are arranged in groups representing the **karyotype** of a species. Individual chromosomes often can be identified by the length of their arms and the location of the centromere. If the centromere is in the middle of the chromosome and the arms (telomeres) are of equal length, the chromosome is said to be **metacentric.** If the centromere is close to one end, the chromosome is **acrocentric,** and if the centromere is between the midpoint and the end, the chromosome is **submetracentric.**

Nuclear Envelope

KEY WORDS: unit membrane, perinuclear space, nuclear pore, nuclear pore complex, lamins, nuclear lamina

The nuclear envelope consists of two concentric **unit membranes,** each 7.5 nm thick, separated by a **perinuclear space** 40 to 70 nm wide. The inner membrane is smooth, whereas the outer membrane often contains numerous ribosomes on its cytoplasmic surface and is continuous with the surrounding endoplasmic reticulum. At irregular intervals around the nucleus, the inner and outer membranes of the envelope become continuous with one another to form small octagonal openings called **nuclear pores** (Fig. 1-19). The pores measure about 10 nm in diameter and are closed by a **nuclear pore complex** that consists of two rings, one of which faces the cytoplasm. Eight radial spokes extend inward from the rings toward a central granule. The number and distribution of nuclear pores depend on the type of cell and its activity.

The nuclear envelope aids in organization of the chromatin and controls the two-way traffic of macromolecules

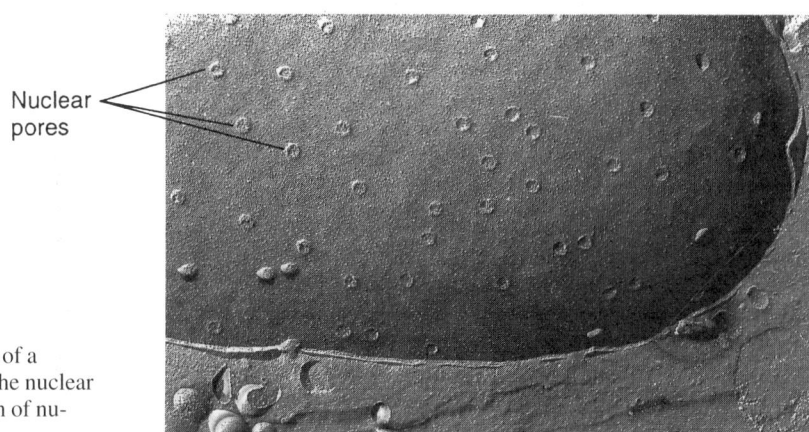

Nuclear pores

Fig. 1-19. A freeze-fracture replica of a human odontoblast that illustrates the nuclear envelope. Note the even distribution of nuclear pores (× 18,500).

between the nucleus and cytoplasm. Molecules less than l0 nm in diameter pass through the nuclear pore complex by passive diffusion, whereas large molecules (entering newly synthesized proteins, exiting ribonucleoproteins) require an energy-dependent transport mechanism. It is thought that a signal sequence of amino acids directs them to the nuclear envelope, and after binding of a signal sequence to a receptor in the nuclear pore complex, the nuclear pore opens much like an iris diaphragm to permit passage of the larger molecules.

A thin meshwork of filaments called **lamins** is made up of three polypeptides and lies along the inner surface of the nuclear membrane to form the **nuclear lamina.** The lamins are structurally similar to intermediate filaments and are classified as types A, B, and C according to their location and chemical properties. Type B lamins lie nearer the outer surface of the nuclear lamina and bind to specific integral (receptor) proteins of the inner nuclear membrane. Types A and C lamins lie along the inner surface of the nuclear lamina and link membrane-bound lamin B to chromatin. The three lamins are thought to function in the formation and maintenance of the nuclear envelope of interphase cells. They also may aid in maintaining the shape of the nucleus.

Nucleolus

KEY WORDS: ribosomal RNA, nucleolus-organizing region, pars granulosa (nucleolonema), pars fibrosa, nucleolus-associated chromatin

A nucleolus appears as a dense, well-defined body, 1 to 3 nm in diameter, contained within a nucleus. Nucleoli are

sites where **ribosomal RNA** (rRNA) is synthesized. Since these sites (**nucleolus-organizing regions**) are located on five different chromosomes, any one cell may contain several nucleoli. Usually only one or two large nucleoli are found, since the nucleolus-organizing regions tend to associate and the RNA produced at these regions aggregates into larger masses.

As seen in electron micrographs, nucleoli lie free in the nucleus, not limited by a membrane. They show two regions, each associated with a particular form of ribonucleoprotein. The dominant region, the **pars granulosa (nucleolonema),** consists of a network of dense granules of RNA 13 to 15 nm in diameter. The second region, the **pars fibrosa,** tends to be centrally placed and consists of dense masses of filaments 5 nm in diameter.

Deoxyribonucleoprotein also is associated with the nucleolus and is present in filaments of chromatin that surround or extend into the nucleolus. This chromatin forms the **nucleolus-associated chromatin;** its DNA constitutes the template for synthesis of rRNA. In humans, rRNA genes represent clusters of DNA segments located near the tips of chromosomes 13, 14, 15, 21, and 22. Ribosomal gene transcription occurs in the fibrillar region of the nucleolus. Following transcription, the rRNAs are processed and then assembled in the pars granulosa. With the addition of the small ribosomal subunit, ribosomes are then transported to the cytoplasm through nuclear pores.

Nucleoli are found only in interphase nuclei and are especially prominent in cells that are actively synthesizing proteins. They are dispersed during cell division but reform at the nucleolus-organizing regions during reconstruction of the daughter nuclei after cell division.

Review Table 1-1. Key Features of the Structural Components of the Cell

	Morphologic Features	Function
Cytoplasmic constituents		
Membranous organelles		
Cell membrane (plasmalemma)	Electron-lucent layer (3.5–4.0 nm) separating two electron-dense layers (2.5–3.0 nm)	Selectively permeable membrane; cell-to-cell communication, receptors for hormones, cell recognition, ion pumps; generates messenger molecules
Granular endoplasmic reticulum	Network of membrane-bound tubules and cisternae; outer surface is covered with ribosomes	Site of synthesis of protein to be discharged from cell, lysosomal proteins
Smooth endoplasmic reticulum	Network of branching, anastomosing, smooth, membrane-bound tubules	Varies with cell type; implicated in synthesis of steroid hormones, cholesterol, triglycerides, lipoproteins; detoxification of lipid-soluble drugs; acquisition or release of Ca^{2+} ions
Golgi complex	Stacked, parallel array of several flattened, membrane-bound saccules (cisternae); convex surface is the forming face; concave surface is the maturing face	Site of concentration, modification, and packaging of secretory products; limited synthetic activity for complex carbohydrates and glycoproteins
Annulate lamellae	Parallel stacks of membranes enclosing cisternal lumina 30–50 nm in diameter; become continuous at regular intervals forming fenestrations (pores) in cisternae; pores often lie in register with one another	Unknown
Mitochondria	Size variable; usually round, oval, or rod-shaped; consist of two membranes: an outer smooth membrane enveloping a thinner, highly folded inner membrane; lamellar or tubular cristae projects into the interior (matrix); matrix granules prominent	Provide energy in form of phosphate bonds (ATP) for cell activities; involved in synthesis of steroid hormones in specific cell types
Lysosomes	Demonstration of acid phosphatases needed for confirmation	
Primary	Round, membrane-bound dense bodies 0.25–0.55 μm in diameter; interior variable in density	Yet to begin digestive activity
Secondary	Membrane-bound bodies, ovoid or irregular in outline, size variable. Usually contain material of some type (crystalloids, debris, organelle remnants, lipofusion pigment)	Currently or have been involved in digestive activity
Multivesicular body	Numerous small vesicles limited by a single, smooth membrane	Unknown
Peroxisomes	Spherical, membrane-bound bodies 0.2–0.4 μm in diameter; matrix may contain dense crystalline structures (nucleoids)	Oxidation of long-chain fatty acids

Review Table 1-1 (continued)

	Morphologic Features	Function
Nonmembranous organelles		
Ribosomes	Dense granules (15–25 nm) comprised of a large and small subunit. Polyribosomes are several ribosomes united by a strand of mRNA	Protein synthesis
Centrioles	Cylindrical structures (0.2 × 0.5 μm) with an electron-dense wall containing nine triplet microtubules surrounding an electron-lucent center; occur in pairs oriented at right angles	Essential for formation of microtubules of mitotic spindle, cilia, and flagella
Components of cytoskeleton		
Microtubules	Long, slender tubules 25 nm in diameter; wall 5–7 nm in width surrounding an electron-lucent lumen	Maintain cell shape and participate in shape changes; forms axonemes of cilia and flagella; intracellular transport
Microfilaments	Actin — 6-nm microfilament Myosin — 15-nm microfilament	Cell movement; control focal movement of plasmalemma or in endocytosis
Intermediate filaments	8–11 nm in diameter, subdivided immunocytochemically into five classes: 1. Keratin filaments — occur in epithelia 2. Desmin filaments — occur in three types of muscle 3. Vimentin filaments — occur in fibroblasts and mesenchymal cell derivatives 4. Neurofilaments — occur in nerve cells 5. Glial filaments — occur in glial cells	Stabilize cell shape and attachments
Cytoplasmic inclusions		
Glycogen	Beta particles (20–30 nm) occur singly and are irregular in outline; alpha particles are variously sized aggregates of the beta form	Storage form of glucose
Lipid	Spheroidal droplets of varying size and density; lack a limiting membrane	Primarily a storage form of short carbon chains (triglycerides)
Pigments		
Melanin	Dark brown; complexed in ellipsoidal melanosomes	Inert
Hemosiderin	Gold-brown; ultrastructurally appears as collections of 9-nm particles	Degradation product of hemoglobin; inert
Lipofuscin	Coarse, irregularly shaped, brown-gold granules	End product of lysosomal activity; inert

Review Table 1-1 (continued)

	Morphologic Features	Function
Crystalline structures	Vary considerably in size and structure; may occur in cytoplasmic matrix, organelles, or nucleus	Most thought to be a storage form of protein
Secretory granules	Membrane-bound; vary in size, density, and internal consistency	Material synthesized to be released by cell
Nuclear constituents		
Nuclear envelope	Consists of inner and outer membranes that become continuous around nuclear pores; outer membrane studded with ribosomes, inner membrane smooth; are separated by a perinuclear cistern; nuclear pore complex is associated with each pore	Specialized segment of endoplasmic reticulum that bounds nucleus; nuclear pores permit communication between cytoplasm and nucleoplasm
Nuclear lamina	Thin network of interwoven filaments	Stabilizes inner nuclear membrane; attachment site for components of nucleoplasm
Heterochromatin	Dense staining; condensed chromatin	Inactive; part of genome not being expressed
Euchromatin	Light staining; dispersed chromatin	Active; part of genome being expressed
Nucleolus	Conspicuous round body in nucleus; nucleolonema consists of dense granules in a matrix of filaments; amorphous component may be present	Synthesis of ribosomes

Functional Review: The Cell

Cells are the fundamental units of structure of all tissues and organs and perform all the activities necessary for the survival, growth, and reproduction of an organism. They carry out energy transformations and biosynthetic activities and are able to replicate themselves. The functional activities of a cell are carried out by specialized structures, the organelles, many of which consist of or are bounded by biologic membranes essential to the organization of the cell. In addition to forming the interface between the cell and its external environment, they also are important in transportation of materials, organization of different energy transfer systems, transmission of stimuli, provision of selectively permeable barriers, and separating various intracellular compartments.

All materials that pass into or out of a cell must cross the plasmalemma; this structure is instrumental in selecting what enters and leaves the cell. Large molecules are taken in by pinocytosis, particulate matter can be incorporated into the cell by phagocytosis, and small molecules enter the cell by diffusion. The rate at which material diffuses into the cell depends on whether the material is soluble in lipid or water. Lipid-soluble materials readily dissolve in the lipid matrix of the plasmalemma and pass through the cell membrane relatively unhindered. The transmembrane proteins of the plasmalemma act as sites for passage of water-soluble substances and may serve as carrier proteins for materials such as glucose.

Although simple in microscopic appearance, the plasmalemma is very complex and variable in its molecular organization and function. Significant physiologic and biochemical differences may occur at specific regions of the plasmalemma according to the function of the cell. The plasmalemma contains special receptor sites that can react with agents such as hormones or neurotransmitters to transfer information to the cell and elicit specific responses. The structure and composition of the plasmalemma are important in determining the immunologic properties of the cell and its relationships and interactions with other cells. The plasmalemma may contain enzyme systems that act as ion pumps or initiate and control cellular activities by generating secondary messenger molecules. Membrane specializations also are involved in cell-to-cell attachments and communication.

The endoplasmic reticulum is associated with many synthetic activities of the cell. Smooth endoplasmic reticulum in liver cells is believed to serve in lipid and cholesterol metabolism, glycogen synthesis and storage, and detoxification of drugs. In endocrine cells, the SER has been implicated in the synthesis of steroid hormones. It plays an important role in muscle contraction, being responsible for the release and recapture of calcium ions. The GER is involved in the synthesis of proteins and provides a membrane for the transport vesicles that carry newly formed protein to the Golgi complex, where it is concentrated and packed into secretory granules. Golgi membranes are able to synthesize carbohydrate, complex it to protein, and form glycoproteins. The Golgi complex also acts in the synthesis of lipoproteins, being responsible for complexing lipid to proteins produced by the GER. Synthesis of protein occurs in ribosomes, and protein formed by ribosomes attached to the GER usually is secreted by the cell, whereas that formed on free polysomes is used by the cell itself.

Mitochondria perform a number of functions, chiefly the production of energy for the cell. Pyruvate, produced by the degradation of glucose in the cytoplasm, enters the mitochondrion, where, along with amino acids and fatty acids, it is processed by the enzymes of the Krebs cycle located in the mitochondrial matrix. Transfer of electrons and hydrogen atoms to oxygen is mediated by enzymes of the electron transport system, which are located on the elementary particles or within the mitochondrial membrane that forms the cristae. Energy released by the oxidations is incorporated into the high-energy bonds of ATP, which then diffuses into the cytoplasm, where it is available for use by the cell. Mitochondria also play a role in the synthesis of proteins, lipids, and steroids.

Lysosomes mainly function in intracellular digestion, which occurs in membrane-enclosed digestion vacuoles that isolate the lysosomal enzymes from the remainder of the cytoplasm. Materials may be taken into the cell during phagocytosis, or portions of cytoplasm containing aged or damaged organelles may be sequestered and digested by lysosomal activity. Usable end products diffuse into the cytoplasm, and the undigested residue remains within the vacuoles to form the residual bodies. Lysosomes might have other functions in normal cells. They have been implicated in the degradation of glycogen, in the removal from cells of excess substances such as unsecreted products, and in the release of thyroid hormone by splitting off the globulin to which the hormone is conjugated during its synthesis. Rupture of lysosomes in some cells may initiate mitosis.

Peroxisomes are similar in morphology to lysosomes but contain a number of enzymes capable of producing and degrading hydrogen peroxide. They play a role in the conversion of noncarbohydrate precursors to glucose and are instrumental in preventing the accumulation of lethal levels of hydrogen peroxide.

Microtubules, intermediate filaments, and microfilaments form the cytoskeleton of cells and perform a number of functions. Generally, intermediate filaments act as supporting elements. Microtubules contribute to the structure of cilia, flagella, and centrioles, provide supporting structures (stiffening rods) in some cells, have been implicated in intracellular transport of materials, and are essential for cell division and motility. Microfilaments (primarily actins) together with myosin function in cell movement, focal movements of cell membranes, and endocytosis and are responsible for the contraction ring during mitosis.

All the information needed to initiate and regulate the activities of a cell is encoded in the nucleus. This structure controls the growth, differentiation, maturation, and metabolic activities of the cell. The nucleus also is equipped to duplicate and pass on DNA to its daughter cells. The nucleolus, which is the site of production of rRNA and tRNA, plays a central role in the control of protein synthesis by the cell. All the nuclei in an individual contain identical genomes, but some portions are actively expressed in some cells and repressed in others. The differential activity of genes is due to differential activation, not to selective removal of genes.

Mitosis

Nearly all multicellular organisms grow by increasing the number of cells. The zygote, which is formed at conception, divides repeatedly and gives rise to all the cells of the body. Every cell in the resulting individual thus contains a nucleus, and each nucleus possesses identical genetic information. In the adult, most cells have a finite life span and must be replaced continuously. Proliferation of somatic cells is the result of *mitosis,* which can be defined as the production of two daughter cells with exactly the same number of chromosomes and the same DNA content as the original parent cell.

Mitosis usually lasts from 30 to 60 minutes and involves division of the nucleus (karyokinesis) and the cytoplasm (cytokinesis). Both events usually take place during mitosis, but karyokinesis may occur without division of the cytoplasm, resulting in formation of multinucleated cells, such as the megakaryocytes of bone marrow. Although mitosis is a continuous process, for descriptive purposes it is convenient to divide it into prophase, metaphase, anaphase, and telophase (Fig. 1-20). The time between successive mitotic divisions constitutes interphase and is the period when the cell performs its usual functions, contributes to the total economy of the body, and makes preparations for the next division.

Interphase

KEY WORDS: replication of DNA, centromere

Replication of DNA occurs during interphase, before the cell visibly enters into mitosis. The double helix of the chromosome unwinds, and each strand acts as a template for the development of a complementary strand of DNA that contains an exact copy of the sequence of molecules in the original DNA (Fig. 1-21). A new double helix is formed that contains a new strand and a parent strand of DNA. Replication begins at the ends of the chromosomes and progresses toward the center, where a small area of the chromosome, the **centromere,** remains unduplicated. Shortly thereafter, the chromosomes begin to coil, shorten, and become visible within the nucleus, and the cell enters the prophase of mitosis.

Prophase

KEY WORDS: chromatid, centriole replication

In prophase, the cell assumes a more spherical shape and appears more refractile, while in the nucleus, chromosomes become visible and appear as threadlike structures. At prophase each chromosome consists of two coiled sub-units called **chromatids** that are closely associated along their lengths. Chromatids are the functional units of chromosomes, and each contains a double strand of DNA. As prophase progresses, the chromatids continue to coil, thicken, and shorten, reaching about one-twenty-fifth of their length by the end of prophase. The chromosomes then appear to approach the nuclear envelope.

As these events occur, nucleoli become smaller and finally disappear, and the nuclear envelope breaks down. When this occurs, the center of the cell becomes more fluid, and the chromosomes move more freely, making their way to the equator of the cell. Simultaneous with the nuclear events, the **centrioles replicate,** and the resulting pairs migrate to the opposite poles of the cell.

Metaphase

KEY WORDS: mitotic spindle, equatorial plate, continuous fiber, centromere (kinetochore), chromosomal fiber, splitting of centromere

The disappearance of the nuclear envelope marks the end of prophase and the beginning of metaphase, which is characterized by formation of the **mitotic spindle** and the alignment of chromosomes along the equator to form the **equatorial plate.** The mitotic spindle is a somewhat diffuse body formed mainly of microtubules; those which pass from pole to pole of the spindle are called **continuous fibers.** Other microtubules extend from the poles of the spindle to attach to the **centromere (kinetochore)** of each chromosome and form the **chromosomal fibers.** The centromere is a specialized region of unduplicated DNA and protein that holds together the chromatids of each chromosome and forms an attachment site for the chromosomal fibers.

The final act of metaphase is duplication of the DNA at the centromere, after which the **centromeres split.** The two chromatids of each chromosome separate and begin to migrate toward the centrioles at the opposite poles of the cell. Duplication of the centromeres and migration of chromatids occur simultaneously in all chromosomes of a given cell.

Anaphase

KEY WORDS: telomere, daughter chromosome

The initial separation of chromatids marks the beginning of anaphase. As the chromatids move toward the opposite poles, the centromeres travel in advance of the **telomeres** (arms) of the chromosomes that trail behind. Movement of chromatids, which now are **daughter chromosomes,** is an active, dynamic process, but the mechanisms by which the movement is effected are not certain.

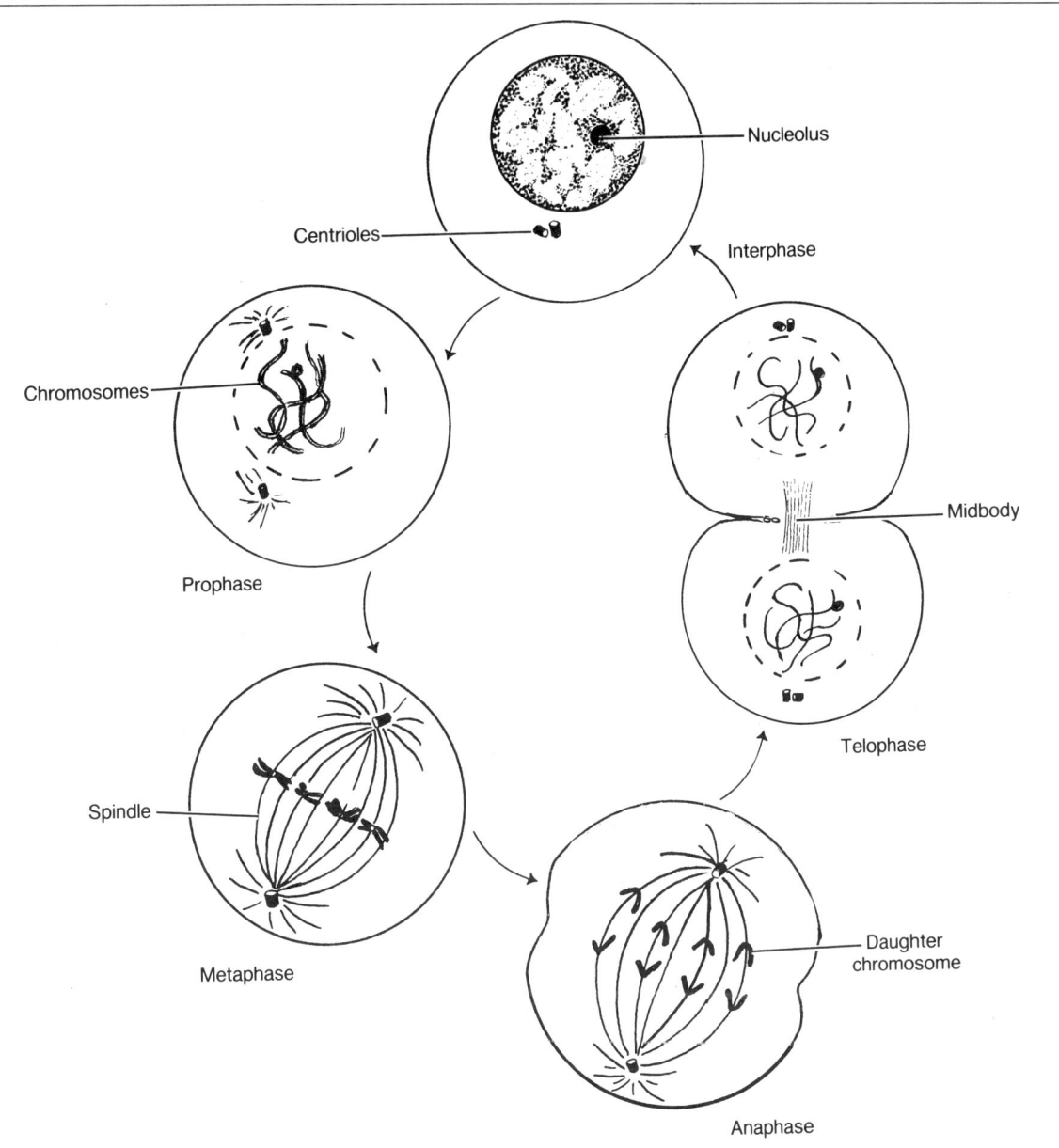

Fig. 1-20. Changes occurring during mitotic division of a cell.

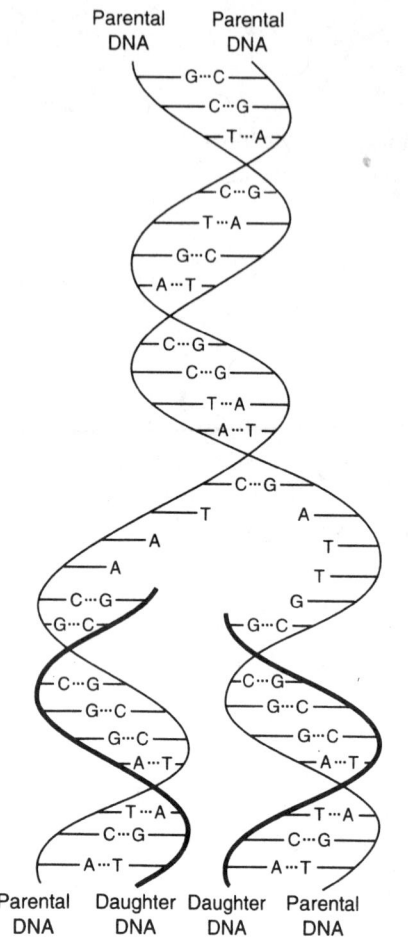

Parental DNA   Parental DNA

Parental DNA   Daughter DNA   Daughter DNA   Parental DNA

Fig. 1-21. Semiconservative replication of DNA. Each replicated DNA contains one of the original DNA strands (parental strands shown as thin black lines) and one newly made strand (daughter strand shown as thick black lines).

Telophase

KEY WORDS: re-forming nuclear envelope, nucleolus-organizing region, karyokinesis

As the daughter chromosomes reach their respective poles, discontinuous portions of endoplasmic reticulum form around each group of chromosomes and begin to **re-form the nuclear envelope.** This event initiates telophase. When the nuclear envelope has been re-formed completely, the chromosomes uncoil and become indistinct, and the two nuclei reassume the interphase configuration. The normal complement of nucleoli also reappears at this time. Their development is associated with specific **nucleolus-organizing regions** present on certain chromosomes. With this event **karyokinesis** is complete.

Cytokinesis

KEY WORDS: interzonal fiber, midbody

During telophase the mitotic spindle begins to disappear. The fibers between the two forming nuclei appear stretched and often are called **interzonal fibers.** Midway between the two nuclei, in the region formerly occupied by the equatorial plate, a constriction of the plasmalemma forms a furrow that extends around the equator of the cell. The constriction extends deeper into the cytoplasm, separating the daughter chromosomes until they are joined only by a thin cytoplasmic bridge, called the **midbody,** that contains interzonal fibers. Eventually the daughter cells pull away from each other by ameboid movement, thus completing the separation of the cells and ending cytokinesis.

Cell organelles are evenly distributed between the two daughter cells. Immediately after division, the daughter cells enter a phase of active RNA and protein synthesis, resulting in an increase in the volume of the nucleus and cytoplasm. The endoplasmic reticulum and the Golgi complex are restored to their original concentrations, mitochondria reproduce by fission, and the centrioles replicate in the daughter cells just before the next division.

Functional Review: Mitosis

Mitosis produces daughter cells that have a genetic content identical to that of the parent cell. What determines if and when a cell divides is not known, but there are certain requirements that must be met before a cell can enter mitosis. There is some relationship between cell mass and cell division. For example, in *Amoeba proteus* mitosis can be prevented by periodic amputation of the cytoplasm. The cell then merely regenerates the lost cytoplasm without entering mitosis. In general, each daughter cell achieves the mass of the parent cell before it can divide. However, the relationship between cell growth and mitosis is not a causal one, and the two events can be separated in time. An example of this is seen in cleavage of ova, where growth of the cell occurs long before division takes place.

Duplication of the entire DNA complement is essential for cell division. Depriving cells of thymidine blocks DNA synthesis and prevents cells from entering mitosis. Provision of thymidine to such cells results in a wave of DNA synthesis followed by a wave of mitosis. Again, however, DNA duplication may be completed long before division occurs, and the existence and maintenance of polyploid cells argue against DNA synthesis being the trigger for mitosis.

The intense coiling and contracting of chromosomes at metaphase results in small, compact units that can be transported more easily to the poles of the cells. The destination of the chromosomes, their orderly arrangement at metaphase, and the plane of cell cleavage are determined by the mitotic spindle. It would seem reasonable that one of the preparations for mitosis must be synthesis of protein specifically for the formation of the spindle. Certain amino acid analogues result in synthesis of faulty spindle proteins, and cells cultured in the presence of these analogues fail to divide. When removed to normal media, the cells enter mitosis but only after a delay, during which the faulty spindle protein is replaced by newly synthesized normal protein.

Centrioles are responsible for polarization of the mitotic spindle and may play a role in assembling the spindle, but how this is achieved is unknown. Centrioles complete their duplication before mitosis begins, and cell division can be prevented by suppressing the replication of centrioles.

Mitosis must be an energy-consuming process, but cell division is not a period of intense respiratory activity. Restriction of energy sources and inhibition of respiration and oxidative phosphorylation do not stop mitosis once it has begun. Thus it can be assumed that the energy needs are met during preparation for cell division.

Cell division often occurs in waves, with patches of cells or whole tissues undergoing synchronous division, implying some kind of cellular or tissue control. Substances that promote or initiate mitosis have been isolated from epidermis and salivary glands in several species, and other substances that appear to suppress mitosis (chalones) have been identified in a number of different tissues. Thus the cell population and the rate of cell division may be governed by the reciprocal action of agents that control the rate of entry into mitosis either by stimulation or suppression of mitotic activity.

Review Questions and Answers

Questions

1. How do organelles differ from inclusions?
2. In what ways do mitochondria differ from other membrane-bound organelles?
3. Briefly discuss the role of the cytoplasmic microtubules.
4. What is a unit membrane? List structures bounded by a unit membrane.
5. What functions do mitochondria serve?
6. Trace the synthetic pathway of a glycoprotein.
7. Describe the cytoplasmic events that occur during mitosis. What is achieved by mitotic division?

Answers

1. Organelles are specialized, dynamic cellular units that perform specific roles in the functional activities, survival, and reproduction of cells; they constitute a part of the living substance of a cell. Inclusions, on the other hand, can be regarded as nonliving components. They make no contribution to the economy of the cell in which they are found and generally are either the end product of activities carried out by the organelles or foreign materials taken into the cell.

2. Mitochondria differ in several respects from other cell organelles. They are limited by two distinct membranes (each of which represents a unit membrane). They are capable of self-replication by fusion and contain their own complement of DNA. The mitochondrial DNA is unusual in that it has a circular rather than a linear (helical) arrangement. Although centrioles also are self-replicating organelles, they are not membrane-bound.

3. Microtubules are involved in the formation of the mitotic spindle, contribute to the cytoskeleton of the cell,

and may be involved in the transport of materials from one part of a cell to another. They constitute the structural elements of centrioles, flagella, cilia, and ciliary basal bodies and have been implicated in the physical movement of cells.

4. The unit membrane is a trilaminar structure representative of all biologic membranes. Morphologically it appears as two dense lines separated by a clear line, reflecting the molecular configuration of a phospholipid-lipid bilayer in which the hydrocarbon chains are directed inward and the polar groups outward. Although all the membranous structures of the cell have the configuration of the unit membrane, they vary in their enzymatic properties and functional activities in different locations. Structures bounded by unit membranes include the cell itself (the plasmalemma), the nucleus, mitochondria, lysosomes, peroxisomes, transport vesicles, and secretion granules. Unit membranes enclose the tubules and cisternae of the smooth and granular endoplasmic reticulum and form the Golgi saccules. Pinocytotic vesicles and phagocytic vacuoles also are bounded by unit membranes. The only organelles not enclosed by unit membranes are ribosomes, centrioles, filaments, and microtubules.

5. Mitochondria have been called the "power house" of the cell because, ultimately, they provide the energy required for the metabolic activity of the cell. The enzymes of the Krebs citric acid cycle, responsible for the oxidation of pyruvate to carbon dioxide and water, are present in the mitochondrial matrix. The energy released by this oxidation is accepted by the electron transfer system of cytochromes, which are capable of fixing the energy in the high-energy bonds of ATP. The enzymes of this system are located in mitochondria either in the elementary particles or in the membranes that form the cristae. Mitochondria also contain several enzymes that implicate them in the synthesis of steroid hormones. Enzymes involved in protein and lipid synthesis also occur in the mitochondria.

6. Synthesis of the protein moiety of glycoprotein occurs on the ribosomes that are attached to granular endoplasmic reticulum. However, the ribosomes are inactive until they become associated with messenger RNA produced in the nucleus. The mRNA transmits the instructions for protein synthesis (encoded in the DNA of the nucleus) to the ribosomes, where it dictates the order in which amino acids are to be incorporated. The amino acids are transported to the ribosomes on yet another form of RNA, the transfer RNA, there being a specific tRNA for each amino acid. The ribosomes assemble the amino acids into the polypeptide chain and release it into the cisternae of the granular endoplasmic reticulum. The peptides are transported to the Golgi complex in small transport vesicles, the membranes of which are contributed by the endoplasmic reticulum. The protein material passes through the stacks of Golgi saccules, entering at the forming face and leaving at the maturing face. During passage through the Golgi complex, the protein is conjugated to carbohydrate synthesized by the Golgi. The newly formed glycoprotein is condensed and packaged into condensing vacuoles, which develop as the Golgi saccules at the maturing face bud off to form small vesicles that contain the glycoprotein product. The material is further condensed, and the vacuoles eventually mature into secretory granules.

7. The nuclear changes taking place during mitosis are so dramatic and visible that the importance of the cytoplasmic events tends to be overlooked. As a cell enters mitosis it appears to round up and the cytoplasm becomes clear and less viscous. During prophase the centrioles replicate and the resulting pairs migrate to the opposite poles. The nuclear envelope breaks down and by the end of prophase has disappeared. The cytoplasm at metaphase is characterized by the appearance of numerous microtubules that form the mitotic spindle. Some of the spindle fibers extend from pole to pole as the continuous fibers; others extend from one pole to the centromere (kinetochore) of a chromosome as the chromosomal fibers. The microtubules of the spindle may contribute to the movement of chromosomes from the equatorial plate to the opposite poles of the dividing cells.

After the daughter chromosomes have completed their migration, discontinuous segments of the nuclear envelope begin to form from surrounding segments of the granular endoplasmic reticulum. The nuclear envelope becomes complete following the reappearance of nucleoli in the daughter nuclei. Simultaneous with these events, a cleavage furrow develops midway between the daughter nuclei as a result of an invagination (contraction) of the plasmalemma. The contraction continues until the daughter nuclei are united only by a slender cytoplasmic bridge that contains numerous spindle microtubules; ameboid movement also aids in separating the daughter cells. Before the appearance of the cleavage furrow, mitochondria migrate in nearly equal numbers to the opposite poles of the dividing cells, and they and the remainder of the cytoplasmic organelles are partitioned nearly equally between the two daughter cells.

Immediately following cytokinesis, the mitochondria replicate by fission to restore their numbers to that of the parent cell. The centrioles, also self-replicating organelles, duplicate in the period following cytoplasmic division. The daughter cells enter a period of metabolic activity following their separation, during which time the endoplasmic reticulum, Golgi membranes, and other organelles are restored to the concentrations found in the parent cell.

The end result of mitosis is the provision of two daughter cells, each with the identical genetic makeup of the parent cell and each with a complement of organelles necessary to carry out the normal cell activities and functions.

2 Epithelium

Epithelium is one of the four basic tissues, the others being connective, muscle, and nerve tissue. A basic tissue is a collection of cells of similar type that, together with their associated extracellular substances, are specialized to perform a common function or functions. Each basic tissue is present, in variable amounts, in the organs that collectively make up the entire organism.

The various epithelia consist of closely aggregated cells with only minimal amounts of intervening intercellular substances.

Structural Organization

KEY WORDS: sheet, basal lamina, lamina lucida, lamina densa, basement membrane, avascular, gland

The cells of an epithelium form **sheets** that cover the external surfaces of an organism and line the digestive, respiratory, cardiovascular, and urogenital tracts. Thus substances that enter or leave the body must pass through an epithelial layer. Cells within an epithelial sheet are bound firmly together and resist forces that tend to separate them. The space between the membranes of adjacent epithelial cells is narrow (about 20 nm) and contains fibronectin and a small amount of proteoglycan that is rich in cations, chiefly calcium. Glycoproteins associated with the plasmalemma are integral (transmembrane) glycoproteins that act as specialized adhesion molecules that aid in holding adjacent epithelial cell membranes in close opposition.

Epithelial cells rest on a **basal lamina** that separates the epithelium from underlying connective tissue. The basal lamina consists of an electron-lucid **lamina lucida** immediately adjacent to the epithelium and a denser layer, the **lamina densa,** next to the connective tissue. The basal lamina, a product of the epithelial cells, consists of the glycoprotein laminin, limited mainly to the lamina lucida, a proteoglycan rich in heparan sulfate, limited mainly to the lamina densa, and type IV collagen. It is reinforced on the connective tissue side by a layer of reticular fibers embedded in proteoglycan. The basal lamina provides support and a surface for the attachment of the overlying epithelium. The basal lamina and the interface of the underlying connective tissue (reticular lamina) together make up the epithelial **basement membrane** that is visible with the light microscope.

With rare exceptions, epithelium is **avascular** and depends on diffusion of substances across the basement membrane for its nutrition. Epithelial cells also occur in groups to form **glands** specialized for the secretion of various substances.

Classification of Epithelia

KEY WORDS: simple epithelium, stratified epithelium, squamous, cuboidal, columnar

The lining and covering forms of epithelium may consist of a single layer of cells, forming a **simple epithelium,** or multiple layers of cells may be present to form a **stratified epithelium.** The epithelial cells themselves can be divided into three types according to their shapes: **squamous** (thin, platelike cells), **cuboidal** (cells in which the height and width are approximately equal), and **columnar** (cells in which the height is greater than the width). Cells intermediate in height between cuboidal and columnar do occur and frequently are referred to as *low columnar.*

Simple Epithelia

KEY WORDS: simple squamous, simple cuboidal, simple columnar, pseudostratified columnar, apical surface, basal surface, lateral surface, polarity, mesothelium, endothelium

Epithelium consisting of a single layer of cells can be classified as **simple squamous, simple cuboidal,** or **simple columnar,** depending on the shape of the constituent cells (Figs. 2-1 through 2-3). A fourth type of simple epithelium, **pseudostratified columnar,** is composed of more than one type of cell whose nuclei are at different levels, falsely suggesting that the epithelium is made up of two or more layers (Fig. 2-4). While all the cells in this type of epithelium are in contact with the basement membrane, not all extend to the surface.

The surfaces of cells within an epithelial sheet show different orientations. The **apical surface** is free, the opposite or **basal surface** is directed toward the underlying connec-

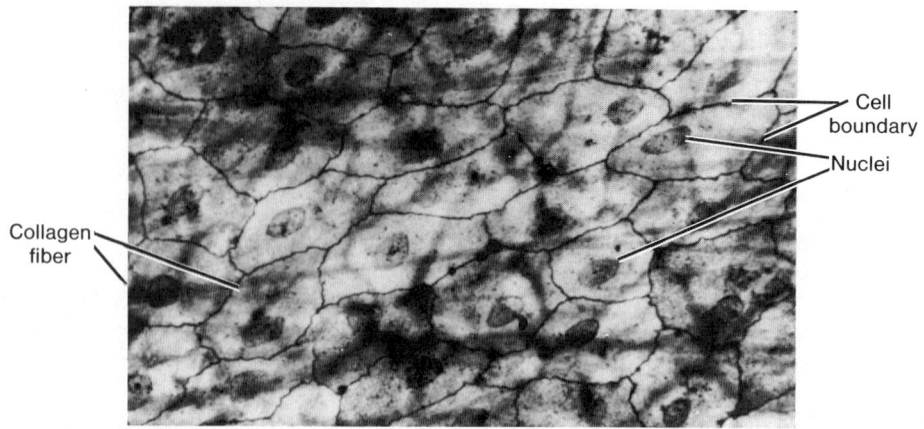

Fig. 2-1. A portion of human mesentery treated with silver nitrate to demonstrate cell boundaries. Note the collagen fibers showing through the translucent cytoplasm of the simple squamous mesothelial cells (LM, × 400).

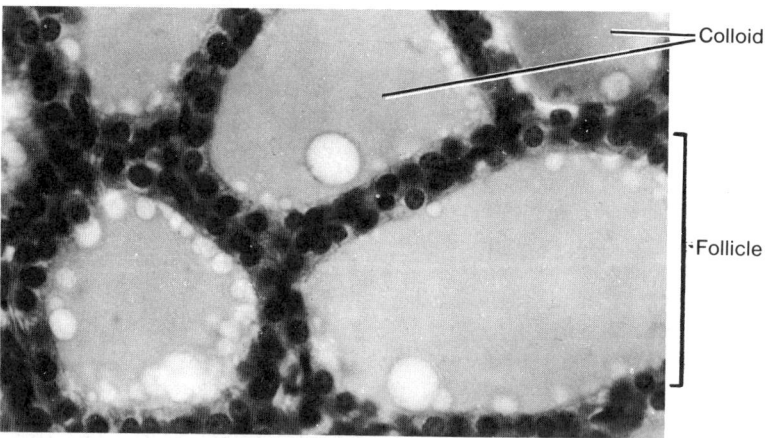

Fig. 2-2. An example of simple cuboidal epithelium taken from a human thyroid. Note that several thyroid follicles lie immediately adjacent to one another (LM, × 300).

tive tissue, and the **lateral surfaces** face adjacent epithelial cells. Differences in the orientation of the plasmalemma (i.e., apical, basal, or lateral) are associated with differences in cell function and morphology. Organelles, including the nucleus, frequently take up preferred locations in the cell, in which case the cell is said to show **polarity.** An epithelial cell may be polarized apically, basally, or later-

ally with respect to the distribution of its organelles, depending on the functional status of the cell.

The term **mesothelium** is the special name given to the simple squamous epithelia that form the serous membranes of the pleura, pericardium, and peritoneum. A specific name, **endothelium,** is given to the simple squamous epithelium that lines the cardiovascular and lymph vascular systems.

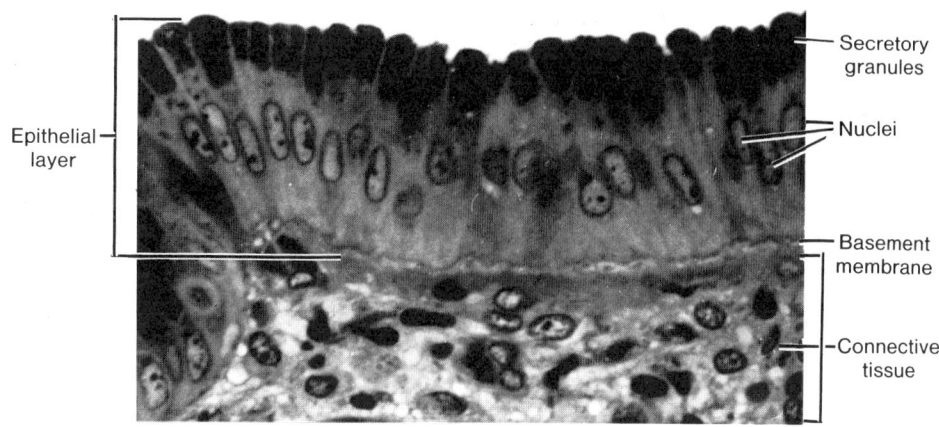

Fig. 2-3. A region of simple columnar epithelium taken from the lining of a human stomach. Note that the epithelium lies on a distinct basement membrane and that the apical region of each cell is filled with secretory granules (LM, × 500).

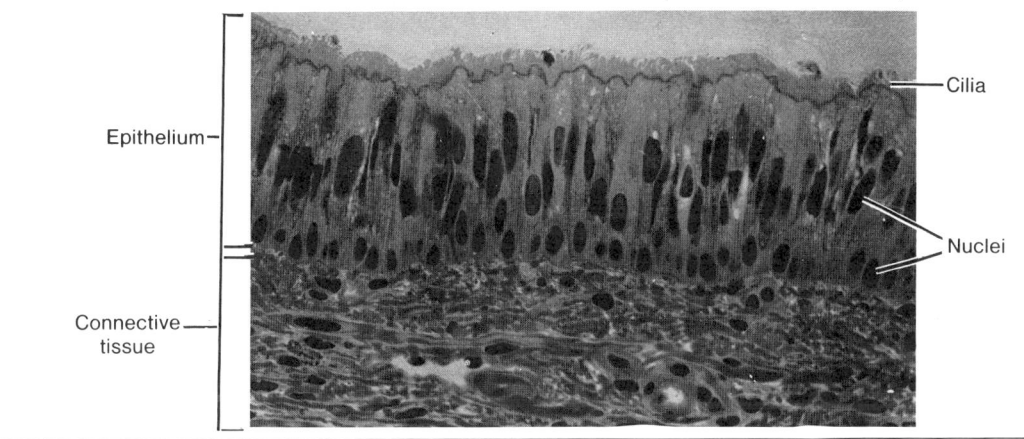

Fig. 2-4. A demonstration of ciliated pseudostratified columnar epithelium taken from the human trachea. Nuclei are found at different levels within this simple epithelium — hence its name (LM, × 400).

Stratified Epithelia

KEY WORDS: stratified squamous, stratified cuboidal, stratified columnar, keratinized stratified squamous, nonkeratinized stratified squamous, transitional epithelium, germinal epithelium

Stratified epithelia are classified according to the shape of the superficial (surface) cells, regardless of the shapes of the cells in the deeper layers. Except for those in the basal layer, cells of stratified epithelia are not in contact with the basement membrane, and only the most superficial cells have a free surface. In the thicker stratified types, the deeper cells often are irregular in shape. The types of stratified epithelia are **stratified squamous, stratified cuboidal,** and **stratified columnar.**

In regions where the surface is dry or subject to mechanical abrasion, the outer layers of cells are transformed and filled with a protein called *keratin*. Keratin is an especially tough protein that is resistant to mechanical injury and is relatively impervious to bacterial invasion or water loss. This type of stratified epithelium is referred to as **keratinized stratified squamous** epithelium (Fig. 2-5). The keratinized layer of cells at the surface is constantly being shed and replaced. In other regions the stratified squamous type of epithelium is moist, and recognizable, intact cells rather than flat, keratin-filled cells are sloughed from the surface. This type of epithelium is called a wet or **nonkeratinized stratified squamous** epithelium (Fig. 2-6).

Transitional and germinal epithelia are not classified according to the cells of the surface layer. **Transitional epithelium** (uroepithelium) is restricted to the lining of the urinary passages and extends from the minor calyces of the kidneys to the urethra. Its appearance varies considerably, depending on the degree of distension to which it is sub-

Fig. 2-5. Keratinized stratified squamous epithelium taken from the skin covering a human thigh. Note the layer of keratin and that cells forming the surface-most layer have lost their nuclei (LM, × 250).

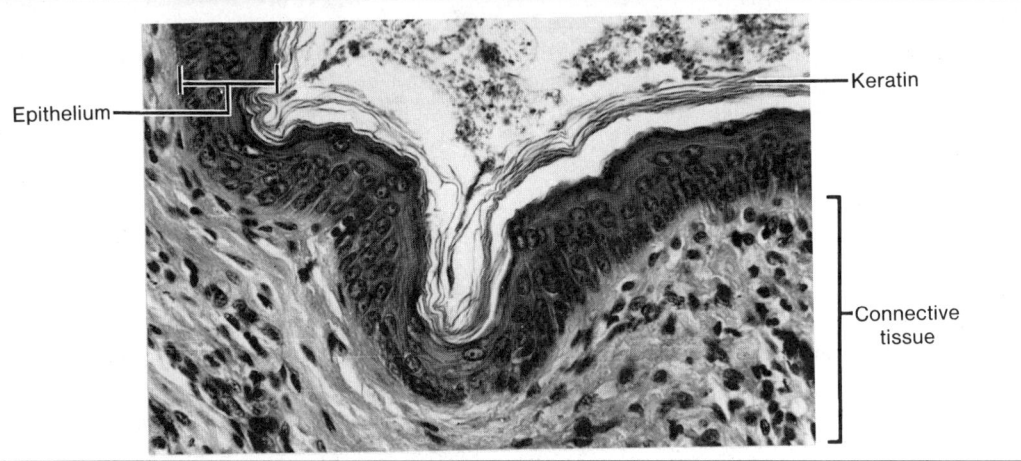

Fig. 2-6. An example of nonkeratinized stratified squamous epithelium. Note that the surface cells retain their nuclei (LM, × 300).

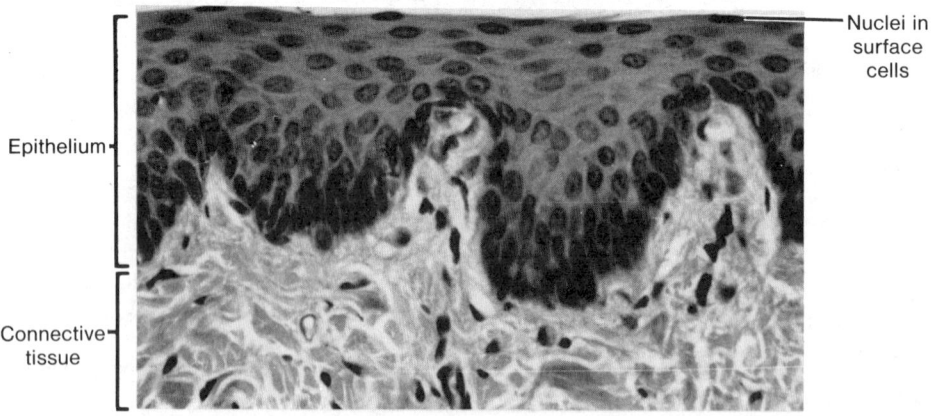

jected. The superficial cells of a nondistended organ appear rounded or dome-shaped and often show a free convex border that bulges into the lumen (Fig. 2-7). In the distended organ, the superficial cells may vary in shape from squamous to cuboidal. **Germinal epithelium** in the seminiferous tubules of the testes is a complex, stratified epithelium that consists of supporting cells and spermatogenic cells.

A classification of epithelia and the locations where each type can be found is presented in Table 2-1.

Fig. 2-7. Transitional epithelium lining the urinary bladder is of considerable depth. The surface-most cells often appear dome-shaped in the nondistended condition, and many surface cells are binucleate (LM, × 300).

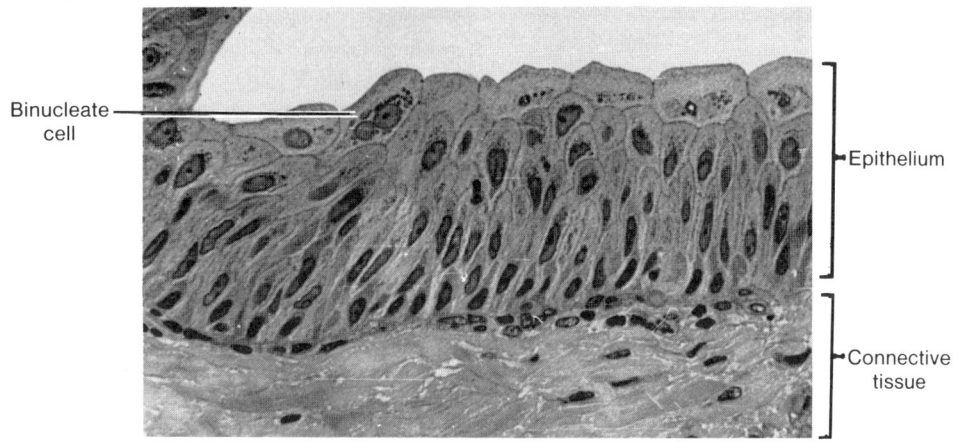

Table 2-1. Classification of Epithelia

Type of Epithelium	Location	Specialization
Simple squamous	Endothelium, mesothelium, parietal layer of Bowman's capsule, thin segment of loop of Henle, rete testis, pulmonary alveoli	
Simple cuboidal	Thyroid, choroid plexus, ducts of many glands, inner surface of the capsule of the lens, covering surface of ovary	
Simple columnar	Surface epithelium of stomach	
	Small and large intestine	Striated border
	Proximal convoluted tubule of kidney	Brush border
	Distal convoluted tubule of kidney	Basal striations
	Gallbladder, excretory ducts of glands	
	Uterus, oviducts	Cilia
	Small bronchi of lungs	Cilia
	Some paranasal sinuses	Cilia
Pseudostratified columnar	Large excretory ducts of glands, portions of male urethra	
	Epididymis	Sterocilia
	Trachea, bronchi	Cilia
	Eustachian tube	Cilia
	Portions of tympanic cavity	Cilia
Stratified squamous	Buccal surface, esophagus, epiglottis, conjunctiva, cornea, vagina	
	Epidermis of skin	Keratin
	Gingiva, hard palate	Keratin
Stratified cuboidal	Ducts of sweat glands	
Stratified columnar	Cavernous urethra, fornix of conjunctiva, large excretory ducts of glands	
Transitional	Urinary system; renal calyces to urethra	
Germinal	Seminiferous tubules of adult testes	

Cell Attachments

In addition to the intercellular matrix, several attachment points or junctions occur between neighboring epithelial cells that aid in holding adjacent cell membranes close together. They also serve as anchoring sites for microfilaments and intermediate filaments of the cytoskeleton, which assists in stabilizing the cell shape.

Junctional Complexes

KEY WORDS: terminal bar, zonula occludens (tight junction), zonula adherens, macula adherens (demosome), plaque, tonofilament, punctum adherens, hemidesmosome, focal adherens

Terminal bars occur along the apical, lateral interfaces of most simple cuboidal and columnar epithelial cells that border on luminal surfaces. They are visible under the light microscope as short, dense lines. In electron micrographs, a terminal bar is seen to form an area of specialization called the *junctional complex*, which consists of three distinct regions (Fig. 2-8). Near the free surface, the external laminae of adjacent plasma membranes fuse, and the intercellular space is obliterated. This region is the **zonula occludens (tight junction),** an area of specialization that forms a zone or belt around the perimeter of each cell to create an occluding seal between the apices of adjacent epithelial cells. Materials that pass through the epithelial sheet must cross the plasma membranes and cannot pass between cells through the intercellular spaces. The zonula occludens junction is important in separating apical from basolateral cell membrane domains by preventing the lateral migration of specialized membrane proteins (e.g., receptors, transport proteins) in the plasmalemma. In most cases, membrane proteins in the apical plasmalemma function quite differently from those occupying the basolateral regions. The plasmalemma of the apical domain is rich in cholesterol and glycolipids and contains ion channels, transporter proteins, hydrolytic enzymes, and hydrogen-ATPase. Regulated secretory products are released from this domain by exocytosis. In contrast, the plasmalemma of the basolateral domain is characterized by receptors for hormones, anion channels, and sodium-potassium ATPase and is where most constitutive secretion occurs. As a result of this distribution, the plasmalemma of epithelial cells functions to regulate the flow of ions and solutes between external environments and body fluids as well as between compartments within the body. Tight junctions also have been reported in simple squamous epithelia and in the superficial layers of stratified squamous epithelia.

The second part of the junctional complex, the **zonula adherens,** also forms an adhering belt or zone that surrounds the apex of an epithelial cell, immediately below the zonula occludens. Adjacent plasma membranes are separated by an intercellular space 15 to 20 nm wide, and

Fig. 2-8. A tripart junctional complex between two gastric lining epithelial cells of a human stomach. This complex is formed by a zonula occludens, a zonula adherens, and a macula adherens (TEM, × 22,500).

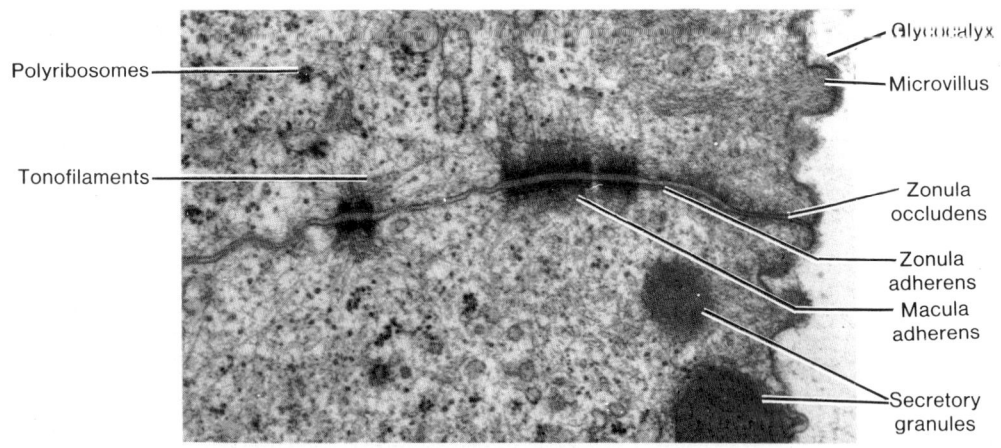

Polyribosomes

Tonofilaments

Glycocalyx

Microvillus

Zonula occludens

Zonula adherens

Macula adherens

Secretory granules

in this region plaques along the internal laminae of the plasma membranes are associated with an actin filament network. The plaques contain α-actinin and vinculin, two actin-binding proteins. Actin is linked to a transmembrane glycoprotein (calherin) that holds cells together in this region in the presence of calcium ions.

The third element of the junction is the **macula adherens,** also called the **desmosome.** These small, elliptical discs form spotlike points of attachment between cells that function to link the intermediate filament network of adjacent cells. The internal laminae of the adjacent plasma membranes appear dense and thick due to the presence of an amorphous material called a **plaque.** Intermediate filaments (usually cytokeratins) extend into the plaque, where they appear to form hairpin loops. A thin, moderately dense lamina often is seen in the intercellular space of each desmosome. These are thought to contain transmembrane linking glycoproteins (desmogleins) that hold adjacent cell membranes in close apposition. Desmosomes are scattered along epithelial surfaces and are not restricted to regions of tight junctions (Fig. 2-9). They are especially abundant in stratified squamous epithelia, where they are associated with numerous **tonofilaments** — intermediate filaments of the keratin type. Desmosome-like junctions that lack attachment plaques and associated tonofilaments also occur between epithelial cells. This type of cell junction has been called a **punctum adherens.**

The basal plasma membrane, although not immediately adjacent to another cell, also shows an attachment, the **hemidesmosome,** which closely resembles one-half of a desmosome (Fig. 2-10). Hemidesmosomes act as points of adhesion between the basal plasma membrane and the underlying basal lamina and link the intermediate keratin filament network of the epithelial cells to the extracellular matrix via transmembrane anchoring glycoproteins. These transmembrane glycoproteins act as laminin receptors and bind to laminin, a glycoprotein comprising the lamina lucida of the basal lamina.

Focal adherens type junctions also occur along the basal plasmalemma and link actin filaments to the extracellular matrix. Intracellular attachment proteins (α-actinin, vinculin, talin) link actin filaments to another of the transmembrane glycoproteins that serves as a fibronectin receptor. The latter binds to fibronectin, a multifunctional extracellular glycoprotein that interacts with various types of collagens and proteoglycans in the extracellular matrix as well as the fibronectin receptor. Additional support is provided by anchoring fibers of type VII collagen that extend from the basal lamina to form loops around collagen fibers (type I) in the subjacent connective tissue prior to terminating in anchoring plaques.

Thus, through a series of transmembrane glycoproteins that constitute a large group of cellular adhesion molecules (CAMs) or integrins, epithelial cells are not only held tightly together in a contiguous layer but also are anchored firmly to the underlying extracellular matrix. Desmosomes

Fig. 2-9. Desmosomes (maculae adherens) between two adjacent human intestinal epithelial cells (TEM, × 30,000).

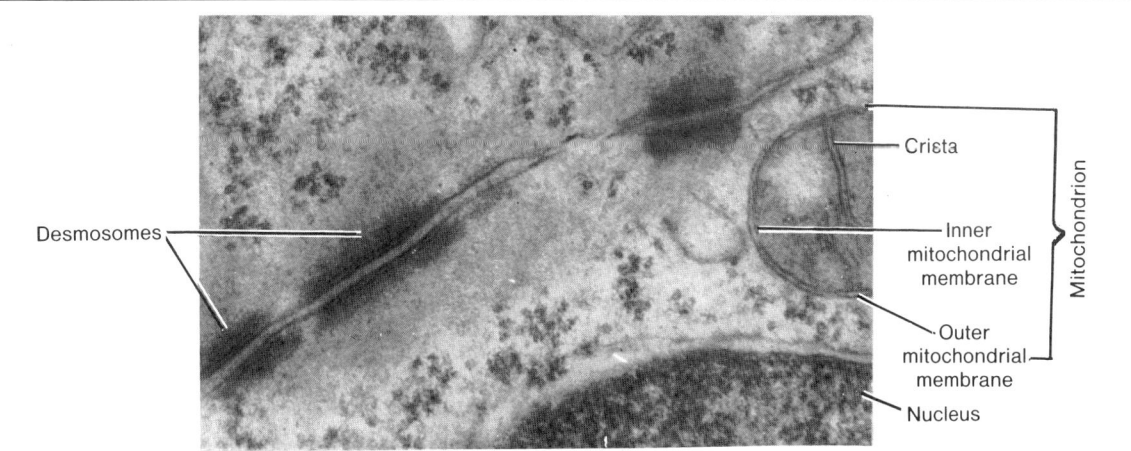

and hemidesmosomes serve as firm anchoring sites for intermediate (keratin) filaments of the cytoskeleton. In addition, desmosomes function to interconnect the keratin filament network between adjacent cells, and hemidesmosomes link this network to the extracellular matrix. Such an arrangement of interconnected intermediate filaments gives mechanical stability to a group of cells so that they can function as a cohesive unit. Likewise, actin microfilaments are linked not only to plasmalemma at zonula adherens and focal adherens junctions but also are interconnected with actin filaments of adjacent cells through these attachment points as well as to elements of the extracellular matrix. This arrangement provides epithelial cells with a degree of mobility and is particularly important in migrating epithelial cells (gastric and intestinal lining epithelial cells) that move continuously along the basal lamina. The integrity of desmosomes and hemidesmosomes appears to depend on the presence of calcium ions. Heparan sulfate–rich proteoglycans in the basal lamina give its surface a strong ionic charge that, acting with calcium ions, mediates the attachment of transmembrane linker glycoproteins to elements in the extracellular matrix.

Nexus Junction

KEY WORDS: nexus (gap junction), connexon, connexin, communicating junction

The **nexus,** or **gap junction,** like tight junctions and desmosomes, forms a relatively firm point of attachment between adjacent cells (Fig. 2-11). At a nexus, the plasma membranes are separated by a gap 2 nm wide that is bridged by particles 8 nm in diameter. These particles, the **connexons,** mainly consist of the protein **connexin.** Each connexon consists of six identical transmembrane connexin proteins that are arranged around a small aqueous channel or pore. Connexon units extend through the plasmalemma and cross the intercellular gap to join a corresponding unit from the opposing cell membrane. The hydrophilic channel within the center of each connexon permits intercellular passage of ions, amino acids, and nucleotides. The nexus also provides an avenue whereby cyclic adenosine monophosphate (cAMP), released by one cell, can pass to adjacent cells. Thus epithelial cells are electrically and metabolically coupled, and the response of a number of cells to a specific stimulus can be coordinated. Because nexus junctions act as sites of communication between epithelial cells, they often are called **communication junctions.** Nexus junctions also appear to be capable of changing from low to high resistance and thereby inhibit cell communication.

Adjacent lateral cell membranes often do not run parallel to one another but show numerous interdigitations. Such interlocking of cell membranes also is thought to aid in maintaining cell-to-cell adhesion in an epithelial sheet.

Fig. 2-10. Hemidesmosomes along the basal cell membrane of cells from the stratum basale of the epidermis. Note the abundance of tonofilaments (keratin filaments) associated with the hemidesmosomes (TEM, × 12,000).

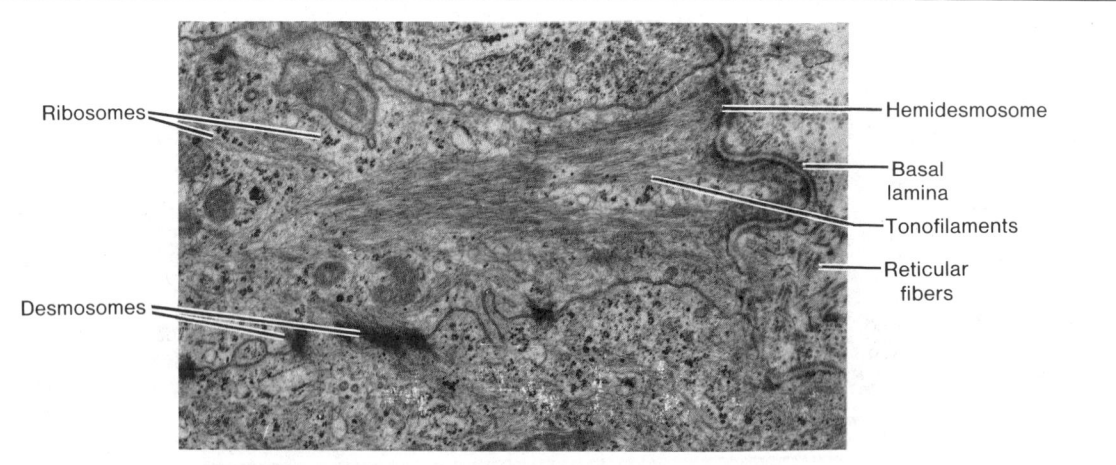

Ribosomes

Hemidesmosome

Basal lamina

Tonofilaments

Reticular fibers

Desmosomes

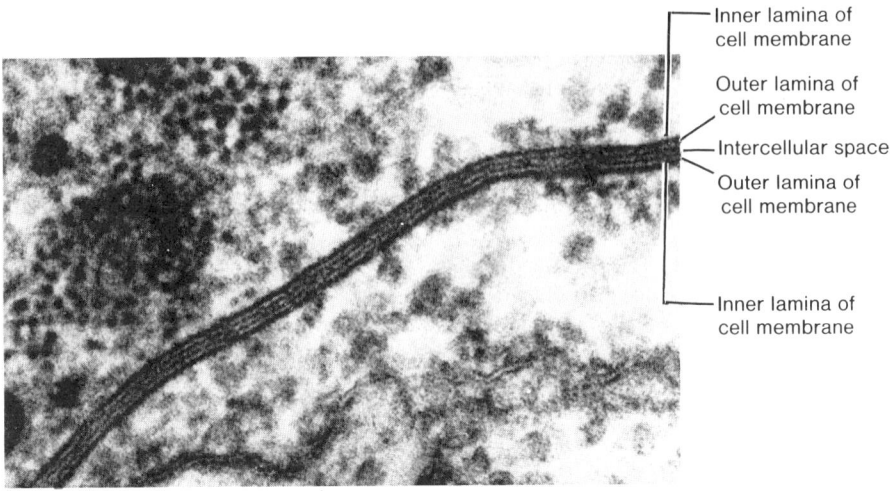

Inner lamina of
cell membrane

Outer lamina of
cell membrane

Intercellular space

Outer lamina of
cell membrane

Inner lamina of
cell membrane

Fig. 2-11. Subcomponents of a nexus (gap) junction between two adjacent epithelial cells (TEM, × 206,000).

Specializations of Epithelial Membranes

The cells in an epithelial sheet often show special adaptations of apical or basal-lateral surfaces that increase the surface area of the cell or constitute motile "appendages" that move material across the surface. The surface specializations include microvilli, cilia, and basal-lateral infoldings.

Microvilli

KEY WORDS: striated border, brush border, stereocilia

Microvilli are minute, finger-like evaginations of the apical plasma membrane that enclose cores of cytoplasm (Fig. 2-12). Scattered microvilli occur on the surfaces of most cells of simple epithelia, but individually they are usually too small to be seen with the light microscope. In

Fig. 2-12. The microvillus border of an intestinal lining epithelial cell. Individual microvilli can be visualized (TEM, × 20,000).

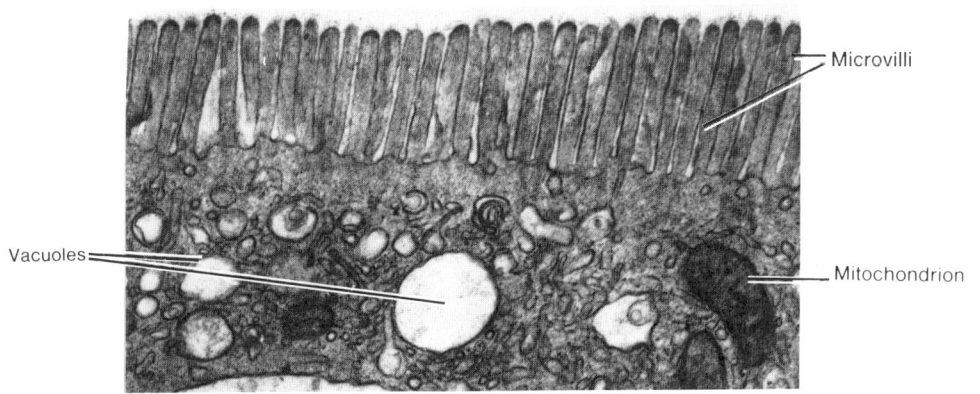

Microvilli

Vacuoles

Mitochondrion

epithelia such as that lining the small intestine, however, microvilli are so numerous, uniform, and closely packed that they form a **striated border** that is visible by light microscopy. Microvilli also are numerous on the cells of the proximal convoluted tubules of the kidney, where they are somewhat longer and less uniform than those in the small intestine and appear as tufts on the cell apices. Because of this appearance, they are called a **brush border.** Microvilli increase the surface area enormously, thereby enhancing the absorptive capacity of epithelial cells. The uniform shape of microvilli forming the striated and brush borders is maintained by a bundle of 25 to 40 actin filaments linked together by the actin-binding proteins fimbrin and fascin. Minimyosin molecules attach the actin filament bundle within the microvillus core to the inner surface of the plasmalemma. The actin filaments extend from the tips to the microvilli to the apical cytoplasm, where just beneath the cell surface they join a transverse zone rich in actin filaments called the *terminal web.*

Epithelial cells that line the epididymis show unusually long, slender, branching microvilli that have been called **stereocilia.** However, stereocilia are not cilia but are true microvilli and also serve to increase the surface area of a cell.

Cilia

KEY WORDS: microtubule, doublet, central pair, (9 + 2), dynein, basal body, triplet, (9 + 0), ciliated simple columnar epithelium, ciliated pseudostratified columnar epithelium

Like microvilli, cilia represent appendages of the apical cell membrane but are relatively large and often are motile (Fig. 2-13). They may be single or occur in large numbers and are seen readily with the light microscope. The limiting plasmalemma of the cilium is continuous with that of the cell and encloses a core of cytoplasm containing 20 **microtubules.** Nine pairs of fused microtubules, the **doublets,** form a ring at the periphery of the cilium and surround a **central pair** of individual microtubules (Figs. 2-14 and 2-15). This arrangement has been called the **(9 + 2)** configuration. Within each doublet, the microtubules are named the A and B microtubules. Microtubule A has two short, divergent arms that project toward tubule B of the adjacent doublet. The arms consist of the protein **dynein,** which has activity like adenosinetriphosphatase (ATPase). Radial spokes project from the doublets to the central pair of microtubules.

The microtubules of a cilium originate from a **basal body** in the apical cytoplasm of the cell. The basal body is similar in structure to the centriole and appears as a hollow cylinder, the wall of which consists of nine **triplet** microtubules. Since there are no central microtubules in a basal body, the arrangement of microtubules is expressed as **(9 + 0).** Each of the nine doublets in the cilium is continuous with the two innermost microtubules of the triplets in the basal body. The two central microtubules of the ciliary shaft terminate as they reach the basal body and do not join with it. Ciliary motion is thought to involve a sliding tubule mechanism, dependent on the ability of the dynein protein arms to generate energy through its ATPase-like activity.

Fig. 2-13. Surface specializations (cilia, microvilli) of the ciliated simple columnar epithelium lining the oviduct (SEM, × 5,000).

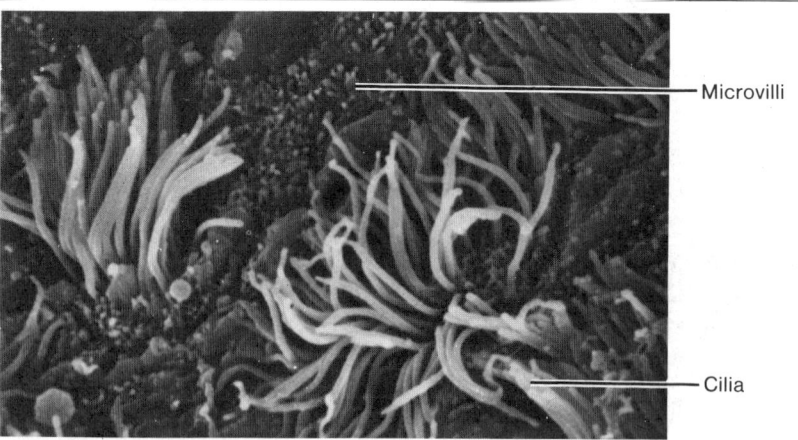

Microvilli

Cilia

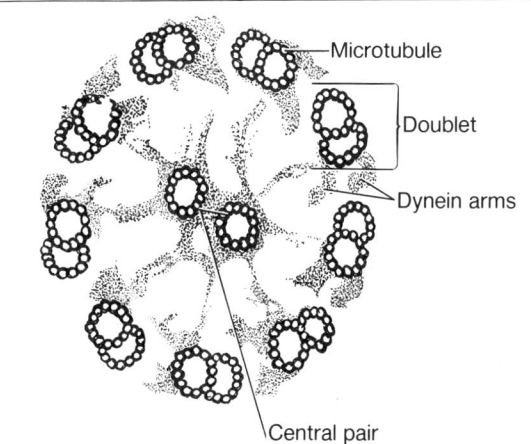

Fig. 2-14. Diagrammatic representation of a cilium in cross section.

A rare type of immotile cilia syndrome (Kartagener's) occurs in which the dynein arms are absent. Patients with this congenital disorder are subject to respiratory infection because clearance from the lungs is impaired due to the absence of coordinated ciliary motion within the respiratory tree. Men with this syndrome are usually infertile because the axonemes of sperm flagella also are affected.

Cilia are numerous on epithelial cells that line much of the respiratory tract and parts of the female reproductive tract. Groups of cilia beat in a coordinated, rhythmic fashion to move a thin layer of fluid in one direction across the epithelial surface. In regions such as the distal tubules of the kidney, where each cell bears a single cilium, or in the olfactory epithelium, where they occur in small groups on cells, the cilia are nonmotile and are believed to act as chemoreceptors. Cilia also are found in the maculae and cristae of the inner ear and act as mechanoreceptors. The most extreme modification of cilia is seen in the formation of the outer segments of the rods and cones of the retina, where the modified cilia serve as photoreceptors.

The presence of cilia contributes to the classification of epithelia. Where cilia are present on simple columnar epithelial cells, as in the oviduct, the epithelium is called a **ciliated simple columnar epithelium.** When cilia are associated with pseudostratified epithelium, as in the trachea, the epithelium forms a **ciliated pseudostratified columnar epithelium.**

Basal-Lateral Infoldings

KEY WORDS: basal-lateral plasmalemma, fluid transport, basal striations

In addition to the apical specializations, some epithelial cells show elaborate infoldings of the **basal-lateral plasmalemma.** These are called *basal-lateral infoldings* and

Fig. 2-15. Cilia and microvilli from epithelial cells lining the trachea as seen in cross section. Compare the diameters and note the central axoneme of microtubules within each cilium. Basal bodies are seen in the apical cell cytoplasm (TEM, × 4,000; inset, × 80,000).

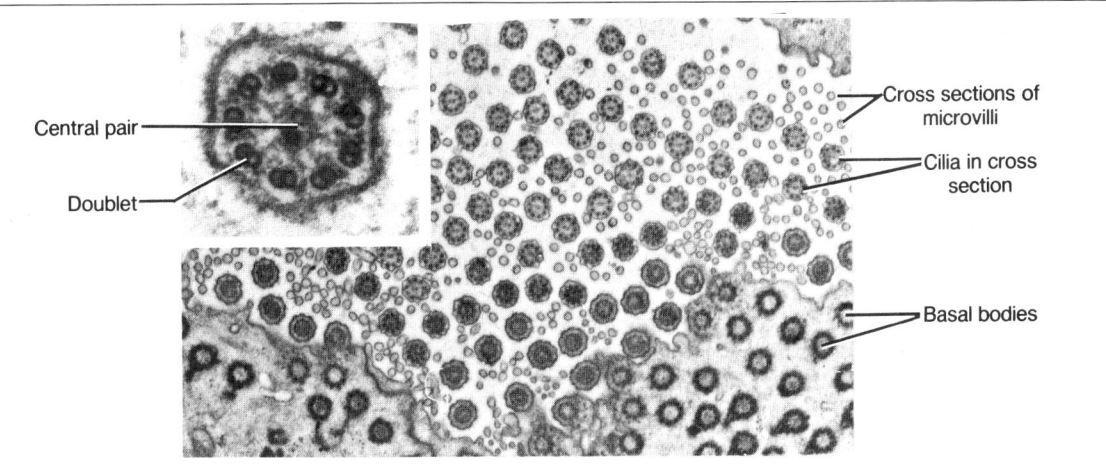

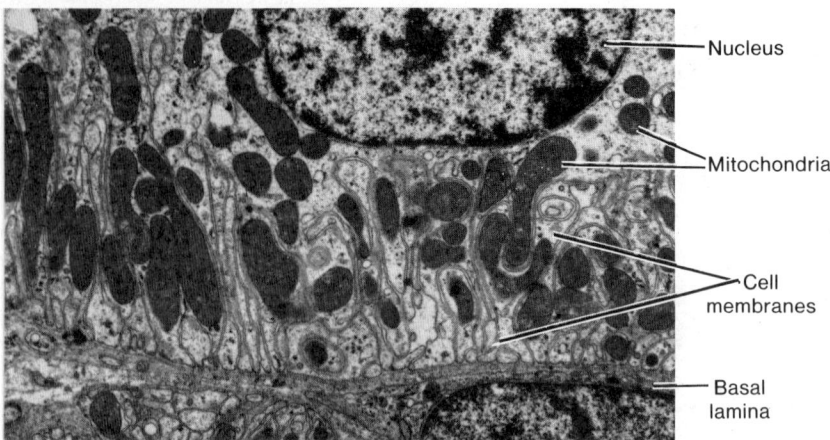

Fig. 2-16. Basal-lateral infoldings of the cell membrane from a cell within the proximal tubule of the kidney. Note the parallel alignment of associated mitochondria (TEM, × 2,000).

usually are prominent in epithelia active in **fluid transport,** such as in the proximal and distal convoluted tubules of the kidneys and in some ducts of salivary glands. Like microvilli, they greatly increase the surface area of the cell membrane. Mitochondria, which often are numerous in this area of the cell, tend to lie parallel to the infoldings of the plasma membrane (Fig. 2-16). Because of this association, the infoldings are visible with the light microscope and appear as **basal striations.**

Glandular Epithelium

In addition to performing functions such as protection and absorption, the cells of an epithelial sheet also may secrete materials. Epithelial cells adapted specifically for secretion constitute the glands.

Classification of Glands

KEY WORDS: unicellular, multicellular, exocrine, endocrine, type and mode of secretion

Glands can be classified according to their histologic organization, possession of ducts, type of material secreted, and manner in which material is secreted. Glands that consist of only a single cell are called **unicellular;** aggregates of secreting cells form **multicellular** glands. The latter can be divided into **exocrine** and **endocrine** types according to whether or not they secrete onto a surface. Exocrine glands

can be subdivided on the basis of the branching of their ducts, the configuration of the secretory end piece, and the **type and mode of secretion.** A classification of glands is shown in Table 2-2.

Table 2-2. Classification of Glands

Type	Example
Unicellular glands	
Exocrine	Goblet cells
Endocrine	Gastrin-cell and other endocrine cells of the gastrointestinal mucosa
Multicellular exocrine glands	
Secretory sheet	Surface epithelium of gastric mucosa
Intraepithelial glands	Urethral (Littre's) glands
Simple tubular glands	Intestinal glands
Simple coiled tubular glands	Apocrine sweat glands
Simple branched tubular glands	Esophageal glands
Simple branched acinar glands	Meibomian glands
Compound tubular glands	Glands of gastric cardia
Compound tubuloacinar glands	Pancreas
Compound saccular glands	Prostate

Unicellular Glands

KEY WORDS: goblet cell, mucin, mucus

The classic example of a unicellular gland is the **goblet cell,** found scattered among columnar epithelial cells of the trachea, small intestine, and colon (Fig. 2-17). The cell has a narrow base and an expanded apex filled with secretory droplets. Goblet cells elaborate **mucin,** which, on hydration, produces a viscous lubricating fluid called **mucus.** Hence a goblet cell can be classified as a unicellular exocrine gland. Unicellular endocrine cells (glands) also occur. They are numerous in the epithelium lining the gastrointestinal tract and produce a number of peptide hormones and/or amines. They also are found in the epithelium lining the respiratory system.

Multicellular Glands

KEY WORDS: secretory sheet, intraepithelial gland

The simplest form of multicellular gland is the **secretory sheet,** exemplified by the gastric lining epithelium, in which the secreting cells form a continuous epithelial layer. **Intraepithelial glands** are small clusters of secretory cells that lie wholly within an epithelial sheet, clustered about a small lumen.

Exocrine Glands

KEY WORDS: compound, simple, tubule, acinus, alveolus, saccule

Exocrine glands secrete onto a luminal epithelial surface either directly, as in secretory sheets and intraepithelial glands, or by a ductal system. If the duct branches, the gland is said to be **compound;** if it does not branch, the gland is **simple** (Fig. 2-18).

The secretory cells may form **tubules, acini, alveolae** (berry-like end pieces), or **saccules** (dilated, flasklike end pieces). Simple and compound glands can be named from the shape of the secretory portion. Thus simple glands can be classed as simple tubular, simple coiled tubular, simple branched tubular, or simple branched acinar (Figs. 2-17 and 2-19). Similarly, compound glands are subdivided into compound tubular, compound saccular, and compound tubuloacinar (tubuloalveolar). The initial subdivision into simple and compound is made according to whether or not the ducts branch. Subsequent classification depends on the shape and configuration of the secreting portion.

Endocrine Glands

KEY WORDS: lack ducts, hormone

Like exocrine glands, the endocrine glands arise from an epithelial sheet, but during their formation, endocrine glands lose their connections with the surface and thus **lack ducts** (Fig. 2-20). Cells of endocrine glands secrete regulatory material — **hormones** — directly into the bloodstream, lymphatics, or intercellular spaces (paracrine and/or autocrine secretion). Scattered, solitary endocrine cells occur in the gastrointestinal tract and along the conducting portion of the respiratory tract.

Fig. 2-17. Intestinal glands within the mucosa of the human jejunum are simple tubular glands (LM, × 250).

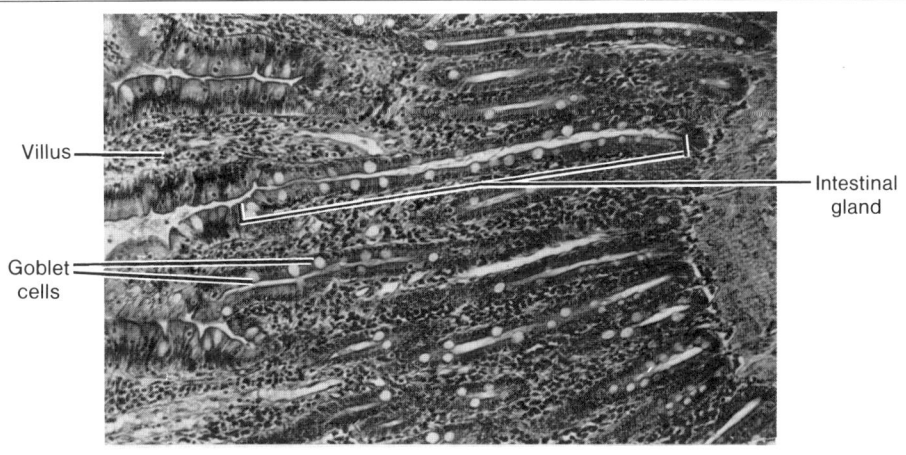

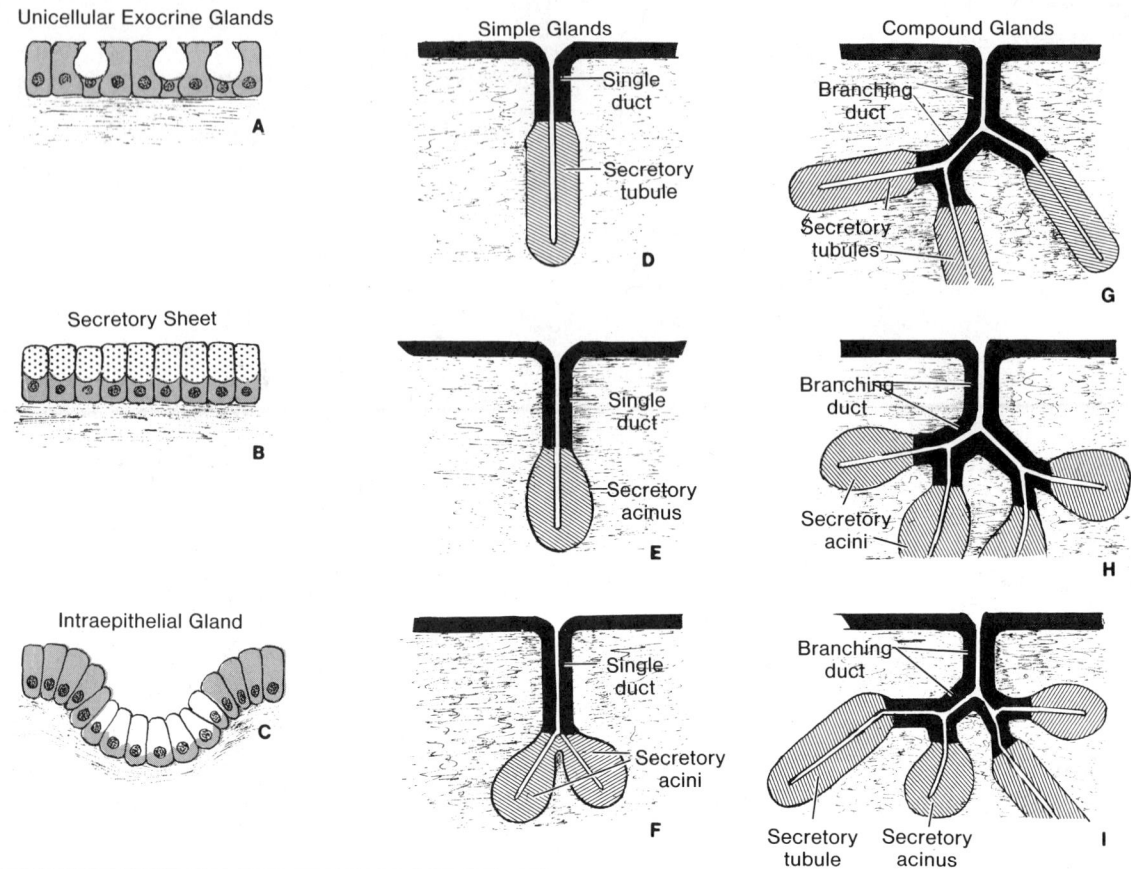

Fig. 2-18. Schematic illustration of several glandular forms. A. Unicellular glands. B. Secretory sheet. C. Intraepithelial gland. D. Simple tubular gland. E. Simple acinar gland. F. Simple branched gland. G. Compound tubular gland. H. Compound acinar gland. I. Compound tubuloacinar gland.

The structure of the large endocrine glands is so diverse that the organs do not readily lend themselves to histologic classification. However, some overall structural patterns can be made out, and the cells may be arranged as clumps, cords, or hollow spheroidal structures called *follicles*.

Type and Mode of Secretion

KEY WORDS: mucous, serous, mixed gland, merocrine, holocrine

Glands can be classified as mucous, serous, or mixed according to the type of material secreted. Cells of **mucous** glands have pale, vacuolated cytoplasm and flattened, dense nuclei oriented toward the base of the cell. **Serous** cells show spheroidal nuclei surrounded by a basophilic cytoplasm that contains discrete secretory granules. **Mixed glands** contain variable proportions of serous and mucous cells.

From the mode of secretion, two types of glands can be distinguished. **Merocrine** secretion refers to release of the product through fusion of secretory vesicles with the apical plasmalemma (Fig. 2-21). This is the process identified as exocytosis (regulated secretion) by electron microscopy, and most exocrine secretion is of this type. **Holocrine** secretion involves release of whole cells (such as sperm from the testes) or the breakdown of an entire cell to form the secretion (e.g., sebum in sebaceous glands). A third type of secretion, *apocrine*, has been described in which a portion of the apical cytoplasm was said to be lost

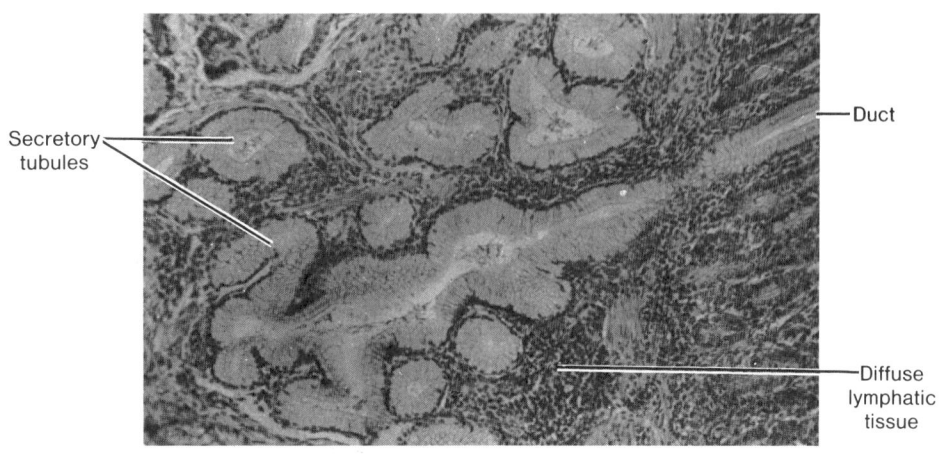

Fig. 2-19. The human duodenal gland shown is an example of a simple branched tubular gland (LM, × 200).

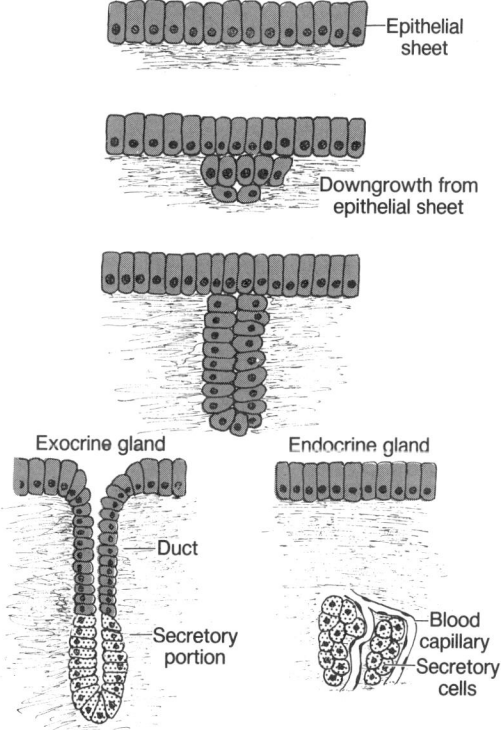

Fig. 2-20. Formation of exocrine and endocrine glands. All glands form as an invagination from an epithelial sheet. Glands that retain a connection (duct) to the epithelial sheet of origin are classed as exocrine glands. Those which lose their connection are called endocrine glands and secrete into the blood vasculature.

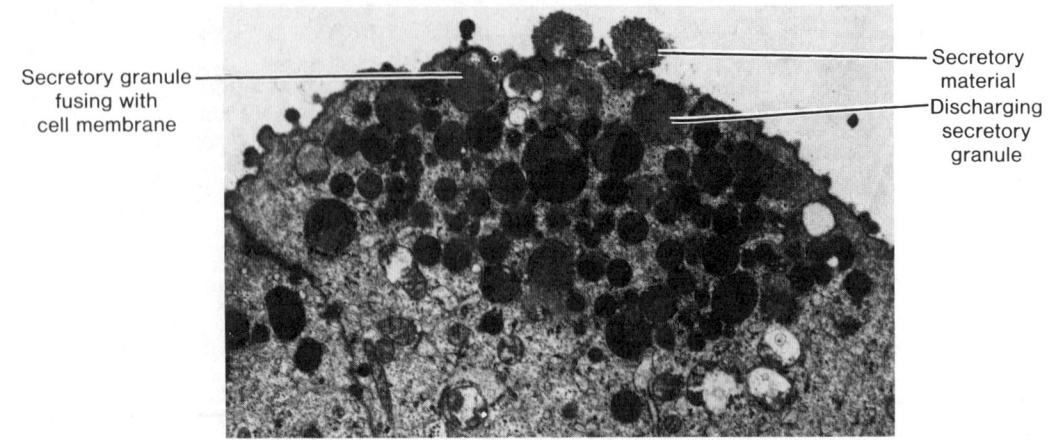

Secretory granule
fusing with
cell membrane

Secretory
material

Discharging
secretory
granule

Fig. 2-21. Exocytosis (merocrine secretion) by a human gastric lining epithelial cell (TEM, ×5,000).

along with the secretory material. Electron microscopy has failed to support this mode of secretion except perhaps for the mammary glands. Even here, however, the existence of an apocrine secretion is dubious.

Myoepithelial Cells

The secretory units of some glands are associated with myoepithelial cells, which are specialized cells located between the glandular cells and the basal lamina. The cells are contractile and their cytoplasm contains numerous actin filaments. Long cytoplasmic processes extend from the body of the cell to course around the secretory unit. The processes, by their contraction, aid in expressing products from the secretory units into the ductal system. Myoepithelial cells occur in sweat, mammary, and salivary glands and glands along the bronchi and esophagus.

Review Table 2-1. Key Features in Epithelial Classification

Number of Layers	Cell Shape	Epithelial Classification*
One	Flat, platelike	Simple squamous
One	Cuboidal, height equals width	Simple cuboidal
One	Columnar, height greater than width	Simple columnar
Cells staggered suggest stratification but all in contact with basement membrane	Majority columnar, cuboidal cells present at base	Pseudostratified columnar
Two or more; superficial cells not in contact with basement membrane	Surface layer squamous	Stratified squamous
Two or more; superficial cells not in contact with basement membrane	Surface layer cuboidal	Stratified cuboidal
Two or more; superficial cells not in contact with basement membrane	Surface layer columnar, deep layer(s) cuboidal	Stratified columnar
Multiple	Variable, surface cells, dome-shaped; in distended organs surface cells may be squamous	Transitional
Multiple	Columnar supporting (Sertoli) cells, round (or polyhedral) spermatogenic cells	Germinal (of male)

* Prefixes such as ciliated, keratinized, or nonkeratinized may be added to the classification.

Functional Review: Epithelium

Epithelium covers the entire body surface and lines the digestive, respiratory, cardiovascular, and urogenital systems. Any substance that enters or leaves the body must pass through an epithelial layer. Epithelia protect the organism from mechanical trauma and invasion by small organisms. They also limit fluid loss from underlying connective tissues and thus aid in preventing dehydration. Other functions they contribute to are absorption, transport of useful materials, and excretion of wastes. In the form of glands, they secrete products such as hormones, which regulate and integrate the activities of many tissues and organs.

Surface specializations — microvilli, stereocilia, and basal-lateral infoldings — provide a greatly increased cell surface for the absorption and transport of materials into or out of the cells. Cilia move materials across the surface of an epithelium. The nonmotile cilia of olfactory epithelia and kidney tubules serve as chemoreceptors, the cilia of the inner ear act as mechanoreceptors, and the modified cilia of retinal rods and cones are involved in photoreception.

The junctions between adjacent epithelial cells hold the cells together to form a continuous sheet. Desmosomes help maintain these cell-to-cell adhesions and also act as sites where keratin filaments of the cytoskeleton are anchored. The zonula occludens provides a tight seal between cells so that material crossing an epithelium must pass through the cell membrane and not through intercellular spaces between cells. Thus the ultimate barrier between the external world and the interior of the body is the plasmalemma. Nexus junctions provide an avenue for intercellular communication and exchange of materials and coordinate responses to stimuli throughout a mass of cells.

All epithelia lie on a basal lamina that provides physical support and serves as a surface for attachment and migration of cells. It also acts as a barrier for some molecules, depending on their size, charge, and shape, and plays an important inductive role in differentiation. It is important in repair processes after injury.

Review Questions and Answers

Questions

8. How are sheet or lining epithelia classified?
9. List three specializations of epithelial cell membranes and briefly discuss their functional significance.
10. List three cell attachments and comment on their function.
11. Briefly discuss how exocrine glands may be classified.

Answers

8. The lining or sheet form of epithelia is classified according to the number of layers of cells and the geometric shape of component cells (squamous, cuboidal, columnar). If an epithelial sheet consists of a single layer of cells, it is simple; if more than one layer of cells is present and the superficial cells are not in contact with the basement membrane, the epithelium is termed stratified. In cases where more than one cell type is present and in which not all the cells reach the surface but all are in contact with the basement membrane, this form of epithelium is termed pseudostratified. The stratified forms of epithelia are classified further by the geometric shape of the most superficial cells (e.g., stratified squamous, stratified columnar). If cells of an epithelial sheet exhibit adaptations or specializations (keratinization, cilia), the specialization is often included in the classification: keratinized or nonkeratinized stratified squamous epithelium, ciliated simple columnar or ciliated pseudostratified columnar epithelium.

9.a. Microvilli are finger-like evaginations of the apical cell membrane. In regions of high concentration (small intestine, proximal convoluted tubule of the kidney), they form a microvillus border that greatly increases the surface area and absorptive capacity of the cell.

b. Cilia are large, elongated evaginations of the apical cell membrane that contain two central and nine peripheral pairs of microtubules associated with an underlying basal body. In some regions they are numerous, motile, and beat in a coordinated manner to move a thin fluid film in a single direction (oviduct, trachea). In other locations where they occur singly or in small groups (olfactory epithelium, collecting tubules), cilia may function as chemoreceptors.

c. Basal-lateral infoldings are elaborate infoldings of the basal and lateral cell membrane found in the proximal and distal tubules of the kidney and the striated ducts of salivary glands. They greatly increase the surface area of the plasmalemma at the base of the cell and are associated with numerous mitochondria. They are observed in regions of high fluid/ion transport.

10.a. The zonula occludens is a fusion of the external laminae of adjacent epithelial cell membranes bordering a lumen. It forms an impermeable seal around the perimeter of each cell, uniting them in an epithelial sheet. The zonula occludens junction also is impor-

tant in separating apical from basal-lateral cell membrane domains by preventing the migration of specialized membrane proteins (receptors, transport proteins) in the plasmalemma.

b. A nexus is a firm attachment point between adjacent cells. The intercellular space of this junction measures about 2 nm, and it is here that exchange of ions and small molecules may occur, allowing direct intercellular communication.

c. The macula adherens (desmosome) acts as a firm attachment point between adjacent-cell membranes and also as an anchoring point for numerous keratin filaments in the cytoplasm.

11. Exocrine glands may be classified as either (a) unicellular or multicellular; (b) by the nature of the duct, simple (unbranched) or compound (branched); (c) by the shape of the secretory unit (acinus, tubule, saccule); (d) by the type of secretory cell (serous, mucous, or, where both are present, a mixed gland); or (e) by the mode of secretion (merocrine, apocrine, holocrine).

3 General Connective Tissue

Connective tissues form a diverse group of tissues derived from mesenchyme and, in keeping with the diversity of types, have a number of different functions. They provide structural elements, serve as a foundation for the support of organs, form a packing material for otherwise unoccupied space, provide an insulating layer (fat) that also acts as a storage depot that can be used to provide energy, and play a vital role in the defense mechanisms of the body and in repair after injury. Some are functions of ordinary (general) connective tissues; others are functions of specialized connective tissues. In organs, connective tissues form the *stroma* and epithelial components make up the *parenchyma*.

Organization

KEY WORDS: cell, intercellular substance, fiber, ground substance, matrix

Like all basic tissues, connective tissues are made up of **cells** and **intercellular substances.** The latter consist of **fibers, ground substance,** and tissue fluid. Unlike epithelium, where the cells are closely apposed with little intercellular material, connective tissue cells are widely separated by the intercellular fibers and ground substance that form the bulk of these tissues. Together, the fibers and ground substance form the **matrix.** With some exceptions, connective tissue is generally well vascularized.

Classification

KEY WORDS: extracellular material, general, loose, dense, regular, irregular

Classification of connective tissues into various subgroups is largely descriptive of **extracellular materials** rather than features of the cellular components. Two large categories can be defined — general and special — from which further subdivisions are made. **General** connective tissues are distinguished as loose or dense according to whether the fibers are loosely or tightly packed. **Loose** connective tissues can be subdivided on the basis of some special properties of their constituents, such as adipose (fatty) tissue, reticular tissue, and so forth. **Dense** connective tissues can be divided according to whether the fibers are randomly

distributed or show an orderly arrangement. Thus dense connective tissues are classed as dense **irregular** or dense **regular** connective tissues. A classification of connective tissues is given in Table 3-1.

Loose Connective Tissue

Loose connective tissue is a common and simple form of connective tissue and can be considered the prototype of connective tissue. Essentially all other types of connective tissues are variants of loose connective tissue in which one or more components have been emphasized to serve specific functions.

Fibers of Connective Tissue

KEY WORDS: collagen fiber, unit fibril, tropocollagen, alpha unit, reticular fiber, elastic fiber, microfibril, elastin, fibrillin

The fibrous intercellular substances in connective tissue consist of collagenous, reticular, and elastic fibers. **Collagen fibers** are present in all connective tissues and run an irregular course with much branching. The fibers vary in thickness from 1 to 10 μm and are of undefined length (Fig. 3-1). Although flexible, collagenous fibers have extremely high tensile strength. Ultrastructurally, the fibers consist of parallel fibrils, each of which represents a structural unit. These **unit fibrils** (Fig. 3-2) show repeating transverse bands spaced at 64-nm intervals along their length and are composed of macromolecules of **tropocollagen** that measure about 260 nm long by 1.5 nm wide. The tropocollagen molecules lie parallel to each other, overlapping by about one-fourth their lengths; the overlap is responsible for the banding pattern. Each molecule of tropocollagen consists of three polypeptide chains called **alpha units** arranged in a helix and linked by hydrogen bonds. The polypeptides are rich in glycine and proline and also contain hydroxyproline and hydroxylysine. However, alpha units isolated from collagens taken from different sites vary somewhat in the composition and sequence of amino acids. Several distinct molecular types of collagen have been identified, all of which consist of three alpha units arranged in a righthand helix and differ mainly in the amino acid constituents of the alpha chain. Not all unit fibrils of the various collagens present a banded,

Table 3-1. Subdivisions of Connective Tissue

Type	Locations
General connective tissues	
Loose connective tissue	
Mesenchyme	Primarily in the embryo and developing fetus
Mucoid	Umbilical cord
Areolar	Most organs and tissues
Adipose	Omentum, subcutaneous tissue
Reticular	Lymph nodes, bone marrow
Dense connective tissue	
Irregular	Dermis, capsules of organs, periosteum, perichondrium
Regular	
Collagenous	Tendon, ligaments, aponeurosis, cornea
Elastic	Ligamentum nuchae, ligamenta flava
Special connective tissues	
Cartilage	
Hyaline	Costal cartilages, trachea
Fibrous	Symphysis pubis, intervertebral disk
Elastic	External ear, epiglottis
Bone	Skeleton
Blood	Cardiovascular system
Hemopoietic	Bone marrow, lymphatic tissue and organs

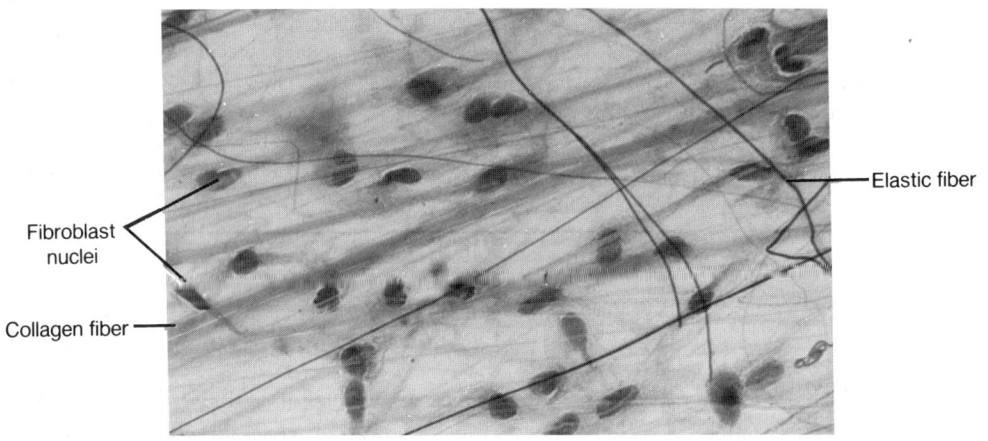

Fig. 3-1. A spread preparation of human loose areolar connective tissue stained with orcein, which selectively stains elastic fibers. Note that collagen fibers and numerous fibroblasts also are shown (LM, × 400).

fibrillar appearance, nor are they arranged in a similar fashion. Thus collagen represents a family (in excess of 12 types) of closely related but genetically distinct proteins. Some of these are listed in Table 3-2. Various types of collagen can be made by cells other than fibroblasts, and collagen is synthesized by chondroblasts, osteoblasts, odontoblasts, epithelial cells, endothelial cells, and smooth muscle cells. Collagen molecules are important components of basement membranes.

Reticular fibers do not gather into bundles as do col-

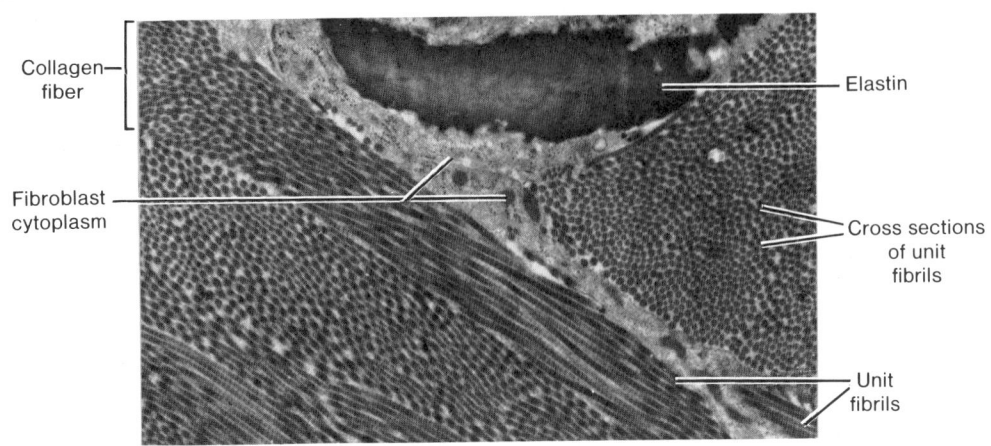

Fig. 3-2. Elastic and collagen fibers from the connective tissue component of the pleura as seen in the electron microscope. Note that each collagen fiber is made up of several unit fibrils that show cross striations (TEM, × 11,000).

Table 3-2. Types of Collagen*

Type	Morphologic Features	Distribution
I	Broad, banded fibrils	Widespread; tendon, bone, dermis, dentin, fascia
II	Small-diameter, banded fibrils	Hyaline cartilage, vitreous body, nucleus pulposus, notochord
III	Small-diameter, banded fibrils	Corresponds to reticular fibers; prominent in organs with a major smooth muscle component; uterus, blood vessels
IV	Feltwork of nonbanded fibrils	Basal laminae of epithelial cells, glomerular epithelium
V	Thick nonbanded fibrils	Widespread; pericellular laminae of smooth and striated muscle cells, tendon sheaths
VI	Thin, banded fibrils	Makes up about 25% of collagen in cornea; small amounts where types I and III are found
VII	Small-diameter, banded fibrils	Form anchoring fibrils that link the basal lamina of many epithelia to the underlying connective tissue
VIII	Unknown	A major component of Descemet's membrane; associated with and produced by endothelial cells
IX	Unknown	Found mainly in cartilage; links type II forms in three-dimensional arrangement
X	Unknown	Cartilage matrix surrounding hypertrophic chondrocytes during endochondral bone formation

* Various types of collagen can be elaborated by cells other than fibroblasts. Collagen is known to be synthesized by epithelial cells, endothelial cells, chondroblasts, osteoblasts, and smooth muscle cells.

lagenous fibers but tend to form delicate networks. They are not seen in routinely prepared sections but can be shown with silver stains (Fig. 3-3) or by the periodic acid–Schiff reagent, which reacts with proteoglycans that bind reticular fibers together. In electron micrographs, reticular fibers show the same banding pattern as collagen. They differ principally in the number, diameter, and arrangement of the unit fibrils.

Elastic fibers appear as thin, homogeneous strands that are similar and of more uniform size than collagen fibers (see Fig. 3-1). They cannot be distinguished in routine sections and require special stains to make them visible. In

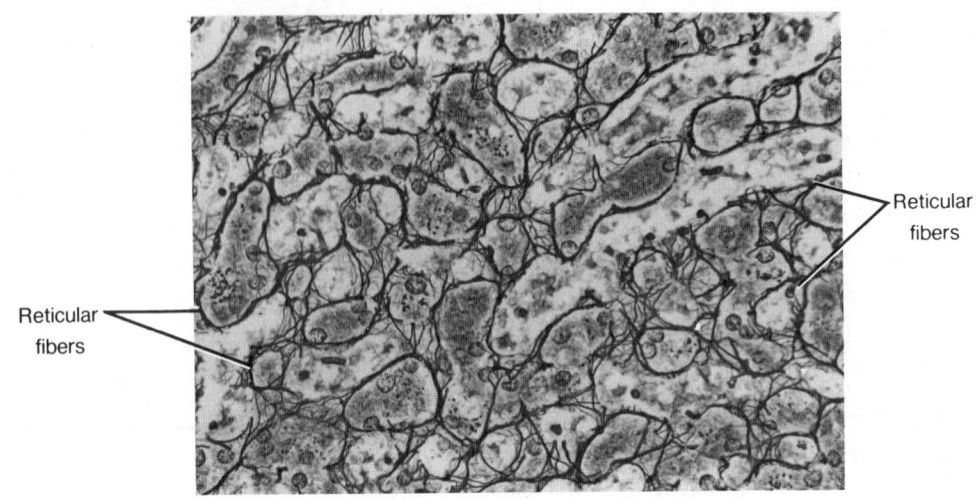

Reticular
fibers

Reticular
fibers

Fig. 3-3. A portion of human liver stained with silver, which is selective for demonstrating reticular fibers (LM, × 200).

electron micrographs, elastic fibers are seen to consist of bundles of **microfibrils** embedded around an amorphous component called **elastin** (see Fig. 3-2). The microfibrils are about 11 nm in diameter and lack crossbanding. During formation of elastic fibers, microfibrils are laid down first and elastin is added secondarily but soon forms the bulk of the fiber. A precursor molecule, tropoelastin, is released by cells and polymerizes extracellularly to form elastin. The microfibrillar component tends to be peripherally located on the fiber. Elastin, like collagen, contains glycine and proline but has little hydroxyproline and lacks hydroxylysine. It has a high content of valine and contains two amino acids, desmosine and isodesmosine, that are specific to elastin. Microfibrils consist of a nonsulfated glycoprotein called **fibrillin.** Similar microfibrils also exist that are associated with the basal lamina of epithelia.

Elastic fibers can stretch more than 130 percent of their original length, and their presence permits connective tissues to undergo considerable expansion or stretching with return to the original shape or size when the deforming force is removed. In addition to forming fibers, elastin may be present in fenestrated sheets as in some arterial walls (Fig. 3-4).

Ground Substance

KEY WORDS: gel, proteoglycan, glycosaminoglycan, bound water, adhesion glycoprotein, laminin, fibronectin, thrombospondin

The fibers and cells of connective tissue are embedded in an amorphous material called *ground substance* that is present as a transparent **gel** of variable viscosity. Ground substance consists of glycoproteins, glycosaminoglycans, and proteoglycans that differ in amount and type in different connective tissues (Table 3-3). **Proteoglycans** have a bottle brush configuration with long protein cores bound covalently to numerous glycosaminoglycan side chains. **Glycosaminoglycans** are linear polymers of repeating disaccharide units that contain hexosamine, D-glucosamine, or D-galactosamine as their most constant feature. Tropocollagen has been extracted from ground substance but cannot be demonstrated histologically.

Ground substance contains a high proportion of water, which is bound to long-chain carbohydrates and proteoglycans; most of the extravascular fluid is in this state. This **bound water** acts as a medium by which nutrients, gases, and metabolites can be exchanged between blood and tissue cells. Hyaluronic acid is the chief glycosaminoglycan of loose connective tissue, and because of its ability to bind water, it is primarily responsible for changes in the permeability and viscosity of loose connective tissue. Proteoglycans have an important role in maintaining the structural organization of the fibrous constituents in the matrix and basal laminae and aid in preventing or retarding the spread of microorganisms and their toxic materials from the site of an infection.

The primary glycoproteins of the matrix are fibronectin, laminin, and thrombospondin. These are **adhesion glyco-**

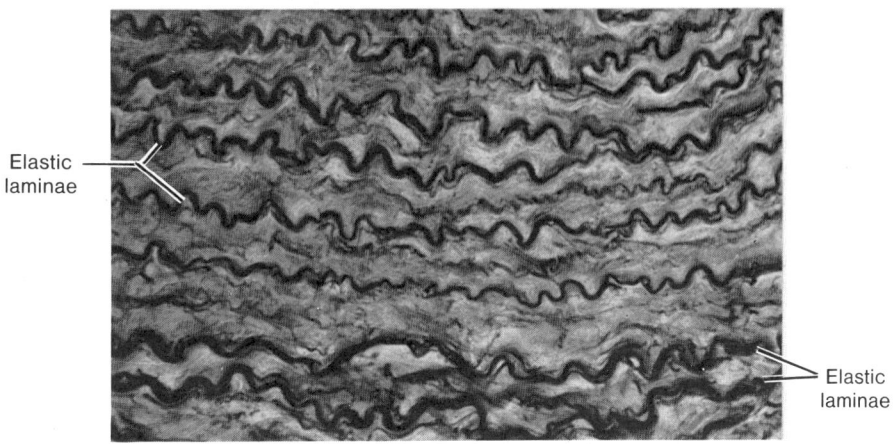

Elastic laminae

Elastic laminae

Fig. 3-4. A region of the human aorta stained with orcein to demonstrate elastic laminae (LM, × 150).

Table 3-3. Types of Glycosaminoglycans

Type of Glycosaminoglycan	Some Locations
Hyaluronic acid	Umbilical cord, vitreous humor, synovial fluid, loose connective tissue
Chondroitin	Cornea
Chondroitin-4-sulfate	Aorta, bone, cartilage, cornea
Chondroitin-6-sulfate	Cartilage, nucleus pulposus, sclera, tendon, umbilical cord
Dermatan sulfate	Aorta, heart valves, ligamentum nuchae, sclera, skin, tendon
Keratan sulfate	Bone, cartilage, cornea, nucleus pulposus

proteins that function to link cell membranes of various cell types to elements comprising the extracellular matrix. **Laminin** is the largest of these adhesion glycoproteins and is most abundant in the lamina lucida of epithelial basal laminae and in the external laminae of muscle cells. It binds the plasmalemma, via a transmembrane glycoprotein laminin receptor, of epithelial and muscle cells to type IV collagen and proteoglycans (heparan sulfate) in the adjacent matrix. **Fibronectin** is a large, flexible glycoprotein of the extracellular matrix synthesized by fibroblasts and some epithelia. It also occurs in the external laminae of muscle cells and Schwann cells as well as on the surfaces of other cell types. Fibronectin mediates adhesion between cells and between cells and substrates in the matrix. It has several receptor-binding domains and links the plasmalemma of epithelial cells to collagen fibers and glycosaminoglycans. Through fibronectin receptors (integrins) in the basal epithelial plasmalemma, fibronectin links actin

filaments in the cell cytoplasm to components of the extracellular matrix. Fibronectin is thought to be involved in cell spreading and locomotion. **Thrombospondin** is another adhesive glycoprotein that binds to collagen and some proteoglycans. It is produced by fibroblasts and smooth muscle and epithelial cells and binds the plasmalemma of these cell types to the extracellular matrix.

Cells of Connective Tissue

KEY WORDS: fibroblast, myofibroblast, macrophage, system of mononuclear phagocytes, mast cell, heparin, histamine, fat cell, plasma cell, leukocyte, neutrophil, eosinophil, lymphocyte, monocyte

Connective tissue contains several different cell types. Some are indigenous to the tissues; others are transients derived from blood. The most common connective tissue cells are **fibroblasts** — large, flattened cells with elliptical nuclei that contain one or two nucleoli. The cell body is irregular and often appears stellate with long cytoplasmic processes extending along the connective tissue fibers (see Fig. 3-1). The boundaries of the cell are not seen in most preparations, and the morphology varies with the state of activity. In active cells the nuclei are plump and stain lightly, whereas the nuclei of inactive cells appear slender and dense. Ultrastructurally, active cells show increased amounts of granular endoplasmic reticulum (GER). Fibroblasts elaborate the precursors of collagen and reticular and elastic fibers, produce the ground substance, and maintain the extracellular materials that are constantly being removed and renewed.

The peptides of collagen are formed on the ribosomes of the GER, from which they are transported to the Golgi complex. Procollagen, a molecular form of collagen, is released at the maturing face of the Golgi body into the surrounding cytoplasm and then released from the cell (Fig. 3-5). An enzyme called *procollagen peptidase* converts the procollagen molecule into tropocollagen, which then polymerizes to form the unit fibril of collagen.

Myofibroblasts resemble fibroblasts but contain aggregates of actin microfilaments and myosin. Normally found only in small numbers, they increase following tissue injury. Myofibroblasts produce collagen, and their contractile activity contributes to the retraction and shrinkage of early scar tissue.

Macrophages (histiocytes) are almost as abundant as fibroblasts in connective tissue. They are actively phagocytic, ingesting a variety of materials from particulate matter to bacteria, tissue debris, and whole dead cells (Fig. 3-6). The material ingested is broken down by lysosomal digestion.

The effectiveness of macrophages is enhanced by the binding of complement and antibodies to the surface of

Fig. 3-5. Formation of collagen by fibroblasts.

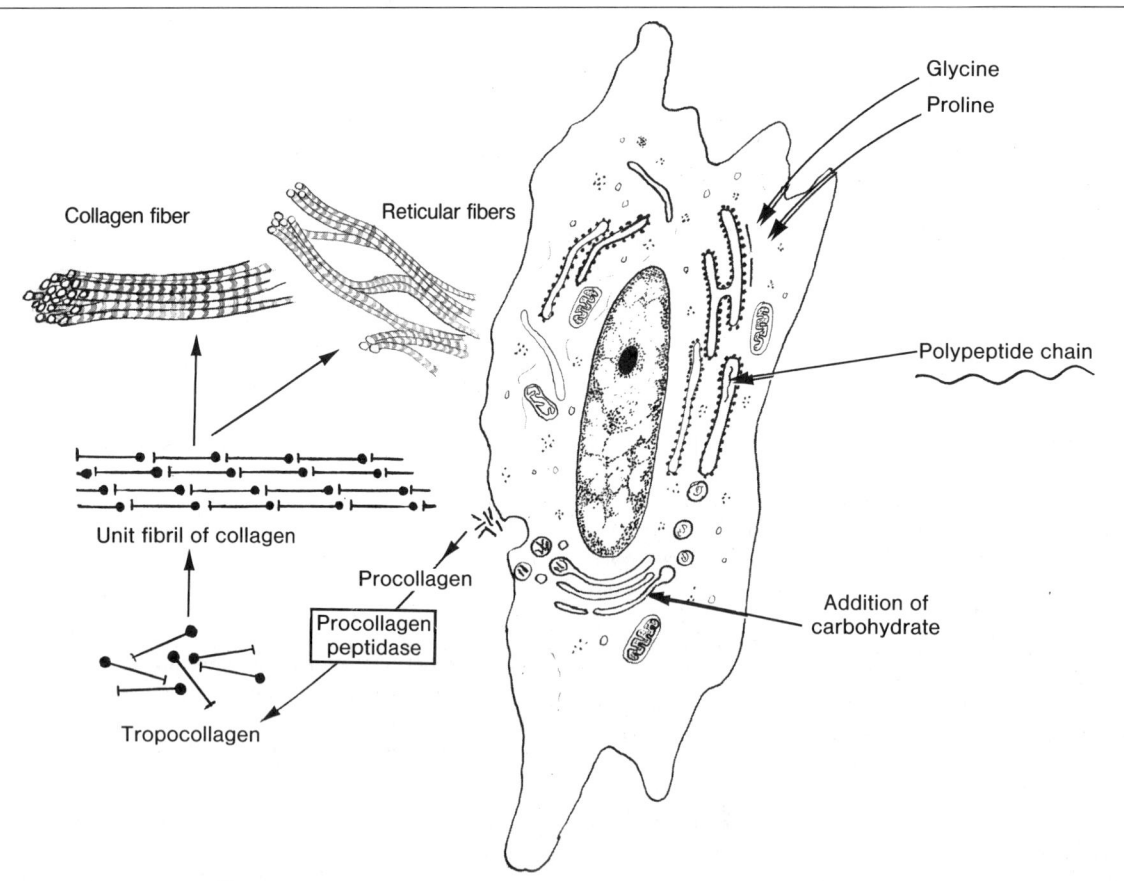

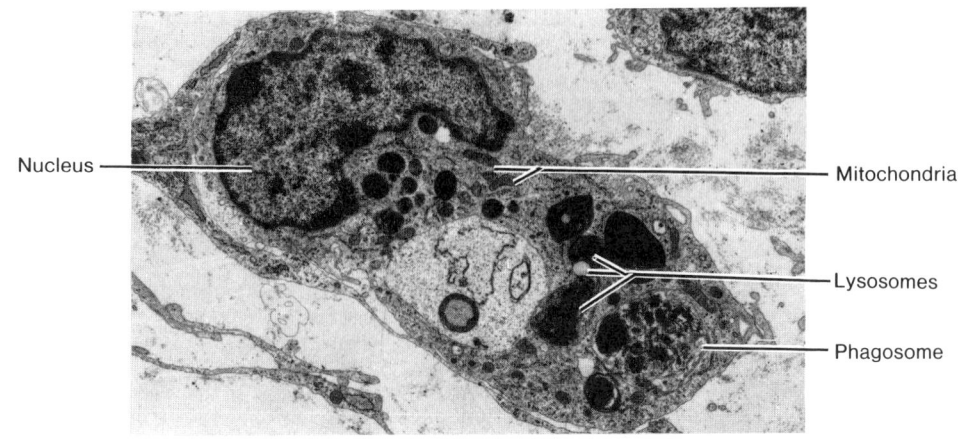

Nucleus — Mitochondria — Lysosomes — Phagosome

Fig. 3-6. An active human macrophage that has recently phagocytosed foreign material (TEM, × 4,000).

bacteria (opsonization). Opsonized bacteria are more vulnerable to phagocytosis. Complement is a group of proteins circulating in the blood plasma that are synthesized and released by the liver. Macrophages also may interact with lymphocytes in combating infections. Macrophages can synthesize and release a number of factors such as interleukins 1 and 4, fibroblast growth factor, transforming growth factor, and tumor necrosis factor, as well as connective tissue enzymes such as collagenase and elastase.

Macrophages commonly are described as irregularly shaped cells with blunt cytoplasmic processes and ovoid or indented nuclei that are smaller and stain more deeply than those of fibroblasts. In fact, macrophages are difficult to distinguish from fibroblasts, especially active fibroblasts, unless the macrophages show evidence of phagocytosis.

The macrophages of loose connective tissue are part of a widespread **system of mononuclear phagocytes** that includes phagocytes of the liver (Kuppfer cells), lung (alveolar macrophages), serous cavities (pleural and peritoneal macrophages), nervous system (microglia), lymphatic tissue, and bone marrow. Regardless of where they are found, macrophages have a common origin from precursors in the bone marrow, and the monocytes of blood represent a transit form of immature macrophages.

Mast cells are present in variable numbers in loose connective tissue and often collect along small blood vessels. They are large, ovoid cells 20 to 30 μm in diameter with large granules that fill the cytoplasm (Fig. 3-7). Two populations of mast cells are known to exist: connective tissue

mast cells and mucosal mast cells. In humans, the granules of connective tissue mast cells are membrane-bound and in electron micrographs show a characteristic tubular pattern. The granules contain **heparin,** a potent anticoagulant; **histamine,** an agent that causes vasodilation and increased permeability of capillaries and venules; and eosinophil chemotactic factor. The granules of mucosal mast cells contain chondroitin sulfate rather than heparin. Both mast cell types arise from bone marrow stem cells. Mast cells are especially numerous along small blood vessels and beneath the epithelia of the intestinal tract and respiratory system. Here they detect the entry of foreign proteins and initiate a local inflammatory response by rapidly discharging their secretory granules. The discharge of granules is unique in that several fuse together and release their contents simultaneously. Mast cells also can promote immediate hypersensitivity reactions (hay fever, asthma, anaphylaxis) following their release of secretory granules, which act as chemical mediators.

Fat cells are specialized for synthesis and storage of lipid. Individual fat cells may be scattered throughout loose connective tissue or may accumulate to such an extent that other cells are crowded out and an adipose tissue is formed. Each fat cell acquires so much lipid that the nucleus is flattened to one side of the cell and the cytoplasm forms only a thin rim around a large droplet of lipid (Fig. 3-8). In ordinary sections, fat cells appear empty due to the loss of lipid during tissue preparation, and groups of fat cells have the appearance of chicken wire.

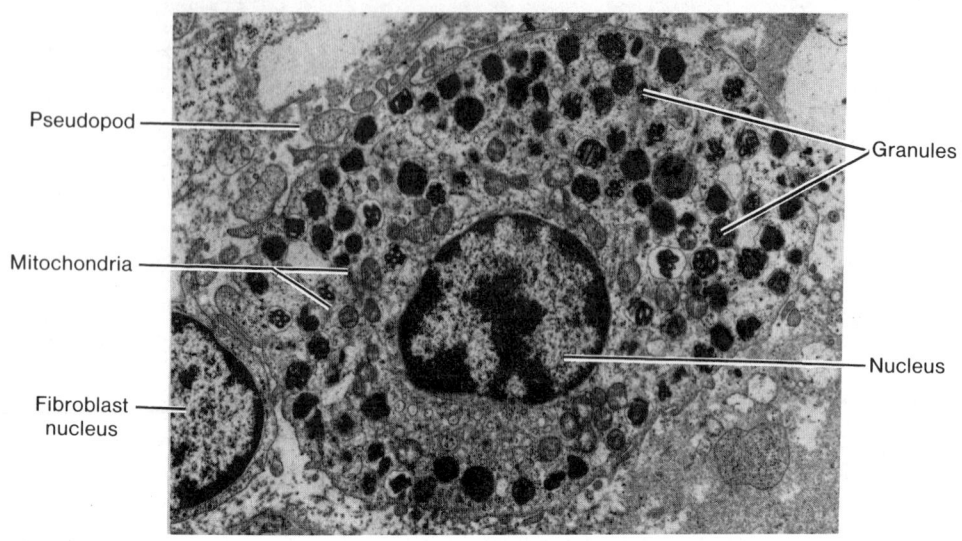

Fig. 3-7. A human mast cell from the lamina propria of the stomach (TEM, × 4,000).

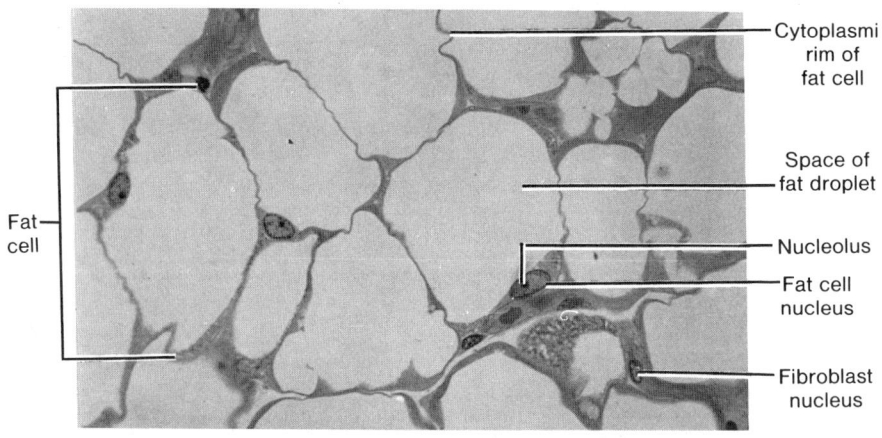

Fig. 3-8. White (unilocular) fat cells taken from a region of human adipose tissue (LM, × 350).

Plasma cells are not common in most connective tissues but may be numerous in the lamina propria of the gastrointestinal tract and are present in lymphatic tissues. Plasma cells appear somewhat ovoid, with small, eccentrically placed nuclei in which the chromatin is arranged in coarse blocks to form a "clock face" pattern. The cytoplasm is deeply basophilic due to the abundance of GER (Fig. 3-9). A weakly staining area of cytoplasm often appears adjacent to one side of the nucleus and corresponds to the site of the Golgi complex. Plasma cells produce immunoglob-

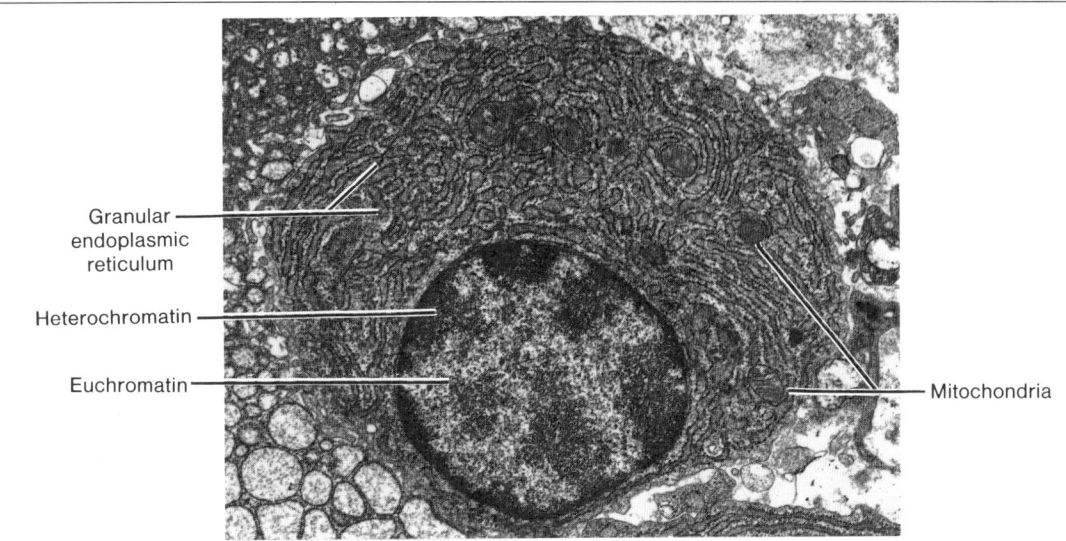

Granular endoplasmic reticulum

Heterochromatin

Euchromatin

Mitochondria

Fig. 3-9. A human plasma cell taken from the lamina propria of the stomach (TEM, × 5,000).

ulins (antibodies) that form an important defense against infections. Various acidophilic inclusions may be found in the cytoplasm of some plasma cells. Plasma cells are a differentiated form of the B-lymphocyte.

Variable numbers of **leukocytes** constantly migrate into the connective tissues from the blood (diapedesis). **Neutrophils** are one type of leukocyte characterized by a lobed nucleus. In this cell type, the granules are said to be neutrophilic, although in sections they stain faintly pink (i.e., acidophilic). The cells are avidly phagocytic for small particles and are especially numerous at sites of infections. They are the major cellular component of pus. **Eosinophils** also are characterized by lobed nuclei, but their cytoplasmic granules are larger, more spherical, and more discretely visible than those of neutrophils. The granules stain intensely with acid dyes such as eosin. These cells are phagocytic and have a special avidity for antigen-antibody complexes. They increase in number as a result of allergic or parasitic types of disease. Eosinophils are attracted to the sites of histamine release, and their cytoplasmic granules contain enzymes capable of breaking down histamine. **Lymphocytes** belong to the class of leukocytes called *mononuclear leukocytes* and are characterized by a single, round, nonlobed nucleus. They are the smallest of the cells that migrate into connective tissues and measure about 7 μm in diameter. These cells form part of the immunologic

defense system and may give rise to antibody-producing cells or elaborate nonspecific factors that destroy foreign cells. They are few in number in normal connective tissue but increase markedly in areas of chronic inflammation. **Monocytes** are the blood-borne forerunners of tissue macrophages. Once the cells have entered connective tissue, it is difficult, if not impossible, to distinguish them from macrophages. Monocytes can fuse with one another to form multinucleated giant cells in an attempt to engulf or wall off objects that are too large or resistant to phagocytosis by a single cell.

Subtypes of Loose Connective Tissue

KEY WORDS: mesenchyme, mucoid tissue, areolar connective tissue

Mesenchyme is a loose, spongy tissue that serves as packing between the developing structures of the embryo. It consists of a loose network of stellate and spindle-shaped cells embedded in an amorphous ground substance with thin, sparse fibers. The cells have multiple developmental potentials and can give rise to any of the connective tissues. The presence of mesenchymal cells in the adult often has been proposed to explain the expansion of adult connective tissue. However, fibroblasts, too, are capable of sequential divisions, and the idea of mesenchymal rests can well be discarded.

Mucoid tissue occurs in many parts of the embryo but is particularly prominent in the umbilical cord, where it is called *Wharton's jelly*. It resembles mesenchyme in that the constituent cells are stellate fibroblasts with long processes that often make contact with those of neighboring cells, and the intercellular substance is soft and jelly-like and contains thin collagen fibers. The fiber content increases with the age of the fetus. Mucoid tissue does not have the developmental potential of mesenchyme.

Areolar connective tissue is a loosely arranged connective tissue that is widely distributed in the body. It contains collagen fibers and a few elastic fibers embedded in a thin, almost fluid-like ground substance. This kind of connective tissue forms the fascia that binds organs and organ components together. It forms helices about the long axes of expandable tubular structures such as the ducts of glands, the gastrointestinal tract, and blood vessels.

Adipose Tissue

KEY WORDS: white fat, unilocular, brown fat, multilocular

Adipose tissue differs in two respects from other connective tissue: Fat cells and not the intercellular substances predominate, and unlike other connective tissue cells, each fat cell is surrounded by its own basal lamina. Reticular and collagenous fibers also extend around each fat cell to provide a delicate supporting framework that contains numerous capillaries.

Adipose tissue can be subdivided into white and brown fat. **White fat** is the more plentiful and is found mainly in the subcutaneous tissue (where it forms the panniculus adiposus), omenta, mesenteries, pararenal tissue, and bone marrow. White fat is an extremely vascular tissue and also contains many nerve fibers. In white fat, the cells are filled by a single, large droplet of lipid; thus it is often referred to as **unilocular** fat. The lipid droplet is composed of glycerol esters and fatty acids. The materials within the fat droplet are in a constant state of flux and are not permanent entities. White fat serves as a storage depot for calories taken in excess of the body's needs and can be used for production of energy as required or as a source of material for steroid hormone synthesis. It also acts as an insulating layer to conserve body heat, acts mechanically as a packing material, and forms shock-absorbing pads in the palms

of the hands, on the soles of the feet, and around the eyeballs.

Brown fat is present in many species, including humans, and is prominent in hibernating animals and newborns. Brown fat has a restricted distribution, occurring mainly in the interscapular and inguinal regions. The brown or tan color is due to the high content of cytochrome enzymes. The cells show round nuclei, and the cytoplasm is filled with multiple small droplets of lipid; hence this type of fat is called **multilocular** fat. Mitochondria here are more numerous and larger than those in the cells of white fat. Each cell of brown fat receives direct sympathetic innervation. During arousal from hibernation, the lipid within the brown fat is rapidly oxidized to produce heat and release substances such as glycerol that are used by other tissues. Because brown fat is even more vascular than white fat, it raises the temperature of the blood significantly, thus increasing the general body temperature.

Reticular Connective Tissue

KEY WORDS: reticular cell, reticular fiber, fibroblast

Reticular connective tissue is characterized by a cellular framework as seen in lymphatic tissues and bone marrow. **Reticular cells** are stellate, with processes extending along the **reticular fibers** to make contact with neighboring cells. The cytoplasm stains lightly and is attenuated, and the large nucleus stains weakly. The reticular cell is equivalent to the **fibroblast** of other connective tissues and is responsible for the production and maintenance of the reticular fibers, which are identical to those found in loose connective tissue.

Dense Connective Tissue

KEY WORDS: dense irregular, dense regular

Dense connective tissue differs from the loose variety chiefly in the concentration of fibers and reduction of the cellular and amorphous constituents. Two types, dense irregular and dense regular, are identified.

Dense irregular connective tissue contains abundant, thick, collagenous bundles that are woven into a compact network. Among the collagen fibers is an extensive network of elastic fibers. An example of dense irregular connective tissue is the dermis of the skin (Fig. 3-10).

Dense regular connective tissue also contains a predominance of collagen fibers arranged in bundles, but these

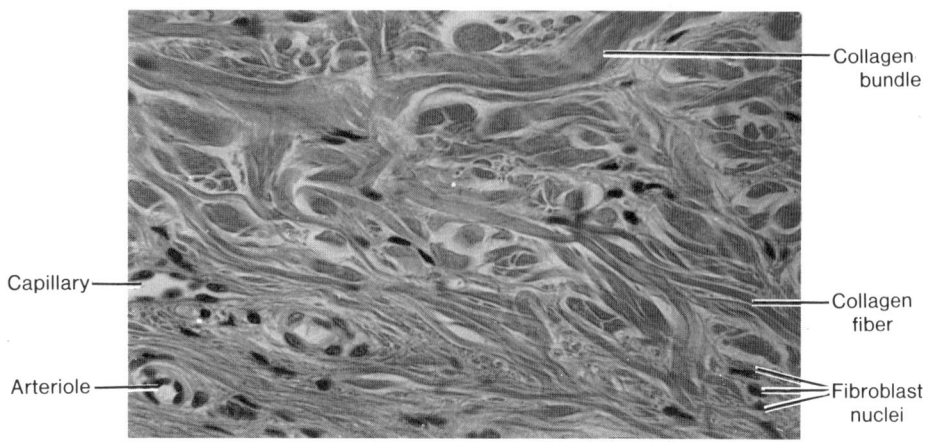

Fig. 3-10. The dermis of human skin is a classic example of dense irregular connective tissue (LM, × 250).

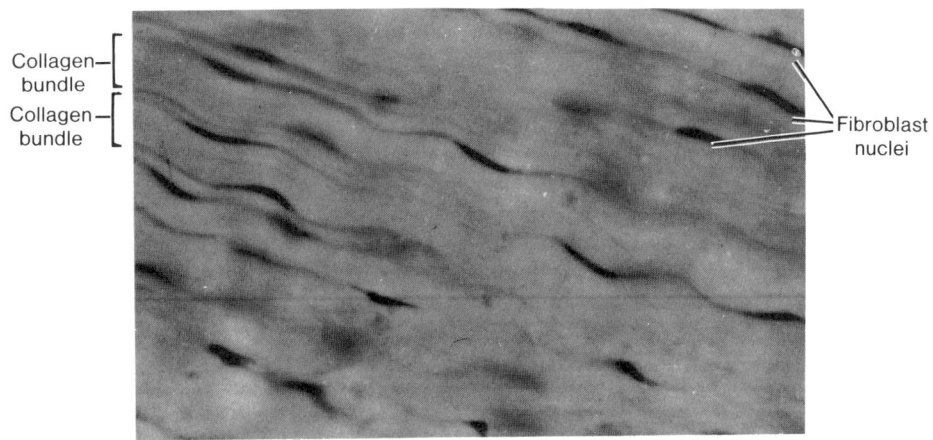

Fig. 3-11. Dense regular connective tissue from a human tendon (LM, × 400).

have a regular, precise arrangement. The organization of the collagen bundles reflects the mechanical needs of the tissue. In tendons and ligaments, for example, they are oriented in the direction of pull. Fibroblasts are the primary cells present and occur in rows parallel to the bundles of collagen fibers (Fig. 3-11).

Special Connective Tissue

Special connective tissues have functions and a histologic organization sufficiently unique to warrant their consideration as distinct and special forms of connective tissue. They are considered individually in succeeding chapters.

Review Table 3-1. Key Features of General Connective Tissue

Key Features of Connective Tissue Cells

Cell Types	Nuclear Characteristics	Cytoplasmic Characteristics	Primary Activity
Resident			
Fibroblasts	Oval, centrally placed, staining variable depending on activity	Elongate, spindle- or stellate-shaped cell; usually not visible in sectioned material	Produce fibers and proteoglycans
Myofibroblasts	Same as above	Same as above; bundles of actin filaments at electron microscopic level	Produce collagen fibers, contract
Adipose cells			
Unilocular fat cells	Usually compressed at edge of cell (signet ring–like), staining variable	Forms a thin rim around single central lipid droplet	Store lipid
Multilocular fat cells	Central, spheroid, light staining	Numerous lipid droplets, abundant mitochondria	Store lipid
Mast cells	Central, spheroid to ovoid, may show abundant heterochromatin	Filled with granules	Store and release histamine, heparin, eosinophil chemotactic factor
Macrophages	Large, ovoid, most frequently indented	Light staining; contains phagocytosed material	Phagocytic for a variety of materials
Plasma cell	Usually eccentric, spheroid; heterochromatin clumps may form "clock face"	Basophilic, slate gray in color; may show negative Golgi image	Antibody production
Migratory			
Neutrophils	Polymorphonuclear, 3–5 lobes common, chromatin dense	Lilac staining	Phagocytic for small particles
Eosinophils	Polymorphonuclear, 2–4 lobes common, chromatin dense	Red-orange granules fill cytoplasm	Phagocytic for antigen-antibody complexes, degrade histamine
Lymphocytes	Single, spheroid, abundant heterochromatin	Thin rim, light blue staining	Antibody production, cytotoxic agents

Key Features of Connective Tissue Fibers

Fiber Type	Light Microscopic Appearance	Electron Microscopic Appearance	Primary Locations
Collagen fiber (type I collagen)	Coarse fibers 0.5–10.0 μm in diameter; indefinite length; stain with protein dyes	Unit fibrils 50–150 nm in diameter; repeating transverse bands every 64 nm; organized into large fibers	Tendon, ligament, dermis, fascia, capsules, sclera, bone, dentin
Reticular fiber (type III collagen)	Delicate network of fine fibers; must be stained specifically to demonstrate, usually by reduction of silver	Banded unit fibrils 50 nm or less in diameter, organized into tiny fibers	Stroma of lymphatic organs, bone marrow, glands
Elastic fiber	Smooth, homogeneous fibers; varying diameter; must be stained specifically to demonstrate (orcein, Verhoeff's stain)	Amorphous core of elastin; microfibrils 11 nm in diameter at periphery of fiber	Dermis, lung, arteries, organs that expand

Functional Review: General Connective Tissue

The functions of connective tissue are mainly passive, but mechanically they are indispensable. They provide the supporting framework for organs and for the body itself and serve to connect distant structures as, for example, in the connection of muscle to bone by tendons. Connective tissues bind organs and unite the organ components into a functioning unit. Adipose tissue forms an insulating blanket to limit heat loss and forms mechanically protective cushions in the hands and feet and about the eyeball. It also stores excess calories in the form of fat that can be called on and used as needed.

Collagenous fibers provide tissues with flexibility and tensile strength; elastic fibers allow considerable deformity of an organ with recovery of the shape and size when the force is removed. Collagen molecules form a major component of basal laminae, which determine vascular and epithelium permeability.

Connective tissue also contributes significantly to the defense of the body. This is the function of cellular components mainly, as in the phagocytic activity of macrophages and granular leukocytes and the antibodies and cytotoxic agents elaborated by plasma cells and lymphocytes. The matrix also plays an important role in the body's defenses by preventing or slowing the spread of microorganisms and their toxic products from sites of infection.

The various connective tissues together form a single vast compartment limited by sheets of epithelia. It is through this compartment that tissue fluid must pass during exchange of nutrients and waste products between cells and tissues and the blood vasculature. Most immune reactions take place in this compartment. Because the ground substance of loose connective tissue binds water, it can act as the medium by which these exchanges are effected.

Review Questions and Answers

Questions

12. How do connective tissues differ from epithelia?
13. What is the difference between collagen and reticular fibers?
14. What conclusions would you draw from the appearance by light and electron microscopy of the plasma cell?
15. In what ways does adipose tissue differ from other loose connective tissues?
16. Briefly define *ground substance* and discuss its role in connective tissue.
17. What role does the fibroblast play in the connective tissues?

Answers

12. Connective tissues differ from epithelia in the proportion of cells and intercellular substances. In epithelia, the cellular component predominates, the cells are closely apposed, and there is little intercellular material between them. Connective tissues, on the other hand, are characterized by a preponderance of intercellular materials, and the cells are widely separated by the ground substance and fibers. With rare exceptions, connective tissues are well vascularized, whereas epithelia are avascular.

13. Collagen and reticular fibers differ in the number and arrangement of the unit fibrils of which the fibers are composed. Typical collagen fibers consist of type I collagen, whereas reticular fibers consist primarily of type III collagen.

14. The deeply basophilic cytoplasm indicates the presence of a large amount of granular endoplasmic reticulum, and the clear area of the cytoplasm corresponds to a prominent Golgi complex. These are consistent with a cell that is actively engaged in protein (glycoprotein) synthesis. Plasma cells produce immunoglobulins (antibodies).

15. Adipose tissue differs from other connective tissues in the predominance of fat cells over intercellular substances, and unlike other connective tissues, the fat cells of adipose tissue are each surrounded only by their own basal laminae.

16. The ground substance is best defined as that amorphous component of connective tissue in which the cells and extracellular fibers are embedded. The ground substance together with extracellular fibers constitutes the matrix of connective tissue. Ground substance usually occurs as a transparent gel and is of variable viscosity. It is made up of a number of proteoglycans, glycosaminoglycans, and glycoproteins that differ in type, amount, and consistency depending on the kind of connective tissue. Most extravascular fluid is bound within the ground substance and serves as a medium in which metabolites, nutrients, and gases can be exchanged between tissue cells and the surrounding vasculature. Ground substance also plays an important role in retarding the spread of microorganisms and toxic materials from the sites of infections.

17. Fibroblasts are the most common and widespread of the connective tissue cells. They elaborate and maintain the extracellular fibers of connective tissue (collagenous, reticular, and elastic fibers) as well as the amorphous ground substance.

4 Special Connective Tissue: Cartilage, Bone, and Joints

Cartilage

Although adapted to provide support, cartilage contains only the usual elements of connective tissue — cells, fibers, and ground substance. It is the ground substance that gives cartilage its firm consistency and ability to withstand compression and shearing forces. Collagen and elastic fibers embedded in the ground substance impart tensile strength and elasticity. Together, the fibers and ground substance form the matrix. Cartilage differs from other connective tissues in that it lacks nerves, a blood supply, and lymphatics and is nourished entirely by diffusion of materials from blood vessels in adjacent tissues. Although relatively rigid, the cartilage matrix has a high water content and is freely permeable, even to fairly large particles.

Classification of cartilage into hyaline, elastic, and fibrous types is based on differences in the abundance and type of fibers in the matrix.

Hyaline Cartilage

KEY WORDS: **chondrocyte, lacuna, isogenous group, proteoglycan, territorial matrix, interterritorial matrix, perichondrium**

Hyaline cartilage is the most common type of cartilage and forms the costal cartilages, articular cartilages of joints, and cartilages of the nose, larynx, trachea, and bronchi. It is present in the growing ends of long bones. In the fetus, most of the skeleton is first laid down as hyaline cartilage.

The cells of cartilage are called **chondrocytes** and reside in small spaces called **lacunae** scattered throughout the matrix (Fig. 4-1). The cells generally conform to the shape of the lacunae in which they are contained. Deep in the cartilage the cells and their lacunae usually appear rounded in profile, whereas at the surface, they are elliptical, with the long axis parallel to the surface. Chondrocytes often occur in small clusters called **isogenous groups** that represent the offspring of a single cell.

In the usual preparations, the cells show an irregular outline and appear shrunken and pulled away from the walls of the lacunae. Electron microscopy reveals that each cell completely fills the lacunar space and sends short processes into the surrounding matrix, but neighboring cells do not touch one another. The nucleus is round or oval and contains one or more nucleoli. The cytoplasm shows the usual organelles as well as lipid and glycogen inclusions. In growing chondrocytes the Golgi body and endoplasmic reticulum are well developed.

In fresh cartilage and in routine histologic sections, the matrix appears homogeneous because the ground substance and the collagen fibers embedded in it have the same refractive index. The collagen in hyaline cartilage rarely forms bundles but is present as a feltwork of slender fibrils in which the banding pattern shows variable periodicities or even is lacking. Collagen in cartilage appears to be less polymerized than in other tissues; cartilage is unique in that only type II collagen is present. Chondronectin, a glycoprotein, mediates the adhesion of type II collagen to chondroycytes. The ground substance consists mainly of **proteoglycans,** the specific glycosaminoglycans of cartilage being chondroitin-4- and chondroitin-6-sulfate, keratan sulfate, and a small amount of hyaluronic acid. The proteoglycans are responsible for the basophilic properties of the ground substance. The matrix around each isogenous group stains more deeply than elsewhere, forming the **territorial matrix.** The less densely stained intervening areas form the **interterritorial matrix.**

Except for the free surfaces of articular cartilages, hyaline cartilage is enclosed by a specialized sheath of connective tissue called the **perichondrium** (see Fig. 4-1). The outer layers of the perichondrium consist of well-vascularized, dense, irregular connective tissue that contains elastic and collagen fibers and fibroblasts. Where it lies against cartilage, the perichondrium is more cellular and passes imperceptibly into cartilage. The slender collagen fibers of the cartilage matrix blend with the wider, banded type I collagen fibers of the perichondrium. Perichondrial cells adjacent to the cartilage retain the capacity to form new cartilage.

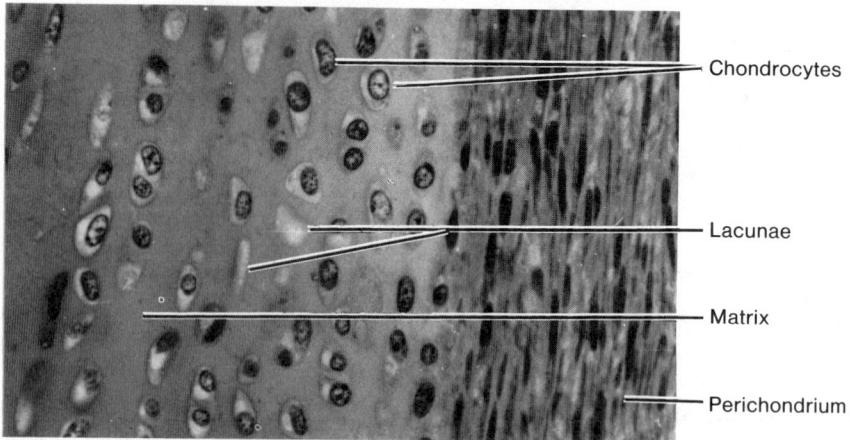

Fig. 4-1. A region of mature hyaline cartilage from a human trachea that illustrates the perichondrium (LM, × 400).

Elastic Cartilage

KEY WORDS: elastic fibers

Elastic cartilage is a variant of hyaline cartilage, differing chiefly in that it contains branched elastic fibers in the matrix. Collagen fibers of the type found in hyaline cartilage also are present but are masked. Deep in the cartilage, **elastic fibers** form a dense, closely packed mesh that obscures the ground substance, but beneath the perichondrium, the fibers form a looser network and are continuous with those of the perichondrium. The chondrocytes are similar to those of hyaline cartilage and elaborate the elastic and col-

lagen fibers. Elastic cartilage is more flexible than hyaline cartilage and is found in the external ear, auditory tube, epiglottis, and some of the smaller laryngeal cartilages.

Fibrous Cartilage

Fibrous (or fibro-) cartilage represents a transition between dense connective tissue and cartilage. It consists of typical cartilage cells enclosed in lacunae, but only a small amount of ground substance is present in the immediate vicinity of the cells (Fig. 4-2). The chondrocytes lie singly, in pairs, or in short rows between bundles of dense collagen fibers that

Fig. 4-2. Fibrous cartilage taken from a human symphysis pubis. Note that collagen bundles dominate the field of view and that chondrocytes in lacunae appear only as scattered entities (LM, × 300).

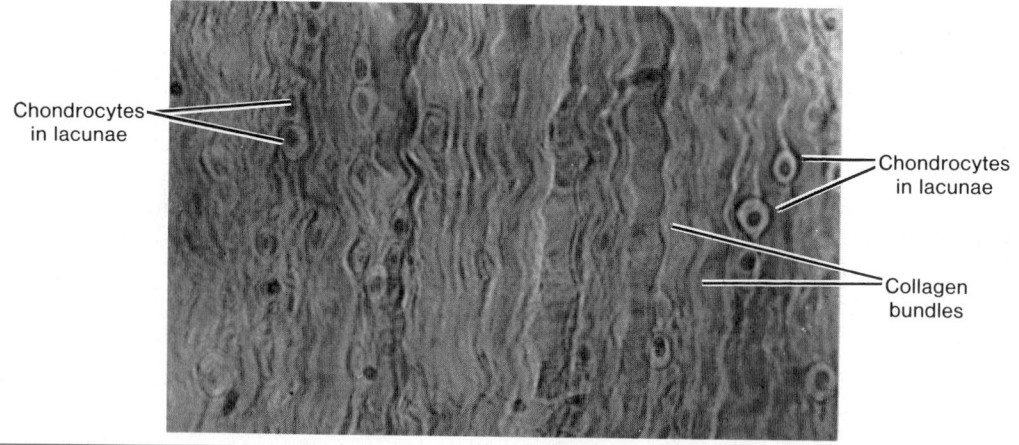

show the 64-nm banding typical of type I collagen. Fibrous cartilage always lacks a perichondrium and merges into hyaline cartilage, bone, or dense fibrous connective tissue. It occurs in the intervertebral discs, in some articular cartilages, in the symphysis pubis, and at sites of attachment of certain tendons to bone.

Bone

Bone is a connective tissue specialized for support. However, in bone the matrix is mineralized and forms a dense, hard, unyielding substance with high tensile, weight-bearing, and compression strength. Despite its strength and rigidity,

bone is a dynamic, living tissue constantly turning over, constantly being renewed and re-formed throughout life.

Macroscopic Structure

KEY WORDS: cancellous (spongy) bone, trabecula, compact (dense) bone, diaphysis, epiphysis, periosteum, endosteum, diploë

Grossly, cancellous and compact forms of bone can be identified. **Cancellous (spongy) bone** consists of irregular bars or **trabeculae** of bone that branch and unite to form a three-dimensional, interlacing network of boney rods, delimiting a vast system of small communicating spaces that in life are filled with bone marrow (Figs. 4-3 and 4-4). Tra-

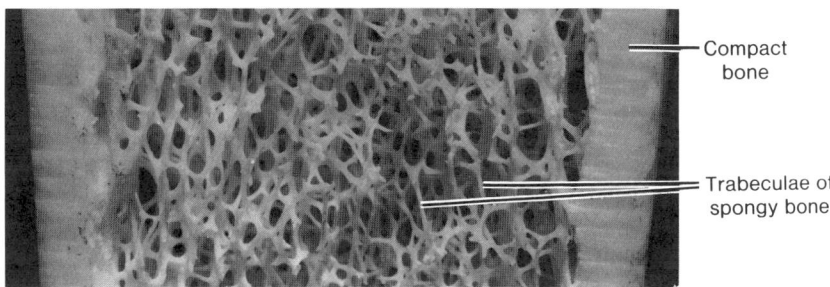

Compact bone

Trabeculae of spongy bone

Fig. 4-3. A cut through the shaft of a human ulna illustrates both compact and spongy bone (× 5).

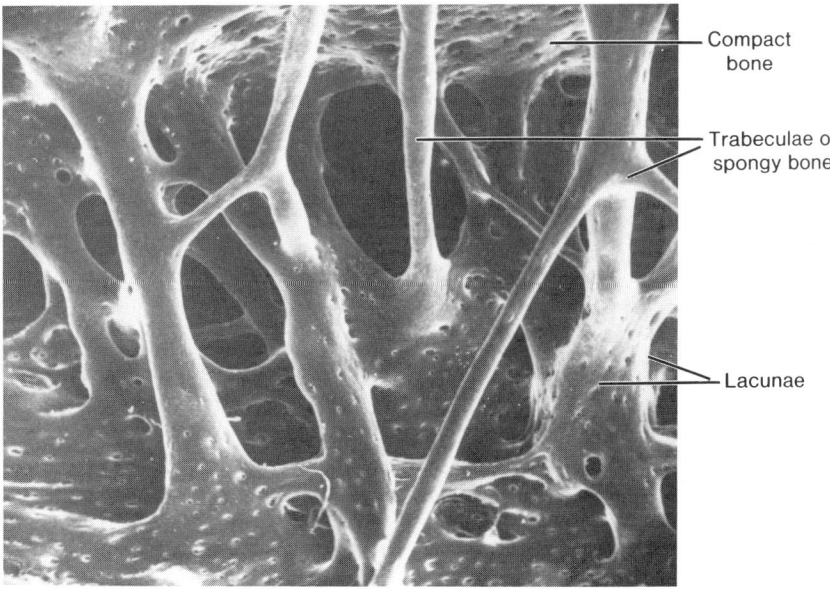

Compact bone

Trabeculae of spongy bone

Lacunae

Fig. 4-4. A scanning electron micrograph illustrates in greater detail the trabeculae of spongy bone (compare with 4-3) (× 50).

beculae of spongy bone do not form a random network but occur in a precisely organized pattern along compression and tension stress lines. This strutlike arrangement of the trabeculae contributes to the strength of bone. **Compact (dense) bone** appears as a solid, continuous mass in which spaces cannot be seen with the naked eye. The two types of bone are not sharply delimited and merge into one another.

In a typical long bone the shaft or **diaphysis** appears as a hollow cylinder of compact bone enclosing a large central space called the *marrow cavity.* The ends of the long bones, the **epiphyses,** consist mainly of cancellous bone covered by a thin layer of compact bone. The small intercommunicating spaces in the spongy bone are continuous with the marrow cavity of the shaft.

Except over articular surfaces and where tendons and ligaments insert, bone is covered by a fibroelastic connective tissue called the **periosteum.** The marrow cavity of the diaphysis and the spaces within spongy bone are lined by **endosteum,** which is similar to periosteum but is thinner and not as fibrous. Periosteum and endosteum have the ability to form bone under appropriate stimulation.

Flat bones, such as those of the skull, also consist of compact and spongy bone. The inner and outer plates (often called *tables*) consist of thick layers of compact bone, while the space between the plates is bridged by spongy bone called the **diploë.**

Microscopic Structure

KEY WORDS: lamella, lacuna, osteocyte, canaliculus, osteon (haversian system), interstitial lamella, cement line, circumferential lamella, haversian canal, Volkmann's canal

A fundamental characteristic of bone is the arrangement of its mineralized matrix into layers or plates called **lamellae.** Small, ovoid spaces, the **lacunae,** occur rather uniformly within and between the lamellae, each occupied by a single bone cell or **osteocyte.** Slender tubules called **canaliculi** radiate from each lacuna and penetrate the lamellae to link up with the canaliculi of adjacent lacunae. Thus lacunae are interconnected by an extensive system of fine canals.

In compact bone, the lamellae show three arrangements. Most are arranged concentrically around a longitudinal space to form cylindrical units that run parallel to the long axis of the bone. These are **osteons** or **haversian systems** and make up the structural units of bone (Fig. 4-5). Osteons vary in size and consist of 8 to 15 concentric lamellae that surround a wide space occupied by blood vessels. In longitudinal sections, an osteon appears as plates of

boney matrix running parallel to the slitlike space of the vascular channel. Throughout its thickness, compact bone contains a number of osteons running side by side. Osteons take an irregular course through the length of the bone and may branch.

Other lamellae appear as angular, irregular bundles of lamellar bone that fill the spaces between osteons. These are called **interstitial lamellae.** Osteons and interstitial lamellae are outlined by a refractile line, the **cement line,** which consists of modified matrix. Cement lines are not traversed by canaliculi.

At the external surface of the bone, immediately beneath the periosteum, several lamellae run around the circumference of the bone. A similar but less well-developed system of lamellae is present on the inner surface, just beneath the endosteum. These two systems of lamellae are the outer and inner **circumferential lamellae,** respectively. The lamellar arrangement of bone is shown in Figures 4-6 and 4-7.

The longitudinally oriented channels at the center of osteons are the **haversian canals.** These communicate with one another by oblique branches and by transverse connections called **Volkmann's canals** that penetrate the bone from the endosteal and periosteal surfaces. Unlike haversian canals, Volkmann's canals are not surrounded by concentric lamellae. Volkmann's canals unite the blood vessels within the haversian canals with vessels of the marrow cavity and periosteum. Thus the nutritional needs of compact bone are met by a vast, continuous network of vascular channels. Canaliculi adjacent to a haversian canal open into the perivascular space, and the canalicular system brings all the lacunae into communication with the canal.

Spongy bone shows a lamellar structure also but differs from compact bone in that it is not usually traversed by blood vessels. Therefore, osteons are rare or lacking in the irregular rods of lamellar bone that comprise spongy bone. It is thought that a critical distance (number of lamellae) exists beyond which osteocytes cannot be maintained by the canalicular system. These needs are met in trabeculae only a few lamellae in width as the canalicular system is linked to the perivascular region of the marrow cavity that resides around all trabeculae. Trabeculae that exceed this distance, like compact bone, must contain haversian systems to meet their nutritional needs.

Periosteum and Endosteum

KEY WORDS: dense irregular connective tissue, Sharpey's fibers

Fig. 4-5. An osteon and its subcomponents taken from a region of compact bone in a human ulna (LM, × 400).

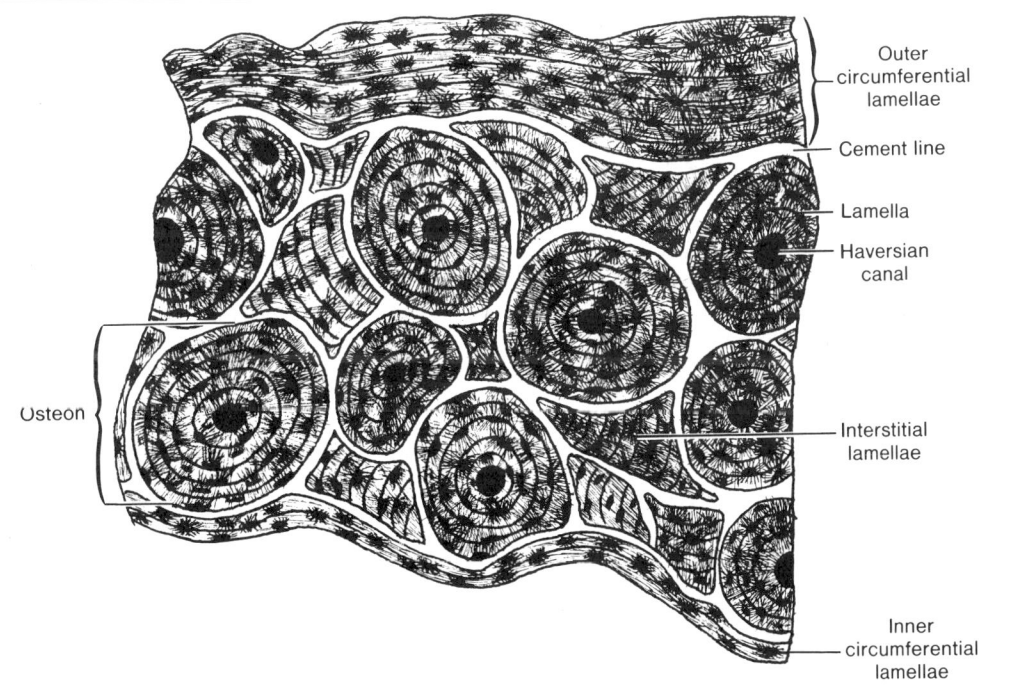

Fig. 4-6. Diagrammatic representation of a cross section of ground bone.

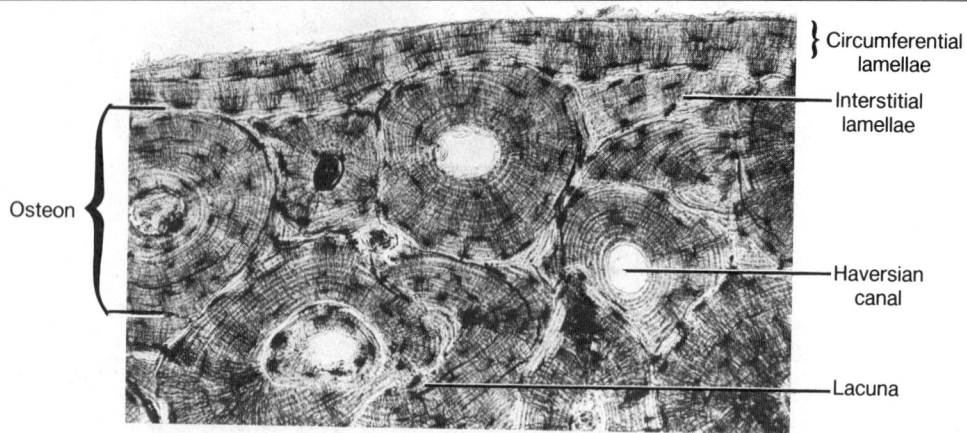

Fig. 4-7. A region of ground, compact bone taken from the peripheral edge of a human ulna illustrates its substructure (LM, × 250).

The outermost layer of the periosteum is a relatively acellular **dense irregular connective tissue** with abundant collagen fibers, a few elastic fibers, and a network of blood vessels, branches of which enter Volkmann's canals. The deeper layers of periosteum (those closest to the bone) are more cellular (osteogenetic layer) and consist of more loosely arranged connective tissue. During development of bone, collagen fibers of the periosteum are trapped in the circumferential lamellae and form **Sharpey's fibers,** which anchor the periosteum to the underlying bone.

Endosteum lines all the cavities of bone, including the marrow spaces of the diaphysis and epiphysis and the haversian and Volkmann's canals. It consists of a single layer of squamous to cuboidal cells with a small and variable content of reticular and collagen fibers. As do the cells of the inner layer of the periosteum, the cells of the endosteum retain osteogenic potential.

Cells of Bone

KEY WORDS: osteocyte, osteoblast, osteoclast, Howship's lacunae, ruffled border

Osteocytes are the chief cells found in mature bone and take the shape of the lacunae in which they are housed. They differ from **osteoblasts,** which are bone-forming cells. Osteocytes are not isolated in their lacunae but maintain extensive communication with nearby osteocytes. Delicate cytoplasmic processes extend from the cell bodies and traverse the canaliculi to contact processes from neighboring cells. At the points of contact, apposed cells form nexus

(communicating) junctions. A thin layer of unmineralized matrix separates the cell and its processes from the walls of the lacunae and canaliculi and provides the means for diffusion of materials throughout the bony matrix.

Osteocytes contain the usual cell organelles, but the Golgi apparatus is relatively inconspicuous, and ribosomes, mitochondria, and endoplasmic reticulum are scarce. Although the osteocyte appears to be not actively elaborating protein, it is not metabolically inert. It is responsible for maintaining bone, plays an active role in regulating calcium concentration in the body fluids, and is implicated in the resorption of bone.

In areas where bone is being resorbed, large multinucleate giant cells are present. These are **osteoclasts** that often lie within shallow depressions along the surface of the bone, most frequently on an endosteal surface. These cavities are called **Howship's lacunae.** Although the cell may contain as many as 30 nuclei, the individual nuclei show no unusual features and usually are located in the part of the cell farthest from the bone. The cytoplasm, usually weakly to moderately acidophilic, contains multiple Golgi bodies and centriole pairs and numerous mitochondria and lysosomes. The cell surface adjacent to the bone shows a **ruffled border,** a surface modification unique to osteoclasts; osteoclasts that lack a ruffled border do not participate in bone resorption. The border consists of elaborate folds and clefts that abut the surface of the bone; the plasmalemma on the ruffled border is covered by small, bristle-like projections. A clear zone containing many actin filaments lies adjacent to the ruffled border, and mitochondria tend to accumulate nearby. The cell surface farthest from the bone shows a smooth contour. The cytoplasm of active osteoclasts has a frothy appearance due to numerous vacuoles that contain fragments of collagen and crystals of matrix. Osteoclasts arise by fusion of monocytes that have emigrated from the blood. Osteoclasts are stimulated by parathyroid hormone.

Matrix of Bone

KEY WORDS: ground substance, collagen fiber, inorganic component

Bone matrix forms the bulk of bone and consists of collagen, ground substance, and inorganic components. The **ground substance** contains proteoglycans similar to those of cartilage but in a lesser concentration. Glycosaminoglycans of bone include chondroitin sulfate, keratin sulfate, and hyaluronic acid. Osteocalcin and osteopontin are glycoproteins of bone that play a role in binding calcium salts to the ground substance. An additional glycoprotein isolated from bone, osteonectin, aids in binding cells of bone to the matrix.

Collagen fibers are the major organic constituent of bone and are striated fibers whose cross-linked molecules form a network on which bone mineral is deposited. The collagen is present in a highly organized arrangement: In any one lamella the fibers are parallel to each other but run a helical course. The pitch of the helix differs in adjacent lamellae such that the fibers are at right angles to those in the next lamellae. The collagen of bone is exclusively type I.

The **inorganic component** is responsible for the rigidity of bone and consists of calcium phosphate and carbonate with small amounts of calcium and magnesium fluoride. The minerals are present as crystals with a hydroxyapatite structure and are present on and within the collagen at regular intervals along the fibers.

Joints

Adjacent bones are united at joints, which show considerable variation in their structure and function. At some joints, the articulating bones are so interlocked that they are immovable, while at others, the joint structure permits a wide range of motion. Based on their structure, joints can be classified as fibrous, cartilaginous, or synovial.

Fibrous Joints

KEY WORDS: suture, syndesmosis, gomphosis

Fibrous joints are formed by dense fibrous connective tissue and permit little or no movement. **Sutures** are immovable fibrous joints found only in the skull. The gap between the articulating bones is filled by a fibrovascular connective tissue and is bridged at the surfaces by periosteum. Sutures are sites of osteogenic activity and allow for growth of the skull. They are temporary joints; with aging the fibrous connective tissue gradually is replaced by bone to form a permanent boney union (synostosis). The adjoining edges of the bones are highly irregular and interlock to create a firm union. A **syndesmosis,** such as occurs at the inferior tibiofibular joint, contains much more connective tissue than does a suture, allowing somewhat more movement. A third type of fibrous joint is the **gomphosis,** a peg-and-socket joint restricted to fixation of the teeth into the jaws.

Cartilaginous Joints

KEY WORDS: epiphyseal plate, synchondrosis, symphysis, amphiarthrosis, intervertebral disc, annulus fibrosus, nucleus pulposus

Cartilaginous joints are present when bones are united by a continuous plate of hyaline cartilage or a disc of fibrocartilage. The hyaline cartilage of an **epiphyseal plate** forms a temporary joint uniting the shaft and epiphysis of a long bone during its development. This type of immovable joint, a **synchondrosis,** sometimes is classed as a primary cartilaginous joint. Permanent or secondary cartilaginous joints are represented by **symphyses,** which include unions such as those between the pubic bones (pubic symphysis) and between the vertebral bodies. The articulating surfaces are covered by thin layers of hyaline cartilage that in turn are joined to a central disc of fibrous cartilage. The bone, hyaline cartilage, and fibrous cartilage are intimately united, and the joint allows only limited movement (**amphiarthrosis**). The degree of movement possible depends on the thickness of the fibrous cartilage. A slitlike cavity may appear in the fibrous cartilage of some symphyses (e.g., symphysis pubis), but the specialized sliding surfaces of a synovial joint are absent.

In the spine, specialized discs of fibrous cartilage form the **intervertebral discs** that make up the major connections between adjacent vertebral bodies. The outer part of the disc, the **annulus fibrosus,** consists of a peripheral zone of collagen fibers and an inner zone of fibrous cartilage that is especially rich in collagen. The central part of the disc, the **nucleus pulposus,** is soft and gelatinous in the young but with aging is gradually replaced by fibrous cartilage. The articular surfaces of each vertebra are covered by thin plates of hyaline cartilage. The annulus fibrosus is firmly attached to the vertebral body by collagen fibers that pierce the cartilaginous end-plates and enter the compact bone of the vertebral body as Sharpey's fibers.

Synovial Joints

KEY WORDS: diarthrosis, capsule, articular disc, meniscus, articular cartilage, tangential zone, transitional zone, radial zone, calcified zone, lamina splendens

Synovial joints, also called **diarthroses,** are complex, freely mobile joints capable of a wide range of motion. Most of the joints of the limbs fall into this category. The bones involved in the joint are contained within and linked by a **capsule** of dense irregular connective tissue that is continuous with the periosteum over the articulating bones. Additional stability may be provided by intra- and extracapsular ligaments that also are dense irregular connective tissue. The opposing boney surfaces are covered by articular cartilages that provide low-friction gliding surfaces. Although it appears smooth to the naked eye, the free surface of the cartilage is beset with various undulations, troughs, and valleys.

The inner surface of the capsule is lined by a synovial membrane that elaborates the lubricating synovial fluid. An **articular disc** may be interposed completely or partially between the articular surfaces; a partial disc is called a **meniscus.** Discs and menisci consist of fibrous cartilage with an abundance of collagen fibers; at the periphery, they are connected to the capsule. The general structure of a synovial joint is shown in Figure 4-8.

The **articular cartilages** of most joints are hyaline in type but show a greater content of collagen than usual. In some joints (e.g., temporomandibular), frank fibrous cartilage is present. Articular cartilage lacks a perichondrium, and joint contact is made between the free, uncovered surfaces of the opposing cartilages.

The collagen fibers of articular cartilage have a unique arrangement. At the surface, the fibers are organized into tight bundles that tend to run at right angles to one another and are parallel to the surface. This forms the **tangential zone.** In the deeper **transitional** and **radial zones,** the collagen fibers are randomly arranged. Where it abuts the underlying bone, the articular cartilage becomes calcified and fuses with the bone, thus forming the **calcified zone** in which the collagen fibers are perpendicular to the surface. The diameters of the fibers increase progressively as they pass from the tangential zone to the calcified zone. The fibers of the calcified layer penetrate the subchondral bone, where they are firmly embedded like Sharpey's fibers. Overlying the tangential zone is the **lamina splendens,** a layer of randomly arranged fine fibrils.

The junction between bone and calcified cartilage is irregular and shows fine ridges, grooves, and interdigitations that help unite the subchondral bone and articular cartilage. The junction between the calcified and radial zones

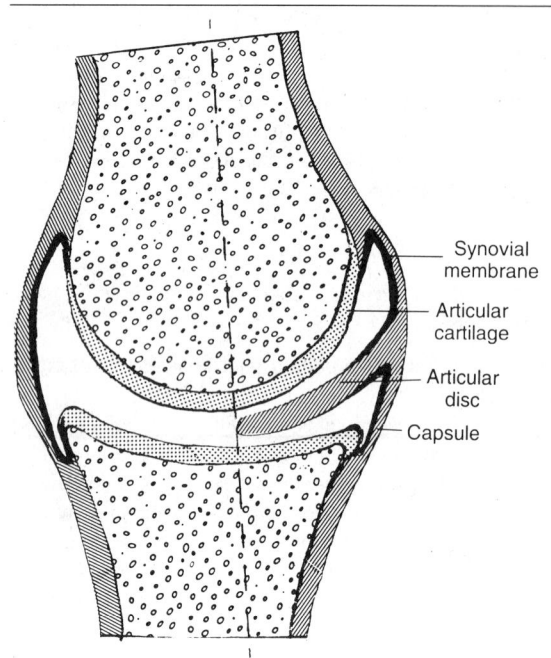

Fig. 4-8. The synovial joint.

Labels: Synovial membrane; Articular cartilage; Articular disc; Capsule

appears, in sections, as a basophilic "tide line" that represents the advancing front of the calcification process.

The cells of the articular cartilage also show zonal differences. The tangential zone contains several layers of flattened, fibroblast-like cells whose long axes, like those of the fibers, are parallel to the surface. Their oval or elongated nuclei usually have smooth outlines, but some are irregular and show a variety of undulations, deep indentations, or clefts; the patchy, clumped chromatin stains deeply. The cytoplasm contains short cisternae of granular endoplasmic reticulum (GER), a small Golgi body, and small, round mitochondria. Inclusions such as lipids and glycogen are rare, but pinocytotic vesicles are plentiful.

In the transitional zone, the cells are rounded and show many long cytoplasmic processes that often bifurcate at their tips. The round, usually eccentric nuclei contain finely granular chromatin and frequently show one or more nucleoli. Granular endoplasmic reticulum is abundant, the Golgi apparatus is well developed, and secretory granules are prominent. The cells are scattered randomly throughout the transitional zone.

The cells of the radial zone also are rounded but tend to form short columns or isogenous groups. The endoplasmic reticulum is less developed, the Golgi body is sparse, and mitochondria are small and dense. Intracellular filaments are increased in number, and lipid droplets and glycogen granules are common.

The calcified zone is characterized by short columns of enlarged, pale-staining cells that are in the advanced state of degeneration. The nuclei are dense and pyknotic, the nuclear envelope is fragmented, and cytoplasmic organelles are lacking.

The organization of the chondrocytes of articular cartilage into successive zones is reminiscent of the arrangement of cells in an epiphyseal plate during endochondral bone formation. Indeed, during growth, articular cartilage does serve as a growth zone for the subchondral bone. When epiphyseal growth is complete, the deep zones of chondrocytes in the articular cartilage are converted to compact bone and incorporated into the subchondral bone layer.

Like other cartilage, articular cartilage is avascular and aneural. The central regions of the cartilage receive their nutrition by diffusion from the synovial fluid, which bathes the cartilages, and, to a lesser extent, from vessels in the subchondral bone. At their edges, the articular cartilages are well nourished from blood vessels in the nearby synovial membrane.

Synovial Membrane

KEY WORDS: synovial cell, A-cell, B-cell, areolar synovial membrane, adipose synovial membrane, fibrous synovial membrane, synovial fluid

The synovial membrane is one of the characteristic features of a synovial joint. It is a loose-textured, highly vascular connective tissue that lines the fibrous capsule and extends onto all intraarticular surfaces except those subjected to compression during movement of a joint. Thus articular cartilages, articular discs, and menisci are not covered by synovial membranes. Occasional finger-like projections, the *synovial villi,* and coarser folds of the synovial membrane project into the joint cavity.

The free surface (synovial intima) of the synovial membrane consists of one to three layers of flattened **synovial cells** embedded in a granular, fiber-free matrix. The surface cells do not form a continuous layer, and in places, neighboring cells are separated by gaps through which the synovial cavity communicates with tissue spaces in the synovial membrane. Where the cells do make contact, their surfaces may be complex and interdigitated. Desmosomal junctions have been described in rat synovial membranes, but their presence in humans has not been confirmed.

Ultrastructurally, two types of synovial cells have been described. **A-cells** are predominant and resemble macro-

phages (hence their alternate name, *M-cells*). The cell membrane shows numerous invaginations, filopodia, and micropinocytotic vesicles, while the cytoplasm contains large Golgi bodies and numerous mitochondria and lysosomes but only scant profiles of GER. **B-cells** resemble fibroblasts and sometimes are called *F-cells.* They have a smooth outline and contain abundant GER and scattered free ribosomes. Both cell types contain much glycogen but only a few lipid inclusions. The relationship between A- and B-cells is unknown, but the existence of intermediate forms (AB-cells) suggests that the two may be merely different functional forms of a single type.

The subintimal tissue varies from place to place within the same joint, and based on the structure of this tissue, the synovial membrane is classified as areolar, adipose, or fibrous. In **areolar synovial membranes,** the underlying tissue is a loose connective tissue with relatively few collagen fibers and an abundant matrix. Cellular constituents include fibroblasts, mast cells, macrophages, and lymphocytes. **Adipose synovial membranes** line the articular fat pads, and the subintimal tissue mainly consists of fat cells. In **fibrous synovial membranes,** the underlying tissue is a dense irregular connective tissue and is found in regions subjected to tension; where such forces are extremely high, fibrous cartilage may be present.

Synovial fluid is a clear or slightly yellow viscous fluid that bathes the joint surfaces. Normally, the volume of fluid is sufficient only to form a thin film over all the surfaces within a joint. In composition, synovial fluid is an ultrafiltrate of plasma to which mucin has been added. Mucin is a product of the surface synovial cells and consists mainly of highly polymerized hyaluronic acid, which gives the viscosity to synovial fluid. Normal synovial fluid contains only 50 to 100 cells per milliliter, and these consist of macrophages, free synovial cells, and an occasional leukocyte. Amorphous particles and fragments of cells and fibrous tissue may be present and are believed to result from slow wearing of the joint surface.

Development of Cartilage, Bone, and Joints

Cartilage

KEY WORDS: chondroblast, chondrocyte, interstitial growth, appositional growth

All cartilage arises from mesenchyme, and at sites where cartilage is to form, the mesenchymal cells lose their processes, round up, proliferate, and crowd together in a dense aggregate (center of chondrification). The cells in the interior of the mass are **chondroblasts,** which lay down fibers and ground substance. As the amount of intercellular material increases, the cells become isolated in individual compartments (lacunae) and take on the characteristics of **chondrocytes.** New chondroblasts are recruited from the more peripherally placed immature cells. The perichondrium is derived by condensation of the mesenchyme that surrounds the developing cartilage. Further growth occurs by expansion of the mass of cartilage from within (interstitial growth) and by formation of new cartilage at the surface (appositional growth).

Interstitial growth results from proliferation of young chondrocytes that divide and lay down a new matrix to produce an internal expansion of cartilage. Such growth occurs only in young cartilage that is plastic enough to allow daughter cells to move away from each other after division. **Appositional growth** occurs as a result of perichondrial chondrogenic activity. The cells immediately adjacent to the cartilage (chondrogenic layer) divide and differentiate into chondrocytes, secrete matrix about themselves, and add new cartilage to the surface. Although adult perichondrium retains some potential to form new cartilage, the capacity for growth and repair is limited by its avascularity. Adult cartilage has the lowest mitotic rate of any skeletal tissue.

Where elastic cartilage forms, chondroblasts at first elaborate fibers that are neither elastic nor collagen in character. These indifferent fibers subsequently mature into elastic fibers. Fibrous cartilage develops much as does ordinary connective tissue, but some fibroblasts transform into chondroblasts and secrete a small amount of cartilage matrix about themselves.

Bone

Bone is mesenchymal in origin and develops by replacement of a preexisting tissue. If bone is formed directly in a primitive connective tissue, the process is called *intramembranous ossification;* when replacement of a preformed cartilaginous model is involved, the process is known as *endochondral ossification.* The essential process of bone formation is the same in both, however.

Intramembranous Ossification

KEY WORDS: osteoprogenitor cell, osteoblast, osteocyte, osteoid, woven bone

The flat bones of the cranium and part of the mandible develop by intramembranous ossification; these bones fre-

quently are called *membrane bones.* In the areas where bone is to form, the mesenchyme becomes richly vascularized, and the cells proliferate actively. Some cells undergo changes that subsequently lead to bone formation and are regarded as **osteoprogenitor cells.** Structurally they closely resemble the mesenchymal cells from which they arise, and their identification depends mostly on their proximity to forming bone.

The osteoprogenitor cells enlarge and transform into **osteoblasts** that lay down bars of matrix. The osteoblasts remain in contact with one another by long tapering processes, and as more and more matrix is deposited, the cells and their processes become entrapped in the matrix. The cells then are called **osteocytes.** Ultimately the cells and cell processes are completely enclosed in the matrix, thus forming the lacunae and canaliculi. The first matrix consists of ground substance and collagen fibers and, being unmineralized, is soft. At this stage, the matrix is called **osteoid;** after a short lag period, it becomes mineralized to form true bone. As osteoblasts become trapped in the newly deposited matrix and become osteocytes, new osteoblasts are recruited from differentiation of osteoprogenitor cells.

Bone development by intramembranous ossification occurs in several foci throughout the mesenchymal field and results in the formation of scattered, irregular trabeculae of bone that increase in size by apposition of new bone to their surfaces (Fig. 4-9). The resulting bone is of the spongy type. In the first formed bone, collagen fibers run in every direction; the bone is called **woven bone** and has randomly scattered osteocytes and lacks a lamellar arrangement. Woven bone is gradually replaced by lamellar bone.

In areas that become compact bone, such as the inner and outer tables of the skull, the trabeculae continue to thicken by appositional growth, and the spaces between them are gradually obliterated. As bone encroaches on vascular spaces, the matrix is laid down in irregular, concentric layers that surround blood vessels and come to resemble osteons, except that the collagen fibers in the lamellae are randomly arranged. These are called *primary osteons.* Between the plates of compact bone, where spongy bone persists, thickening of the trabeculae ceases and the intervening connective tissue becomes the blood-forming marrow.

The connective tissue surrounding the developing bone condenses to form the periosteum. The osteoblasts on the outer surface of the bone assume a fibroblast-like appearance and are incorporated into the inner layer (cambium) of the periosteum, where they persist as potential bone-forming cells. Similarly, osteoblasts on the inner surface and those covering the trabeculae of spongy bone are incorporated into the endosteum and also retain their potential for producing bone.

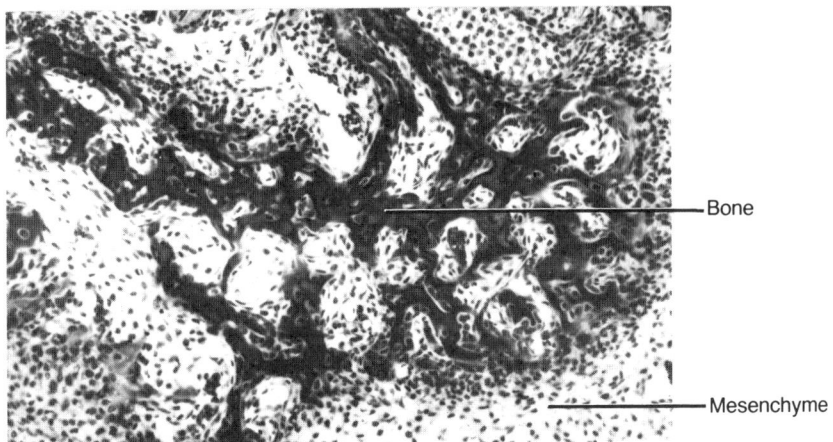

Fig. 4-9. An example of intramembranous bone formation taken from a human skull (LM, × 100).

Endochondral Bone Formation

KEY WORDS: primary ossification center, periosteal collar, periosteal bud, zone of reserve cartilage, zone of proliferation, zone of maturation and hypertrophy, zone of calcification and cell death, zone of ossification, zone of resorption, epiphyseal (secondary) ossification center, epiphyseal plate, primitive osteon, definitive osteon

The bones of the vertebral column, pelvis, extremities, face, and base of the skull develop by formation of bone in a cartilage model or precursor that must be removed before bone can be laid down. Endochondral bone formation involves simultaneous formation of the bone matrix and removal of the cartilage model. The cartilage does not contribute directly to the formation of bone, and much of the complex process involved in endochondral bone formation is concerned with removal of cartilage.

The first indications of ossification of a long bone appear in the center of the cartilage model in the region destined to become the shaft or diaphysis. In this area, called the **primary ossification center,** the chondrocytes hypertrophy and their cytoplasm becomes vacuolated. Lacunae enlarge at the expense of the surrounding matrix, and the cartilage between adjacent lacunae is reduced to thin, fenestrated partitions. The matrix in the vicinity of the hypertrophied chondrocytes becomes calcified by deposition of calcium phosphate. Diffusion of nutrients through the calcified matrix is reduced, and the cells undergo degenerative changes leading to their death.

While these events take place in the cartilage, the vascular perichondrium in this area assumes an osteogenic func-

tion. A layer of bone, the **periosteal collar,** forms around the altered cartilage and provides a splint that helps maintain the strength of the shaft. The osteoblasts responsible for development of the bone are derived from perichondrial cells immediately adjacent to the cartilage. In the region of the boney collar, the perichondrium now has become a periosteum. Blood vessels from the periosteum invade the altered cartilage forming the **periosteal bud.** Contained within the connective tissue sheath that accompanies the blood vessels are cells with osteogenic properties.

With death of the chondrocytes there is no longer a means of maintaining the matrix, which softens. With the aid of the erosive action of invading blood vessels, the thin plates of matrix between lacunae break down. As the adjoining lacunar spaces are opened up, narrow tunnels are formed that are separated by thin bars of calcified cartilage. Blood vessels grow into the tunnels, bringing in osteogenic cells that align themselves on the surface of the cartilaginous trabeculae, differentiate into osteoblasts, and begin to elaborate bone matrix. At first the matrix is laid down as osteoid, but it soon becomes mineralized to form true bone. The early trabeculae thus consist of a core of calcified cartilage with a shell of bone. These primitive trabeculae soon are removed both by continued resorption of the calcified cartilage and through the activity of osteoclasts that appear on the boney shell. An expanding cavity is opened up within the developing shaft. Support for the bone is provided by extension of the periosteal collar, which also becomes thicker as the periosteum lays down new bone at the surface.

This entire process continues as an orderly progression

of activity that extends toward both ends of the developing bone. Several areas or zones of activity can be distinguished in the cartilage. Beginning at the ends farthest from the primary ossification center, these are the following:

Zone of reserve cartilage. This area consists of hyaline cartilage and initially is a long zone but shortens as the process of ossification encroaches on it. The cells and their lacunae are randomly arranged throughout the matrix, and there is slow growth in all directions.

Zone of proliferation. In this region the chondrocytes actively proliferate, and as the cells divide, the daughter cells become aligned in rows separated by a small amount of matrix (Fig. 4-10). Within the rows, the cells are separated only by thin plates of matrix. Growth occurs mainly in the regions farthest from the reserve cartilage, and the length is increased more than the diameter.

Zone of maturation and hypertrophy. Cell division ceases and the cells mature and enlarge, further increasing the length of the cartilage. Lacunae expand at the expense of the intervening matrix, and the cartilage between adjacent cells in the rows becomes even thinner.

Zone of calcification and cell death. The matrix between the rows of chondrocytes becomes calcified and the cells die, degenerate, and leave empty spaces (Fig. 4-11). The thin plates between the cells break down, and the lacunar spaces are opened up to form irregular tunnels bounded by the remains of calcified cartilage that had been present between the rows of cells.

Zone of ossification. Vascular connective tissue invades the tunnel-like spaces and provides osteogenic cells that differentiate into osteoblasts. These gather on the surface of the calcified matrix and lay down bone.

Zone of resorption. Osteoclasts appear on the trabeculae and begin to resorb the bone. Ultimately all the calcified cartilages and their boney coverings are resorbed, and the marrow cavity increases in size.

As these changes occur within the cartilage, the periosteal collar increases in length and thickness, gradually extending toward the ends of the developing bone to provide a continuous splint around the area of changing and weakening cartilage.

At about the time of birth, new centers of ossification, **epiphyseal,** or **secondary, ossification centers,** appear at the ends of long bones. The cartilage of the epiphyses passes through the same sequences as in the diaphysis, but cartilage growth and subsequent ossification spread in all directions and result in a mass of spongy bone. Ultimately, all the cartilage of the epiphysis is replaced by bone except for that at the free end, which remains as articular cartilage. Cartilage also persists between the epiphysis and diaphysis and forms the **epiphyseal plate,** whose continued growth permits further elongation of the bone. Ossification

Fig. 4-10. A region of epiphyseal plate during endochondral bone formation demonstrates several regions of activity (sample from fetal human finger) (LM, × 250).

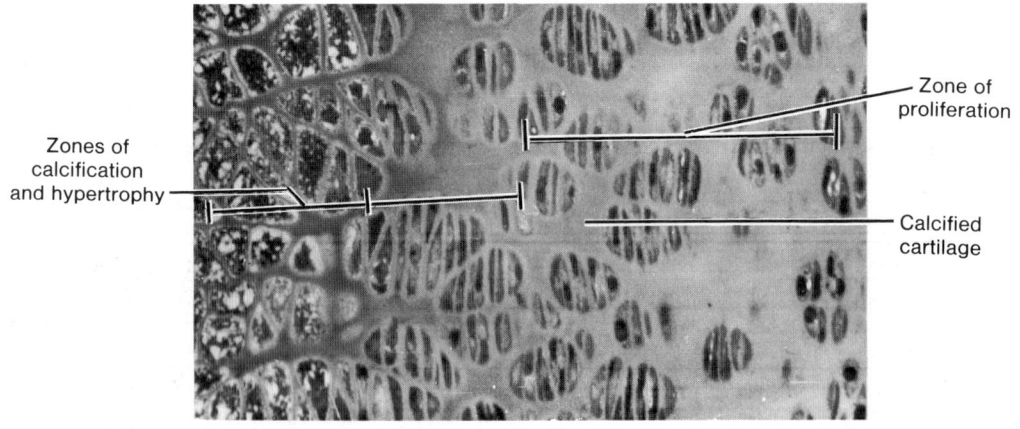

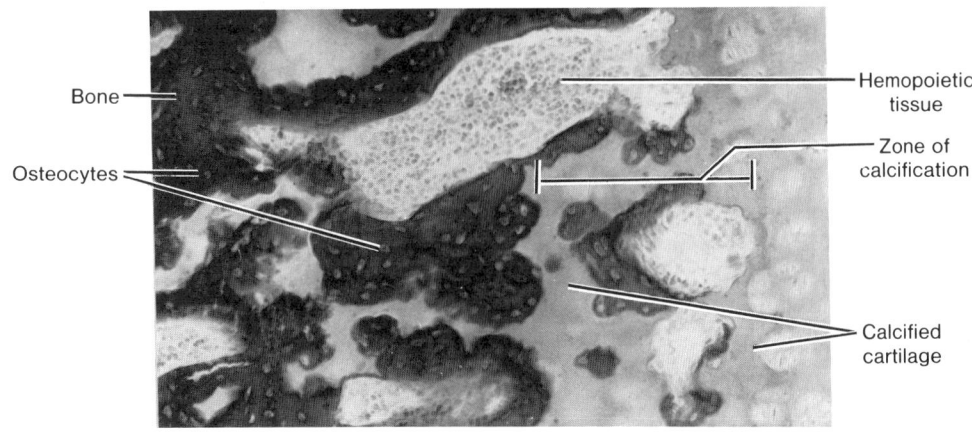

Bone

Osteocytes

Hemopoietic tissue

Zone of calcification

Calcified cartilage

Fig. 4-11. A region of epiphyseal plate during endochondral bone formation clearly demonstrates formed bone and the penetration of calcium resulting in the zone of calcification (sample from fetal human finger) (LM, × 250).

occurs at the diaphyseal side, and for a time formation of new cartilage and its replacement by bone take place at the same rate so that the thickness of the epiphyseal plate remains constant. Eventually, growth in the plate ceases, the cartilage is replaced by bone (closure of the epiphysis), and further increase in length of the bone is no longer possible. Increase in diameter of the bone results from deposition of bone at the outer surface as a result of the activity of the periosteum, while at the same time osteoclasts resorb lesser amounts of bone at the endosteal surface. Thus not only is the gross diameter increased, the thickness of the compact bone and the size of the marrow cavity also increase.

Bone formed by the periosteum represents intramembranous bone and results in a lattice of irregular, boney trabeculae. This is converted to compact bone by a gradual filling of the spaces between trabeculae. Bone is laid down in ill-defined layers, and where they surround and enclose blood vessels, the first haversian systems or **primitive osteons** are formed, in the same way that they form during development of membranous bone. In the primitive osteons the collagen is laid down at random in each lamella. Constant remodeling and reconstruction of bone occur throughout life, with the primitive osteons being replaced by **definitive osteons;** these in turn are replaced by successive generations of osteons.

Replacement of osteons begins with formation of tun-

nels within the compact bone as a result of the osteolytic activity of osteocytes and osteoclasts. The cavities widen and enlarge to reach considerable lengths. Osteolytic activity ceases, osteoblasts become active, and bone is deposited on the walls of the cavity until it is refilled as a definitive osteon. The process continues throughout life; resorption cavities continue to form and to be filled with third, fourth, and higher generations of osteons. The resorption cavities do not always coincide with the osteon being replaced, and fragments of older osteons remain as interstitial lamellae. Replacement of osteons and the remodeling of bone is shown in Figure 4-12.

The successive changes occurring during development of a long bone are summarized in Figure 4-13. The cartilage model (A) becomes encircled by a periosteal collar in the region that will become the shaft (B), and the cartilage in this area calcifies and degenerates (C). A vascular bud of connective tissue and blood vessels penetrates the altered cartilage, which is resorbed, and the cartilage remnants become covered by bone to form irregular trabeculae. These in turn are soon resorbed, and an expanding marrow-filled cavity develops as the process extends toward each end (D). Secondary ossification centers appear in the epiphyses, gradually expanding in all directions (E). Cartilage persists at the ends of the bone and, for a time, as the epiphyseal plate from which all further lengthening will occur (F). Cartilage in the epiphyseal plate ceases its growth and

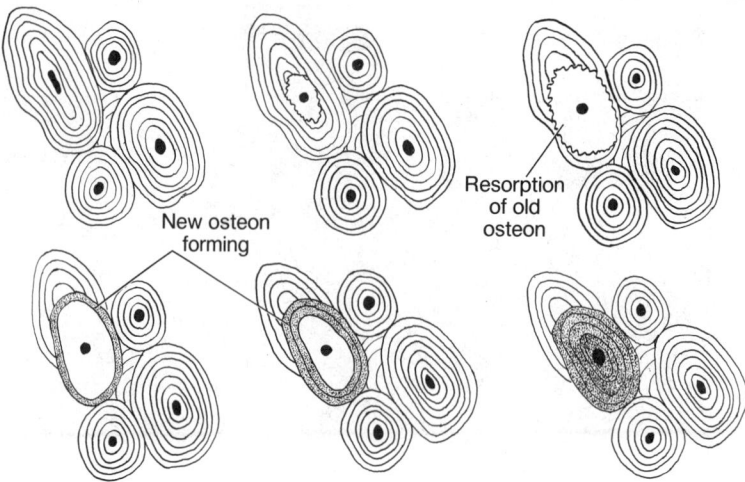

New osteon
forming

Resorption
of old
osteon

Fig. 4-12. Replacement of osteons during remodeling of bone.

is replaced by bone. The marrow cavity then becomes continuous throughout the length of the bone (G).

Cartilage, because of its capacity for rapid growth as compared with bone, is the ideal rigid, lightweight tissue for the embryonic skeleton because it can keep pace with the rapidly growing fetus. Likewise, it is growth of the hyaline cartilage in the epiphyseal plate that allows for the rapid growth of long bones during adolescence. Ultimately, cartilage of the provisional skeleton, as well as that of the epiphyseal plate, is replaced by the slower-growing bone.

Joints

Fibrous Joints

Fibrous joints form simply by the condensation and differentiation of the mesenchyme that persists between developing bones, resulting in formation of a dense irregular connective tissue that binds the bones together.

Cartilaginous Joints

In the early stages of development, a symphysis is represented by a plate of mesenchyme between separate cartilaginous rudiments of bone; the mesenchyme differentiates into fibrous cartilage. When the adjacent cartilaginous bone models have ossified, thin layers of hyaline cartilage are left unossified and persist as laminae of hyaline cartilage between the bone and fibrous cartilage.

The nucleus pulposus of an intervertebral disc develops from the notochord. Early in embryonic life, notochordal cells in the regions between future vertebrae proliferate rapidly but later undergo mucoid degeneration to give rise to the gelatinous nucleus pulposus. The surrounding fibrocartilage of the annulus fibrosus is derived from the mesenchyme between adjacent vertebrae.

Synovial Joints

The formation of appendicular bones is heralded by development of cartilage in the center of each future skeletal element. As the cartilage models expand, their ends approach one another, and the intervening undifferentiated tissue condenses to form the interzonal mesenchyme or midzone from which the joint will form. Simultaneously, the outer layer of mesenchyme also condenses to form the primitive joint capsule, which is continuous with the mesenchyme that gave rise to the perichondrium over the adjacent cartilage model. The mesenchyme enclosed by the primitive joint capsule forms three distinct zones: a central, loose layer of randomly arranged cells lying between two dense layers in which the cells are parallel to the surface of the underlying cartilage. The dense zones give rise to the articular cartilages.

Development of the joint cavity begins in the loose, central region of the interzone with the appearance of small, fluid-filled spaces between the mesenchymal cells. The spaces coalesce and come to form a continuous cavity

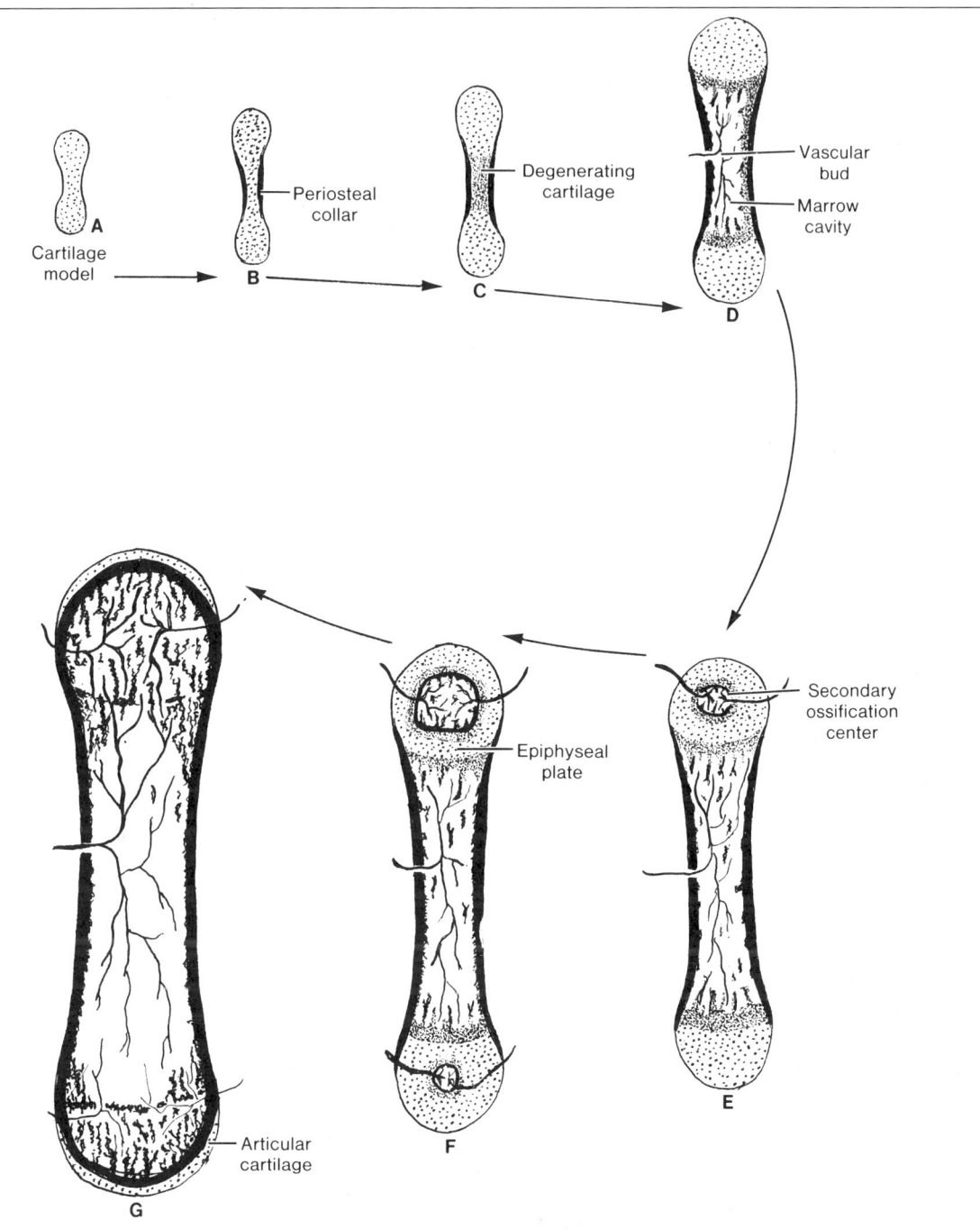

Fig. 4-13. Diagrammatic representation of the successive changes during the development of a long bone.

separating the dense chondrogenic layers and extending along the margins at the ends of the cartilage models. As the spaces unite, the primitive joint capsule completes its differentiation and forms two layers. The outer layer becomes the dense, irregular connective tissue of the joint capsule proper, while the inner layer becomes cellular and gives rise to the synovial membrane. Menisci and articular discs develop from the mesenchyme that gave rise to the fibrous joint capsule.

Repair of Cartilage and Bone

Cartilage

Because of its avascularity, mammalian cartilage has a limited capacity to restore itself after injury. Damaged regions of cartilage become necrotic, and these areas then are filled in by connective tissue from the perichondrium. Some of the connective tissue may slowly differentiate into cartilage, but most remains as dense irregular connective tissue that may later calcify or even ossify.

Bone

Repair of a broken bone is influenced by the size of the bone, the thickness of its compact bone, and the complexity of the fracture. However, the underlying process of repair remains the same and in general recalls the events of bone formation.

A fracture ruptures blood vessels in the marrow, periosteum, and bone itself, and bleeding may be extensive. As a result of the vascular damage, bone dies for some distance back from the fracture site, as do the periosteum and marrow. However, the latter have a greater blood supply than bone tissue itself, so the area of cell death in these tissues is not as great. The blood clot that forms is removed by lysis and phagocytosis. Fibrovascular invasion of the clot immediately around and within the fracture and its conversion to a fibroconnective tissue are not prominent phenomena in humans. Repair occurs by activation of osteogenic cells in the viable endosteum and periosteum near the fracture site. The cells proliferate and form new trabeculae of bone within the marrow cavity and beneath the periosteum.

Repair occurring within the marrow cavity can be very

important in human fractures. Medullary healing begins as foci of vascular and fibroblastic proliferations in the viable tissue that borders the damaged marrow, followed by osteogenic activity and bone formation. Ultimately, a network of fine bony trabeculae extends across the marrow cavity from cortex to cortex on either side of the fracture and finally across the fracture line, providing an internal scaffold until union of the fractured ends can be effected. All this new bone arises by intramembranous bone formation and is woven bone. Medullary bone healing is particularly important for the union of fractures in cancellous bones such as the spinal bodies and lower end of the radius and fractures through the metaphysis of long bones.

On the periosteal surface, the repair process arises from the inner osteoblastic layer of the periosteum (osteogenetic layer) beginning a short distance from the fracture zone. Periosteal proliferation occurs on both sides of the fracture gap, resulting in collars of bony trabeculae that grow outward and toward each other, ultimately fusing to span the gap in a continuous arch. The new trabeculae are firmly attached to the old bone surface, including the dead bone. Growth of the osteogenic cells outstrips their vascular supply so that in the midzone of the fracture site the cells differentiate into chondroblasts rather than osteoblasts and lay down a callus of hyaline cartilage. This also bridges the fracture gap to form a stabilizing splint around the fracture. The cartilage is converted to bone by endochondral bone formation, but the process is self-limiting, and all the cartilage disappears without continuous formation of new cartilage as in an epiphyseal plate. Cartilage always appears during the repair of long bones, whereas flat bones heal without cartilage formation.

Union of the compact cortical bone occurs from sources arising in the medullary cavity or from the periosteum. Since the ends of the bone at the fracture line consist of dead bone, direct union of the fractured ends is very rare if it occurs at all. Occasionally the gap is filled by formation of hyaline cartilage, which then undergoes endochondral ossification to achieve cortical union. More frequently, the bone that initially unites the broken ends is a network of woven bone formed by intramembranous bone formation. The last act in repair is the resorption of excess bone and the remodeling of newly formed and dead bone that is replaced by lamellar bone.

Review Table 4-1. Key Features in Identifying Cartilage, Bone, and Tendon

Type	Arrangement of Cells and Lacunae	Other Features
Hyaline cartilage	Glasslike matrix; lacunae randomly arranged, slitlike in appearance near perichondrium	Isogenous groups; territorial matrix
Elastic cartilage	Matrix more fibrous in appearance; elastic fibers (need to be stained selectively)	Large isogenous groups
Decalcified bone	Lacunae show organization within lamellae of osteons; haversian canals	Tide marks; bone marrow
Ground bone	Haversian systems; interstitial and circumferential lamellae	Canaliculi between lacunae
Fibrous cartilage	Dense fibrous connective tissue a dominant feature with a small amount of ground substance; lacunae few in number and scattered between fibers	Round-shaped chondrocytes
Tendon	Fibroblast nuclei; dense staining and elongate; lie in parallel rows between collagen fibers	Lacunae absent

Functional Review: Cartilage and Bone

Cartilage serves as a rigid yet lightweight and flexible supporting tissue. It forms the framework for the respiratory passages to prevent their collapse, provides smooth "bearings" at joints, and forms a cushion between the vertebrae, acting as a shock absorber for the spine. Cartilage is important in determining the size and shape of bones and provides the growing areas in many bones. Its capacity for rapid growth while maintaining stiffness makes cartilage suitable for the embryonic skeleton. About 75 percent of the water in cartilage is bound to proteoglycans, and these compounds are important in the transport of fluids, electrolytes, and nutrients throughout the cartilage matrix.

Bone forms the principal tissue of support and is capable of bearing great weight. It provides attachment for muscles of locomotion, carries the joints, serves as a covering to protect vital organs, and houses the hemopoietic tissue. Bone is the major storehouse of calcium and phosphorus in the body.

Osteocytes are the dominant cells of mature bones and are responsible for maintaining the matrix. They also aid in regulating the calcium and phosphorus levels of the body and play a role in the resorption of bone. Osteoclasts are specifically implicated in bone resorption. They destroy the ground substance and collagen and release minerals from the matrix. These cells elaborate lysosomal enzymes and contain high concentrations of citrate, which is involved in mobilizing calcium from bone. The initial stage of bone resorption by osteoclasts is extracellular: glycosaminoglycans of the matrix are degraded, permitting fragmentation of the bone. The fragments are phagocytosed by osteoclasts and digested intracellularly. The ruffled border of the osteoclasts increases the surface area and may seal off the area of resorption and allow a local environment conducive to the digestion of bone.

Longitudinal growth of a developing bone depends on the interstitial growth of cartilage in the zone of proliferation of the epiphyseal plate and on the enlargement of chondrocytes in the zone of maturation and hypertrophy nearby. Cartilage of the epiphyseal plate is under the influence of growth hormone secreted by cells in the anterior pituitary gland. Thyroid hormone modulates the activity of growth hormone, and at puberty, androgens and estrogens contribute to masculinization or feminization of the skeleton, respectively. Increased width of the bone is achieved by appositional growth and the deposition of bone at the boney collar. The first primitive trabeculae consist of cores of cartilage covered by bone. They are formed in the region where the boney collar is thinnest and serve as internal struts to support the developing bone until the collar is thick enough to provide support. As the collar increases in thickness, osteoclasts remove bone at the endosteal surface, thereby enlarging the marrow cavity.

Although most joints serve to allow movement, sutures are immobile and act only to unite bones. Sutures in the young provide areas where bone can grow, and it is only after they have been replaced by boney unions that their uniting functions become paramount. Even the articular cartilages of synovial joints serve as areas of growth for subchondral bone during development. The cartilaginous joints of the spine, although permitting some degree of motion, serve primarily as shock absorbers to dampen and distribute mechanical forces acting on the spine. This is mainly the function of the nucleus pulposus, which, because of its high water content, can act as a water cushion.

The joints specifically designed for movement are the synovial joints. The fibrous capsule and ligaments provide a tough, unyielding binding for the articulating bones yet remain flexible enough to allow movement of the joint. The synovial membrane mainly produces the lubricant of the joint, the synovial fluid, but also may participate in removing particulate matter from the joint cavity. Synovial villi increase the surface area for secretion and absorption, while the larger folds serve as flexible pads that can accommodate to the changing shape and size of the joint space and its recesses during movement. The function of the articular discs and menisci is uncertain, but they may act as shock absorbers, improve the fit between the joint surfaces, spread lubricant, distribute weight over a larger area, or protect the edges of the articular cartilages.

Cartilages form the bearings on which the articulating surfaces of the joint move and offer a wear-resistant surface of low friction — about half that of Teflon, the most slippery of synthetic materials. The arrangement of fibers in the tangential zone places a dense mat of collagen fibers right at the articulating surface. The cartilage is anchored to the underlying bone by collagen fibers (Sharpey's fibers) that are embedded in subchondral bone. The irregular and interlocking margin between the calcified zone and bone further stabilizes the cartilage. The calcified zone also may help regulate diffusion of materials between bone and cartilage.

Synovial fluid provides for the nutrition of articular cartilages, menisci, and articular discs and also lubricates the joint surfaces. Exactly how the synovial fluid performs its lubricating function is unknown. It has been suggested that the fluid forms a thin film between the bearing surfaces and that the lubricating activity depends on internal properties of the fluid such as cohesiveness, shear rate, and flow rate. Another mechanism, "weeping lubrication," has been proposed as a method of joint lubrication. Because of the content of proteoglycans, articular cartilage avidly binds water that, during compression, exudes ("weeps") from the cartilage to lubricate the surface. When compression is removed, fluid is reabsorbed into the cartilage. Thus weeping lubrication also could serve as a method for pumping nutrients and wastes into and out of the cartilage. A mechanism of "boosted lubrication" takes into account the irregularities normally seen on the surface of articular cartilage. As the cartilage is compressed, pools of synovial fluid are trapped in the furrows and valleys of the articular surfaces. As compression increases, a mobile component of the fluid containing small molecules is forced into the cartilage at the areas of pressure. The fluid in the valleys thus becomes increasingly rich in hyaluronidate and more viscous so that as the pressure increases, the synovial fluid progressively becomes more effective as a lubricant.

Review Questions and Answers

Questions

18. What would be the advantages and disadvantages of entering the thoracic cavity by an incision made through the costal cartilages? (Ignore the structures in the intercostal spaces.)

19. In what ways are bone and cartilage similar? How do they differ? What do these similarities and differences tell you about possible reparative processes in bone and cartilage?

20. What contributes to lengthening of a bone during endochondral ossification?

21. In which dimension will activity in the secondary ossification center result in an increase in bone size? Growth from the epiphyseal plate increases bone size in which dimension and of which part of the bone?

22. Describe the unit structure of compact bone.
23. What are interstitial lamellae and what is their origin?

Answers

18. Being cartilaginous, the costal cartilages are without a nerve or blood supply of their own. Section through the cartilages therefore would be bloodless and would not sacrifice nerves. However, because cartilage has no blood supply, healing is poor.

19. Bone and cartilage are similar in that the intercellular substance forms the bulk of the tissues, and their constituent cells reside in spaces (lacunae) within the intercellular substance. They are also similar in that both bone and cartilage (except for fibrous cartilage and articular cartilage) are covered by connective tissue membranes (periosteum and perichondrium, respectively) that are able to lay down new bone or cartilage by the process of appositional growth. The tissues differ in that the cartilage cells are isolated and do not communicate, whereas bone cells contact neighboring cells by means of cytoplasmic processes that traverse the bone matrix through the canaliculi. In bone the matrix becomes calcified, whereas under normal conditions cartilage does not calcify; if it does, the chondrocytes degenerate. Cartilage is avascular, and nutrition is supplied totally by diffusion through the matrix; (compact) bone is a relatively vascular tissue, and diffusion throughout the tissue is accomplished through the canaliculi. Because of its rich vascular supply, bone heals readily following injury, whereas cartilage rarely heals once fractured due to its poor vascularization.

20. Lengthening of bone during its development results from events occurring within the cartilage. Increase in size occurs as a result of interstitial growth in the zone of cell proliferation, with increase in length occurring because the cells tend to align themselves in rows oriented in the long axis. Further increase in length occurs in the region of cell hypertrophy as the cells enlarge.

21. Since bone formation in the secondary centers occurs in all directions, increase in bone size (epiphysis) will likewise be in all directions. Replacement of cartilage by bone occurs at the diaphyseal side of the epiphyseal plate and therefore will result in lengthening of the shaft. It will not contribute to an increase in size of the epiphysis.

22. The unit structure of compact bone is the osteon (or haversian system), which consists of 8 to 12 concentric lamellae of bone arranged about a central canal called the haversian canal. In life the haversian canal is occupied by blood vessels. Arranged rather uniformly within and between lamellae are lacunae that contain the osteocytes. Extending from each lacunae are fine channels or canaliculi, which join similar structures from adjacent lacunae; the canaliculi that border on a haversian canal make contact with the space of the canal. Thus all the lacunae are continuous with each other and ultimately with the haversian canal by means of the canalicular system. Osteons are separated from neighboring osteons or other lamellar structures by the cement line. Canaliculi do not penetrate the cement line, so each osteon represents a discrete unit.

23. Interstitial lamellae are angular or irregular groups of lamellae that fill in the spaces between osteons. They represent the remnants of earlier osteons that have been replaced during remodeling of bone.

5 Special Connective Tissue: Blood

Blood often is defined as a special connective tissue in which the intercellular substance is a fluid. Unlike other connective tissues, however, the intercellular substance of blood lacks a fibrous component, and most of the intercellular protein is produced by cells in other tissues (chiefly the liver) and not by the blood cells. Many of the formed elements of blood consist of anucleate elements (erythrocytes) and bits of cytoplasm (platelets). True cells — leukocytes — make up only a small part of the formed elements and are present only as transients that use the blood for transportation to other organs and tissues into which they migrate to carry out their functions. Only erythrocytes and platelets function while in the bloodstream. None of the formed elements normally replicate within the blood; as they are lost, new elements are added from special blood-forming tissues outside the circulation. Thus blood can be considered the secretory product of several organs and tissues rather than a tissue per se.

The total quantity of blood forms a rather constant proportion of the body mass. In humans the volume is about 5 liters, or approximately 7.5 percent of body weight.

Components of Blood

KEY WORDS: formed elements, erythrocyte, platelet, leukocyte, plasma, albumin, fibrinogen, globulin, serum

Blood consists of **formed elements** that include erythrocytes, platelets, leukocytes, and a fluid intercellular material called *plasma.* These elements can be separated by centrifugation, and when done in calibrated tubes, the result (hematocrit) gives an estimate of the volume of the formed elements. The heaviest components, **erythrocytes,** form the lower layer and in humans make up about 45 percent of the blood volume. **Platelets** and **leukocytes** are present in the buffy coat, a grayish white layer immediately above the erythrocytes, and form about 1 percent of the total blood volume. The uppermost layer consists of **plasma,** which contains three main types of protein: albumin, globulin, and fibrinogen. **Albumin,** the most abundant and smallest of the plasma proteins, is formed in the liver, as is **fibrinogen,** an essential component for blood clotting. The **globulins** (alpha, beta, and gamma) include several proteins of different sizes. The gamma globulins

Table 5-1. A Partial List of Normal Values of Human Blood

pH (arterial)	7.35–7.45
Osmolality	280–296 mOsm/kgH$_2$O
O$_2$, saturated (arterial)	96–100%
CO$_2$	24–30 mEq/liter
PCO$_2$ (arterial)	35–45 mm Hg
Total protein	6.0–8.4 g/100 ml
Albumin	3.5–5.0 g/100 ml
Globulin	2.3–3.5 g/100 ml
Cholesterol	120–220 mg/100 ml
Triglycerides	40–150 mg/100 ml
Lactic acid	0.6–1.8 mEq/liter
Ascorbic acid	0.4–1.5 mg/100 ml
Uric acid	3.0–7.0 mg/100 ml
Glucose (fasting)	70–110 mg/100 ml
Sodium	135–145 mEq/liter
Potassium	3.5–5.0 mEq/liter
Chloride	100–106 mEq/liter
Calcium	8.5–10.5 mg/100 ml
Phosphorus	3.0–4.5 mg/100 ml
Magnesium	1.5–2.0 mEq/liter
Iron	50–150 μg/100 ml

are immunoglobulins (antibodies) synthesized by cells of the lymphatic organs and tissues. A partial list of normal values of other substances found in human blood is provided in Table 5-1.

Plasma is obtained from blood after treatment with an anticoagulant and contains all the components of the fluid portion of the blood. **Serum** is obtained from clotted or defibrinated blood and does not contain fibrinogen; it does contain other components elaborated during the process of blood clotting.

Erythrocytes

At rest, the average human uses about 250 ml of oxygen and produces almost 200 ml of carbon dioxide per minute. With activity, these quantities can increase 10- to 20-fold. Oxygen and carbon dioxide are carried by the erythrocytes, which transport the gases with great efficiency. Exchange of gases occurs in the blood capillaries of the lung, through which the erythrocytes pass in less than 1 second, yet gaseous exchange is complete in approximately the first third of this time.

Structure

KEY WORDS: anucleate, hemoglobin, absence of organelles

In blood smears and tissue sections, erythrocytes appear as **anucleate,** uniformly acidophilic bodies devoid of any internal structure. The pigment **hemoglobin** makes up about 95 percent of the dry weight. *Erythroplastid* more aptly describes these elements, but custom and common use have given the terms *erythrocyte* and *red cell* the status of proper terminology. Electron microscopy confirms the **absence of organelles** in the interior of mature red cells, which are unable to synthesize protein or renew constituents of their cell membrane. The cholesterol of the plasmalemma, important for the flexibility of the cell, is controlled by the plasma concentration of cholesterol, not by cell metabolism.

Carbohydrate chains extending from glycoproteins and glycolipids of the erythrocyte plasmalemma contain a number of antigenic determinants that are the basis for the ABO blood-group system. The cell membrane limiting erythrocytes of some individuals may have either antigen A (type A), antigen B (type B), or both antigens A and B (type AB). Erythrocytes of other individuals may have neither antigen, and such individuals are classified as having type O blood. Most individuals have antibodies in their blood plasma against erythrocyte antigens, with the exception of their own, and if a different blood type is transfused, massive intravascular agglutination and lysis of erythrocytes in the recipient of the transfusion will occur. Therefore, it is essential to determine what antigens are present on the erythrocytes of donor blood and what antibodies are present in the plasma of the recipient before giving a transfusion.

Shape and Size

KEY WORDS: biconcave disc, central depression, cell membrane, spectrin, ankyrin

The human red cell usually is described as a **biconcave disc.** Face on, the normal erythrocyte presents a smooth, rounded contour with a **central depression** (Figs. 5-1 and 5-2). In smears, human erythrocytes average 7.6 μm in diameter, are slightly smaller in tissue sections (6.5–7.0 μm), and are slightly larger (about 8 μm) in fresh blood. The thickness of the red cell is about 2 μm. Apparently this thickness allows hemoglobin to maintain a distance from the cell surface that is optimal for its function.

Erythrocytes are enclosed in a typical **cell membrane,** whose flexibility and elasticity allow the red cell to accommodate its passage through the small capillaries. In vivo, the cells often assume a cup shape as they pass through blood vessels. A subplasmalemmal network of protein (spectrin) helps to maintain the biconcave shape and still allows flexibility. **Spectrin,** linked by actin, appears to form a web immediately beneath the plasmalemma and

Fig. 5-1. Erythrocytes from human peripheral blood (SEM, × 2,000).

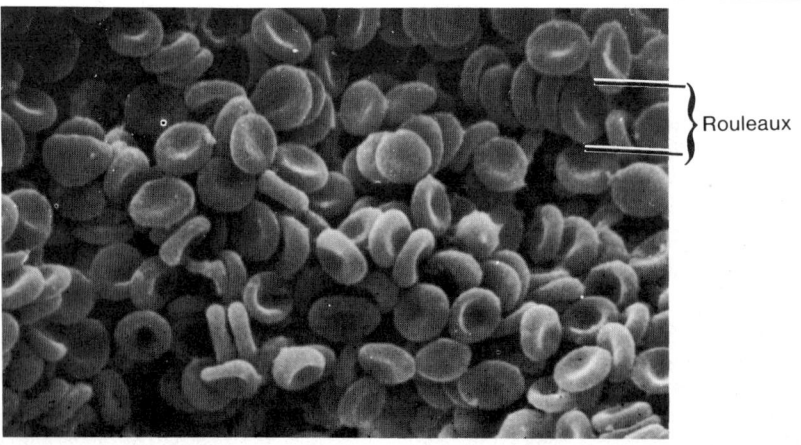

Rouleaux

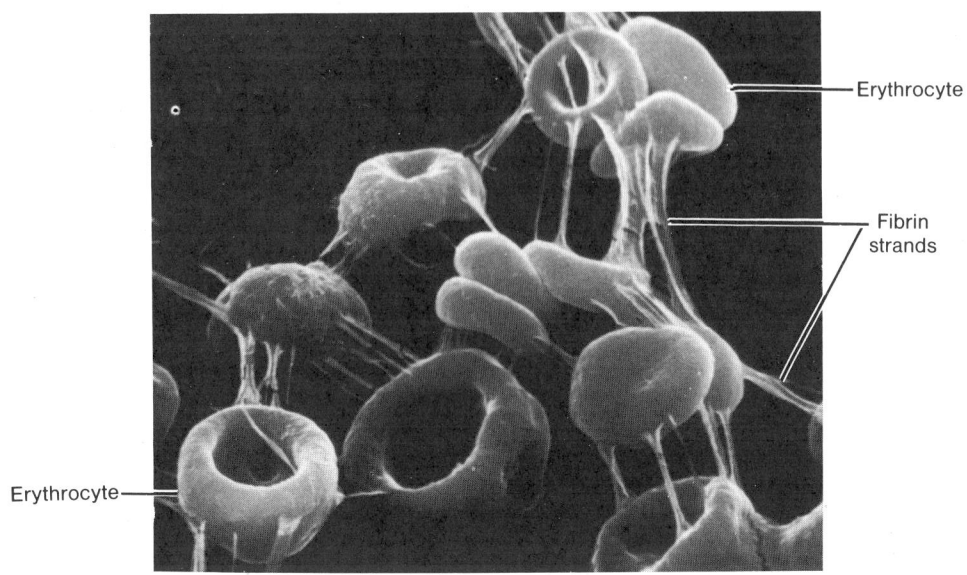

Fig. 5-2. Blood clot of human peripheral blood (SEM, × 5,000).

may act as a cytoskeleton. The network is attached to the cell membrane by the protein **ankyrin** so that the cytoskeleton and cell membrane are linked to act as a unit. Hemoglobin also may play a role in maintaining cell shape, since marked changes in shape are associated with the abnormal hemoglobin of sickle cell anemia.

Number and Survival

Consistent sex differences in the number of red cells are seen, with lower values occurring in women. Values of 4,500,000 to 6,000,000/mm³ in men and 3,800,000 to 5,000,000/mm³ in women are considered normal.

In humans the survival time of red cells is about 120 days. Although red cells generally are removed as they age and wear out, a certain amount of random destruction also occurs; many red cells are destroyed in the marrow without ever being released.

As erythrocytes age, they use up most of the enzymes associated with adenosine triphosphate (ATP) and are unable to maintain themselves, synthesize protein, or renew their cell membranes. By the end of their life span, red cells have become rigid due to loss of cholesterol from the cell membrane and to degradation of protein and its cross-linkage with calcium. These aged red cells are trapped and destroyed by phagocytes mainly in the spleen but also in

the liver and bone marrow. Iron in hemoglobin is recovered, stored, and recycled into new red cells.

Reticulocytes

KEY WORDS: polychromatophilic cell, ribonucleoprotein

Most erythrocytes stain an orange-red with the usual blood stains, but a few take on a bluish or slate gray tint. These **polychromatophilic cells** are erythrocytes that are not fully mature and contain a small amount of **ribonucleoprotein** that, when stained with brilliant cresyl blue or new methylene blue, precipitates as a network or web. These cells are called *reticulocytes*. Their numbers in peripheral blood provide a rough index of erythrocyte production. Normally, reticulocytes make up only 1 to 2 percent of the red cells in humans.

Rouleaux

Erythrocytes tend to adhere to each other by their broad surfaces to form stacks called *rouleaux* (see Fig. 5-1). Rouleaux depend on changes in the blood plasma rather than in the cell. Any condition that increases the net positive charge in the plasma produces changes in the surface charge of erythrocytes, allowing them to adhere to each other more readily. Rouleaux are temporary phenomena, may occur intravascularly, and appear to do no harm to the red cells. Increased rouleaux are reflected in an increase in the rate at which red cells settle out or sediment.

Abnormalities

KEY WORDS: anisocytosis, macrocyte, microcyte, poikilocytosis, crenated, sickling, hypochromia, target cell, Howell-Jolly bodies, siderocyte, basophilic stippling, spherocyte

Departures from normal size, shape, or staining properties of erythrocytes can be important indicators of disease, but to a much lesser degree, some of these abnormalities may be found in healthy individuals also. **Anisocytosis** describes abnormal variations in the size of red cells, which may be **macrocytes** (larger than normal) or **microcytes** (smaller than normal). In macrocytes the central pale area is less marked than in normal cells, but the concentration of hemoglobin is not increased.

Irregularity in shape is called **poikilocytosis;** the cells may show blunt, pointed, or hook-shaped projections from their surfaces. Under various conditions, red cells in vitro may become shrunken and show numerous projections on their surfaces; these cells are said to be **crenated** and are called *echinocytes.* Crenated cells can be produced by subjecting normal red cells to fatty acids, anionic compounds, elevated pH, or hypertonic media. The changes are re-versible. One of the most severe changes in shape occurs during **sickling** of red cells in sickle cell anemia, in which erythrocytes appear as crescents, holly leaves, or even tubes (Fig. 5-3).

Hypochromia denotes cells that stain poorly due to a decrease in hemoglobin; it frequently accompanies microcytosis. In extreme forms, staining may consist simply of a narrow peripheral band. Cells that are thinner than normal (leptocytes) often appear as **target cells** in which the staining is disposed as a central disc and an external band separated by an unstained zone.

Howell-Jolly bodies are nuclear fragments left over from the nucleated precursors of the red cell. They appear as one or two rounded or rod-shaped basophilic granules.

Siderocytes contain clusters of small, darkly stained granules that react positively for iron. They are rare in normal peripheral blood but are quite common in bone marrow. Normally the iron-containing granules are removed by the spleen without harming the cell.

Allied to polychromatophilia is **basophilic stippling,** which appears as blue-black dots that may be coarse or fine and consists of precipitated ribonucleoprotein. It is found occasionally in leukemia and severe anemias but is of diagnostic importance in chronic lead poisoning.

In some pathologic conditions, globular rather than biconcave red cells are produced. These are called **spherocytes** and appear as small, deeply stained cells with a sharp, distinct outline.

Fig. 5-3. Erythrocytes of peripheral blood from a case of human sickle cell anemia (SEM, × 5,000).

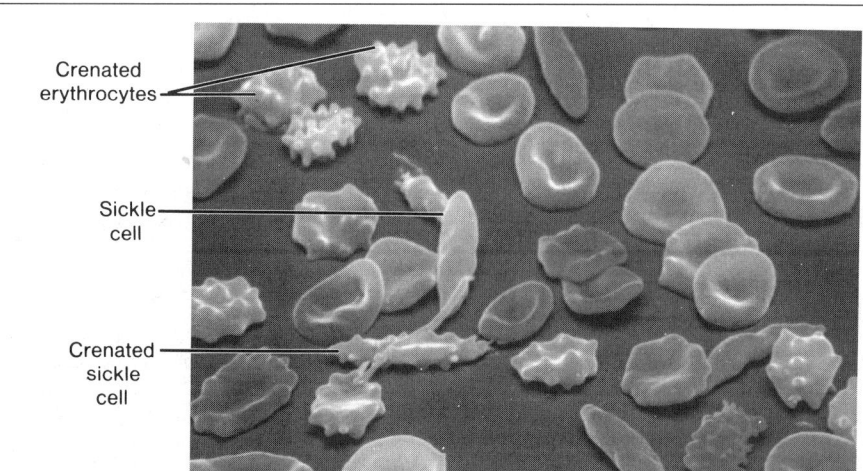

Hemoglobin

Gas transport is carried out by hemoglobin, an iron-containing respiratory pigment that gives the red color to human blood.

Structure

KEY WORDS: heme, globin, HbA, HbA2, HbF, sickle hemoglobin (HbS)

Hemoglobin is one of a group of catalytic compounds that has an iron porphyrin group, heme, attached to a protein, globulin. The **heme** group consists of four pyrrole rings combined with iron through their N groups. Structurally, hemoglobin is a symmetrical molecule formed by two equal mirror-image halves. Each half molecule contains two different peptide chains, each bearing a heme group around which the chain is coiled. The heme groups form the functional component, and the peptide chains make up the **globin** part of the molecule. The amino acid composition and sequences in the globin confer species specificity to the whole molecule and also determine the type of hemoglobin present. Humans synthesize four structurally different globin chains: alpha, beta, gamma, and delta. About 95 percent of normal adult hemoglobin is hemoglobin A (**HbA**), which contains two alpha and two beta chains: about 2 percent is **HbA2,** which consists of two alpha and two delta chains. The remainder, fetal hemoglobin (**HbF**), is composed of two alpha and two gamma chains. Fetal hemoglobin predominates in fetal life but is replaced by the adult type postnatally.

Even minor modifications of the peptide chain can result in gross changes in the entire hemoglobin molecule, as shown by the **sickle hemoglobin (HbS)** associated with sickle cell anemia. In this disease the only fault in the molecule is in the sixth position of the beta chain, where the glutamic acid of normal hemoglobin has been replaced by valine.

Platelets

Platelets are the second-most numerous of the formed elements of blood. Platelets are not cells but merely fragments of cytoplasm derived from a large precursor cell (megakaryocyte) in the bone marrow.

Structure

KEY WORDS: granulomere (chromomere), hyalomere, actin, myosin, marginal bundle

Platelets are small, anucleate bodies 2 to 5 μm in diameter (Fig. 5-4). In stained smears, platelets show a granular part, the granulomere, and a pale, granule-free hyalomere. The **granulomere (chromomere)** often fills the central region of the platelet and may be so compact as to suggest a nucleus. Separation into two zones is not seen in circulating platelets nor in those fixed immediately by drawing blood into fixative. In these conditions, the granulomere remains evenly distributed. Electron micrographs show that the granulomere consists of lysosomes, mitochondria, dense-cored granules that contain serotonin, adenosine

Fig. 5-4. A. Human platelet from peripheral blood (SEM, × 10,000). B. Platelets of peripheral blood as seen in section (TEM, × 8,000).

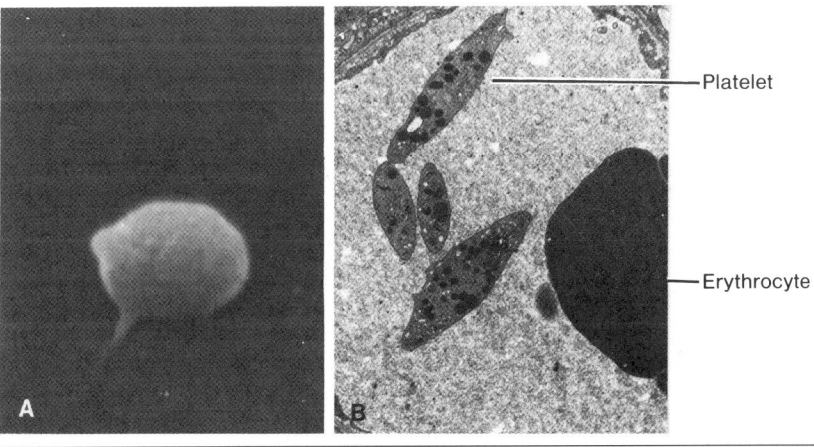

diphosphate (ADP), ATP, and calcium, and alpha particles that contain platelet-specific proteins, fibrinogen and other clotting factors, actin, myosin, adenosinetriphosphatase (ATPase), and agents that increase vascular permeability. Mitochondria are small and few in number and have few cristae. Variable numbers of glycogen granules also are present.

The **hyalomere** represents the cytoplasmic matrix and appears as a homogeneous, finely granular background. A crisscross arrangement of **actin** and **myosin** filaments is present just beneath the plasmalemma. A narrow peripheral zone remains free of granulomere elements and is the site of a system of microtubules that forms the **marginal bundle.** This structure has been described as a single, coiled microtubule that acts as a stiffening element to help maintain the discoidal shape.

Each platelet is bounded by a typical cell membrane that invaginates the cytoplasm and becomes continuous with an elaborate system of channels.

Number

The number of platelets varies widely depending on the species and in individuals also. The variations reflect difficulties in obtaining accurate counts due to physiologic factors that affect the numbers and to inherent properties of the platelets, such as their tendency to stick to foreign surfaces or to each other to form clumps. When in contact with a foreign surface, platelets spread to cover an area several hundred times that of their initial surface. Although platelet levels in humans are given as 250,000 to 400,000/mm^3 of blood, some estimates extend the range to 900,000/mm^3. Transient fluctuations in platelets have been associated with variation in oxygen concentration, exposure to periods of cold, and food intake. Curiously, highly spiced foods have been reported to cause a significant reduction in circulating platelets. Emotional states such as fear and rage result in markedly elevated counts. A progressive decrease in the number of platelets occurs in women during the 2 weeks prior to menstruation, with a rapid return to normal values thereafter.

Leukocytes

Leukocytes (white blood cells, WBCs) are true cells that have nuclei and cytoplasmic organelles and are capable of ameboid movement. They migrate from the blood into tissues, where they perform their functions.

Classification

KEY WORDS: granular, agranular, polymorphonuclear, mononuclear, neutrophil (heterophil), eosinophil, basophil, lymphocyte, monocyte

Leukocytes can be classed as **granular** or **agranular** on the basis of cytoplasmic granulation or as **polymorphonuclear** or **mononuclear** according to the shape of the nucleus. Granular leukocytes contain multilobed nuclei and thus the terms *granular leukocyte* and *polymorphonuclear leukocyte* denote the same class of cell.

Granular leukocytes (or granulocytes) can be subdivided into **neutrophils (heterophils), eosinophils,** or **basophils** according to the color of their cytoplasmic granules after staining with standard blood stains. Heterophil granules do not stain the same in all species, so the term is used inclusively for this type of granular leukocyte, regardless of staining reaction or species. The term *neutrophil* refers to the heterophil of humans.

Agranular leukocytes consist of **lymphocytes** and **monocytes,** which have inconspicuous or no granules and a single, nonlobed nucleus. Thus the terms *mononuclear leukocytes* and *agranular leukocytes* are used to denote the same class of cells. The distinction between granular and agranular leukocytes is not absolute: Monocytes regularly show fine, dustlike granules that are barely resolvable by light microscopy, and lymphocytes occasionally may show a few coarse granules. The distinction is an old one based on appearances with less refined dyes but remains useful because the granules of granular leukocytes are specific, distinctive, and prominent.

Neutrophil (Heterophil) Leukocytes

KEY WORDS: lobe, filament, band form, drumstick, specific granule, lysozyme, lactoferrin, phagocytin, azurophil granule, lysosome

The neutrophil granulocyte varies from 12 to 15 μm in diameter and is characterized by the shape of the nucleus, which contains small **lobes** connected by fine **filaments.** The nucleus stains deeply, and the chromatin is aggregated into clumps that form a patchy network. Nucleoli are absent. The number of lobes varies, but in humans, two to four are usual. A single, elongated, nonlobed nucleus is seen in a small number of granulocytes; called **band forms,** they represent cells recently released from the bone marrow. Excessive lobation occurs in some diseases or as an inherited anomaly in humans.

In addition to nuclear lobation, nuclear appendages in the shape of hooks, hand racquets, clubs, or **drumsticks** may be present (Fig. 5-5). Only the drumstick is significant: It represents the female sex chromatin and usually is found on a terminal lobe in about 2 to 3 percent of the neutrophils.

The cytoplasm of neutrophils contains two types of granules. Most numerous are the **specific granules** (secondary, type B granules), which are small and take on a pinkish hue. These granules contain **lysozyme,** an enzyme complex that acts against components of bacterial cell walls, collagenase, and **lactoferrin,** another antibacterial substance. Several other rather poorly characterized basic proteins (**phagocytins**) are present and also have antibacterial properties. **Azurophil granules** (primary, type A granules) are less numerous, somewhat larger, and stain a reddish purple. They contain lysozyme as well as a battery of acid hydrolases and are considered to be modified **lysosomes.** Azurophil granules are large and homogeneous, whereas the smaller, less dense-specific granules may contain a crystalloid body. A myeloperoxidase that complexes with peroxide to produce activated oxygen, a potent bactericide, also is present in the cytoplasm.

The granules are scattered throughout the cytoplasm, except in a narrow peripheral zone, that contains fine filaments and microtubules that appear to function in cell movement. A small Golgi complex and a pair of centrioles are located centrally in the cell. Rare profiles of endoplasmic reticulum, a few mitochondria, and ribosomes are present also.

Neutrophils produce a protein that converts plasminogen to plasmin, the proteolytic enzyme responsible for fibrinolysis. Whereas the release of granule-associated factors is associated with death of the cell, release of plasminogen-activating substance is a secretory activity directly related to cell viability.

Neutrophils are the chief leukocytes in humans, forming 55 to 70 percent of the circulating white blood cells.

Eosinophil Leukocytes

KEY WORDS: diurnal variation, nuclear lobation, crystalloid, lysosome, myeloperoxidase

Eosinophils form only a small proportion of the white cells and in humans normally make up 1 to 3 percent of the circulating leukocytes. Their numbers in blood bear some relation to the activity of the adrenal glands; a decrease in eosinophils is seen in the alarm reaction. This relationship may account for the wide range of values reported. Distinct **diurnal variations** related to activity cycles also occur.

The eosinophil is about the same size as the neutrophil and also is characterized by **nuclear lobation.** Although often described as having a bilobed nucleus, eosinophils with three or four lobes are not uncommon (Fig. 5-6A). Conditions that produce hypersegmentation in neutrophils also affect the eosinophil granulocyte.

The distinctive feature of eosinophils is the closely packed, uniform, spherical granules that stain a brilliant red or orange-red. The granules are membrane-bound and

Fig. 5-5. Erythrocytes and leukocytes of human peripheral blood (LM, × 1,000).

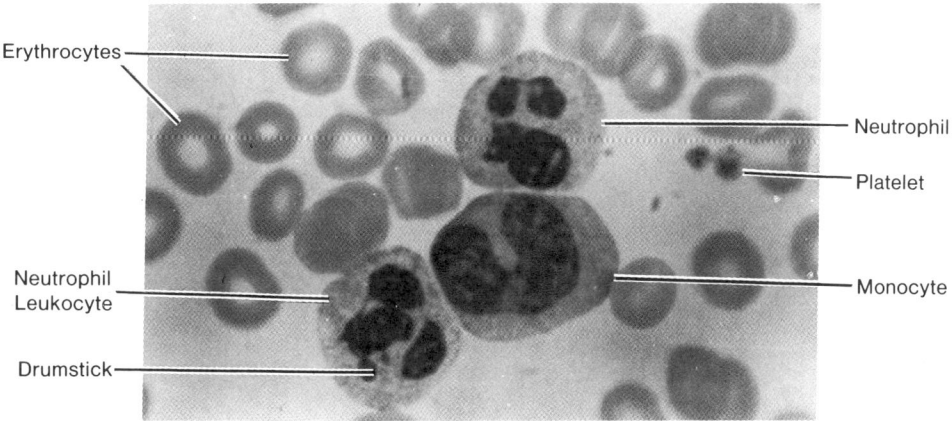

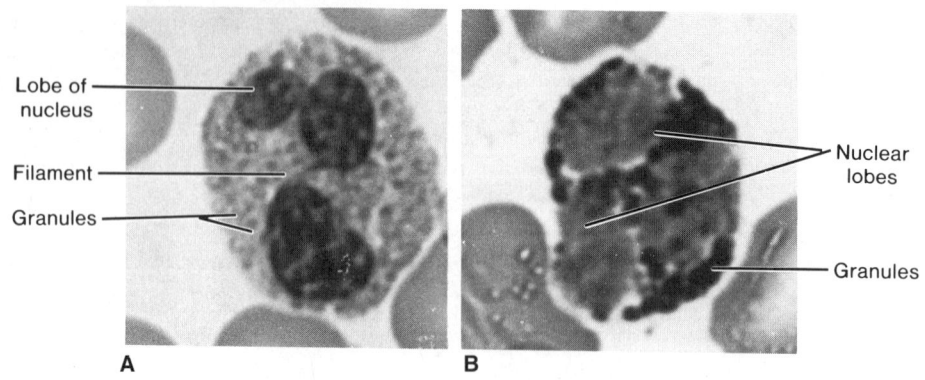

Lobe of nucleus

Filament

Granules

Nuclear lobes

Granules

A   B

Fig. 5-6. A. Eosinophil from human peripheral blood (LM, × 1,500). B. Basophil from human peripheral blood (LM, × 1,500).

show an internal structure called a **crystalloid,** or internus (Fig. 5-7). In humans the crystalloid assumes various forms and may be multiple. The crystalloid is embedded in a finely granular matrix.

Eosinophil granules are **lysosomes** and contain the usual lysosomal enzymes. They show a higher content of **myeloperoxidase** than do the azurophil granules of neutrophils and lack lysozyme and phagocytin. The granules also contain lipid and a major basic protein (MBP) that is responsible for the eosinophilia.

Basophil Leukocytes

KEY WORDS: nuclear lobe, filament, metachromatic

In humans basophils make up only about 0.5 percent of the total leukocytes. They are slightly smaller than neutrophil granulocytes and in blood smears measure 10 to 12 μm in diameter.

Nuclear lobes are less distinct than in the neutrophil or eosinophil, and nuclei with more than two or three lobes are rare. **Filaments** between lobes tend to be short and broad and rarely form the threadlike structures seen in neutrophils. The chromatin is relatively homogeneous and stains less deeply than in the other granulocyte types. Nucleoli are absent.

The cytoplasm of basophils contains prominent, rather coarse granules that stain a deep violet with the standard blood stains and **metachromatically** with toluidine blue or thionin. In the human, well-preserved granules are spherical and uniform but, being soluble in water or glycerin, frequently appear irregular in size and shape in fixed preparations. The granules are scattered unevenly throughout

the cytoplasms, often overlie and obscure the nucleus (Fig. 5-6B), and are neither as numerous nor as densely packed as the granules of eosinophil leukocytes. Electron microscopy shows an internal striated pattern to the granules. The granules contain myeloperoxidase, glycosaminoglycans, eosinophilic chemotactic factors, prostaglandins, and a platelet-activating factor. The basophilia of the granules is due to the presence of heparin. There is no evidence that the granules have lysosomal activity.

Lymphocytes

KEY WORDS: agranulocyte, small lymphocyte, mononuclear leukocyte, B-lymphocyte (B-cell), humoral immunity, antibody, T-lymphocyte (T-cell), cell-mediated immunity, T-helper cell, T-suppressor cell, lymphokine, cytotoxic T-cell, null cell, large granular lymphocyte, natural killer cell, mitogen, blastogenic transformation, recirculation, memory cell, motility

Lymphocytes are the predominant **agranulocyte** and in the adult human account for 20 to 35 percent of the circulating leukocytes.

Classification

Lymphocytes comprise a family of cells, each differing in functions, life span, background, and size. They are found in blood, lymph, lymphatic tissues, and organs. The most common type is the **small lymphocyte** (Fig. 5-8A). In blood smears this cell has a diameter of 6 to 8 μm and appears to be little more than a nucleus surrounded by a thin rim of pale blue cytoplasm. At one aspect of the cell, the cytoplasm may be somewhat expanded. The round or

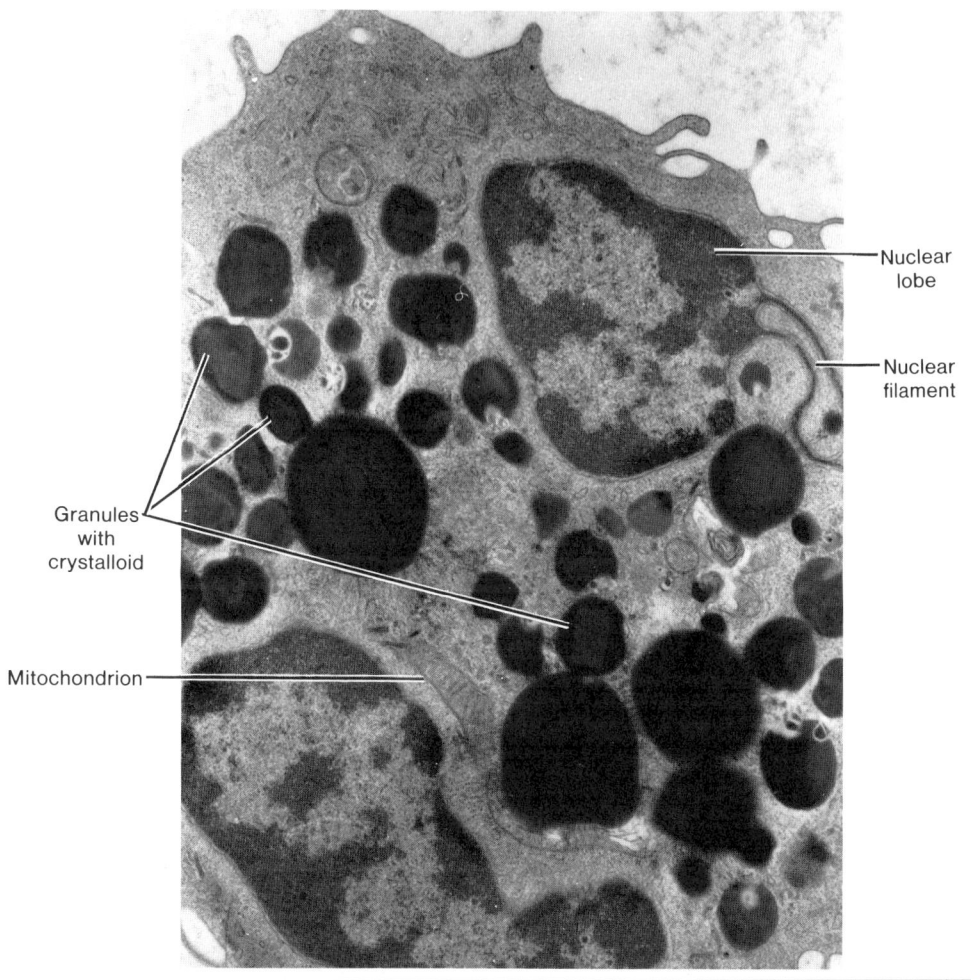

Nuclear
lobe

Nuclear
filament

Granules
with
crystalloid

Mitochondrion

Fig. 5-7. An electron micrograph of a portion of a human eosinophil. The granules contain crystalloids. Note the nuclear filament (TEM, × 15,000).

slightly indented nucleus is heterochromatic and stains deeply. Except in living cells or in electron micrographs, nucleoli usually are not visible. Because of the round nucleus, lymphocytes are classified as **mononuclear leukocytes.** Electron micrographs reveal a small Golgi body and a few mitochondria and lysosomes; a small amount of SER is present, but GER is very scarce. While lymphocytes are classified as agranular leukocytes, a few nonspecific azurophil granules may be present and represent lysosomes.

Although morphologically indistinguishable by ordinary means (Fig. 5-9), small lymphocytes are a mixture of functional types that can be divided into two major classes, B- and T-lymphocytes, by means of distinctive surface markers. **B-lymphocytes (B-cells)** play a central role in **humoral immunity** and represent cells that have been conditioned to transform into plasma cells and secrete **antibody** when stimulated by foreign antigens. The antibody reacts specifically with the antigen that induced its formation. The organ responsible for conditioning B-cells in mammals appears to be the bone marrow. In aves, B-cells are processed in the bursa of Fabricius, a lymphoreticular, appendix-like diverticulum of the cloaca. B-cells form only 5 to 10 percent of the blood lymphocytes in humans.

T-lymphocytes (T-cells) have been processed by the thymus, where they acquire distinctive T-cell antigens and become immunologically competent. However, immuno-

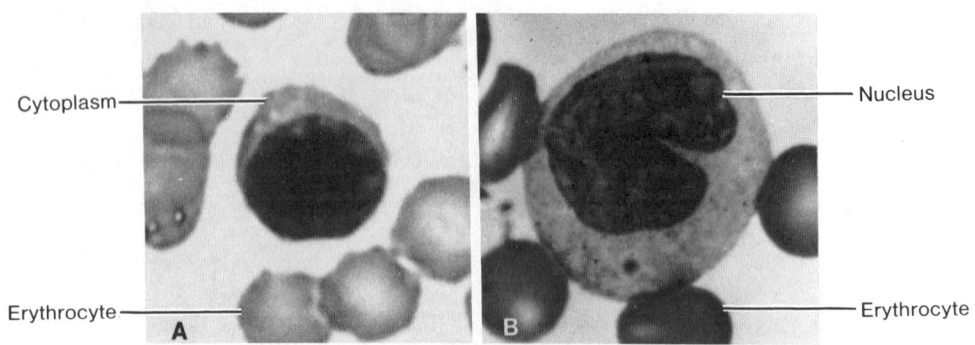

Fig. 5-8. A. Lymphocyte from human peripheral blood (LM, × 1,200). B. Monocyte from human peripheral blood (LM, × 1,200).

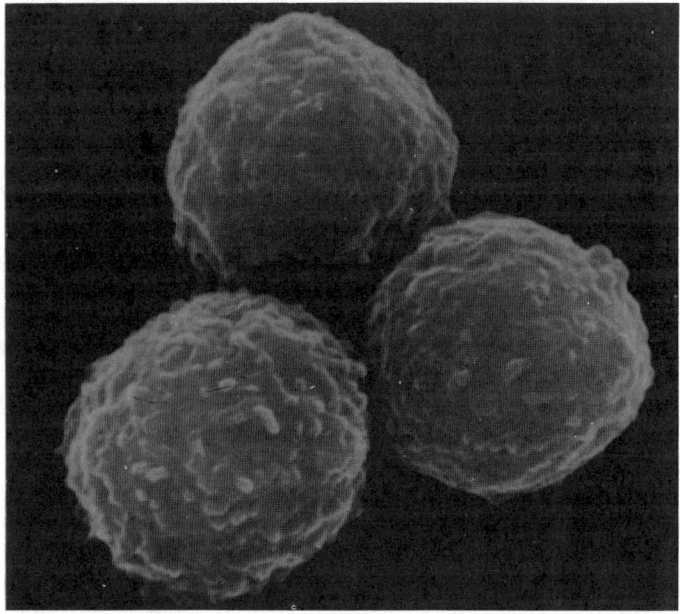

Fig. 5-9. Small lymphocytes from human peripheral blood (SEM, × 10,000).

logic capacities are not fully developed until the cells are exported to peripheral lymphatic organs, especially the spleen. T-cells contribute to **cell-mediated immunity** and perform a variety of functions within the immune system. When appropriately stimulated, some T-cells interact with B-cells and, as **T-helper cells,** amplify the production of antibody; as **T-suppressor cells,** they dampen antibody formation. Other T-cells produce **lymphokines** that have important roles in body defenses. Of the many lymphokines described, the best characterized are agents that damage or destroy foreign cells or interfere with their replication, such as lymphotoxin, interleukin, interferon, and tumor necrosis factor; agents that inhibit migration of macrophages (MIF) and thus concentrate macrophages at the site of an antigen; macrophage-activating factor, which stimulates macrophage activity; lymphocyte blastogenic factor (LBF), which induces blastic transformation, growth, and differentiation of small lymphocytes; factors that attract neutrophil, eosinophil, or basophil granulocytes to the sites of antigens (i.e., chemotactic agents); and factors that increase vascular permeability. Thus lymphokines represent a large class of lymphocyte-derived substances that have a role in controlling immunologic and inflammatory responses or have cytotoxic activities.

Some activated T-cells, the **cytotoxic T-cell** or killer (K) lymphocytes, take part in antibody-dependent, cell-to-cell interactions with foreign cells that result in lysis of the target cell. Only target cells that share antigens with the foreign cell originally inciting the appearance of cytotoxic cells are killed. Cytolysis of the foreign cell requires intimate contact between the effector and its target, the latter then undergoing a series of membrane changes that result in rupture of the cell. The K lymphocyte is unharmed by the interaction and remains free to react with other target cells.

A third class of small lymphocyte, the **null cell,** has been described. These cells lack B- or T-cell surface markers. Null cells also appear to be a diverse group among which may be circulating hemopoietic stem cells.

Larger forms of lymphocytes, 10 to 14 μm in diameter, also are present in peripheral blood in small numbers. These cells have been called **large granular lymphocytes.** The cytoplasm of these cells is more abundant and more basophilic than that of the small lymphocyte. Electron micrographs show an increased number of ribosomes and mitochondria, a greater amount of endoplasmic reticulum, and an increase in the size of the Golgi apparatus. Scattered azurophil granules are present. These cells are directly cytotoxic and also lyse target cells by contact, but they appear to function in the absence of antibody and without known prior exposure to antigen. Thus they have been named **natural killer (NK) cells.** Commonly considered to be null cells, there is growing evidence that they are a subset of T-cells and may be related to immune-dependent cytotoxic T-cells. The mechanism of NK cell activity is unknown, but it appears to involve the azurophil granules, which have been identified as lysosomes.

Stimulation by Antigens and Mitogens

Small lymphocytes are not simply terminal effector cells engaged in T- or B-cell activities; they also have the properties of "stem" cells and can propagate additional small lymphocytes when stimulated by antigens or nonspecific agents called **mitogens.** The best studied mitogens are plant extracts — phytohemagglutinin (PHA), concanavalin A (con A), pokeweed mitogen (PWM) — but other agents such as liposaccharide from *Escherichia coli,* antilymphocyte serum, and various pollens also have mitogenic activity. These agents have a common ability to bind to surface groups on small lymphocytes, thereby triggering **blastogenic transformation.**

During transformation, small lymphocytes progressively enlarge; the nuclei increase in size and become euchromatic, nucleoli become visible, the Golgi body enlarges, and the cell synthesizes deoxyribonucleic acid (DNA) and ribonucleic acid (RNA). Within 36 to 48 hours, the small lymphocyte has assumed the appearance of a lymphoblast that divides and gives rise to new T- or B-cells. The new cells can produce lymphokines (T-cells) or become plasma cells and synthesize antibody (B-cells).

The reactions to mitogens show a degree of specificity: PHA and con A stimulate T-cells but not B-cells; *E. coli* lipopolysaccharide stimulates only B-cells; PWM activates both types of small lymphocyte. The mechanism of the reaction is not understood, but the reaction is specific to lymphocytes in that although these agents bind to the surface of other cells, they do not induce their transformation. The response of lymphocytes to mitogens resembles that after exposure to antigen or lymphoblastic lymphokine.

Lymphocytes **recirculate** repeatedly between the blood, lymphoid organs, and lymph. Only small lymphocytes recirculate, and most of these (80–85 percent) are long-lived cells of T-cell origin, with a life span measured in years. A small proportion (12–15 percent) of the circulating cells are short-lived B-cells with a life expectancy of several months. Until the lymphocytes make contact with an antigen, the cells continue to recirculate. A small number of long-lived cells, of T- and B-origin, are present as **memory cells.** These are small lymphocytes that have been stimulated by antigen but have not responded with production of antibody or lymphokines. They remain for years as conditioned cells that can react immediately to the same antigen on subsequent exposure to it.

Lymphocytes are highly **motile** cells that leave the bloodstream to enter lymphatic organs, migrate through connective tissues of the body, or penetrate into epithelia. They have little ability to adhere to or to spread on surfaces of any kind.

Monocytes

KEY WORDS: **mononuclear leukocyte, azurophil granule, primary lysosome, mononuclear phagocyte system**

Monocytes make up the second type of **mononuclear leukocyte** normally found in the blood. Their proportions are fairly consistent, and they regularly form 3 to 8 percent of the circulating leukocytes. They vary in size from 12 to 20 μm in diameter and contain a fairly large nucleus that may be rounded, kidney-shaped, or horseshoe-shaped (see Figs. 5-5 and 5-8B). Segmentation of the nucleus and formation of filaments never occur, but coarse constrictions with blunt, broad lobes may be present. The chromatin is more loosely dispersed than in lymphocytes and therefore stains much less densely. Two or three nucleoli may be seen. The blue-gray cytoplasm is abundant and contains numerous fine **azurophilic granules** that impart an opacity to the cytoplasm, giving it a "ground glass" or "dusty" appearance.

Ultrastructurally, the nucleus shows one or more nucleoli and the cytoplasm contains a small amount of GER, ribosomes, and polyribosomes. Small, elongated mitochondria are present, and the Golgi apparatus is well formed. The azurophil granules represent **primary lysosomes** and in electron micrographs appear as dense, homogeneous structures.

Monocytes are part of the **mononuclear phagocyte system** and represent the cells of this system in transit. They serve little function while in the blood but migrate into various organs and tissues throughout the body, where they differentiate into macrophages. In addition to serving as tissue scavengers, monocytes also have a role in processing antigen in the immune response and are able to fuse with one another to form various phagocytic giant cells. Macrophages do not possess true storage granules for their enzymes as do granulocytes, and the cell behaves mostly as a secretory cell.

Abnormalities

KEY WORDS: leukocytosis, leukopenia, neutrophilia, neutropenia, lymphocytosis, lymphopenia, Döhle's bodies, toxic granulation, Alder's anomaly, hypersegmentation, macropolycyte, Pelger-Huët anomaly

As with erythrocytes, the total number of circulating leukocytes tends to fall within a relatively narrow range in normal individuals. In humans, this range is 5,000 to 12,000/mm³. An increase above normal is called a **leukocytosis** and may be due to disease or emotional or physical stress. In humans, a leukocytosis with counts of 25,000 to 35,000/mm³ has been reported after severe exercise. The increase represents the flushing of leukocytes sequestered in capillary beds and marginated at the edges of the bloodstream. A decrease in leukocytes is called **leukopenia**. Diurnal variations of variable degree are associated with activity cycles.

Fluctuations in leukocyte numbers in humans usually involve neutrophil granulocytes and lymphocytes. If neutrophil granulocytes are in excess, a **neutrophilia** is said to be present; if decreased, a **neutropenia** is present. An increase or decrease in lymphocytes is a **lymphocytosis** or **lymphopenia,** respectively. Increases in eosinophils, basophils, and monocytes are referred to as an *eosinophilia, basophilia,* and *monocytosis,* respectively. Decreases in these cells are difficult to establish from smears because of their normally low numbers, but they do exist.

Various cytologic abnormalities may occur, especially among the granulocytes. **Döhle's bodies** are small, irregular blue clumps or patches, 0.5 to 2.0 μm in diameter, that may appear in the cytoplasm of neutrophils during some infections, after burns, in cancer, or after treatment with some oncolytic drugs. They appear to consist of altered ribonucleoprotein. **Toxic granulation** is seen as coarse, black or purple granules scattered in the cytoplasm of neutrophils and represents altered azurophil granules. Toxic granulation often accompanies Döhle's bodies and, with cytoplasmic vacuolation, may occur during severe bacterial infections. **Alder's anomaly** is a rare abnormality characterized by large, coarse granules in the cytoplasm of neutrophils. An apparently inherited anomaly, it has no pathologic significance.

Hypersegmentation of granulocyte nuclei occurs as an inherited anomaly of no significance and also accompanies anemia resulting from deficiency of vitamin B_{12}. In such anemia, the cells are abnormally large and have been called **macropolycytes.** At the other extreme, poorly segmented granulocytes are present in the **Pelger-Huët anomaly,** an inherited condition. All types of granulocyte are involved, and the mature cells rarely show more than two lobes; more frequently, the cells have round nuclei with no filaments. In humans the anomaly is without significance.

Atypical lymphocytes may show vacuolated cytoplasm, folded (monocytoid) nuclei, nuclear vacuolation, or prominent nucleoli. Most of the atypical cells are medium lymphocytes. Dark, basophilic cytoplasmic granules have been reported in the cytoplasm of lymphocytes in cases of gargoylism (Hurler's syndrome), and inclusions similar to Alder's granules occur in the lymphocytes of children as a familial disorder.

Blood in Children

During infancy and childhood, the composition of blood differs significantly from that of adults. Hematologically, a child is not merely a small adult, and children and infants have their own specific blood pictures. In general, the younger the individual, the greater is the deviation from normal adult values and the greater the instability of the blood picture during disease. At birth, the hemoglobin content of red cells is higher than at any subsequent period, and compared with adult values, the number of red cells is increased (a polycythemia) and the cells are macrocytic. An increase in reticulocytes and the presence of nucleated red cells (normoblasts) are characteristic of the normal neonate. Within a few weeks after birth, macrocytes disappear and normal, adult-sized red cells are present. By about the third month of life, hemoglobin levels drop, accompanied by a small decrease in the number of red cells so that the red cells (as judged against adult standards) are hypochromic. This picture is maintained until about the second year, after which the values for red cells and hemoglobin gradually rise to reach adult levels at puberty. It is only after puberty that the sex differences in red cells and hemoglobin become apparent.

The number of leukocytes is much greater in the neonate than in the adult, and neutrophil granulocytes predominate. Most of these cells have only two lobes, many are band forms, and myelocytes and even premyelocytes (immature forms normally found only in bone marrow) may be present. The leukocyte count drops quickly and reaches adult levels by about the third month. From the second week to about the second year, lymphocytes dominate the blood picture rather than neutrophils as in the adult.

In disease states, a child's blood is apt to show far greater deviations and variations than occur in adult blood. An infection that in the adult would produce a mild leukocytosis and an increase in band form leukocytes in a child may evoke a leukocytosis of leukemic proportions (leukemoid reaction) with the appearance of myelocytes and promyelocytes. Similarly, anemia in a child may be marked by the presence of nucleated red cells where in the adult only polychromatophilia would occur. Lymphocytosis develops much more readily in children, and platelets are apt to fluctuate wildly.

Review Table 5-1. Key Features of Cells in Peripheral Blood

Cell	Diameter (mm)	Number	Cytoplasm	Nucleus	Nucleolus
Erythrocyte					
Male	7.6	4,500.000 to 6,000,000/mm^3	Biconcave disc, eosinophilic	—	—
Female	7.6	3,800.000 to 5.000,000/mm^3		—	—
Platelet	2.1–5	200,000 to 400,000/mm^3	Chromomere light blue, Hyalomere pale	—	—
Neutrophil	10–15	55–70% of leukocytes	Numerous fine granules, lilac color	2–4 lobes connected by fine filament	—
Eosinophil	10–14	1–3% of leukocytes	Numerous round, red-orange granules	2 lobes connected by filament usual; 4–5 lobes may occur	—
Basophil	10–12	0.5% of leukocytes	Coarse, blue-black granules	Less distinct lobulation, broad nuclear filament	—
Lymphocytes					
Small	6–8	20–35% of leukocytes	Thin, blue rim	Round, deep staining	+
Large	10–14	Small number	Abundant, basophilic	Round, lighter staining than small form	+
Monocytes	12–20	3–8% of leukocytes	Abundant, bluish gray, opaque in appearance	Large, kidney or horseshoe shape	+

Functional Review: Peripheral Blood

Blood is important in the transport of materials throughout the body, maintaining the acid-base balance, and providing defense mechanisms. Transport functions include carriage of oxygen to all cells of the body, transport of nutrients, and removal of waste products of cell metabolism. It aids in regulating body temperature by dissipating heat formed during metabolism and distributes hormones, thus integrating the functions of the endocrine system. Through its buffering capacities, blood helps maintain the acid-base balance and ensures an environment in which cells may function normally.

Transport of oxygen and carbon dioxide is a function of erythrocytes and their hemoglobin. Packaging hemoglobin into a cell allows for a local environment in which hemoglobin can function most effectively and not be converted to useless forms. The red cell plasmalemma provides mechanical protection as well as enzymes, which prevent irreversible oxidation of hemoglobin. Retention of hemoglobin in a cell permits increased concentration of the pigment without increasing the viscosity of blood. The absence of a nucleus allows for a smaller cell and a more efficient distribution of hemoglobin within the cell. The biconcave shape gives hemoglobin optimal access to the surface for gas exchange and allows the cell to undergo considerable increase in volume without rupturing. An iron atom on each of the four polypeptide chains of the hemoglobin molecule serves as the binding site for oxygen molecules. In contrast, carbon dioxide binds to the amino acid portion of the hemoglobin molecule and does not interfere with oxygen transport.

Platelets have several diverse functions concerned with maintaining the integrity of the blood vasculature. Because of their ability to stick to one another and to foreign surfaces, platelets can cover and temporarily plug small gaps in blood vessels. They play an important part in blood clotting, releasing factors that initiate the clotting process; they also contain fibrinolytic factors that dissolve clots and thus aid in maintaining the fluidity of blood. They are necessary for clot retraction, which results in a firm, dense clot and reduces its bulk to prevent obstruction of the vessel. Platelets also are important in maintaining the integrity of endothelium, which becomes more permeable when platelets are decreased. Although not synthesized by platelets, the vasoconstrictors epinephrine and 5-hydroxytryptamine (serotonin) are contained in platelets. While probably contributing little to the body's overall defenses, platelets can phagocytize small particles, viruses, and bacteria.

Neutrophils are avidly phagocytic and part of the first line of defense against bacterial infections. Azurophil granules are lysosomes, and phagocytosis by granulocytes is associated with fusion of the granules with the ingested material in the usual manner of lysosomal digestion. The cells also contain lysozyme, which hydrolyzes glycosides in bacterial cell walls, and peroxidase, which complexes with hydrogen peroxide to release activated oxygen, an antibacterial agent. Specific granules contain a number of antibacterial agents including lysozyme; lactoferrin, which binds iron (required by bacteria for growth); and cationic compounds with bactericidal properties.

Eosinophils have a special affinity for antigen-antibody complexes, which tend to bind complement and induce cell lysis. Phagocytosis of the complexes may suppress cell destruction in normal tissues. Enzymes such as histaminase released by eosinophils may dampen allergic responses by degrading histamine and histamine-like substances released from mast cells and basophils. Eosinophils also have a major role in controlling parasitic infections: Eosinophils selectively interact with the larvae of some helminthic parasites and damage them by oxidative mechanisms. A major basic protein (MBP) in the crystalloid core of the eosinophil granules enhances the antibody-mediated destruction of parasites.

Basophil granules have a high content of heparin (an anticoagulant), histamine, vasodilating agents, and slow-reacting substance (SRS). Histamine induces a prompt and transient vasodilation, whereas that induced by SRS is more sustained and occurs after a latent period. Both increase vascular permeability. Basophils play a role in acute systemic allergic reactions; antigens that bind to specific sites on the cell membrane cause degranulation with release of agents that cause smooth muscle spasm, mucous secretion, hives, itching, and rhinitis.

Lymphocytes are concerned primarily with the two major types of immune responses, humoral and cellular. The basis of humoral immunity is the production of antibodies and their diffusion throughout the body fluids.

As an antigen enters the body, it is complexed on the surface of B-cells, whose surface antibodies fit determinants on the surface of the antigen. The antigen is internalized and triggers cell proliferation and the differentiation of the lymphocytes into antibody-producing cells. Some of the stimulated cells do not go on to produce immunoglobulins but remain as memory cells. The process of humoral immunity is highly specific: Only those antigens that fit receptors on the surface of B-cells elicit an immune response, and the antibody produced will react only with the antigen that induced its formation.

Cellular immunity depends on T-cells. In response to an antigen, T-cells proliferate and release a variety of lymphokines that may inhibit macrophage migration from the site of the antigen, increase the activity of macrophages, interfere with virus replication, or lyse bacterial walls. For most antigens that stimulate a humoral response, cooperation between T-cells and B-cells is essential, but the nature of this interaction is not known. T-cells may recognize the antigen as foreign and concentrate it on their surfaces before presenting it to the B-cell, or they may elaborate substances that activate the B-lymphocyte.

Monocytes leave the blood and differentiate into tissue macrophages. They not only serve as tissue scavengers, ingesting and removing particulate matter, tissue debris, and infective agents, they also play a role in the immune response. Their exact role in immune reactions is unknown, but macrophages may retain antigen on their surfaces to build up a critical amount that can stimulate B-cells, hold it long enough to stimulate B-cells, or in some way alter the antigen to expose antigenic determinants that are recognizable by lymphocytes. Macrophages also liberate antiviral agents and a number of enzymes that digest collagen, elastin, and fibrin.

Review Questions and Answers

Questions

24. What comprises the complex fluid called blood?
25. How do serum and plasma differ?
26. What is the significance of the "drumstick" observed in polymorphonuclear leukocytes?
27. What is the functional significance of the two types of cytoplasmic granules found in neutrophil granulocytes?
28. Briefly discuss the two major functional types of small lymphocytes.

Answers

24. Most of the intercellular protein of blood is produced by cells in other tissues and not by blood cells. Albumin, the most abundant and smallest of the plasma proteins, is formed in the liver. Fibrinogen, an essential component for blood clotting, also is formed in the liver. Globulins, which include several proteins of different sizes, among which are the immunoglobulins (antibodies), are produced by cells of the lymphatic tissues and organs. The formed elements of blood consist of anucleate elements, erythrocytes, bits of cytoplasm, and the platelets, important in the blood clotting process. Leukocytes, the true cells, make up only a small part of the formed elements and are present as transients that use the blood for transportation to other organs. Only erythrocytes and platelets function while in the bloodstream.

25. Serum is obtained from defibrinated (clotted) blood and does not contain fibrinogen; it does contain other components formed during the clotting process. Plasma, on the other hand, is obtained from blood after treatment with an anticoagulant and contains all the components of the fluid portion of blood.

26. The drumstick of polymorphonuclear leukocytes usually is found on a terminal nuclear lobe and represents female sex chromatin.

27. The cytoplasm of neutrophilic granulocytes contains two types of granules: specific granules and azurophil granules. Specific granules contain lysozyme, collagenase, and lactoferrin. They also contain several other basic proteins called phagocytins that have antibacterial properties. Azurophil granules are considered to be modified lysosomes and contain numerous acid hydrolases as well as myeloperoxidase, a potent bactericide.

28. Lymphocytes are concerned primarily with the immune responses, of which there are two major types, humoral and cellular. The process of humoral antibody production is highly specific: only those antigens that fit receptors on the surface of B-lymphocytes

elicit an immune response, and the antibody produced will react only with the antigen that induced its formation. The antigen is internalized and stimulates the proliferation and differentiation of B-lymphocytes into antibody-producing cells. For most antigens that stimulate a humoral response, cooperation between B-lymphocytes and T-lymphocytes is essential; however, the nature of this interaction is poorly understood. Cellular immunity depends on T-lymphocytes. In response to an antigen, T-cells proliferate and release a variety of lymphokines that may inhibit macrophage migration from the site of antigens, increase macrophage activity, lyse foreign cells, or interfere with virus replication. T-lymphocytes may concentrate antigen on their surfaces before presenting it to B-lymphocytes or they may release substances that activate B-lymphocytes. Some T-lymphocytes may suppress responses to foreign (or self) antigens; others may remain as memory cells.

6 Special Connective Tissue: Hemopoietic Tissue

The formed elements of blood are not self-replicating; their numbers in circulation are maintained by continuous replacement from specialized blood-forming (hemopoietic) tissues. These consist of bone marrow, spleen, lymph nodes, and thymus.

Embryonic and Fetal Hemopoiesis

In the adult blood formation is restricted to the bone marrow and lymphatic tissues, but in embryonic and fetal life, hemopoiesis occurs first in the yolk sac and then successively in the liver, spleen, and bone marrow. In some pathologic conditions, the human liver and spleen may resume a role in hemopoiesis.

Yolk Sac

KEY WORDS: blood island, primitive erythroblast

Blood development in the embryo begins soon after formation of the germ layers. It first occurs in the walls of the yolk sac with the appearance of **blood islands.** Discrete foci of mesenchymal cells of the yolk sac proliferate to form solid masses of cells that soon differentiate along two lines. The peripheral cells flatten and become primitive endothelial cells; the central cells round up, acquire a deeply basophilic cytoplasm, and detach from the peripheral cells to become the first hemopoietic precursors. Most of the first blood-forming cells differentiate into **primitive erythroblasts** that synthesize hemoglobin and become nucleated red cells characteristic of the embryo. The isolated blood islands eventually coalesce to form a network of vessels that ultimately join with intraembryonic vessels. Few or no hemopoietic foci develop in the vessels of the embryo proper.

Yolk sac hemopoiesis begins 19 days after fertilization in humans and continues until the end of the twelfth week. Subsequent hemopoiesis in the liver, spleen, and bone marrow develops as the result of migration of stem cells from the yolk sac. The cells gain access to the embryonic circulation via vitelline and allantoic vessels.

Hepatic Hemopoiesis

KEY WORDS: erythropoiesis, definitive erythroblast

The liver is the second important blood-forming organ of the embryo. Hepatic hemopoiesis begins during the sixth week of gestation. At this time, hemopoietic cells already make up about 10 percent of the total cells present in the liver. Hemopoiesis reaches a peak at 7 to 8 weeks and remains steady until about the fifteenth week, after which it slowly declines. **Erythropoiesis** dominates throughout the entire period of hepatic hemopoiesis, and while there is significant production of macrophages (monocytes) and platelets during the first 2½ weeks, granulocyte formation is minor. Some neutrophils and eosinophils are produced, almost always in the connective tissue around portal spaces.

The liver remains the principal site of red cell formation from the third to sixth month of gestation; erythropoiesis continues, in decreasing amounts, until birth. Hemopoiesis in the liver resembles that in bone marrow, and the red cell precursors in the liver have been called **definitive erythroblasts** because they give rise to nonnucleated red cells. The developing cells form islands of hemopoietic tissue between the cords of developing hepatic cells, but there is no basement membrane between the hepatocytes and adjacent blood-forming cells.

Splenic Hemopoiesis

A low level of splenic hemopoiesis overlaps that of the liver, contributing mainly to the production of erythrocytes, although some granulocytes and platelets are formed also. Erythroblastic islands and some megakaryocytes are present by the twelfth week of gestation. Splenic hemopoiesis wanes as the bone marrow becomes active, but the spleen produces lymphocytes throughout life.

Myeloid Phase

The myeloid phase of hemopoiesis begins when ossification centers develop in the cartilaginous models of the long bones. A few hemopoietic cells can be found in the clavicle of the human fetus at 2½ months. Foci of erythropoietic cells are present in many bones by 4 months, and by 6 months the bone marrow is an important source of circulating blood cells. During the last 3 months of pregnancy, the bone marrow is the main blood-forming organ of the fetus.

Bone Marrow

In the adult bone marrow is the major organ for production of erythrocytes, platelets, granular leukocytes, and monocytes. Many lymphocytes also are produced in the marrow and reside in the lymphatic tissues secondarily. In toto, the bone marrow constitutes an organ that rivals the liver in weight and in humans is estimated to account for 4 to 5 percent of body weight.

Types

KEY WORDS: red marrow, active hemopoietic marrow, yellow (fatty) marrow, extramedullary hemopoiesis

Bone marrow is an extremely cellular connective tissue that fills the medullary cavities of bone. On gross inspection, it may have a red or yellow color. **Red marrow** is actively engaged in the production of blood cells and represents the **active** or **hemopoietic marrow.** The red color is due to the content of red cells and their pigmented precursors. **Yellow (fatty) marrow** is inactive, and its principal cellular components are fat cells. Fat cells also are scattered sparingly throughout the red marrow.

The amount and distribution of fatty marrow vary with age and the need for blood cells. All bones contain active marrow in the late fetus and neonate, but active marrow gradually is replaced by fatty marrow during development and aging. Beginning in the shafts of long bones, fatty marrow gradually replaces red marrow until, by adulthood, almost all the marrow in the limbs is the fatty type. Active marrow remains at the ends of the long bones and in the ribs, sternum, clavicles, vertebrae, pelvis, and skull. In humans fatty marrow forms about 50 percent of the total marrow.

Fatty marrow is very labile and easily replaced by active marrow. Yellow marrow serves as a reserve space for the expansion of active marrow to meet increased demands for blood. Active marrow first replaces the fat cells scattered within the red marrow itself, but if demands for blood remain high or are increased, red marrow gradually encroaches into the gross areas of fatty marrow. Prolonged or increased demands then are met by expansion of hemopoiesis into other organs or tissues such as the spleen, liver, or lymph nodes. Blood formation in tissues other than the bone marrow is called **extramedullary hemopoiesis** and occurs in some pathologic states.

Structure

KEY WORDS: reticular connective tissue, reticular fiber, hemopoietic cell, free cell, reticular cell, fixed cell

Bone marrow is a specialized connective tissue that, on the basis of its fiber content, can be classed as a **reticular connective tissue.** A loose, spongy network of **reticular fibers** and associated cells fills the medullary cavities of bone and provides a framework for the **hemopoietic cells.** The network of fibers and cells is continuous with the endosteum of the bone and is intimately associated with blood vessels that pervade the marrow. Within the meshes of the reticular network are all the cell types normally found in blood, their precursors, fat cells, plasma cells, and mast cells. These constitute the **free cells** of the marrow. The **reticular cells** are **fixed cells** that have no special phagocytic powers and do not give rise to precursors of hemopoietic cells. They are fibroblasts responsible for the formation of reticular fibers. Reticular cells have large, palely stained nuclei and irregularly branched cytoplasm that extend long slender processes along the reticular fibers.

Vasculature

KEY WORDS: nutrient artery, central nutrient (longitudinal) artery, sinusoid (venous sinus), adventitial reticular cell, central longitudinal vein

In long bones, the **nutrient artery** enters the marrow cavity through the nutrient canal and gives off **central nutrient (longitudinal)** branches that run centrally in the marrow cavity toward the ends of the bone. These are supplemented by branches from epiphyseal and periosteal arteries. Along its course, the central nutrient artery provides numerous branches that pass toward the boney wall. Some of these branches enter the bone to supply osteons; others turn back to the marrow and join the venous system directly by uniting with sinusoids.

Sinusoids (venous sinuses) are thin-walled vessels, 15 to 100 μm in diameter, that form an extensive and complex network throughout the marrow. They are lined by thin, flattened endothelial cells that are closely apposed and joined by zonula adherens and gap junctions. A basal lamina is absent or discontinuous, with only scattered patches present on the basal surface. The endothelium is incompletely wrapped by a loose coat of **adventitial reticular cells** whose broad, branching processes merge with the reticular mesh that supports the hemopoietic cells. The adventitial reticular cells contain ribosomes, granular endoplasmic reticulum, and bundles of microfilaments located just beneath the plasmalemma. The proportion of endothelium covered by these cells varies with the functional state of the marrow and may be reduced to one-third when there is heavy cell traffic across the sinusoidal wall.

Sinusoids are the first venous elements. They pass toward the center of the marrow cavity and join the **central longitudinal vein** either directly or after union with other sinusoids to form collecting sinusoids. Thus, in the marrow, blood flows from the center to the periphery and back to the center. The central longitudinal vein also is thin-walled and consists of a low endothelium, a thin but complete basal lamina, and a thin outer layer of supporting reticular cells. The venous blood drains from the marrow by a main vein that exits through the nutrient canal, where it narrows abruptly. The blood supply appears to be "closed" — that is, there is endothelial continuity between venous and arterial components.

Distribution of Marrow Elements

KEY WORDS: **hemopoietic compartment, vascular compartment, extravascular**

The hemopoietic elements form irregular cords between the sinusoids, and collectively, these cords of blood-forming cells make up the **hemopoietic compartment;** the blood vascular components are referred to as the **vascular compartment.** In smears, marrow cells appear to be randomly distributed, but in serial sections, some ordering of the cells can be seen. Fat cells, though generally scattered, tend to concentrate toward the center of the marrow, where they cluster about the larger blood vessels. The vascular arrangement of the marrow results in a concentration of small vessels, especially sinusoids, at the periphery, and the hemopoietic cells are largely confined to the area near the vessels, close to the endosteum.

Erythrocytes develop in small islets close to the sinusoids with some ordering of the cells within a cluster according to their stage of development. The most mature cells occupy the outer rim of the islet. Macrophages are commonly associated with developing red cells and often lie in the center of an erythroblastic islet surrounded by several layers of red cells. Cells that give rise to platelets (megakaryocytes) are closely applied to the walls of sinusoids and frequently project small cytoplasmic processes through apertures in the sinusoidal wall. Developing gran-

ulocytes form nests of cells that tend to locate at some distance from the sinusoids.

Formation of blood cells occurs **extravascularly:** The cells form outside the blood vessels and enter the circulation through the sinusoids. The blood cells must penetrate the outer investment of reticular cells and cross the endothelial lining to gain access to the lumen. In areas of active passage of cells, the reticular sheath is much reduced, and the vessel appears to consist only of an endothelium. Blood cells penetrate the endothelium through apertures in the cytoplasm and not by insinuating themselves between adjacent endothelial cells. The openings are relatively large, 1 to 3 μm in diameter. They are not permanent structures and develop only in relation to and during the actual passage of cells.

Stem Cells

KEY WORDS: **pluripotent stem cell, spleen colony, colony-forming unit (CFU), restricted stem cell, candidate stem cell**

The hemopoietic cells of the marrow encompass various developmental stages, from a primitive stem cell to the mature circulating elements in the blood. A *stem cell* can be defined as a cell that can maintain itself through self-replication and also can differentiate into a more mature cell type. There is much evidence that bone marrow contains a **pluripotent stem cell** capable of giving rise to all the different types of blood cells. Mice given a lethal dose of irradiation soon die from loss of blood cells due to destruction of the hemopoietic organs. However, transfusion of bone marrow soon after exposure to radiation averts death, and both the marrow and the lymphatic tissues are reconstituted by functional cells derived from the transfused marrow. During recovery, macroscopic nodules appear in the spleen; these **spleen colonies** represent islets of proliferating hemopoietic cells and contain undifferentiated and maturing blood cells. The cells in the marrow inoculum that give rise to the spleen colonies are called **colony-forming units (CFU).**

Some colonies consist only of cells of the erythrocyte line, others contain developing granulocytes or developing megakaryocytes, but many are mixed colonies that contain developing cells of all three lines. The unicellular origin of mixed colonies has been shown by transfusing donor cells that have been irradiated just sufficiently to form unique chromosome markers but not enough to destroy their abilities to replicate. When these cells form spleen colonies, all the constituent cells bear the same marker as the inoculated cells and therefore must have arisen from the same CFU.

The unique chromosome marker also appears in cells that are repopulating the thymus and lymph nodes, indicating that the CFU has the potential for forming cells of lymphatic lineage as well. Cultures of cells from granulocytic and mixed colonies produce monocytes and macrophages that also carry the marker chromosome. The reconstituted marrow of these "rescued" animals, when injected into new recipient animals, produces spleen colonies that still carry the marker chromosomes.

Cells from spleen colonies give rise to new spleen colonies when injected into irradiated animals. Even colonies that are purely erythrocytic or granulocytic can give rise to colonies of all types. Thus the CFU is not only capable of differentiation but also of self-renewal, therefore fulfilling the definition of a stem cell. The bone marrow must contain pluripotent stem cells capable of feeding into the erythrocyte, granular leukocyte, megakaryocyte, monocyte, and lymphocyte lines.

In addition to pluripotent stem cells, bone marrow contains stem cells that have a more restricted capacity for development. These **restricted stem cells** are committed to the development of one specific cell line. Experimental evidence supports the existence of restricted stem cells for erythrocytes (CFU-E), granular leukocytes (CFU-G), and megakaryocytes (CFU-Meg). Most studies suggest a common precursor for granulocytes and monocytes (CFU-GM). Restricted stem cells arise from pluripotent stem cells, are rapidly proliferative, but have limited capacity for self-renewal. Pluripotent stem cells are capable of extensive replication but are only slowly proliferative and actually represent a reserve cell. It is the restricted stem cells that provide for the immediate, day-to-day replacement of blood cells.

Based on size, sedimentation velocities, and response to erythropoietin, three categories of restricted erythropoietic stem cells have been defined, and by their growth rates, size, and sedimentation characteristics, two CFU-GM are known. Within a given cell line, the various categories of committed stem cells represent progressions of development from more primitive, but restricted, stem cells to cells that are the immediate precursors of the classic, cytologically recognizable precursor cells.

While the existence of a pluripotent stem cell is certain, its morphology is in doubt; there is evidence, however, that it may be similar in appearance to a lymphocyte. Fractionation of bone marrow on sedimentation columns has shown that CFUs (stem cells) are contained in a fraction whose cells have a size and weight comparable with those of lymphocytes. The number of marrow stem cells can be increased by administering antimitotic drugs that destroy hemopoietic cells capable of division. Since pluripotent stem cells are only slowly proliferative, they are spared, and their relative numbers in the marrow are greatly increased. Stem cells can be enriched further by density gradient centrifugation, and as shown by their ability to produce spleen colonies, stem cells increase in number in proportion to an increase in cells with lymphoid appearance. These lymphocyte-like cells have been called **candidate stem cells.** While the morphology generally is similar to that of a small lymphocyte, the nucleus is more irregular and less deeply indented than in most lymphocytes, and the chromatin is more finely dispersed. Mitochondria are few in both types of cells but are smaller and more numerous in the candidate stem cell. Free ribosomes but no granular endoplasmic reticulum or lysosomes are present.

The earliest hemopoietic cells are those in the blood islands of the yolk sac, and there is evidence that undifferentiated cells from the yolk sac circulate in the fetus and successively seed the liver, spleen, and marrow. As hepatic hemopoiesis declines, the number of circulating stem cells increases, suggesting a large-scale migration of stem cells into the marrow. Cells of blood islands do contain CFUs and can restore all the hemopoietic organs in an irradiated animal.

Development of Erythrocytes

KEY WORDS: erythropoiesis, proerythroblast, basophilic erythroblast, polychromatophilic erythroblast, acidophilic erythroblast (normoblast), reticulocyte

The process of red cell production is called **erythropoiesis,** during which the erythrocyte undergoes progressive changes that involve the cytoplasm and nucleus. The cell progressively becomes smaller, and the cytoplasm stains increasingly acidophilic as it accumulates hemoglobin and loses organelles. The nucleus shrinks and becomes more heterochromatic and condensed until ultimately it is lost from the cell.

Although it is possible to describe various "stages" in the developmental sequence, the process of erythropoiesis does not occur in stepwise fashion. The process is a continuous one in which, at several points, the cells show distinctive, recognizable features. Unfortunately, the nomenclature of the red cell precursors is confused by the multiplicity of names given by various investigators to stages in the maturational series. The terms used here are commonly used, but alternative nomenclature is provided.

The **proerythroblast** (pronormoblast, rubriblast) is the earliest recognizable precursor of the red cell line and is derived from the pluripotent stem cell through a series of restricted stem cells. The proerythroblast is relatively large, with a diameter of 15 to 20 µm. The nongranular, basophilic cytoplasm frequently stains unevenly, showing patches that are relatively poorly stained, especially in a zone close to the nucleus. The cytoplasmic basophilia is an important point in the identification of this early form of red cell. Synthesis of hemoglobin has begun, but its presence is obscured by the basophilia of the cytoplasm. The nucleus occupies almost three-fourths of the cell body, and its chromatin is finely and uniformly granular or stippled in appearance. Two or more nucleoli are present and may be prominent. In electron micrographs the endoplasmic reticulum and Golgi apparatus are poorly developed, but free ribosomes are abundant and polysomes are scattered throughout the cytoplasm. The proerythroblast undergoes several divisions to give rise to basophilic erythroblasts.

The **basophilic erythroblast** (basophilic normoblast, prorubricyte) generally is smaller than the proerythroblast, ranging from 12 to 16 µm in diameter. The nucleus still occupies a large part of the cell, but the chromatin is more coarsely clumped and deeply stained. Nucleoli usually are not visible. The cytoplasm is evenly and deeply basophilic, sometimes more so than in the proerythroblast. Electron microscopy shows only a few or no profiles of endoplasmic reticulum, but free ribosomes and polyribosomes are abundant. Hemoglobin can be recognized as fine particles of low electron density but, as in the proerythroblast, is not always seen by light microscopy because of the intense cytoplasmic basophilia. The basophilic erythroblast also is capable of several mitotic divisions, its progeny forming the polychromatophilic erythroblasts.

The nucleus of the **polychromatophilic erythroblast** (polychromatophilic normoblast, rubricyte) occupies a smaller part of the cell and shows a dense chromatin network with scattered coarse clumps of chromatin. Nucleoli are absent, and the cell is incapable of division. Cytoplasmic staining varies from bluish gray to light slate gray, reflecting the changing proportions of ribosomes and hemoglobin. When nucleoli disappear, no new ribosomes are formed, and the change in staining characteristics is the result of a decrease in the concentration of ribosomes (which stain blue) and a progressive increase in hemoglobin (which stains red). Cell size varies considerably but generally is less than that of the basophilic erythroblast. The polychromatophilic erythroblasts encompass several generations of cells, the size reflecting the number of previous divisions that have occurred in the basophilic erythroblast. It is sometimes convenient to divide these cells into "early" and "late" stages on the basis of their size and on the intensity of the cytoplasmic basophilia.

Acidophilic erythroblasts (acidophilic normoblast, metarubricyte) are commonly called **normoblasts.** At this stage the cytoplasm is almost completely hemoglobinized and takes on a distinctly eosinophilic tint. Electron micrographs show a uniformly dense cytoplasm devoid of organelles except for a rare mitochondrion and widely scattered ribosomes. The nucleus is small, densely stained, and pyknotic and often is eccentrically located. Ultimately, the nucleus is extruded from the cell along with a thin film of cytoplasm.

Reticulocytes are newly formed erythrocytes that contain a few ribosomes, but only in a few cells (less than 2 percent) are they in sufficient number to impart color to the cytoplasm. After the usual blood stains, these cells have a grayish tint instead of the clear pink of the more mature forms and hence are called *polychromatophilic erythrocytes.* When stained with brilliant cresyl blue, the residual ribosomal nucleoprotein appears as a web or reticulum that decreases as the cell matures and varies from a prominent network to a few granules or threads. Reticulocytes are about 20 percent larger in volume than normal mature red cells.

After loss of the nucleus the red cell is held in the marrow for 2 to 3 days until fully mature. Unless there are urgent demands for new erythrocytes, the reticulocytes are not released except in very small numbers. In humans, these young red cells form a marrow reserve equal to about 2 percent of the number of cells in circulation. Figure 6-1 illustrates the changes in cell morphology during formation of erythrocytes.

Loss of Nucleus

Ordinarily, the nucleus assumes an eccentric position in the cell (late normoblast) and is lost just before the cell enters a marrow sinusoid. Active expulsion of the nucleus by the normoblast has been observed in vitro and may involve some contractile protein, possibly spectrin. Enucleation of erythrocytes also has been described as the cells pass through the pores in the sinusoidal endothelium. The flexible cytoplasm is able to squeeze through the pore, but the rigid pyknotic nucleus is held back and stripped from the cell along with a small amount of cytoplasm. Free nuclei are rapidly engulfed and destroyed by macrophages.

Development of Granular Leukocytes

KEY WORDS: granulocytopoiesis, myeloblast, promyelocyte, azurophil granule, myelocyte, specific

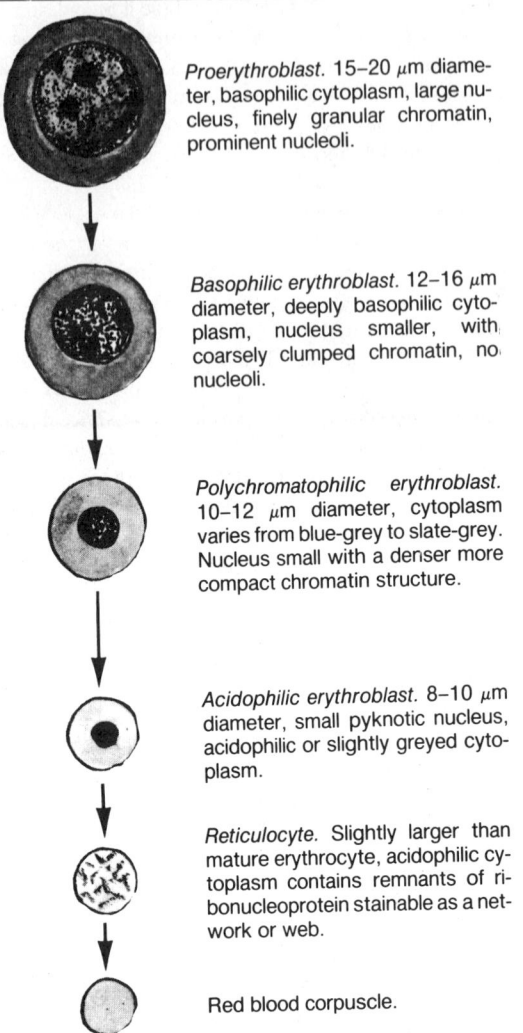

Proerythroblast. 15–20 μm diameter, basophilic cytoplasm, large nucleus, finely granular chromatin, prominent nucleoli.

Basophilic erythroblast. 12–16 μm diameter, deeply basophilic cytoplasm, nucleus smaller, with coarsely clumped chromatin, no nucleoli.

Polychromatophilic erythroblast. 10–12 μm diameter, cytoplasm varies from blue-grey to slate-grey. Nucleus small with a denser more compact chromatin structure.

Acidophilic erythroblast. 8–10 μm diameter, small pyknotic nucleus, acidophilic or slightly greyed cytoplasm.

Reticulocyte. Slightly larger than mature erythrocyte, acidophilic cytoplasm contains remnants of ribonucleoprotein stainable as a network or web.

Red blood corpuscle.

Fig. 6-1. Cytologic changes during development of erythrocytes.

granule, metamyelocyte, neutrophil band, eosinophil granulocyte, basophil granulocyte

The maturational process that leads to production of mature granular leukocytes is called **granulocytopoiesis.** During this process the cells accumulate granules and the nucleus becomes flattened and indented, finally assuming the lobulated form seen in the mature cell. During maturation, several stages can be identified, but as in red cell development, the maturational changes form a continuum, and cells of intermediate morphology often can be found. The stages commonly identified are myeloblast, promyelocyte, myelocyte. metamyelocyte, band form, and polymorphonuclear or segmented granulocyte. An alternative nomenclature substitutes the stem *granulo* for *myelo,* and the series of stages becomes granuloblast, progranulocyte, granulocyte, metagranulocyte, band form, and polymorphonuclear granulocyte.

Myeloblasts are the first recognizable precursors of granular leukocytes and represent a restricted stem cell committed to granulocyte and monocyte production. It is present in bone marrow only in low numbers. The myeloblast is relatively small, ranging from 10 to 13 μm in diameter. The cytoplasm is rather scant and distinctly basophilic, but much less so than in the proerythroblast, and lacks granules. Electron microscopy reveals abundant free ribosomes but relatively little granular endoplasmic reticulum; mitochondria are numerous and small. The round or oval nucleus occupies much of the cell, stains palely, and presents a somewhat vesicular appearance. Multiple nucleoli are present.

The **promyelocyte** is somewhat larger, measuring 15 to 20 μm in diameter. The nucleus may be slightly flattened, show a small indentation, or retain the round or oval shape. Chromatin is dispersed and lightly stained, and multiple nucleoli still are present. The basophilic cytoplasm contains purplish red **azurophil granules,** which increase in number as the promyelocyte continues its development. Electron micrographs show abundant granular endoplasmic reticulum, free ribosomes, numerous mitochondria, and a well-developed Golgi apparatus. Azurophil granules are formed only during the promyelocyte stage and are produced at the inner (concave) face of the Golgi complex by fusion of dense-cored vacuoles.

Divergence of granulocytes into three distinct lines occurs at the **myelocyte** stage with the appearance of **specific granules.** Thus neutrophil, eosinophil, and basophil myelocytes can be distinguished. Myelocytes are smaller than promyelocytes and measure 12 to 18 μm in diameter. The nucleus may be round or indented, and the chromatin is more condensed. Some myelocytes still may show a nucleolus outlined by a condensation of chromatin, whereas in others the nucleolus is poorly defined. The myelocyte is the last stage capable of mitosis.

The cytoplasm of the neutrophil myelocyte contains two populations of granules: azurophil granules produced at the promyelocyte stage and smaller specific granules formed by the myelocyte. Specific granules also arise by fusion of dense-cored vacuoles but are produced at the outer (convex) face of the Golgi body rather than the concave face.

As the myelocyte undergoes successive divisions, the number of azurophil granules in each cell is progressively reduced, and specific granules soon outnumber the azurophil type. The cytoplasm becomes less basophilic, and free ribosomes and granular endoplasmic reticulum are decreased.

Specific neutrophil granules stain lightly in routine blood smears, taking up a delicate lilac-pink color, and are too small to be resolved individually with the light microscope. Azurophil granules are larger and stain purplish red with the usual blood stains but are less numerous. With the electron microscope, azurophil granules appear more dense than neutrophil granules. The contents of the granules differ: azurophil granules are lysosomes and possess a complex of enzymes, among which aryl-sulfatase, acid phosphatase, beta-galactosidase, beta-glucuronidase, esterase, and nucleotidase have been identified. Specific neutrophil granules contain alkaline phosphatase and proteins with antibacterial properties.

Myelocytes eventually reach a state at which they no longer can divide and then mature into **metamyelocytes.** These cells show most of the features of the myelocyte except that the nucleus is deeply indented to form a horseshoe or kidney shape, nucleoli are lacking, and the chromatin forms a dense network with many well-defined masses of chromatin. Specific granules make up more than 80 to 90 percent of the granules present; the remainder are azurophil.

The **neutrophil band** has the same general morphology as the mature polymorphonuclear cell except that the nucleus forms a variously curved or twisted band. It may be irregularly segmented but not to the degree that definite lobes and filaments have formed.

The stages of maturation of **eosinophil granulocytes** are the same as for the neutrophil. Specific eosinophil granules appear at the myelocyte stage and usually are identifiable soon after they appear. Occasionally, the granules may have a slightly purple-blue color, becoming progressively more orange as the cell matures. A rare eosinophil granule may be present as early as the promyelocyte stage. The granules are much larger than the neutrophil type, stain brilliantly with eosin, and in electron micrographs are only slightly less dense and smaller than azurophil granules.

As the cells mature, the cytoplasm becomes less basophilic and the nucleus more and more indented to form lobes. In the late myelocyte and the metamyelocyte, the contents of the eosinophil granules crystallizes. Some granules show a crystal of variable shape occupying the center of the granule, surrounded by a matrix of lower density; others remain dense and homogeneous. Eosinophil granules are lysosomes and contain the usual battery of lysosomal enzymes.

Basophil granulocytes also pass through the same maturational sequences. The definitive granules usually appear at the myelocyte stage and rarely are seen in the promyelocyte. Initially, the granules are truly basophilic but become metachromatic and with toluidine blue or methylene blue stain violet rather than blue-black. In addition to chemotactic and platelet factors, the granules contain heparin, histamine, and several enzymes including diaphorase, dehydrogenases, peroxidases, and histidine carboxylase, which converts histidine to histamine.

The maturational changes and stages of development in granular leukocytes are depicted in Figure 6-2.

Development of Platelets

KEY WORDS: thrombopoiesis, megakaryocyte, demarcation membrane, megakaryoblast, endomitosis, polyploidy, promegakaryocyte

Thrombopoiesis (thrombocytopoiesis) refers to the formation of platelets. Platelets are derived from giant cells, the **megakaryocytes,** which measure 100 μm or more in diameter. Megakaryocytes are found only in bone marrow.

The nucleus of the megakaryocyte is large and convoluted and contains multiple irregular lobes of variable size interconnected by constricted regions. The coarsely patterned chromatin stains deeply. The cytoplasm is abundant and irregularly outlined and often has blunt pseudopods projecting from the surface. In smears, the cytoplasm appears homogeneous and contains numerous azurophil granules. Ultrastructurally, a variable degree of zoning is apparent. Immediately around the nucleus a narrow perinuclear zone contains a few mitochondria, the Golgi complex, granular endoplasmic reticulum, numerous polyribosomes, and some granules. A large intermediate zone is indistinctly separated from the perinuclear zone and contains granules, vesicles of different sizes and shapes, mitochondria, ribosomes, and components of the Golgi element. Depending on the degree of development of the megakaryocyte, the granules may be distributed uniformly or in small clusters outlined to a variable degree by a system of membranes. The outermost marginal zone is finely granular, varies in width, and lacks organelles; it does contain packets of microfilaments.

Platelets are formed by segmentation of the megakaryocyte cytoplasm via a system of **demarcation membranes.** The azurophil granules form small clusters, and simultane-

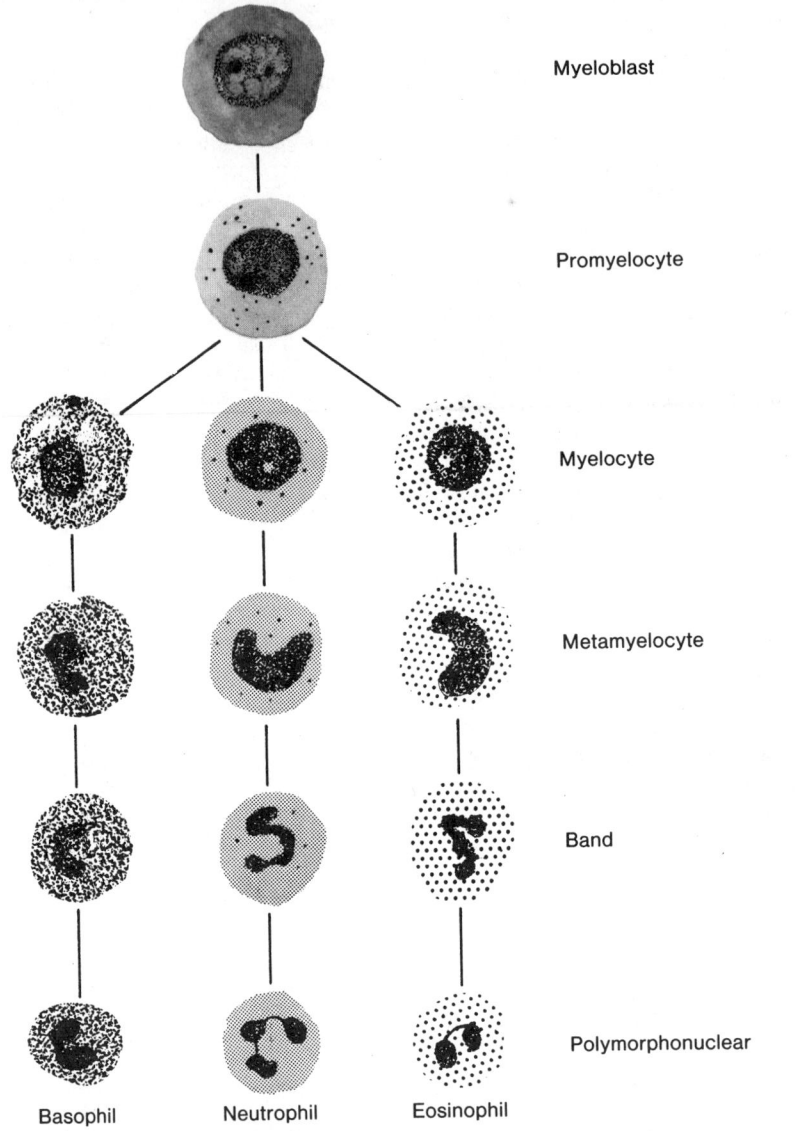

Fig. 6-2. Maturational sequences of granular leukocytes.

ously, small vesicles appear that become aligned in rows between the groups of granules. The vesicles initially are discontinuous but subsequently elongate and fuse to create a three-dimensional system of paired membranes. The narrow spaces between the membranes form clefts that surround each future platelet; as they are shed, the platelets separate along the narrow clefts. Demarcation membranes are continuous with the plasmalemma, and thus each platelet is bounded by a typical cell membrane.

The megakaryocyte delivers platelets through openings in the walls of the sinusoids, either as individual platelets or as ribbons of platelets that separate into individual elements within the sinusoidal lumen. After shedding its platelets, the megakaryocyte consists only of a nucleus surrounded by a thin rim of cytoplasm with an intact cell membrane. It generally is assumed that such megakaryocytes are unable to restore their cytoplasm and degenerate, with new generations of megakaryocytes being formed to replace them. Degenerate megakaryocytes can be found in the circulation, especially in the capillaries of the lung, where they may remain for some time.

Megakaryocytes arise from stem cells, the first recognizable precursor being a large cell, 25 to 45 μm in diameter, with a single round or oval nucleus. The chromatin has a finely granular pattern, and the basophilic cytoplasm is free of granules. These cells, which are rare, have been called **megakaryoblasts.** The cell undergoes a series of divisions in which the nucleus is replicated but the cytoplasm does not divide (**endomitosis**). At metaphase, the chromosomes become aligned in several planes on an increasingly complex multipolar spindle. With subsequent reconstitution, groups of chromosomes are incorporated into a huge lobulated nucleus. Thus a series of **polyploid** cells arises that may reach 64n in some megakaryocytes, although 16n nuclei are the more common. In general, cell and nuclear size are proportional to the degree of ploidy. The intriguing suggestion has been made that segmentation of the cytoplasm by demarcation membranes represents a delayed and modified cytokinesis. Cells of 4n and 8n ploidy, measuring 30 to 40 μm in diameter, frequently are called **promegakaryocytes.** The fully formed but not yet functional megakaryocyte consistently shows a clear marginal zone, whereas in platelet-forming megakaryocytes this zone disappears.

Development of Monocytes

KEY WORDS: promonocyte, azurophil granule, macrophage

Monocytes originate from a pool of precursors in the bone marrow, most likely the same restricted stem cells that give rise to granular leukocytes. In cultures, cells from granulocytic spleen colonies yield monocytes as well as granulocytes.

Immature monocytes of the bone marrow (**promonocytes**) are rare and difficult to distinguish. They range from 8 to 15 μm in diameter and possess large round to oval nuclei with evenly dispersed chromatin and several nucleoli. The cytoplasm is fairly abundant and contains numerous free ribosomes but only scant endoplasmic reticulum. The prominent Golgi complex is associated with numerous small granules that represent the formative stages of azurophil granules.

The mature monocytes of the bone marrow closely resemble those of the blood, are somewhat smaller than promonocytes (9 to 11 μm in diameter), and contain fewer ribosomes and larger and more abundant **azurophil granules.** Azurophil granules contain a variety of hydrolytic enzymes and are primary lysosomes. Following release into the blood, the monocyte continues its maturation in the circulation, and additional azurophil granules are formed.

Monocytes migrate into various tissues where they complete their maturation by transforming into **macrophages,** such as those of the peritoneum, alveoli of the lungs, or Kupffer cells of the liver.

Marrow Lymphocytes

Lymphocytes arise from pluripotent stem cells in the marrow and then emigrate to populate the lymphatic organs and tissues. Restricted stem cells that reach the thymus proliferate there and are released as T-lymphocytes (T-cells). After being released from the thymus, the T-cells migrate to the spleen, where they complete their maturation and are released as long-lived lymphocytes. B-lymphocytes (B-cells) also originate in the marrow. In birds they migrate to the bursa of Fabricius, where they differentiate into B-cells. The mammalian equivalent of the bursa has not been identified, but the bone marrow itself may be the organ for differentiation of B-cells. These cells also appear to complete their development in the spleen. B- and T-cells circulate and recirculate through the blood and lymphatic organs. In humans, B-cells have a life span of at least several months.

Control of Blood Formation

KEY WORDS: hypoxia, erythropoietin

Erythropoiesis is under humoral control and is regulated mainly by a specific erythropoiesis-stimulating factor called *erythropoietin.* However, the fundamental stimulus for red cell production is **hypoxia,** since the rate of production of erythropoietin is inversely related to the oxygen supply of the tissues. **Erythropoietin** is a glycoprotein hormone mainly produced by the kidney, though this is not the sole source of erythropoietin, since it can be produced in animals whose kidneys have been removed. The sites of extrarenal production of erythropoietin are not known, but small amounts of an erythropoietic-active substance have been detected in liver perfusates.

Erythropoietin increases synthesis of several species of ribonucleic acid (RNA) and protein, followed by the initiation of hemoglobin synthesis and differentiation of the restricted stem cell. Erythropoietin also appears to accelerate hemoglobin synthesis in those more differentiated cells which still are capable of synthesizing RNA and promotes release of reticulocytes from the marrow. Hormones other than erythropoietin have been implicated in blood formation. Testosterone stimulates red cell development and appears to be important for normal maintenance of red cell formation; estrogen has the opposite effect.

Whether there are individual hemopoietic hormones for each of the cell lines in bone marrow is unknown. There is evidence for circulating poietins that control the differentiation of granulocytes and megakaryocytes. These factors, granulopoietin and thrombopoietin, have not been characterized, nor is their tissue of origin known. Agents capable of controlling the proliferation and maturation of specific CFUs have been demonstrated in T-lymphocytes.

Lymphatic Tissues

Scattered throughout the body are various lymphatic elements that form part of the body's defense system. Generally distributed so that noxious agents entering the body soon encounter them, lymphatic tissues are prominent in the connective tissue coats of the gastrointestinal, respiratory, and urogenital tracts. Lymphatic structures also are inserted into the lymph drainage and blood circulation, where they serve to filter lymph and blood. Eventually, all the body fluids are filtered through some form of lymphatic structure.

Lymphatic elements may be spread throughout connective tissue and form the lymphatic tissues, or they may be organized into discrete structures, the lymphatic organs.

Classification

KEY WORDS: reticular tissue, hemopoietic tissue, diffuse lymphatic tissue, nodular lymphatic tissue

Lymphatic tissue represents a specialization within the connective tissue compartment and, as for all connective tissues, can be classified according to the type of fiber that is present. On this basis, lymphatic tissue can be defined as a **reticular tissue.** It also contributes cells (lymphocytes) to the blood, so it is a **hemopoietic tissue.** In common with bone marrow, the cellular content exceeds the intercellular material. Subdivision of lymphatic tissue into **diffuse** and **nodular** depends on the arrangement and concentration of the cells, not on differences in fiber types.

Cells of Lymphatic Tissues

KEY WORDS: fixed cell; reticular cell; free cell; small, medium, large lymphocyte; plasma cell, macrophage

The cells of lymphatic tissue are present as fixed and free cells. **Fixed cells** are the **reticular cells** responsible for the formation and maintenance of reticular fibers. The fixed cells of lymphatic tissue and reticular fibers, the two elements of the reticular network, are intimately related, and the fibers often reside in deep grooves in the cytoplasm of the reticular cells. By light microscopy, reticular cells appear as elongated or stellate elements with round or oval, palely stained nuclei and scant, lightly basophilic cytoplasm. The cytoplasm is more voluminous than is apparent under the light microscope, since the bulk of it is spread thinly along the reticular fibers. Electron microscopy reveals a variable amount of endoplasmic reticulum and a moderately well-developed Golgi element; other organelles are relatively inconspicuous. As in bone marrow, lymphatic reticular cells are incapable of giving rise to other cell types and show no special capacity for phagocytosis.

The remaining cells of lymphatic tissue are contained within the spaces of the reticular network and constitute the **free cells.** The bulk of these are lymphocytes, but macrophages and plasma cells also are present in variable numbers. The term *lymphocyte* encompasses a spectrum of cells that possess some general common features. As a class they show a rounded, centrally placed nucleus; their cytoplasm shows variable degrees of basophilia and lacks specific granules. The lymphocyte population customarily is divided into small, medium, and large lymphocytes on the basis of their size, nuclear morphology, and intensity of cytoplasmic staining. Although useful for purposes of description, such a subdivision is somewhat artificial because between the small and large types there is a continuum of cell sizes and morphologies.

Small lymphocytes are the most numerous. As seen in lymphatic tissues, these cells are 4 to 8 μm in diameter and contain a deeply stained nucleus surrounded by a thin rim of cytoplasm that may be expanded slightly at one side of the cell. Although rounded in shape in the blood or lymph, in tissues the cells are crowded together and assume various polyhedral shapes due to compression. The rounded or slightly indented nucleus contains a nucleolus that is barely discernible unless thin sections are examined. The chromatin is present as scattered masses of heterochromatin with some small patches of euchromatin. Electron microscopy shows a small Golgi complex, a few mitochondria and centrioles located in the region of the nuclear indentation, a moderate number of ribosomes, and a few lysosomes scattered throughout the cytoplasm. Granular endoplasmic reticulum is sparse.

Medium lymphocytes (prolymphocytes) range from 8 to 12 μm in diameter. Their nuclei are somewhat larger and more palely stained because of greater dispersion of the chromatin, and the nucleoli are larger and better defined. The cytoplasm is more voluminous and shows a greater basophilia because of an increased number of ribosomes.

Large lymphocytes (lymphoblasts) range in size from 15 to 20 μm. Occasionally even larger forms (25 to 30 μm) may be found, and some authors reserve the term *lymphoblast* for these cells. The nucleus is rounded, palely stained, and contains one or two prominent nucleoli, and the fine chromatin is more uniformly dispersed. The basophilic cytoplasm is abundant and contains free ribosomes and polysomes. The centrosomal region and Golgi body are more highly developed and mitochondria are abundant, but granular endoplasmic reticulum remains scanty.

In the development of lymphocytes, it generally is accepted that the direction of maturation is from large (lymphoblast) to small. However, this progression is complicated by the ability of small lymphocytes to respond to antigenic agents, reassume a blastlike appearance and character, and reproduce additional small lymphocytes and plasma cells.

Plasma cells (plasmocytes) vary in size from 6 to 20 μm in diameter. By light microscopy, the cell presents a rounded, somewhat elongated or polyhedral form, depending on its location. The nucleus is round to oval and usually is eccentrically placed. This is most obvious in elongated cells, but in rounded cells the nucleus may be only slightly eccentric in its placement. The chromatin is dispersed in coarse heterochromatic blocks along the inside of the nuclear envelope, giving the nucleus a clock face or spoke wheel appearance. Binucleate cells may be seen. The cytoplasm is deeply basophilic except for a prominent pale area adjacent to the nucleus. This area represents a negative image of the Golgi body.

Occasionally, plasma cells contain globular or crystalline inclusions in their cytoplasm, the best known of which are Russell's bodies. These are round or ovoid acidophilic inclusions of variable size that represent incomplete antibody.

In electron micrographs the characteristic feature of plasma cells is the extensive development of the granular endoplasmic reticulum, which almost completely fills the cytoplasm. The cisternae may be flat and parallel or distended with flocculent material. Free ribosomes are numerous, and a well-developed Golgi apparatus and centrioles are present in the paranuclear area. Russell's bodies, when present, are contained within distended cisternae of the granular endoplasmic reticulum. Plasma cells are sessile and ultimately degenerate. Despite the massive amounts of endoplasmic reticulum, most antibody production is carried out by forms that are intermediate between lymphocytes and plasma cells. The precursors of plasma cells are B-lymphocytes.

The **macrophages** of lymphatic tissue are part of a system of mononuclear phagocytes widely spread throughout tissues and organs. The cells vary from 10 to 20 μm in diameter and have oval-, kidney-, or horseshoe-shaped nuclei in which one or more nucleoli are present. The cytoplasm is abundant and lightly basophilic and may be vacuolated or contain ingested material. Unless the cytoplasm does contain phagocytosed matter, the macrophage is difficult to distinguish from other large mononuclear cells including active fibroblasts. In electron micrographs, the cytoplasm shows numerous folds, processes, and invaginations of the surface and contains the usual assembly of organelles. The Golgi apparatus is conspicuous, lysosomes are numerous, and residual bodies may be prominent. As are macrophages elsewhere, those in lymphatic tissue are derived from blood monocytes.

Diffuse Lymphatic Tissue

KEY WORDS: reticular fiber, reticular cell, lymphocyte

Diffuse lymphatic tissue is a constituent of lymphatic organs and also is widely dispersed along mucous membranes. It appears as a rather loose aggregate of cells and shows no distinct demarcation from surrounding tissue with which it gradually merges. Basically, diffuse lymphatic tissue consists of a three-dimensional array of **reticular fibers** and their associated **reticular cells.** The two form a spongelike framework pervaded by large numbers of cells, chief of which are **lymphocytes.**

Diffuse lymphatic tissue is particularly prominent in the connective tissue that underlies the epithelium of the intestine. Here the lymphatic tissue, in association with the lining epithelium, produces antibody that bathes the luminal surface. Any antigen that does penetrate the epithelial lining induces an immune response in the lymphatic tissue, the cellularity of which is related to the bacterial content of the intestines.

Nodular Lymphatic Tissue

KEY WORDS: solitary nodule, primary nodule, germinal center, secondary nodule, light pole, dark pole, cap, follicular dendritic cell, confluent lymphatic nodule, Peyer's patch

Nodular lymphatic tissue contains the same structural elements as diffuse lymphatic tissue, differing only in that the components are organized into compact, somewhat circumscribed structures. It may occur anywhere in connective tissue but is prominent along the digestive and respiratory tracts. Lymphatic nodules (also called *follicles*) may be present as **solitary nodules** or in masses forming confluent nodules, as occur in the appendix and the Peyer's patches of the ileum. Lymphatic nodules are prominent in organs such as the tonsils, lymph nodes, and spleen but are absent from the thymus.

In ordinary histologic sections, some lymphatic nodules appear as rounded collections of densely packed small lymphocytes; this type of nodule is called a **primary nodule.** Other lymphatic nodules contain a lightly staining central area surrounded by a deeply stained cuff or cap of closely packed small lymphocytes. The pale region has been called a **germinal center** and the whole structure a **secondary nodule.**

There are few reticular fibers within the germinal center, and the free cells are supported by a cellular framework consisting of stellate reticular cells. The numerous processes of the reticular cells are joined by desmosomes. Reticular fibers are present in the cuff of cells at the periphery of the center, where they form concentric layers around the structure.

Germinal centers have dark and light poles. The **light pole** is sparsely populated by scattered small lymphocytes and reticular cells, whereas the **dark pole** is densely packed with large and medium lymphocytes, cells in transition to plasma cells, macrophages, mitotic cells, and the pyknotic nuclei of degenerating lymphocytes. Surrounding the center is a zone of small lymphocytes that usually is thicker at the light pole, where it forms the **cap.** A thinner layer of lymphocytes covers the dark pole. The structure shows definite polarity: The light pole always is directed toward a surface — the capsule in lymph nodes, red pulp in the spleen, and the epithelium of a tonsil.

Germinal centers appear to be sites where lymphocytes are formed, but many of the newly formed cells die there. They also are the sites of antibody formation, and each germinal center appears to represent a clone of cells derived from an antigen-stimulated lymphocyte and active in the production of one specific antibody. Germinal centers produce B-cells that can migrate through the cap to leave the center and eventually pass to other lymphatic tissues. Areas where germinal centers develop contain **follicular dendritic cells,** which show numerous complex cytoplasmic processes that interdigitate with follicular lympho-

cytes. The cytoplasm contains many mitochondria, but ribosomes, lysosomes, and secretory granules are scarce. The nucleus is irregular and shows peripherally placed bars of heterochromatin and a distinct nucleolus. These cells are considered to be antigen-presenting cells that retain antigen on their surface for the stimulation of T-cells.

Germinal centers develop only after birth and in response to antigenic stimuli. They are absent in animals born and raised in a germ-free environment. Following a first exposure to antigen, germinal centers form de novo and then regress in the absence of antigen and eventually disappear. Upon a second stimulation by the same antigen, there is a rapid and marked production of germinal centers, which precedes a rise in circulating antibody. However, germinal centers are not essential for antibody formation, and it has been suggested that germinal centers arise only after repeated contact with antigen and may be involved in long-term (memory) antibody responses.

Aggregate or **confluent lymphatic nodules** occur in the appendix and are exemplified by Peyer's patches found in the lower half of the small intestine. **Peyer's patches** consist of many solitary nodules massed together to form grossly visible structures that underlie the intestinal epithelium. The light poles of their germinal centers are directed toward the epithelium. Individual nodules may be quite discrete and well defined, but usually they coalesce and can be distinguished only at their apices and by their germinal centers. Peyer's patches are prominent in children but undergo gradual regression with aging.

Interrelationships of Hemopoietic Cells

The current understanding of the interrelationships of the various blood cells is illustrated in Figure 6-3. A pluripotent stem cell, present in the marrow and capable of unlimited self-renewal, gives rise to several types of unipotent (restricted) stem cells. The latter have limited capacity for self-renewal, and each is committed to one line of development. Thus there is a committed stem cell for each of the cell types seen in blood, and these give rise to the morphologic stages that can be recognized in any of the developmental lines. Several orders of restricted stem cells have been recognized, representing increasing stages of maturity. The morphologic identity of the restricted and pluripotent stem cells is undetermined.

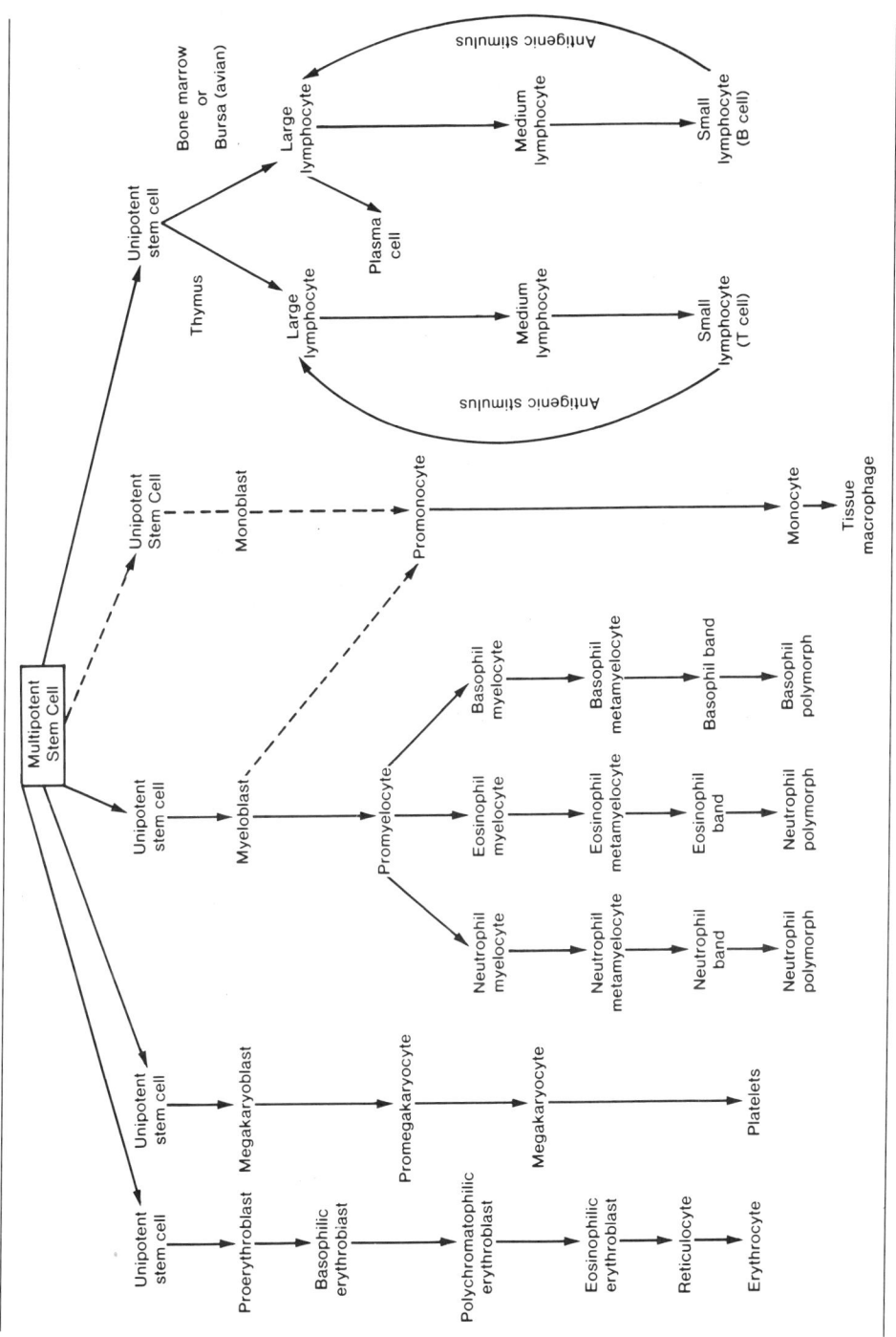

Fig. 6-3. Interrelationships of hemopoietic cell lines.

Review Table 6-1. Key Features of Bone Marrow Hemopoietic Cells

Cell	Diameter (μm)	Cytoplasm	Nucleus	Nucleolus
Myeloblast	10–13	Light blue, transparent; no granules	Round or oval with pale staining; fine vesicular chromatin	Yes
Promyelocyte	15–22	Light blue, transparent; scattered azurophil granules	Round or oval, may have slight indent, fine, lightly stained, dispersed	Yes
Myelocyte	12–18	Light blue, obscured by granules; specific granules are present (neutrophil, eosinophil, basophil)	Round or oval, often with slight indent; chromatin more deeply stained and condensed	No
Metamyelocyte	12–16	Light blue, completely obscured by specific granules; a few azurophil granules present	Horseshoe- or kidney-shaped; coarse masses of deeply stained chromatin	No
Band	10–14	Filled with specific granules; a few azurophil granules present	Ribbon-like, twisted or coiled, no filaments; coarse, deeply stained chromatin	No
Megakaryocyte	50–80	Light blue, transparent; scattered azurophil granules	Large, convoluted, irregular; stains deeply; coarsely patterned	No
Proerythroblast	15–20	Deep blue, opaque; no granules	Round, occupies about three-fourths of cell; texture deeply stained, uniformly "granular"	Yes
Basophil erythroblast	12–16	Deep blue, opaque; may have blotchy appearance; no granules	Round, occupies about half of cell; coarsely clumped, deeply stained	No
Polychromatophilic erythroblast	10–12	Variable — slate gray to blue gray	Round, dense, compact, deeply stained	No
Acidophilic erythroblast (normoblast)	8–10	Pink or pink with slight blue-gray	Round, dense, pyknotic, often eccentric	No

Functional Review: Hemopoietic Tissues

The obvious function of hemopoietic tissues is to provide blood with continuous replacement of cells. Separation of the replicating cells from their end products in the blood and sequestering them in specific hemopoietic tissues allows the establishment of environments best suited to each tissue. It ensures that replicating cells are not unduly influenced by the activities of other organs and tissues, as would occur if replication occurred within the bloodstream. It also provides the blood with small cells that can more easily circulate through fine capillary beds. Since there is no need to set aside, within the circulation, a population of cells concerned with replication, all the blood cells are functional and can serve the body's needs. Subdivision of hemopoietic tissue into lymphatic and bone marrow components allows separate conditioning of cells (such as T-lymphocytes) without, at the same time, exposing the remainder of the hemopoietic cells to the same conditioning factors.

Within bone marrow, yellow marrow fills in spaces unoccupied by active marrow, but being very labile, this fatty marrow can quickly be replaced by hemopoietic marrow when demands for blood become insistent. The sinusoids of the marrow provide sites where newly formed blood cells can gain access to the circulation. Additional maturation of cells occurs within the sinusoids. and the cells held here serve as a reserve that can be mobilized as needed.

Review Questions and Answers

Questions

29. Describe the formation of blood in the embryo and fetus. What are the main cells formed at each stage?
30. What is the difference between reticular cells and reticulocytes?
31. What are the main cytologic changes that occur during erythropoiesis and granulocytopoiesis? (Do not describe specific stages.)
32. Describe the structure of the bone marrow sinusoids.
33. From what cells are platelets formed? How are platelets produced from these cells? What peculiar feature do these platelet-producing cells show?
34. What are colony-forming units?
35. Give a classification of lymphatic tissues.
36. Describe the germinal center. What is its function?
37. What type of granules are present in neutrophil granulocytes?

Answers

29. Blood formation in the embryo occurs first in the blood islands of the yolk sac and commences shortly after the formation of the germ layers. The main product of this activity is nucleated red cells. The second organ involved in hemopoiesis is the liver, which mainly produces nonnucleated erythrocytes; lesser numbers of granulocytes and megakaryocytes also are produced. During the splenic phase, the spleen produces erythrocytes, platelets, granulocytes, and lymphocytes. The bone marrow becomes the major blood-forming organ during the last 3 months of gestation and produces all types of blood cells.
30. The reticular cells are fixed cells that constitute part of the framework of both bone marrow and lymphatic tissues. They are large connective tissue cells responsible for the formation of the reticular fibers, which form the fibrous component of the hemopoietic tissues. Reticulocytes are erythrocytes in the last stages of development. They contain small amounts of residual ribonucleoprotein that, with special stains, can be precipitated and appears as a network within the cell.
31. During the development of red cells (erythropoiesis), the cell as a whole becomes progressively smaller; the cytoplasm first becomes more basophilic and then increasingly acidophilic as hemoglobin increases, and the content of ribosomes diminishes. The nucleus decreases in size, becomes more condensed, and ultimately is lost. During the process of granulocytopoiesis, the cells accumulate cytoplasmic granules of specific types. The nucleus becomes flattened and indented and finally assumes a lobulated form. The overall size of the cell decreases and nucleoli are lost.
32. The sinusoids of the bone marrow are thin-walled vessels of wide bore. The wall consists of three coats: (a) an endothelial lining of attenuated squamous cells united by junctions of the zonula adherens type; (b) a second coat consisting of a thin, discontinuous basal lamina on which the endothelial cells rest; and (c) reticular cells loosely applied to the exterior of the vessel to form an external coat of variable thickness.
33. Platelets are formed from megakaryocytes by segmentation of their cytoplasm. Small cytoplasmic vesicles appear between clusters of granules, coalesce, and elongate to form the demarcation membranes. The membranes surround each future platelet, which separate along the narrow clefts between them and are shed. The megakaryocytes are peculiar in that they arise from precursor cells by a series of nuclear divisions that occur without intervening periods of cytokinesis. This results in polyploid cells that may reach 64n.
34. Colony-forming units (CFUs) are the cells of bone marrow that give rise to spleen colonies when bone marrow is transfused into irradiated mice. The CFUs represent pluripotent and unipotent stem cells and are capable of repopulating all hemopoietic tissues. The CFU has the attributes of a stem cell — self-renewal and the capacity for differentiation.
35. Lymphatic tissue may be classified as a reticular connective tissue on the basis of its content of reticular fibers or as a hemopoietic tissue because it contributes cells to the blood. It may be subdivided into diffuse and nodular types on the basis of its cellular organization.
36. Germinal centers are highly organized components of secondary lymphatic nodules. They show two distinct structural regions called the dark pole and the light pole. The dark pole contains a dense concentration of large and medium lymphocytes, macrophages, and plasma cells. The light pole is less densely populated and contains small lymphocytes and reticular cells. The dark pole is an area of active proliferation of

lymphatic cells, but many of the newly formed cells appear to die there. Reticular fibers are scarce within both poles of the germinal center, and the cells are supported by a cellular framework formed by reticular cells. The germinal center contains B-lymphocytes, is closely associated with humoral immunologic responses, and may represent a clone of cells committed to responding to a single antigen.

37. The neutrophil granulocyte contains two types of granules: the specific granules and the azurophil granules. The azurophil granules are produced only at the promyelocyte stage and are formed at the concave face of the Golgi complex. Specific granules are produced during the myelocyte stage, are formed at the convex face of the Golgi, and become the dominant type of granule. The azurophil granules are lysosomes and contain the usual acid hydrolases associated with lysosomes. The specific granules contain different enzymes and possess proteins with antibacterial properties.

7 Muscle

Muscle is the basic tissue in which the property of contractility is preeminent. The purposeful movements of the body and the maintenance of posture are the result of contractions of muscles attached to the skeleton. Muscular contraction also is responsible for beating of the heart, breathing, constriction of blood vessels, movements of the intestines, emptying of the bladder, and other vital processes. The unit of structure of muscle is the muscle cell, which, because of its elongated shape, also is called a *fiber*. Functionally, the shape of the cell is important, because a greater unidimensional contraction can be achieved by an elongated cell than by a globular cell of the same volume. Within a muscle mass, the fibers are oriented in the direction of movement.

Classification

KEY WORDS: **striated muscle, skeletal muscle, cardiac muscle, smooth muscle, visceral muscle, voluntary muscle, involuntary muscle**

Three kinds of muscle can be distinguished based on their morphology and function. Some muscles show a regular pattern of alternating light and dark bands that extend across the width of the fiber and have been called **striated muscles.** This is the kind of muscle associated with the skeleton and heart and thus forms **skeletal** and **cardiac muscle. Smooth muscle** lacks cross-striations, occurs in the walls of viscera, and often is referred to as **visceral muscle.** Functionally, muscle is either **voluntary** (under control of the will) or **involuntary** (not controlled by the will). Striated skeletal muscle is voluntary; smooth muscle is involuntary; and cardiac muscle is involuntary striated muscle.

Skeletal Muscle

Skeletal muscle is the most abundant tissue in the body and forms those structures generally referred to as "the muscles." It permits a wide range of voluntary activities from the intricate and fine movements involved in writing to the rapid, powerful forces associated with jumping or throwing a ball. The power that can be generated by skeletal muscle is illustrated by the speed attained by a thrown ball which, at the moment it leaves the fingertips, may reach speeds of over 90 mph.

Organization

KEY WORDS: **fiber, multinucleated cell, fascicle, muscle, epimysium, perimysium, endomysium**

The smallest independent unit of striated muscle visible with the light microscope is the **fiber,** which is a long **multinucleated cell.** Groups of parallel fibers form **fascicles** that can be seen with the naked eye. In turn, groups of fascicles make up an entire **muscle.** Fascicles vary in size with different muscles and in general are small in muscles associated with fine movements and large in muscles that perform actions demanding greater power.

At all levels of organization, muscle is associated with connective tissue (Fig. 7-1). The entire muscle is surrounded by a connective tissue sheath called the **epimysium.** Septa pass from the deep surface of the epimysium to invest each fascicle as the **perimysium;** delicate extensions of the perimysium wrap each fiber as the **endomysium.** Although given different names according to its association with different structural units of muscle, the connective tissue is continuous from one part to another. It consists of collagenous, reticular, and elastic fibers and contains several cell types including fibroblasts, macrophages, and fat cells. The endomysium is especially delicate and consists mainly of reticular fibers and thin collagen fibers; it carries blood capillaries and small nerve branches (Fig. 7-2). The larger blood vessels and nerves lie within the perimysium. The connective tissue not only binds the muscle units together but also acts as a harness and aids in integrating and transmitting the force of their contractions. The amount of connective tissue varies from muscle to muscle, being relatively greater in muscles capable of finely graded movements. Elastic tissue is most abundant in muscles that attach to soft parts such as the muscles of the tongue and face.

The Skeletal Muscle Fiber (Cell)

KEY WORDS: **banding, cross-striation, A band, I band, Z line, H band, M line, sarcomere**

117

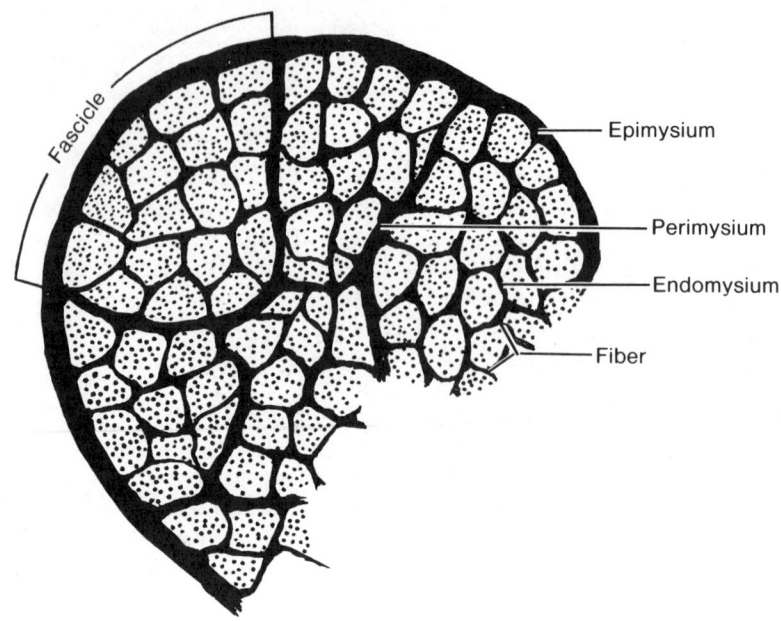

Fig. 7-1. Portion of a muscle in transverse section to show the structural organization and the relationships of the connective tissue.

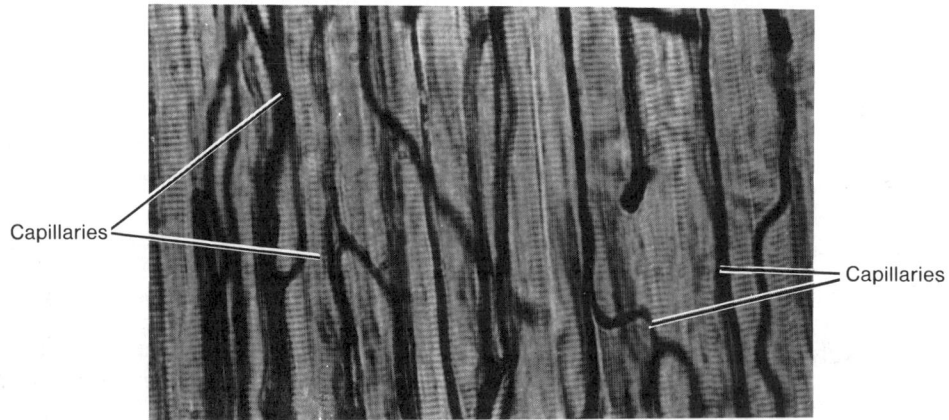

Fig. 7-2. A longitudinal section of human skeletal muscle. The specimen had its vasculature injected previously with colored gelatin. Note that each skeletal muscle fiber is enveloped by capillaries. Striations remain visible on individual skeletal muscle cells (LM, × 250).

Generally, skeletal muscle fibers do not branch, but the muscles of the face and tongue may do so where the fibers insert into mucous membrane or skin. The fibers of skeletal muscle are elongated tubes that vary in shape and length. Some are cylindrical with rounded ends and appear to extend throughout the length of a muscle; others are spindle-shaped with narrowly tapering ends and apparently do not run the length of the muscle. One end of the fiber may attach to a tendon, while the other unites with connective tissue in the belly of the muscle, or both ends may lie within the muscle mass. In transverse sections of fresh muscle the fibers are round or ovoid, but in fixed material they appear as irregular polyhedrons.

The fibers range between 10 and 100 μm in diameter. The size varies from muscle to muscle and even within the same muscle. Longer fibers have the greater diameters and are associated with more powerful muscles. Some spacial arrangement of the fibers within a muscle has been noted, fibers of large diameter tending to be located more centrally. The size of the fibers changes with use, and fibers may hypertrophy (increase in size) in response to continued use or atrophy (decrease in size) with disuse.

The most outstanding structural feature of skeletal muscle fibers is the presence of alternating light and dark segments that result in the **banding** or **cross-striations** that are seen when the fiber is viewed in longitudinal section (Figs. 7-3, 7-4, and 7-5). Under polarized light, the dark bands are anisotropic and are called **A bands,** whereas the light bands are isotropic and thus are called **I bands.** Running transversely through the center of the I band is a narrow, dense line, the **Z line** or Z disc. In good preparations, a pale narrow region, the **H band,** can be seen transecting the A band, with a dark **M line** within it. The relative length of each band depends on the state of contraction of the fiber. The length of the A band remains constant, whereas the I band is prominent in a stretched muscle and short in a contracted muscle (see Fig. 7-10).

The contractile unit of the fiber is a **sarcomere,** defined as the distance between two successive Z lines. Within the sarcomere, as the I band becomes shorter, the Z lines approach the ends of the A bands.

Structure of Skeletal Muscle Fibers

KEY WORDS: sarcolemma, multinucleated cell, peripheral nuclei, sarcoplasm, satellite cell, myofibril, sarcoplasmic reticulum, sarcotubules, terminal cisternae, T-tubule, triad, T-system, junctional feet

Each skeletal muscle fiber is an elongated cell invested by a delicate membrane that, by electron microscopy, can be resolved into the plasmalemma and an adherent outer coat of glycoprotein. The plasmalemma of muscle is referred to as the **sarcolemma** but is no different from the limiting membrane of any cell. A peripheral membrane protein

Fig. 7-3. A longitudinal section of human skeletal muscle cells (LM, × 1,000).

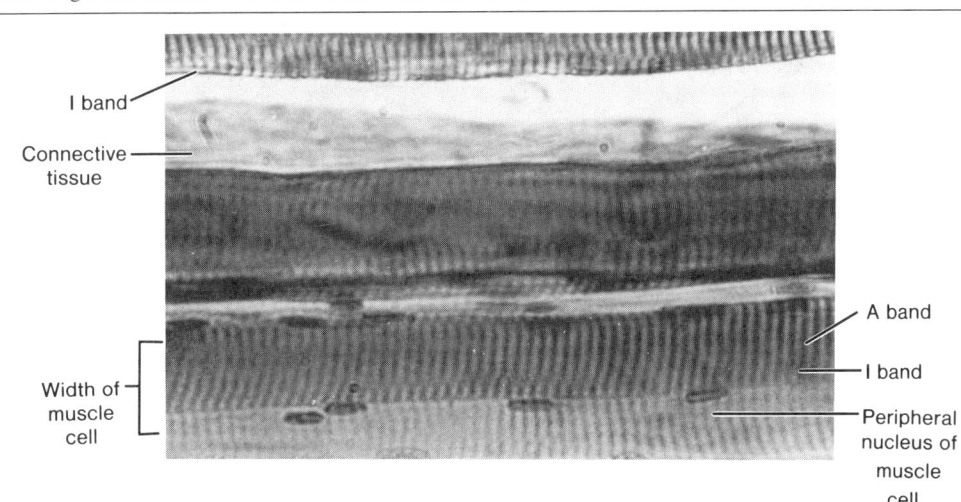

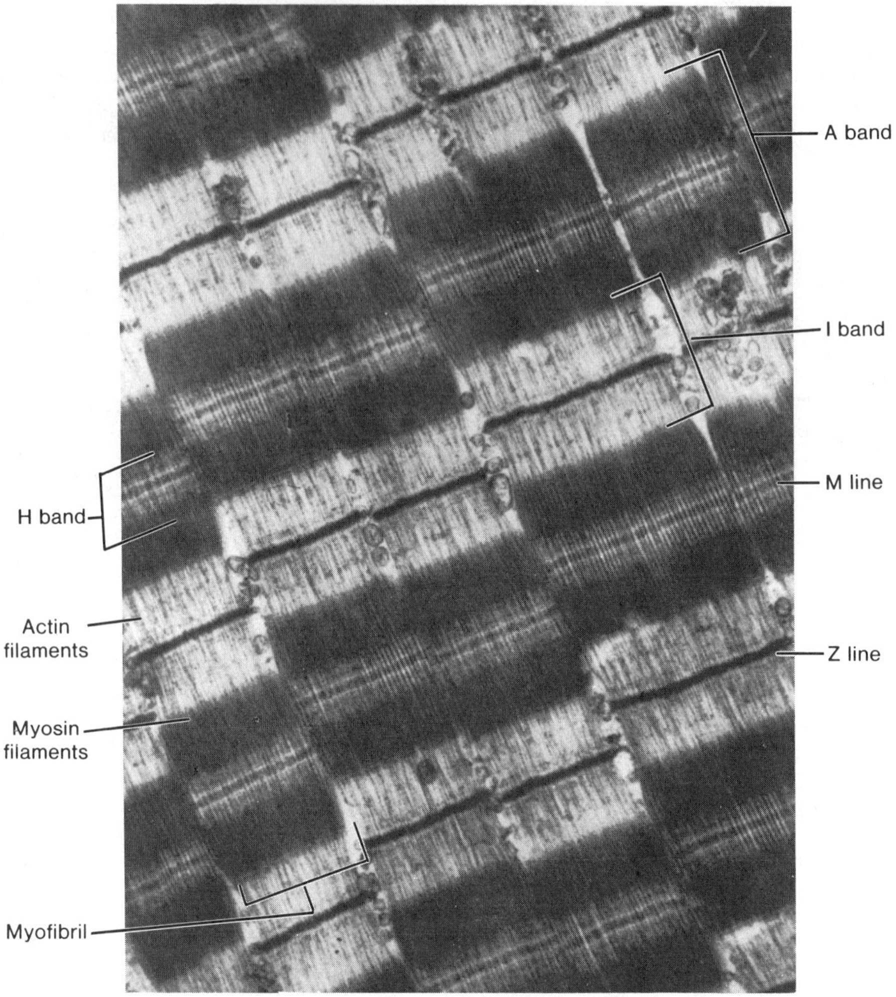

A band

I band

M line

H band—

Z line

Actin
filaments

Myosin
filaments

Myofibril—

Fig. 7-4. Details of the banding on each myofibril shown in a skeletal muscle cell (TEM, × 26,000).

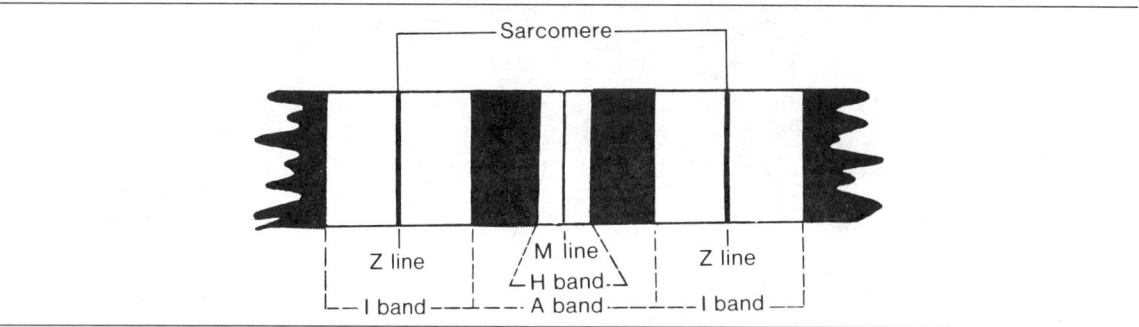

Sarcomere

Z line   M line   Z line
└ H band ┘
I band   A band   I band

Fig. 7-5. Banding pattern of striated skeletal muscle.

called *dystrophin* provides mechanical reinforcement on the cytoplasmic surface of the sarcolemma during muscle contraction. The outer glycoprotein layer corresponds to a basal lamina and is called the *external lamina*. It is associated with delicate reticular fibers that mingle with the reticular fibers of the endomysium.

The striated muscle fiber is a **multinucleated cell.** The nuclei are elongated in the direction of the long axis of the fiber and, in adult muscle, are **peripheral,** located immediately below the sarcolemma. The chromatin tends to be distributed along the inner surface of the nuclear envelope, and one or two nucleoli usually are present. The nuclei are fairly evenly spaced along the fiber but become more numerous and irregularly distributed in the area of attachment of the muscle to a tendon. Although the cytoplasm of the muscle cell is called the **sarcoplasm,** it corresponds to that of any cell. Many small Golgi bodies are present near one pole of a nucleus; lysosomes also usually take up a juxtanuclear position.

Closely associated with the muscle fiber are **satellite cells** that lie flattened against the fiber covered by the same external lamina that invests the muscle cell. Although difficult to distinguish from muscle cells by light microscopy, satellite cells are more fusiform, their nuclei are more condensed, the chromatin is less dispersed, and nucleoli are lacking. The cytoplasm is scanty but may form a clear boundary between the nucleus of the satellite cell and the muscle fiber. In electron micrographs, satellite cells show centrioles (not seen in muscle cells), scant endoplasmic reticulum, a small Golgi complex, and a few mitochondria near the ends of the nuclei. Satellite cells may represent survivors of the primitive myoblasts; they account for less than 4 percent of the muscle-associated nuclei.

Myofibrils are elongated, threadlike structures in the sarcoplasm and run the length of the muscle fiber. At 1 to 2 μm in diameter, myofibrils are the smallest units of contractile material that can be identified with the light microscope. In cross sections, myofibrils appear as small dots (Fig. 7-6), while in longitudinal sections they give a longitudinal striation to the fiber. Each myofibril shows a banding pattern identical to that of the whole fiber (see Fig. 7-4). Indeed, the banding of the fiber results from the bands on consecutive myofibrils being in register. Cross-striations are restricted to the myofibrils and do not extend across the sarcoplasm between fibrils. The striations on adjacent myofibrils are kept in alignment by a system of intermediate filaments composed of the protein desmin. Desmin also links the myofibrils to the cell membrane.

Associated with each myofibril is the **sarcoplasmic reticulum,** a modification of the smooth endoplasmic reticulum seen in other cells. In muscle, the sarcoplasmic reticulum serves as a store for calcium ions. The membrane of the sarcoplasmic reticulum contains numerous transmembrane proteins, including calcium-ATPase, that actively transport calcium from the cytosol into its lumen. Here, calsequestrin and other proteins bind and store the internalized calcium ion. This organelle consists of an extensive and continuous system of membrane-bound tubules called **sarcotubules** that form a mesh around each myofibril. At each junction of A and I bands, a pair of dilated sarcotubules, the **terminal cisternae,** pass around each myofibril and are continuous with the terminal cisternae of

Fig. 7-6. A section through the musculature of the human tongue demonstrates cross-sectional and longitudinal profiles of skeletal muscle cells (LM, × 300).

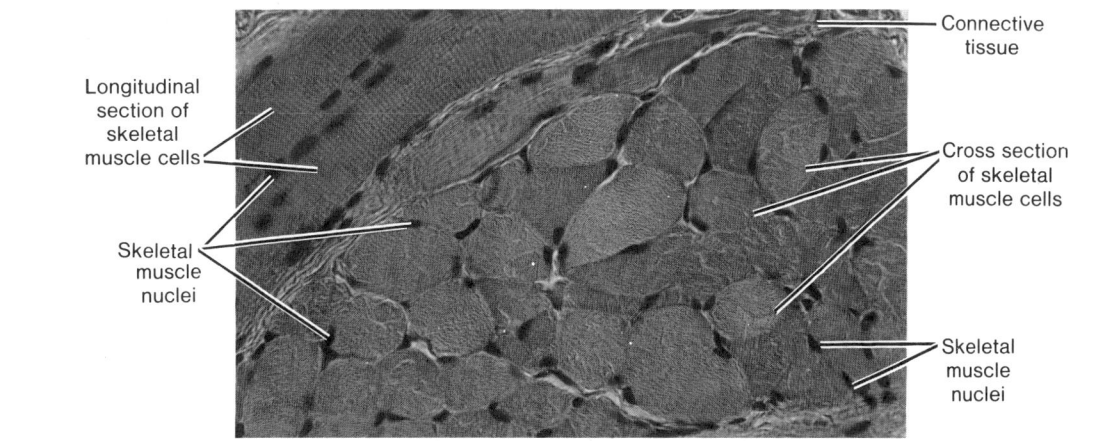

adjacent myofibrils. One of the pair of terminal cisternae serves an A band and the other serves an I band. From each of the cisternae, narrow, longitudinal sarcotubules extend over the A and I bands, respectively. Over the A band, the tubules form an irregular network in the region of the H band, while in the I band a similar confluence occurs at the region of the Z line (Fig. 7-7). Thus each A and I band is covered by a "unit" or segment of sarcoplasmic reticulum. These units consist of the terminal cisternae at the A–I junctions, joined by longitudinal sarcotubules that anastomose in the region of the H and Z lines of their respective A and I bands. The sarcoplasmic reticulum shows the same structure regardless of fiber type.

The pairs of cisternae at the A–I junctions are separated by a slender **T-tubule,** and the three structures — the T-tubule and two terminal cisternae — form the **triad** of skeletal muscle. T-tubules are inward extensions of the sarcolemma and penetrate into the muscle fiber to surround each myofibril. The T-tubules of one myofibril communicate with those of adjacent myofibrils to form a complete network across the fiber. The lumen of the T-tubule does not open into the cisternae but does communicate with the extracellular space at the surface of the sarcolemma. T-tubules are quite distinct from the sarcoplasmic reticulum and collectively make up the T-system.

The **T-system** rapidly transmits impulses from the exterior of the fiber to all the myofibrils throughout the cell, thereby producing a coordinated response. Passage of electrical impulses to the sarcoplasmic reticulum results in release of free calcium ions from the reticulum into the

Fig. 7-7. Diagrammatic representation of the sarcoplasmic reticulum and system of T-tubules in skeletal muscle.

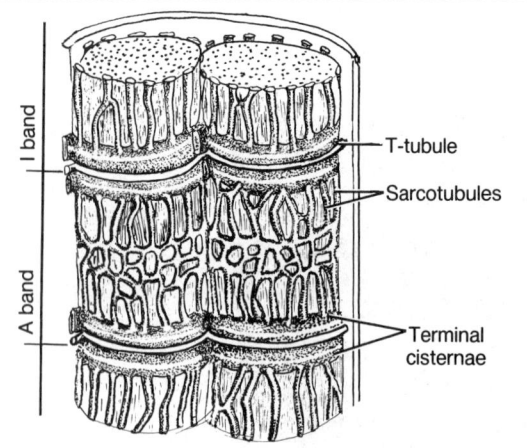

neighboring contractile elements, bringing about their contraction.

T-tubules and terminal cisternae are closely apposed. **Junctional feet** span the narrow gap between T-tubules and terminal cisternae, forming areas of low resistance through which impulses pass to the sarcoplasmic reticulum. In electron micrographs, the junctional feet appear as regularly spaced densities extending from the T-tubules to the terminal cisternae. These are matched by evenly spaced dimples on the cisternal membranes, corresponding to sites where the feet are located. The junctional feet are believed to be calcium ion channel proteins that extend from the terminal cisternae to voltage-sensing calcium ion channel proteins of the T-tubules. When depolarization occurs, the T-tubule channel proteins undergo a conformational change and (because of their intimate association with calcium ion channel proteins of the sarcoplasmic reticulum), the latter open, releasing calcium ions into the cytosol, initiating contraction.

Structure of Myofibrils

KEY WORDS: myofilament, thick filament, myosin, thin filament, actin, tropomyosin, troponin complex

Under the electron microscope, the myofibril is seen to consist of longitudinal, fine **myofilaments,** of which two types have been identified, differing in size and chemical composition. The thick filaments are 10 nm in diameter; thin filaments have a diameter of only 5 nm.

The **thick filaments** consist largely of **myosin** and are 1.5 μm long with a slightly thickened midportion and tapering ends. The midportion is smooth, whereas the ends are studded with many short projections. Myosin filaments can be dissociated into their constituent molecules, of which there are about 180 per filament, each consisting of a head and a tail. The tails of the molecules lie parallel and are so arranged that the smooth midportion of the filament consists of the tails only. The heads project laterally from the filament along its tapered ends and form a helical pattern along the filament. Most of the tail consists of light meromyosin, whereas the heads, with parts of the tails, consist of heavy meromyosin. **Thin filaments** are about 1 μm in length and are composed of globular subunits of **actin** attached end to end to form two longitudinal strands wound about one another in a loose helix. A second protein, **tropomyosin,** lies within the groove between the two actin strands and gives stability to the actin filament.

A protein complex consisting of three polypeptides designated as troponin-T, troponin-I, and troponin-C is bound to tropomyosin at regular intervals of 40 nm. The **troponin**

complex regulates actin and myosin binding. Troponin-T binds the troponin complex to tropomyosin and positions the complex at a site on the actin filament where actin can interact with myosin. Troponin-I prevents actin from binding to myosin. Troponin-C binds to calcium ion. When this occurs, a conformational change occurs in the troponin complex, and myosin can now bind and interact with the actin filament. The energy needed for the cross-bridging between myosin and actin is obtained from the breakdown of adenosine triphosphate (ATP) by ATPase contained in the myosin filaments. As a result of forming and breaking bonds between the myosin and actin filaments, actin filaments slide past the myosin filaments, and the length of each sarcomere is reduced, resulting in an overall shortening of each myofibril.

The arrangement of the thick and thin filaments is responsible for the banding pattern on the myofibrils (Figs. 7-8 and 7-9). The I band consists only of thin filaments, which extend in both directions from the Z line. The filaments on one side are offset from those of the other, and connecting elements appear to run obliquely across the Z line to create a zigzag pattern. The structure and chemical composition of the Z line and the attachments of the thin filaments to it are not well understood. As the actin filament nears the Z line, it becomes continuous with four slender threads, each of which appears to loop within the Z line and join a thread from an adjacent actin filament. Several accessory proteins (alpha-actinin, Z protein, filamin, and amorphin) have been identified in the Z line. Alpha-actinin binds the actin filaments to the Z line. A large filamentous accessory protein called *nebulin* extends from the Z line and lies in close apposition to the actin filaments. Nebulin is thought to organize the actin filaments of the sarcomere and maintain them in three-dimensional space.

The A band consists chiefly of thick filaments with slender cross-connections at their midpoints, which give rise to the M line. A protein, myomesin, in the region of the M line holds the myosin filaments in register. Titin, a large accessory protein with elastic properties, links the myosin filaments to the Z line and helps maintain their position within the sarcomere. Actin filaments extend into the A bands between the thick filaments; the extent to which they penetrate determines the width of the H band (which consists only of thick filaments) and depends on the degree of contraction of the muscle. At the ends of the A bands, where the thick and thin filaments interdigitate, the narrow space between filaments is traversed by cross-bridges formed by the heads of the myosin molecules.

The arrangement of the filaments also establishes the ultrastructural basis of the contractile mechanism. During contraction, the A band remains constant in length, the length of the I and H bands decreases, and the Z lines approach the ends of the A bands. The changes result from alterations in the relative positions of the thick and thin filaments. No change in the lengths of the filaments is involved; the actin filaments slide past the myosin filaments to penetrate more deeply into the A band. As a result, the I and H bands become shorter. The Z lines are drawn closer to the ends of the A bands, thereby decreasing the length of each sarcomere (the distance between successive Z lines) to produce an overall shortening of the myofibrils (Fig. 7-10).

Fig. 7-8. Details of a sarcomere from skeletal muscle (TEM, × 49,000).

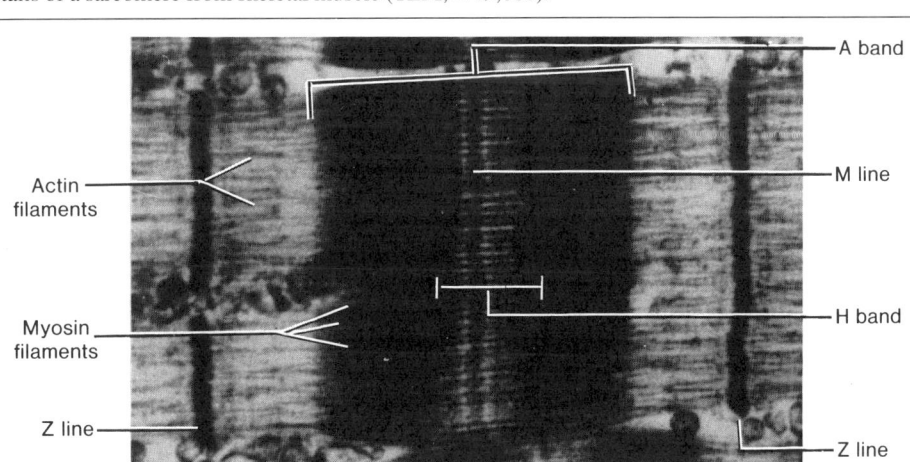

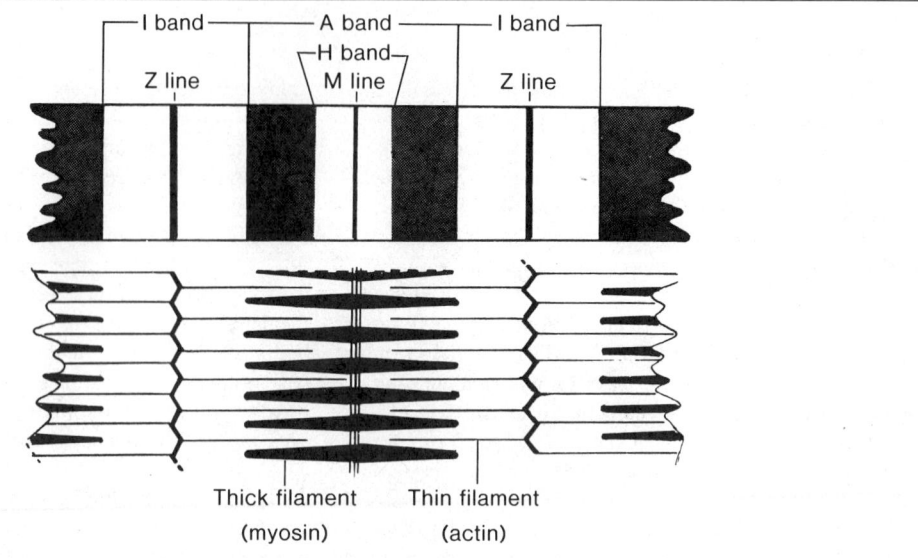

Fig. 7-9. Relationship of cross-banding to the arrangement of thin and thick filaments.

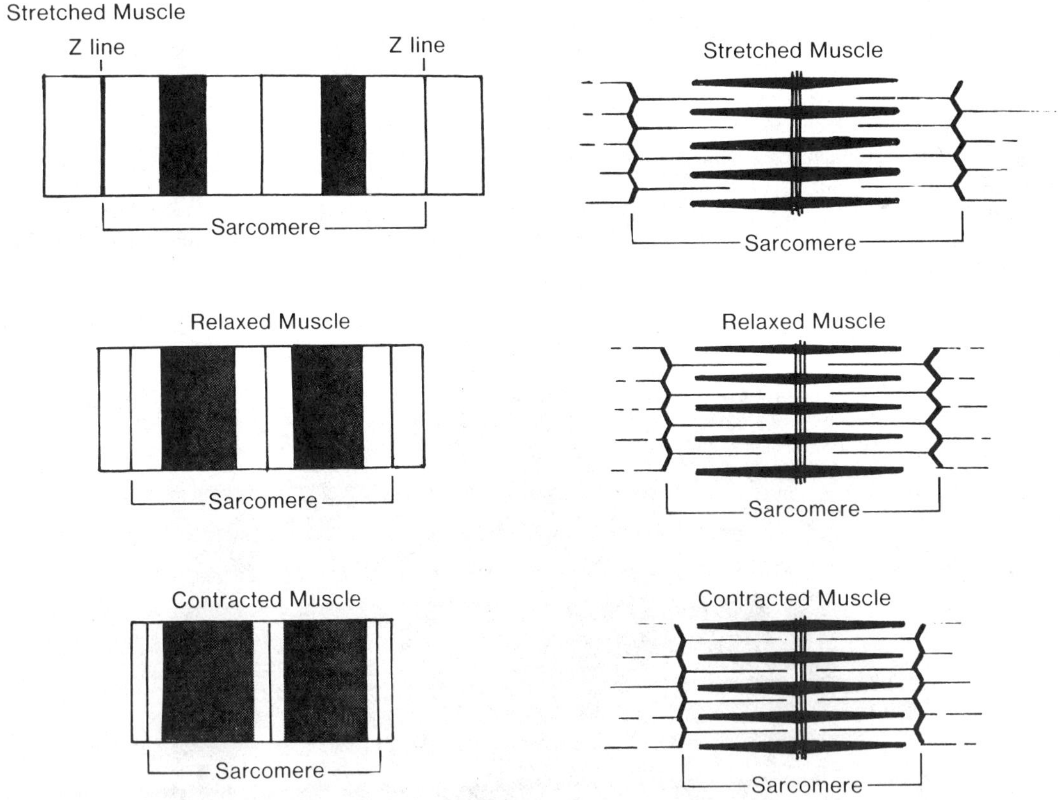

Stretched Muscle

Z line   Z line

—————— Sarcomere ——————

Stretched Muscle

—————— Sarcomere ——————

Relaxed Muscle

—————— Sarcomere ——————

Relaxed Muscle

—————— Sarcomere ——————

Contracted Muscle

—— Sarcomere ——

Contracted Muscle

—————— Sarcomere ——————

Fig. 7-10. Banding pattern and arrangement of thick and thin filaments in stretched, relaxed, and contracted muscle.

124

The sliding action of the filaments results from repeated "make and break" attachments between the heads of the myosin molecules and the neighboring actin filaments. The attachments are made at sites progressively further along the actin filaments, causing the filaments to slide past one another. The force resulting in movement of the filaments appears to be generated as the heads of the myosin change their angle of attachment to the actin filaments.

To translate the contraction of the sarcomere to contraction of an entire fiber, each sarcomere in the fiber must contract simultaneously and not in sequence. This simultaneous contraction — not only in a single fiber but throughout an entire muscle — is brought about by the T-system.

Fiber Types in Skeletal Muscle

KEY WORDS: type I, red fiber, slow twitch, type IIA, aerobic and anaerobic, type IIB, white, anaerobic, fast twitch, type IIC

Some skeletal muscles appear redder than others, reflecting the type of fibers present. Muscle fibers have been classed as red, white, and intermediate from their gross appearance or as type I, IIA, IIB, and IIC on the basis of their histochemical reactions.

Type I fibers are **red fibers.** These contain abundant myoglobin (a pigment similar to hemoglobin) and fat and are surrounded by many capillaries. They have numerous small mitochondria, are rich in oxidative enzymes, are low in myophosphorylase and glycogen, and metabolically are aerobic. The mitochondria are aggregated beneath the sarcolemma and form rows between myofibrils; they possess many closely packed cristae. The Z line is wide. The fibers do not stain when reacted for adenosinetriphosphatase (ATPase) at pH 9.4 but show "reversal" and stain well (dark) when reacted at pH 4.6 or 4.3. Physiologically, type I fibers represent **slow twitch** fibers adapted to slow, repetitive contractions and are capable of long, continued activity. Type I fibers tend to be located deep in a muscle, closer to the bone.

Type IIA fibers also are red due to abundant myoglobin but contain less fat than type I fibers. They contain much glycogen and show both **aerobic and anaerobic** metabolism. Oxidative enzymes are present, but their activity is slightly lower than in type I fibers. Myophosphatase activity is high, mitochondria are large but few in number, and Z lines are wide. ATPase activity is high (the fibers stain darkly) at pH 9.4 but is poor at pH 4.6 or 4.3.

Type IIB fibers are **white** in color and contain moderate amounts of glycogen but little myoglobin or fat. They are distinguished by their large diameters, poorer blood sup-

ply, decreased number of large mitochondria, and narrow Z lines. The fibers are rich in myophosphorylase but poor in oxidative enzymes. Metabolically, they are **anaerobic.** After preincubation at pH 9.4, type IIB fibers show high ATPase activity but fail to stain at pH 4.3 and show only moderate reactivity at pH 4.6 (Fig. 7-11). Type IIB fibers represent **fast twitch** fibers adapted for rapid, short-lived activity. In comparison with types I and IIA fibers, type IIB fibers tire quickly.

Type IIC fibers also have been described. They appear to be similar to type IIB fibers except that their ATPase activity is inhibited only at a pH lower than 4.0. They have narrow Z lines and correspond to the intermediate fiber type. Type IIC fibers represent a fast twitch muscle fiber that is more resistant to fatigue than the type IIB fiber.

Type IID fibers also have been characterized.

Motor Nerve Endings

KEY WORDS: motor unit, motor end-plate, myoneural junction, synaptic trough, primary synaptic cleft, secondary synaptic cleft, junctional fold, synaptic vesicle

Each skeletal muscle is supplied by one or more nerves that contain both motor and sensory fibers. A motor neuron and the muscle fibers supplied by it make up a **motor unit.** The size of the motor unit is determined by the number of muscle fibers served by a single nerve fiber and varies between muscles. Small motor units occur where precise control of muscle activity is required; large units are found in muscles used for coarser activities. In the intrinsic muscles of the eye, every muscle fiber is supplied by a nerve (forming small motor units), whereas in the major muscles of the limb, a single nerve fiber may supply 1,500 to 2,000 muscle fibers — a large motor unit. All the muscle fibers contained in one motor unit appear to be of the same histochemical type; thus there are type I and type II motor units.

Branches of a motor nerve end on the muscle fiber at a specialized region called the **motor end-plate** or **myoneural junction** (Fig. 7-12). The junction appears as a slightly elevated plaque on the muscle fiber associated with an accumulation of nuclei. As the nerve approaches the end-plate it loses its myelin sheath and ends in a number of bulbous expansions, the outer surfaces of which are covered by a continuation of the Schwann's cells. The nerve endings lie in shallow recesses, the **synaptic troughs (primary synaptic clefts),** on the surface of the muscle fiber. The sarcolemma that lines these clefts invaginates to form numerous **secondary synaptic clefts** or **junctional folds.** Although the nerve and muscle fibers are intimately

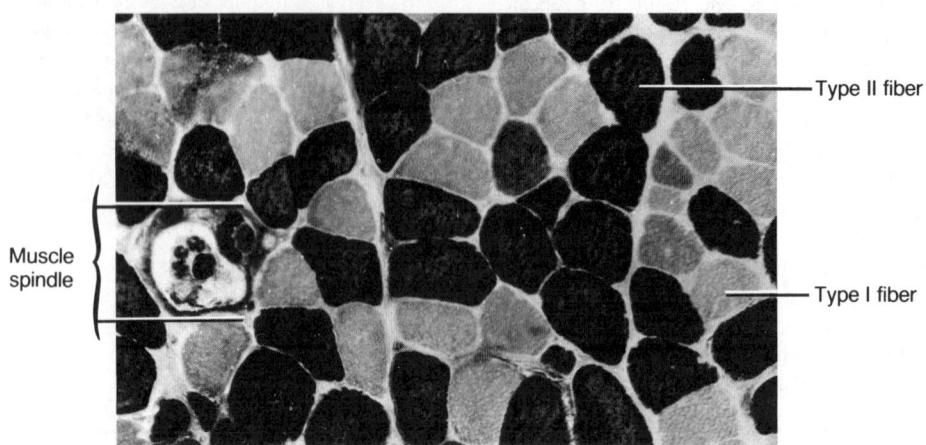

Fig. 7-11. A cross section of skeletal muscle stained with adenosinetriphosphatase at pH 10.4 demonstrates type I and type II skeletal muscle cells (fibers). Note the muscle spindle (LM, × 250).

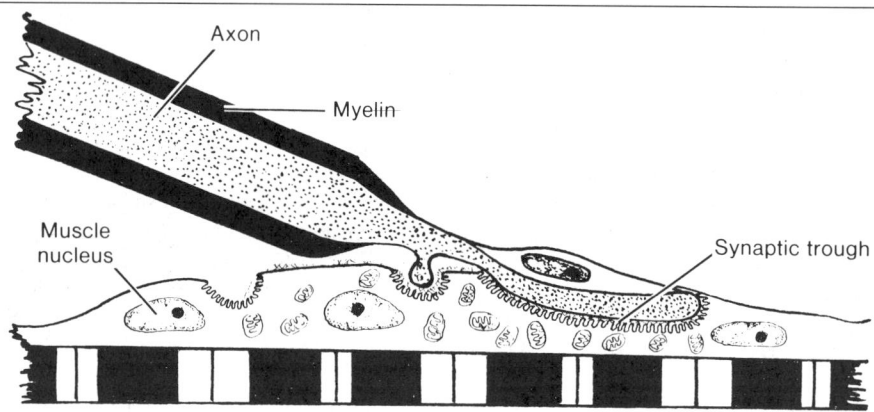

Fig. 7-12. Motor end-plate.

related at the junctions, their cell membranes remain separated by a glycoprotein layer derived from fusion of the basal laminae of the muscle fiber, nerve fiber, and Schwann's cells. This layer extends along the primary synaptic cleft and dips into each secondary cleft. Numerous large, vesicular nuclei and abundant mitochondria, ribosomes, and granular endoplasmic reticulum are present in the sarcoplasm that underlies the nerve terminal. The nerve terminals contain many mitochondria and a large number of **synaptic vesicles** that contain acetylcholine, a neurotransmitter agent.

A nerve impulse results in depolarization of the axon terminal, which permits an influx of calcium ions. This event triggers the release of acetylcholine into the synaptic cleft. The synaptic vesicles release acetylcholine by exocytosis at specific sites in the presynaptic membrane called *active zones*. Acetylcholine receptors are present in the sarcolemma on the outer edges of the secondary clefts, while acetylcholinesterase is located along the inner surface deep within the secondary clefts and in the basal lamina of this area. The acetylcholine receptors contain ion channels made up of a complex of five similar subunits which when bound to acetylcholine undergo a conformational change opening the channel and permitting the entry of sodium ions. This influx results in a depolarization of the sarcolemma, and the action potential created travels over the sarcolemma to the T-tubule system and initiates contraction.

Sensory Nerve Endings

KEY WORDS: muscle spindle, intrafusal fiber, extrafusal fiber, nuclear bag fiber, nuclear chain fiber, annulospiral ending, flower spray ending, motor end-plate, trail ending

Skeletal muscle contains many sensory endings, some of which are simple spiral terminations of nonmyelinated nerves that wrap about the muscle fibers and others that are highly organized structures called **muscle spindles** or neuromuscular spindles (Fig. 7-13). Each muscle spindle consists of an elongated, ovoid connective tissue capsule that encloses a few thin, modified skeletal muscle fibers associated with sensory and motor nerves. The modified muscle fibers traverse the capsule from end to end and form the **intrafusal fibers** to distinguish them from the ordinary muscle fibers (**extrafusal fibers**) outside the capsule (Fig. 7-14).

Two distinct types of intrafusal fibers are present. Both are striated over most of their lengths, but midway along the fibers the striations are replaced by aggregates of nuclei. One type of intrafusal fiber is larger, and its nonstriated region is occupied by a cluster of nuclei that produces a slight expansion of the fibers in this region; these form the **nuclear bag fibers**. The second type of fiber is thinner,

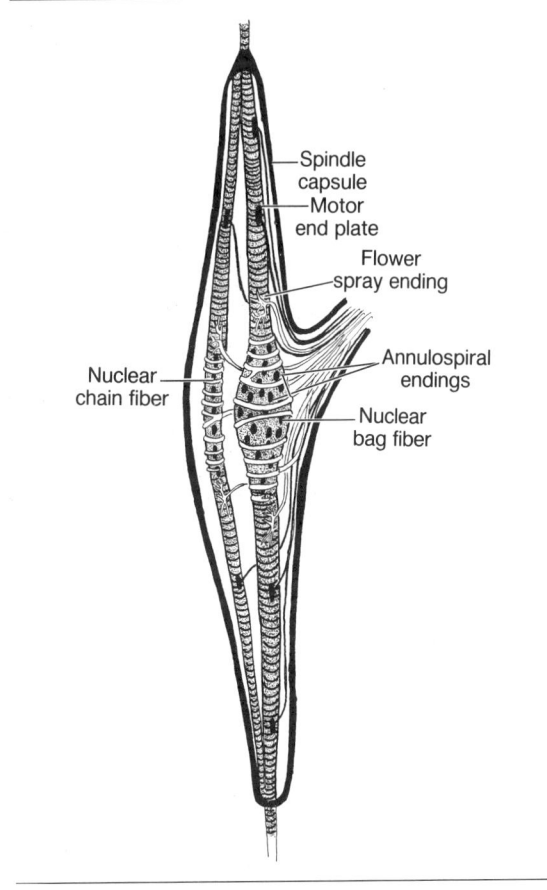

Fig. 7-13. Muscle spindle.

Fig. 7-14. A cross section of a muscle spindle. Compare the diameters of extrafusal and intrafusal fibers (LM, × 400).

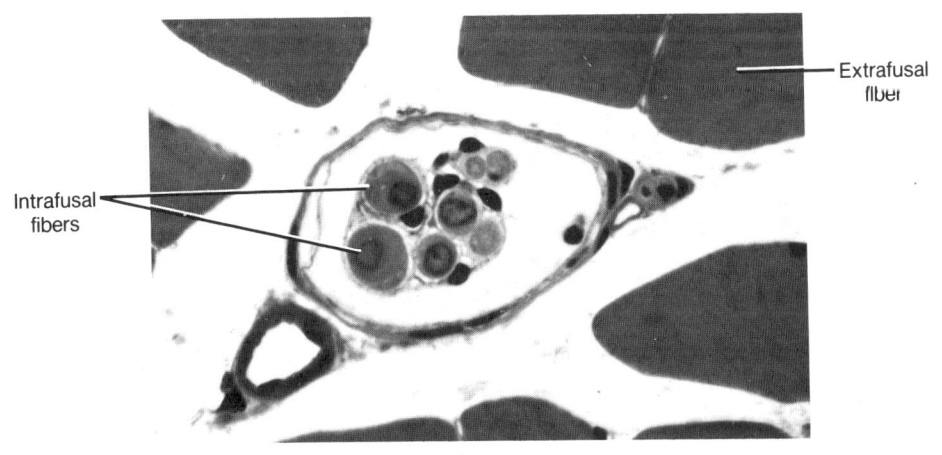

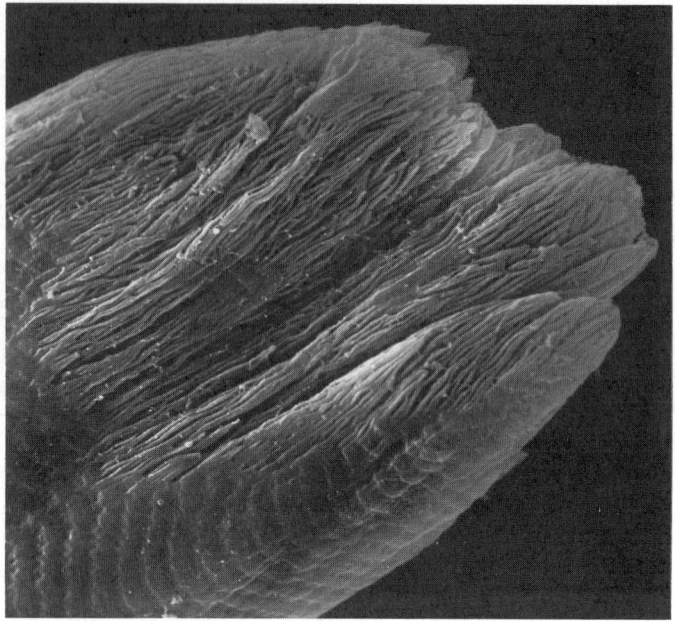

Fig. 7-15. The tapering end of a skeletal muscle fiber at the muscle-tendon junction. The junctional surface is characterized by extensive microvillus-like foldings. The latter greatly increases the surface area for union with connective tissue elements (SEM, × 400).

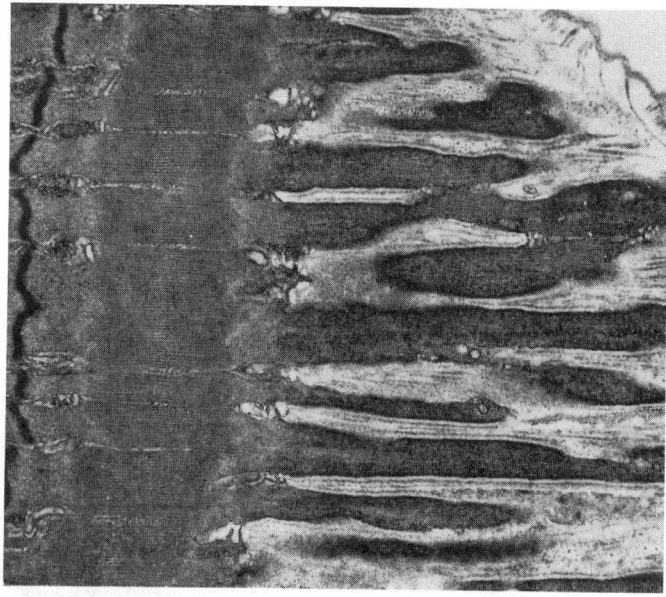

Fig. 7-16. A region of the muscle-tendon junction similar to that shown in 7-15. Note the infolding of the cell membrane and the associated external lamina (TEM, × 20,000).

and the nuclei in the central region form a single row, hence their name, **nuclear chain fiber.** Nuclear bag fibers extend beyond the capsule to attach to the endomysium of nearby extrafusal fibers. Nuclear chain fibers are the more numerous and shorter. They attach to the capsule at the poles of the spindle or to the sheaths of nuclear bag fibers. Since the central areas of the intrafusal fibers lack myofibrils, these regions are noncontractile.

Each spindle receives a single, thick sensory nerve fiber that gives off several nonmyelinated branches that end in a complex system of rings and spirals at the central regions of the intrafusal fibers. These **annulospiral endings** run beneath the basal lamina in grooves in the sarcolemma. **Flower spray endings** are cluster- or spraylike terminations of smaller sensory nerves that occur mainly on nuclear chain fibers. Intrafusal fibers also receive small motor nerve fibers that end either as **motor end-plates** or as long, diffuse **trail endings** that ramify over the intrafusal fibers, making several contacts with them.

Muscle spindles serve as stretch receptors and are found mainly in slow-contracting extensor muscles involved in maintaining posture. They are present also in muscles used for fine movements. Stimulation by motor nerves maintains tension on the intrafusal fibers, resulting in a stretching of their nonstriated segments. The degree of stretch is sufficient to maintain the sensory nerve endings in this region in a state of excitation that is close to their thresholds. When a whole muscle is stretched, it causes increased tension on the intrafusal fibers and stretching of their nonstriated regions, resulting in the discharge of impulses by the sensory nerve. When the muscle contracts, tension on the spindle is reduced and sensory nerve endings cease firing. The frequency of the discharges by the nerve endings is proportional to the tension on the intrafusal fibers; together with the number of active spindles, a degree of muscle contraction appropriate to the stimulus is ensured.

Tendon and Tendon-Muscle Junctions

KEY WORDS: collagen bundle, endotendineum, fascicle, peritendineum, epitendineum, fibroblast, tendon organ

Tendons consist of thick, closely packed **collagen bundles.** Surrounding each bundle is a small amount of loose, fibroelastic connective tissue, the **endotendineum.** Variable numbers of collagen bundles are collected into poorly defined **fascicles** wrapped in a somewhat coarser connective tissue called the **peritendineum.** Groups of fascicles form the tendon itself, which is wrapped in a thick layer of dense irregular connective tissue, the **epitendineum.** The only cells present in the tendon are **fibroblasts** arranged in

columns between the collagen fibers. Tendons are examples of dense regular connective tissue.

Where muscle and tendon join, the connective tissue of the endomysium, perimysium, and epimysium becomes strongly fibrous and blends with the connective tissue of the tendon. On the muscle side of the junction, the connective tissue fibers extend into indentations of the sarcolemma and are firmly attached to the external lamina, to which the sarcolemma also attaches (Figs. 7-15 and 7-16). Within the muscle fiber, the actin filaments of the last sarcomere are anchored to the sarcolemma at the end of the fiber. Thus contraction of the muscle fiber is passed to the sarcolemma, the external lamina, and, by way of the connective tissue sheaths, to the tendon.

Tendon organs are encapsulated sensory receptors found at tendon-muscle junctions. The organ consists of a capsule of dense irregular connective tissue traversed by specialized collagen bundles that are continuous with collagen fibers outside the capsule. A single large sensory nerve pierces the capsule and gives rise to several nonmyelinated branches whose terminations entwine among the collagen bundles. The tendon organ senses stresses produced by muscle contractions, preventing them from becoming excessive.

Cardiac Muscle

Cardiac muscle occurs only in the heart, where it forms the muscle wall (myocardium). It resembles both smooth and skeletal muscle; both skeletal and cardiac muscle are striated, but like smooth muscle, cardiac muscle is involuntary. However, cardiac muscle is unlike either in many of its structural details and is unique in that its contractions are automatic and spontaneous, requiring no external stimulus.

Organization

KEY WORDS: branching fiber, endomysium, perimysium

As in skeletal muscle, the histologic unit of structure of cardiac muscle is the cell (fiber), but in cardiac muscle the cells are associated end to end to form long tracts, each about 80 μm long by 50 μm wide. The cells may divide at their ends before joining to adjacent fibers and thus form a network of **branching fibers.** A web of collagenous and reticular fibers is present between the muscle fibers and corresponds to an **endomysium,** but because of the branching fibers, it is more irregular than in skeletal muscle. Large bundles of fibers are wrapped in a coarser connective tissue of collagen and reticular fibers, forming a **perimysium,** as in skeletal muscle.

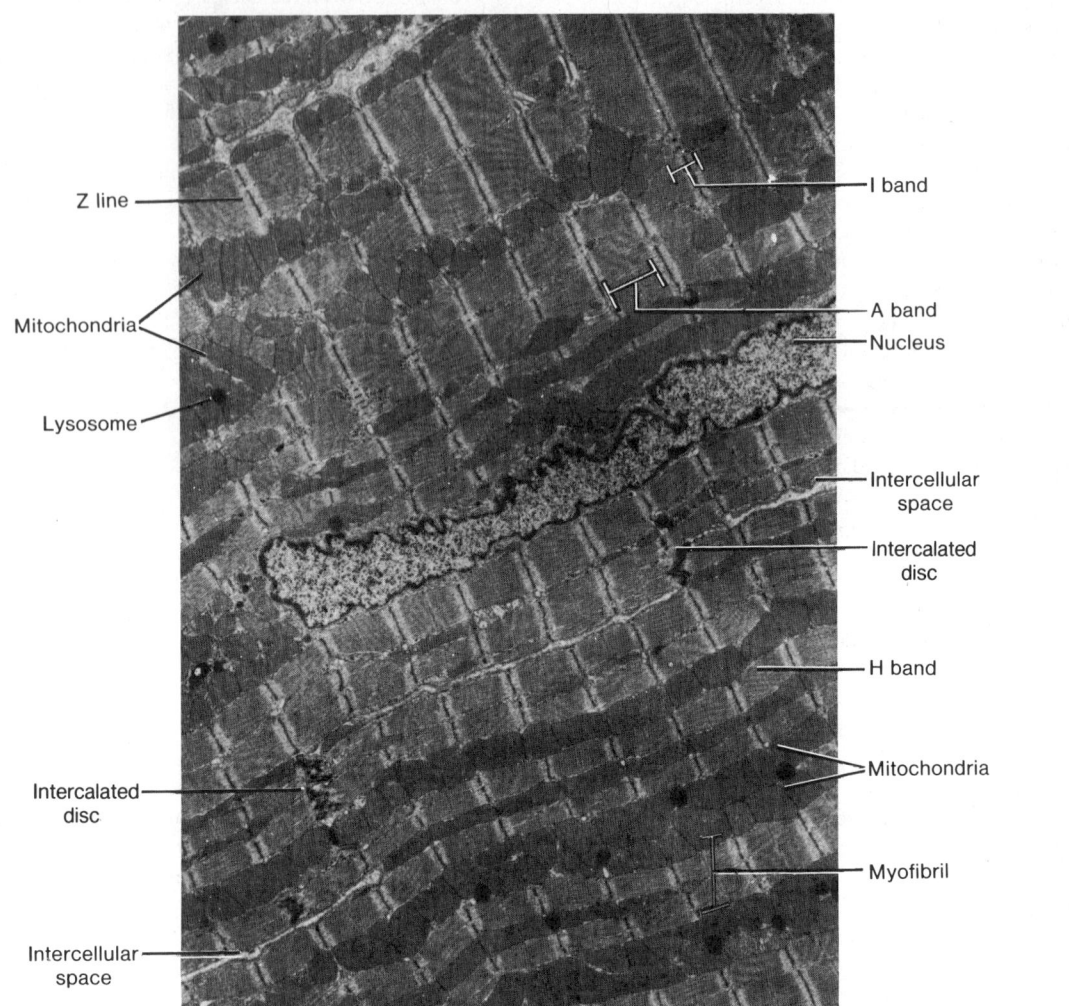

Z line

Mitochondria

Lysosome

I band

A band

Nucleus

Intercellular
space

Intercalated
disc

H band

Mitochondria

Intercalated
disc

Myofibril

Intercellular
space

Fig. 7-17. Two adjacent cardiac muscle cells illustrating individual banded myofibrils and portions of intercalated discs. Note the abundance of mitochondria between myofibrils (TEM, × 6,000).

Fibers

KEY WORDS: sarcolemma; central nuclei; A, I, M, H, and Z bands; actin and myosin filaments; sarcoplasmic reticulum; T-tubule; diad; Z line; peripheral coupling; subsarcolemmal cisternae

Each cell (fiber) is enclosed in a **sarcolemma** external to which is an external lamina associated with fine reticular fibers. Each fiber contains one or two **central nuclei.** The same banding pattern seen in skeletal muscle is present on cardiac fibers, and **A, I, M, H,** and **Z bands** can be distinguished but are not as conspicuous as in skeletal muscle (Fig. 7-17). The banding pattern is due to the arrangement of **actin and myosin filaments,** but the aggregation of filaments into myofibrils is not well defined, and bundles of myofibrils often become confluent with those of adjacent bundles or are separated only by rows of mitochondria. The bundles of myofibrils diverge around the nuclei to leave a fusiform area of sarcoplasm at each pole. These areas are occupied by small Golgi elements and numerous large mitochondria that have closely packed cristae. Glycogen is a prominent feature of cardiac muscle.

The **sarcoplasmic reticulum** is neither as extensive nor as well developed as that of skeletal muscle. The sarcotubules are continuous over the length of the sarcomere and anastomose freely to give a plexiform pattern with no special anastomosis at the H band. There are no terminal cisternae, and the reticulum makes contact with the **T-tubules** via irregular expansions of the sarcotubules. Since most T-tubules are apposed to only one cisterna at any point, the couplings of T-tubule to sarcoplasmic reticulum

have been called **diads** (Fig. 7-18). The T-tubules of cardiac muscle are located at **Z lines** rather than at A–I junctions, and their lumina are much wider than in the T-tubules of skeletal muscle. The luminal surfaces of the cardiac T-tubules are coated by a continuation of the external lamina of the sarcolemma. Sarcotubules at the periphery of the cell may be closely applied to the sarcolemma, and these couplings have been called **peripheral couplings** or **subsarcolemmal cisternae.** As in skeletal muscle, the functional unit of cardiac muscle is the sarcomere, demarcated by two successive Z lines.

Intercalated Discs

KEY WORDS: Z line, desmosome, fascia adherens, nexus (gap) junction

At their end-to-end associations, cardiac muscle cells are united by special junctions that are visible under the light microscope as dark lines. These are the intercalated discs that cross the fibers in stepwise fashion at the level of **Z lines** and thus have transverse and horizontal parts. Along the transverse parts, the opposing cells are extensively interdigitated and at points are united by **desmosomes.** For the most part, the junction between cells is more extensive than that of a desmosome and thus is called a **fascia adherens** (Fig. 7-19). The sarcoplasm immediately adjacent to the cytoplasmic surface of the cell membrane contains a dense mat of filaments into which the actin filaments insert. The fascia adherens contains the actin-binding proteins alpha-actinin and vinculin, which function to anchor actin and intermediate filaments to the plasmalemma. At

Fig. 7-18. Diagrammatic representation of the sarcoplasmic reticulum and system of T-tubules in cardiac muscle.

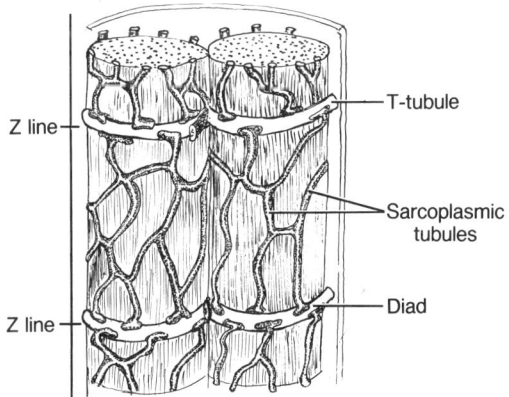

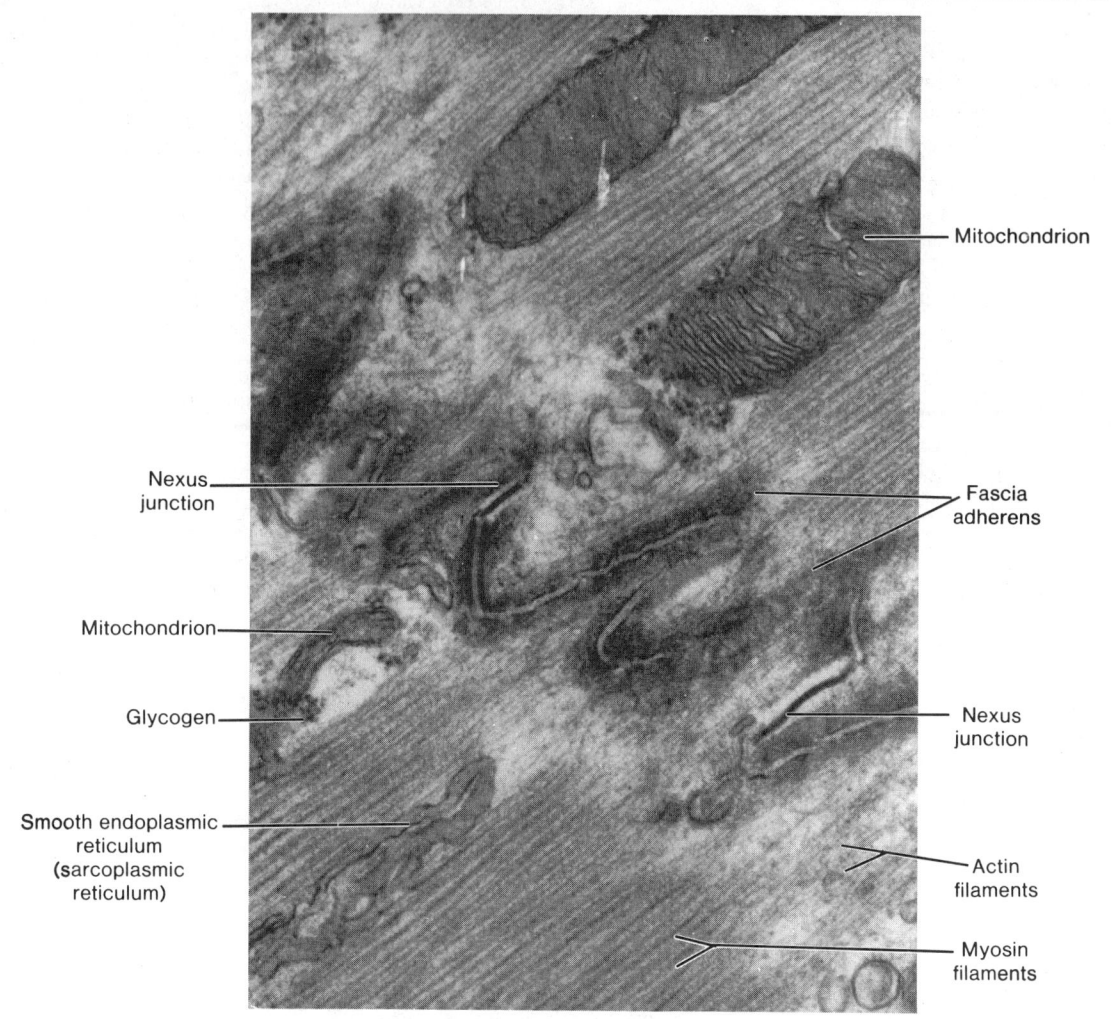

Fig. 7-19. An intercalated disc uniting ends of two cardiac muscle cells (TEM, × 52,800).

irregular sites along the junctions, the membranes of opposing cells are united by **nexus (gap) junctions.** Transmembrane glycoproteins A-CA and plakoglobin function as cell-adhesion molecules and tightly unite cardiac muscle cells end to end at the intercalated disc.

The longitudinal part of the intercalated disc is continuous with the transverse portion and also shows gap junctions that, however, are much more extensive. The gap junctions represent areas of low electrical resistance and allow rapid spread of excitation impulses from cell to cell.

Smooth Muscle

Smooth muscle is widely distributed throughout the body and plays an essential role in the functions of organs. It forms the contractile portion of the walls of blood vessels and of hollow viscera such as the digestive, respiratory, and reproductive tracts. Smooth muscle also is present in the skin, where it forms small muscles attached to hair follicles, and in the iris and ciliary body of the eye, the erectile tissue of the penis and clitoris, and the stroma of the ovary and prostate.

Smooth Muscle Fibers

KEY WORDS: cell, single central nucleus, myofibril, filaments, dense body, basal (external) lamina, gap junction (nexus), attachment plaque

Each smooth muscle fiber is an elongated **cell** with an expanded central region and tapering ends. The fibers vary in length in different organs, from 20 μm in small blood vessels to 500 to 600 μm in the pregnant uterus. A **single central nucleus** occupies the wide portion of the fiber about midway along its length, elongated in the long axis of the fiber (Fig. 7-20). In a contracted fiber, the nucleus has a wrinkled or pleated outline. The fibers lack cross-striations and in the usual histologic preparations appear homogeneous. However, longitudinal striations can be seen after maceration of the fibers in acid; these striations represent the **myofibrils.**

In electron micrographs, the sarcoplasm in the region of the nucleus shows long, slender mitochondria, a few tubules of granular endoplasmic reticulum, clusters of free ribosomes, and a small Golgi body at one pole of the nucleus. Unlike cardiac and skeletal muscle, only a rudimentary sarcoplasmic reticulum is present and a system of T-tubules is absent. The bulk of the cytoplasm contains closely packed, fine **filaments** arranged in bundles that run in the long axis of the fiber. Mitochondria and glycogen granules are interspersed between the myofilaments. Scattered throughout the sarcoplasm are a number of oval **dense bodies** into which the myofilaments (actin) appear to insert (Fig. 7-21).

The cytoplasmic dense bodies contain alpha-actinin, an actin-binding protein. Desmin, the most abundant intermediate filament, and vimentin also insert into these anchoring points.

The sarcolemma is covered externally by a thick **basal** or **external lamina** consisting of proteoglycan associated with numerous fine collagenous and reticular fibers that blend with the surrounding connective tissue. In some regions the basal lamina is lacking, and the cell membranes of adjacent smooth muscle cells are closely apposed in **gap junctions (nexuses)** through which excitation impulses spread from one fiber to another. Areas of increased density, similar to the dense bodies, occur at intervals along the inner aspect of the sarcolemma, becoming more numerous along the ends of the fibers. These dense regions appear to be sites of attachment of myofilaments and intermediate filaments to the cell membrane and have been called **attachment plaques.** The subplasmalemmal attachment plaques have been shown to contain the actin-binding proteins vinculin and talin. Between these areas the sarcolemma may show numerous caveolae that form as a result of invaginations of the cell membrane.

Myofilaments

KEY WORDS: thick filament, thin filament, actin, myosin

Thick and **thin filaments** have been demonstrated in smooth muscle by electron microscopy, and **actin** and **myosin**

Fig. 7-20. Smooth muscle cells from the wall of the oviduct as viewed in longitudinal and cross section (LM, × 400).

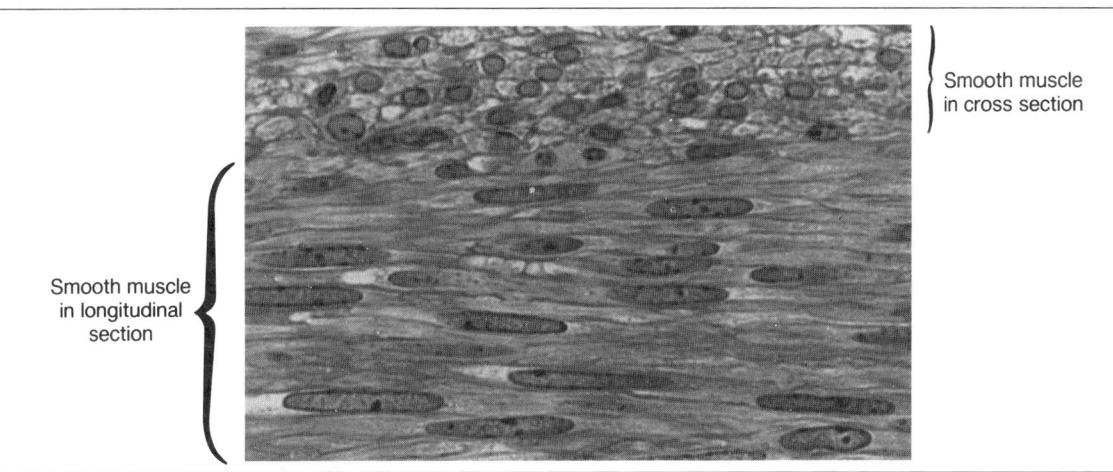

Smooth muscle in cross section

Smooth muscle in longitudinal section

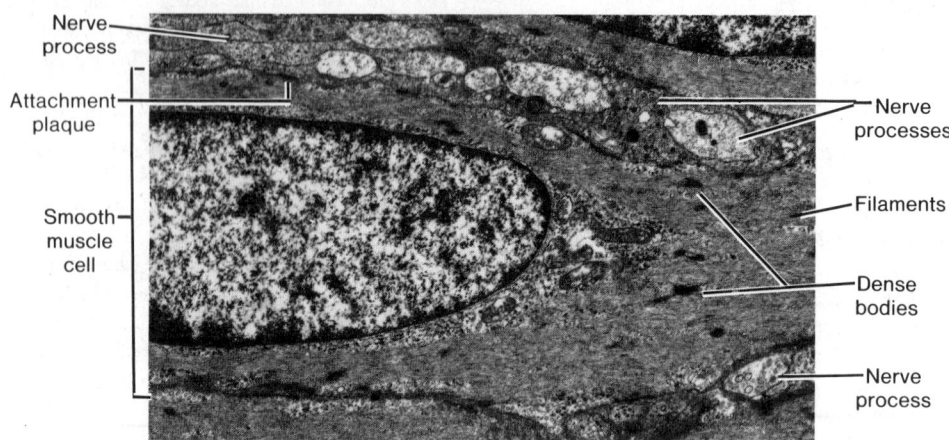

Fig. 7-21. The nuclear region of a smooth muscle cell from a human stomach (longitudinal profile) (TEM, × 15,000).

have been identified biochemically. However, myosin filaments have been difficult to identify in standard electron microscopic preparations, and it has been suggested that in smooth muscle, myosin is labile, aggregating into filaments only on initiation of contraction. Filamentous myosin has been shown in unfixed, rapidly frozen smooth muscle where regular arrays of thick myosin filaments have been seen surrounded by rosettes of actin. As in striated muscle, release of calcium ion in smooth muscle initiates contraction. In smooth muscle calcium ion is complexed to calmodulin (a calcium-binding protein), and this complex activates myosin light-chain kinase, an enzyme necessary for the phosphorylation of myosin, and permits it to bind to actin. Contraction then occurs. Myosin and actin of smooth muscle interact as in striated muscle, and a sliding filament mechanism appears to account for contraction of smooth muscle also.

Contraction of smooth muscle cells is associated with formation of blebs of the plasmalemma; the blebs are lost as the cells relax. The evaginations increase in size and number as the cell continues to shorten, and when the cell has reached 55 percent of its initial length, it is completely covered by blebs. In electron micrographs, the blebs are free of myofilaments, but filaments are prominent in the nonevaginated areas between attachment plaques.

In the contracted cell, the thick and thin filaments and the cytoplasmic dense bodies are oriented obliquely to the long axis, criss-crossing the cell, whereas in the relaxed cell these components generally are parallel to the cell

axis. The contractile elements unite with the attachment plaques and the dense bodies. During contraction, the contractile elements become obliquely oriented. As the cell shortens, its width must increase to accommodate the displaced volume, but this is opposed by the contraction pulling inward at the site of the attachment plaques. Those areas not directly subjected to the inward force bulge outward as blebs. The filaments between the dense bodies represent a contractile unit, several being strung together and anchored at either end in an attachment plaque. In addition to the actin myofilaments, the dense bodies and subplasmalemmal attachment plaques serve as attachment sites for the intermediate filaments (desmin, vimentin), which provide strong cytoskeletal support for the smooth muscle cell.

Organization

Smooth muscle cells may be present as isolated units or small bundles in ordinary connective tissue, as in intestinal villi, the tunica dartos of the scrotum, and the capsule and trabeculae of the spleen. They also form prominent sheets in the walls of the intestines and arteries. In any one sheet of muscle, the fibers tend to be oriented in the same direction but are offset so that the thick portions of the fibers lie adjacent to the tapering ends of neighboring fibers. Thus, in transverse sections, the outlines of the fibers vary in diameter according to where along their lengths the fibers were cut. Nuclei are few and present only in the largest profiles, which represent sections through the expanded areas of the cells.

A thin connective tissue with few fibroblasts and consisting of fine collagenous, reticular, and elastic fibers extends between the smooth muscle cells and becomes continuous with a more dense connective tissue that binds the muscle cells into bundles or sheets. The latter connective tissue contains more abundant fibroblasts, collagenous and elastic fibers, and a network of blood vessels and nerves. The traction produced by contracting fibers is transmitted to the reticuloelastic sheath about the cells and then to the denser connective tissue, permitting a uniform contraction throughout the muscle sheet. The reticular and elastic fibers of the reticuloelastic sheath are products of the muscle fibers.

Review Table 7-1. Key Features of Muscle Types

Type	Cell Shape	Nuclei	Diameter	Striations	Other
Smooth	Small spindles	Central, single	Small with marked variation	Absent	Packed tightly, little connective tissue between fibers
Cardiac	Short branching, anastomosing cylinders	Central, usually single, may be double	Large, moderate variation	Present	Intercalated discs
Skeletal	Long cylinders	Peripheral, multiple	Large, uniform	Present	

Functional Review: Muscle Tissue

Nerve impulses reaching the myoneural junctions of skeletal muscles cause release of acetylcholine contained within the synaptic vesicles. This neurotransmitter diffuses across the synaptic trough, the surface area of which is greatly increased by the junctional folds. Acetylcholine causes depolarization of the sarcolemma at the end-plate, from which an excitation wave sweeps over the surface of the muscle fiber and, by means of the T-system, is delivered to all the myofibrils throughout the fiber. At the triads, the impulse is passed to the sarcoplasmic reticulum, causing release of calcium ions from the terminal cisternae. Free calcium ions activate the myosin-actin interaction through the intervention of tropomyosin and troponin. Troponin is the calcium receptor that, in the presence of increased calcium ion, is thought to undergo structural changes, bringing about the interaction of myosin and actin. The energy for the cross-bridging between myosin and actin is obtained from the breakdown of adenosine triphosphate (ATP) by ATPase contained in the myosin filaments. As the actin filaments slide past the myosin filaments to penetrate more deeply between them, the length of each sarcomere is reduced, resulting in an overall shortening of each myofibril and thus of each fiber. The tension generated is transmitted in succession to the sarcolemma, the external lamina, the connective tissue sheath, and then a tendon or other muscle attachment. Muscle spindles serve as stretch receptors and coordinate the degree of muscle contraction with the strength of the stimulus. Tendon organs sense the stresses produced by the muscle contractions and prevent them from becoming excessive.

The function of skeletal muscle depends on the precise alignment of actin and myosin filaments within each myofibril. A complex skeleton of accessory proteins (myomesin, alpha-actin, nebulin, titin) supports the myofilaments, maintains their alignment, and holds them in register to each other. Intermediate filaments of the cytoskeleton (desmin) link adjacent myofibrils and maintain their register as well as linking them to the lateral aspects of the sarcolemma. Individual skeletal muscle fibers are organized and harnessed by several extracellular components (external lamina, endomysium, perimysium, epimysium) to complete the organization of each cell into a muscle.

The duration of the excitation wave at the myoneural junction is limited by the rapid breakdown of acetylcholine by the enzyme acetylcholinesterase, which is associated with the basal lamina of the junctional folds. With the end of the excitation wave, calcium is rebound in the sarcoplasmic reticulum, and ATPase activity and actin-myosin interactions cease, resulting in relaxation of the muscle.

Skeletal muscle fibers have limited capacity to regenerate but can repair minor injuries. It is thought that repair occurs through activation of satellite cells. These cells, which separate from the parent muscle fiber, divide and fuse with other satellite cells to form myotubules and new fibers that bridge small defects in the parent fiber. Adult muscle increases in mass by increase in the size of existing fibers, not by increase in the number of fibers. A shift occurs from the small- and medium-sized fibers of normal muscle to the large fiber groups, but few of the hypertrophied muscle fibers become larger than the largest normal fibers. Physiologic hypertrophy at first consists of an increase in the amount of sarcoplasm followed by an increase in the number of myofibrils.

Contraction of cardiac muscle occurs spontaneously, with no need for an external stimulus. However, the mechanism of contraction is identical to that of skeletal muscle and involves interaction of actin and myosin filaments. Cardiac muscle depends more on calcium and shows a relatively slower contraction time than does skeletal muscle; these characteristics may relate to the relatively poorer development of the sarcoplasmic reticulum. The high content of mitochondria and glycogen is related to the constant rhythmic expenditure of energy typical of cardiac muscle. Because of the numerous gap junctions between cardiac muscle cells, the fibers of cardiac muscle have an inherent capacity to conduct impulses, and electrically, the cells act as though there were no membranes between them. Rapid distribution of impulses throughout the cardiac muscle is aided by the many branchings that provide unions between adjacent fibers.

Thick and thin myofilaments have been identified in smooth muscle cells, but a sarcomere organization has not been demonstrated clearly. Contraction of smooth muscle is slower than in striated muscle, but the energy cost of an equivalent force is much less. Contraction of smooth muscle is somewhat unique in that forceful contractions can be sustained for long periods without fatigue of the muscle. An example of such contractions is those of the uterine muscle during parturition. This type of contraction differs from that of skeletal muscles, which are unable to sustain contraction without becoming fatigued, and cardiac muscle, which shows inherent intermittent contractions. Absence of an organized sarcoplasmic reticulum is related to the slow contractions exhibited by smooth muscle.

Nexus junctions permit free passage of impulses, and, as in cardiac muscle, smooth muscle frequently acts electrically as though there were no cell boundaries. Thus, a few motor neurons can control large numbers of smooth muscle cells. The connective tissue between and around smooth muscle bundles transmits the forces generated by individual cells as a coordinated force throughout the entire tissue. Smooth muscle cells of different organs may show differences with regard to size, organization (individual cells, bundles, or sheets of cells), and response to different stimuli.

Review Questions and Answers

Questions

38. Briefly outline the arrangement of the connective tissue components of a skeletal muscle and relate it to the gross organization of a muscle. What function does the connective tissue serve?
39. Describe the banding pattern of striated muscle on the basis of the actin and myosin filaments.
40. What is the contractile unit of skeletal muscle? Describe it in terms of the banding pattern on the fiber.
41. Beginning with a whole muscle, outline the various levels of organization of its structural components.
42. What is the function of the sarcoplasmic reticulum? How does that of cardiac muscle differ from that of skeletal muscle?
43. How is an excitation stimulus transmitted from the sarcolemma throughout the muscle cell?
44. What is an intercalated disc?
45. Discuss the type (not mechanism) of contraction of smooth muscle and compare it to that of skeletal and cardiac muscle.

Answers

38. A whole muscle is wrapped in a connective tissue sheath called the epimysium, from which septa pass into the muscle to surround fascicles that consist of groups of muscle fibers. The connective tissue that surrounds fascicles is called the perimysium and in

turn provides slips of connective tissue that surround the individual muscle fibers as the endomysium. The connective tissue forms a continuous harness that binds the muscle units together and integrates the contractions of the individual cells of the muscle to provide continuous force on a tendon or other structures.

39. The light I bands of the muscle fiber consist only of the actin filaments, which extend in both directions from the Z line, offset from one another on each side. The H band is a pale central region of the A band and consists only of myosin fibers joined at their midpoints by slender cross-connections. These cross-connections form the M line, which transects the H band. On either side of the H band, the dark portions of the A band consist of myosin filaments between which actin filaments are interdigitated. The width of the H band and the I band is determined by the degree to which the actin filaments are inserted between the myosin filaments. The width of the A band remains constant.

40. The contractile unit of a skeletal muscle is the sarcomere, defined as the distance between two successive Z lines. The midportion of the sarcomere consists of an entire A band. At either end, it will consist of half of each I band immediately adjacent to the A band.

41. A whole skeletal muscle consists of a variable number of fascicles that in turn are formed by groups of fibers. Each fiber is a cell and contains within its sarcoplasm a number of threadlike structures called myofibrils. In turn, the myofibrils are composed of myofilaments, which are of two types: thick and thin. The thick myofilaments consist of the linearly arranged tails of myosin molecules from which extend the globular heads of the molecule. The thin filaments consist of actin and are composed of globular subunits that form two longitudinal strands.

42. The sarcoplasmic reticulum stores the calcium ions required for the activation of myosin-actin interactions. Upon stimulation, calcium ion is released by the sarcoplasmic reticulum and is recaptured following the termination of the stimulus. In skeletal muscle the sarcoplasmic reticulum is a more extensive organelle, and the terminal cisternae and T-tubules are located at the A–I junctions. The longitudinal sarcotubules form anastomoses at the level of the H bands and Z lines. In cardiac muscle the T-tubules are located at the levels of the Z lines, and terminal cisternae of the sarcoplasmic reticulum are lacking. The T-tubules and sarcoplasmic reticulum make contact at irregular intervals only via small expansions of the sarcotubules. The sarcoplasmic reticulum is not as extensive as in skeletal muscle, and the sarcotubules unite at random over the sarcomere with no special anastomosis at the H bands.

43. Impulses are transmitted from the sarcolemma by the system of T-tubules, which are invaginations of the sarcolemma. The T-tubules surround each myofibril and communicate with those of adjacent myofibrils, forming a complete network across the fiber and establishing a communication with the sarcolemma that extends throughout the whole fiber. The T-tubules are distinct from, and not part of, the endoplasmic reticulum.

44. Intercalated discs are the light microscopic representations of the end-to-end unions of cardiac muscle cells. They contain desmosomes and nexus (or gap) junctions.

45. Smooth muscle is capable of powerful contractions sustained for a considerable period of time without fatigue. An example of this is the contractions of the uterine musculature during parturition. Skeletal muscle, though capable of forceful contractions, is unable to sustain them for any length of time and soon becomes fatigued. The contractions of cardiac muscle, unlike either smooth or skeletal, are inherent and rhythmic and occur in the absence of external stimuli. However, cardiac muscle does not show sustained contractions, and the intervals between contractions represent periods of rest.

8 Nervous Tissue

Nervous tissue, one of the four basic tissues, consists of nerve cells (neurons) and supporting cells (neuroglia). Nerve cells are highly specialized to react to stimuli and conduct the excitation from one region of the body to another. Thus the nervous system shows both irritability and conductivity, properties that are essential to the functions of nervous tissue — to provide communication and to coordinate body activities.

The nervous tissue of the brain and spinal cord makes up the central nervous system; all other nervous tissue constitutes the peripheral nervous system. Nervous tissue specialized to perceive external stimuli is called a *receptor*, from which sensory stimuli are carried by the peripheral nervous system to the central nervous system, which acts as an integration and communications center. Other neurons, the *effectors,* conduct nerve impulses from the central nervous system to other tissues, where they elicit an effect.

The specialized contacts between neurons are called *synapses;* the impulses are transferred between nerve cells by electrical couplings or chemical transmitters. Some neurons of the brain secrete substances (hormones) directly into the bloodstream, and therefore, the brain may be considered a neuroendocrine organ. Neurons also can stimulate or inhibit other neurons they contact.

Neurons

KEY WORDS: perikaryon, dendrite, axon, neurofibril, neurofilament, neurotubule, chromophilic (Nissl) substance, axon hillock, melanin pigment, lipofuscin granule

The nerve cell, or neuron, is the structural and functional unit of nervous tissue. Usually large and complex in shape, it consists of a cell body, the **perikaryon,** and several cytoplasmic processes. **Dendrites** are processes that conduct impulses *to* the perikaryon and usually are multiple. The single process that carries the impulse *from* the perikaryon is the **axon,** or axis cylinder.

The perikarya of different types of neurons vary markedly in size (4 to 140 μm) and shape and usually contain a large, central, spherical nucleus with a prominent nucleolus. The cytoplasm is rich in organelles and may contain a variety of inclusions. Bundles of **neurofibrils** form an anastomosing network around the nucleus and extend into the dendrites and axon (Fig. 8-1). Neurofibrils must be stained selectively to be seen with the light microscope. Ultrastructurally, neurofibrils consist of aggregates of slender **neurofilaments,** 10 nm in diameter, and microtubules that, although called **neurotubules,** appear to be identical to microtubules found in other cells. Neurofila-

Fig. 8-1 Multipolar neurons from the ventral horn of a human spinal cord. Note the abundance of surrounding nerve processes (LM, × 400).

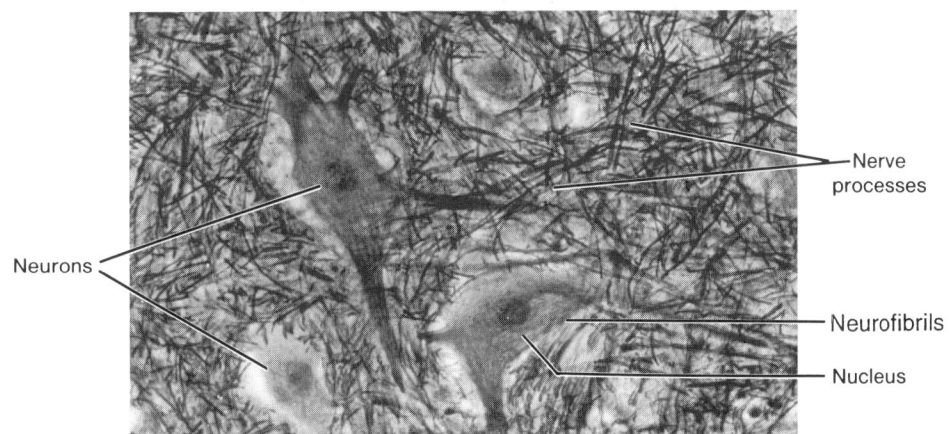

ments act as an internal scaffold for the perikaryon and its processes and function to maintain the shape of neurons. Specialized membrane proteins anchor the neurofilaments to the plasmalemma. The perikarya of most neurons contain the characteristic **chromophilic (Nissl) substance,** which, when stained with dyes such as cresyl violet or toluidine blue, appears as basophilic masses within the cy-

toplasm of the perikaryon and dendrites (Fig. 8-2A). It is absent from the axon and **axon hillock,** the region of the perikaryon from which the axon originates (Fig. 8-2B). In electron micrographs, Nissl substance is seen to consist of several parallel cisternae of granular endoplasmic reticulum (Fig. 8-3).

Small slender or oval mitochondria are scattered through-

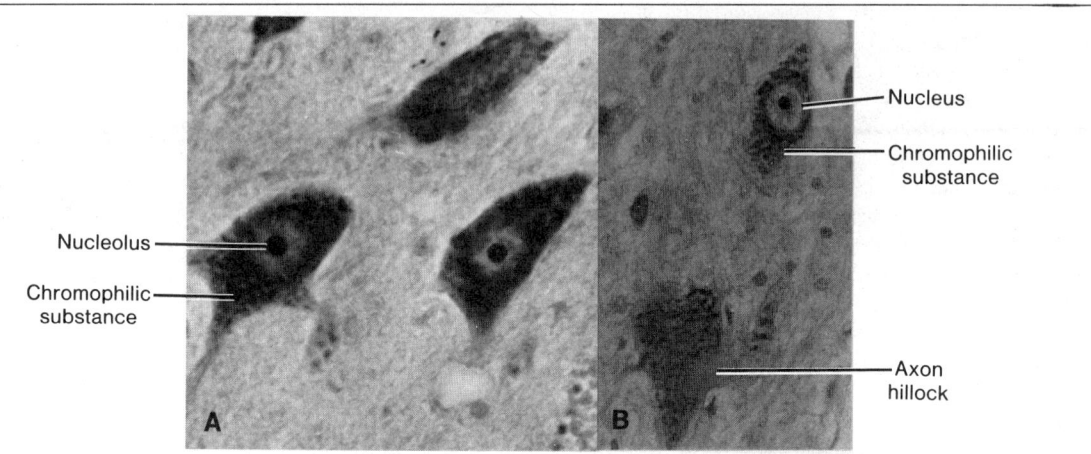

Fig. 8-2. A. Ventral horn of human spinal cord stained with crystal violet to demonstrate chromophilic (Nissl) substance (LM, × 400). B. Similarly stained neurons in human spinal cord, one of which shows an axon hillock (LM, × 400).

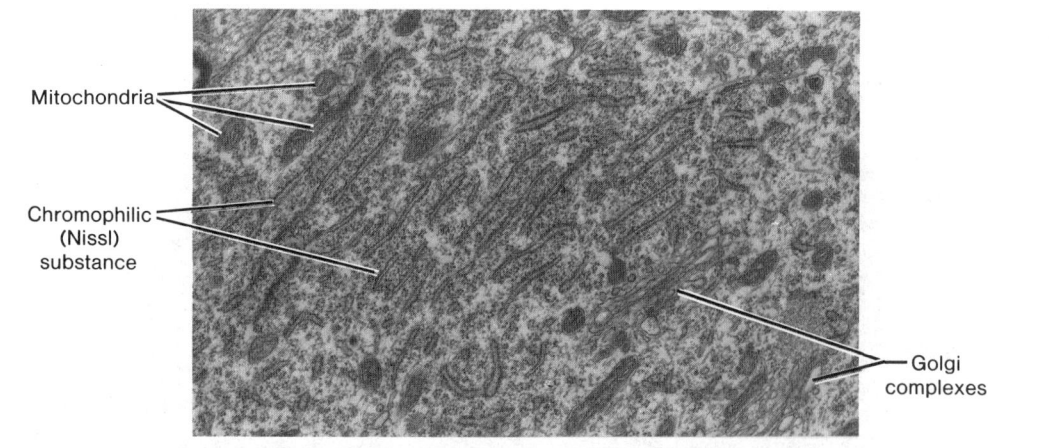

Fig. 8-3. A Nissl body in a neuron from the inferior olivary nucleus illustrates that Nissl consists in fact of parallel arrays of granular endoplasmic reticulum (TEM, × 20,000).

out the cytoplasm of neurons and contain lamellar and tubular cristae. Well-developed Golgi complexes have a perinuclear position in the cell. Although centrioles are seen occasionally, neurons in the adult do not divide.

Inclusions also may be found in the perikarya. Dark brown granules of **melanin pigment** occur in neurons from specific regions of the brain, such as the substantia nigra, locus ceruleus, and dorsal motor nucleus of the vagus nerve, and in spinal and sympathetic ganglia. More common inclusions are **lipofuscin granules,** which increase with age and are the by-products of normal lysosomal activity. Lysosomes are abundant in most neurons due to the high turnover of plasmalemma and other cellular components. The lipid droplets seen in many neurons represent storage material or may occur as the result of pathologic metabolism.

Nerve Processes

KEY WORDS: dendrite, dendritic spine (gemmule), axon (axon cylinder), axon hillock, collaterals, telodendrion, axoplasm, kinesin, dynein, action potential

Nerve processes are cytoplasmic extensions of the perikaryon and occur as dendrites and axons. Each neuron has several **dendrites** that extend from the perikaryon, dividing repeatedly to form branchlike extensions. Although tremendously diverse in the number, size, and shape of dendrites, each variety of neuron has a similar branching pattern. The dendritic cytoplasm contains elongate mitochondria, Nissl substance, scattered neurofilaments, and parallel-running microtubules. The cell membrane of most dendrites forms numerous minute projections called **dendritic spines** or **gemmules** that serve as areas for synaptic contact between neurons; an important function of dendrites is to receive impulses from other neurons. Dendrites provide most of the surface area for receptive synaptic contact between neurons, although in some neurons the perikaryon and initial segment of the axon also may act as receptor areas. The number of synaptic points on the dendritic tree varies with the type of neuron but may be in the hundreds of thousands. Dendrites play an important role in integrating the many incoming impulses.

A neuron has only one **axon (axis cylinder),** which conducts impulses away from the parent neuron to other functionally related neurons or effector organs. The axon arises from the **axon hillock,** an elevation on the surface of the perikaryon that lacks Nissl substance. Occasionally, axons may arise from the base of a major dendrite. Axons usually are much longer and more slender than dendrites and may or may not give rise to side branches called **collaterals,** which, unlike dendritic branches, usually leave the parent axon at right angles. Axons also differ in that the diameter is constant throughout most of the length and the external surface generally is smooth. Axons end in several branches called **telodendria,** which vary in number and shape and may form a network or basket-like arrangement around postsynaptic neurons.

The cytoplasm of the axon, the **axoplasm,** contains numerous neurofilaments, neurotubules, elongate profiles of smooth endoplasmic reticulum, and long, slender mitochondria. Since axoplasm does not contain Nissl substance, protein synthesis does not occur in the axon. Protein metabolized by the axon during nerve transmission is replaced by that synthesized in cisternae of granular endoplasmic reticulum in the Nissl substance of the perikaryon. Neurotubules function to transport the new protein as well as other materials within the axon. Microtubules (neurotubules) are assembled near the axon ending and disassembled at the parikaryon. Thus there is a net flow of microtubules in neurons toward the cell body.

Antegrade transport involves movement of material from the perikaryon to axon terminals and may be slow or fast. Fast transport (410 mm/day) involves transfer of membranous organelles, neurosecretory vesicles, calcium, sugars, amino acids, and nucleotides: Slow transport (1–6 mm/day) includes movement of proteins, metabolic enzymes, and elements of the cytoskeleton. Fast antegrade transport is mediated by **kinesin,** a microtubule-associated protein. One end of the kinesin molecule attaches to a transport vesicle or organelle. The remaining end undergoes cyclic interaction with the microtubule wall, resulting in movement toward the axon terminal. Retrograde transport is the transfer of exhausted organelles, proteins, small molecules, and recycled membrane from the axonal endings to the perikaryon and provides a route by which some toxins (tetanus) and viruses (herpes, rabies) can invade the central nervous system. Retrograde rapid transport occurs at a rate of about 310 mm/day and depends on another microtubule-associated protein called **dynein.** Dendritic transport also occurs but is less well understood.

Axons transmit responses as electrical impulses called **action potentials,** which begin in the region of the axon hillock and the initial segment of the axon. An impulse behaves in an "all or none" fashion. Only when the threshold of activity is reached is the action potential transmitted along the axon to its terminations, which synapse with adjacent neurons or end on effector cells. An axon always induces a positive activity — for example, causing muscle to contract or stimulating epithelial cells to secrete. Impulses directed to another neuron may be excitatory or inhibitory.

Distribution of Neurons

KEY WORDS: gray matter, white matter, nuclei, ganglia

The central nervous system is divided into gray matter and white matter, reflecting the concentration of perikarya. **Gray matter** contains the perikarya of neurons and their closely related processes, whereas **white matter** is composed chiefly of bundles or masses of axons and their surrounding sheaths. In the spinal cord, gray matter has a central location surrounded by white matter; in the cerebral and cerebellar cortices, gray matter is at the periphery and covers the white matter. Aggregates of perikarya occur in the gray matter of the central nervous system and act as distinct functional units called **nuclei.** Similar aggregates or individual neurons located outside the central nervous system are called **ganglia.**

Types of Neurons

KEY WORDS: Golgi type I neuron, Golgi type II neuron, unipolar neuron, bipolar neuron, pseudounipolar neuron, multipolar neuron

Neurons may be divided into types according to their location, number of dendrites, and length and position of axons. In humans, most neurons lie in the gray matter of the central nervous system. Some neurons have numerous, well-developed dendrites and very long axons that leave the gray matter to enter the white matter of the central nervous system and ascend or descend in the major fiber tracts of the brain or spinal cord or leave the central nervous system and contribute to the formation of peripheral nerves. Neurons of this type conduct impulses over long distances and are called **Golgi type I neurons.** Other neurons whose axons are short and do not leave the region of the perikaryon are called **Golgi type II neurons.** These are especially numerous in the cerebellar and cerebral cortices and retina of the eye. They vary greatly in size and shape and have several branching dendrites. Golgi type II neurons primarily serve as association neurons — that is, they collect nerve impulses and disseminate them to surrounding neurons.

Individual neurons also can be classified by their number of dendrites. **Unipolar neurons** lack dendrites and have only a single process, the axon. Although common in the developing nervous system, they are rare in adults. **Bipolar neurons** have a single dendrite and an axon, usually located at opposite poles of the perikaryon. They are found in the retina, olfactory epithelium, and cochlear and vestibular ganglia. Neurons of all craniospinal ganglia originate as bipolar neurons in the embryo, but during de-

velopment, the dendrite and axon migrate to a common site in the cell body, where they unite to form a single process. In the adult, this type of neuron is said to be **pseudounipolar.** The combined process, often called a *dendraxon,* may run for a short distance and then divide into two processes, one of which serves as a dendrite and receives stimuli from peripheral regions of the body and the other of which acts as an axon and enters the gray matter of the central nervous system to synapse with other neurons. Although the process directed toward the periphery acts as a dendrite, it is unusual because, morphologically, it resembles an axon. This functional dendrite is smooth and unbranched and usually receives nervous input from a receptor organ. **Multipolar neurons** are the most common type. They vary considerably in size and shape and are characterized by multiple dendrites. Internuncial and motor neurons are of this type. Of these, the internuncial neuron is the most numerous.

Ganglia

KEY WORDS: cranial ganglion, spinal ganglion, satellite cell, sensory ganglion, pseudounipolar neuron, glomerulus, autonomic ganglion, visceral motor efferent, multipolar neuron

Ganglia are classified as cranial, spinal, and autonomic. Macroscopically, **cranial** and **spinal ganglia** appear as globular swellings on the sensory roots of their respective nerves. Each ganglion is enveloped by a connective tissue capsule and may contain perikarya of only a few neurons or as many as 55,000. A delicate network of collagenous and reticular fibers, accompanied by small blood vessels, extends between individual neurons and together with bundles of nerve processes often separates the perikarya into groups. The perikarya are surrounded by two distinct capsules. The inner capsule is made up of a single layer of low cuboidal supporting cells called **satellite cells.** The outer capsule lies immediately outside the basal lamina of the satellite cells and consists of a vascular connective tissue (Fig. 8-4).

Cranial and spinal ganglia are **sensory ganglia** and contain **pseudounipolar neurons.** The perikarya appear spherical or pear-shaped and are 15 to 100 μm in diameter. They often show a large, central nucleus and a distinct nucleolus. The dendraxon of the pseudounipolar neuron may become convoluted to form a **glomerulus.** This nerve process then divides: One branch, a functional dendrite, passes to a receptor organ; the other, a functional axon, passes into the central nervous system. The perikarya of these pseudounipolar neurons do not receive synapses from other neurons.

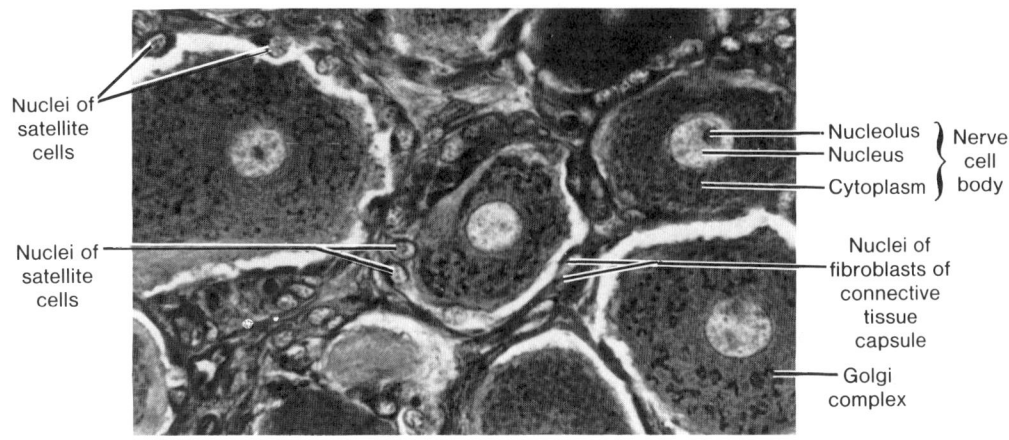

Fig. 8-4. A dorsal root ganglion associated with a human spinal cord (osmium stain) (LM, × 300).

Autonomic ganglia consist of collections of perikarya of **visceral efferent motor neurons** and are located in swellings along the sympathetic chain or in the walls of organs supplied by the autonomic nervous system. They generally have a connective tissue capsule. The neurons usually are **multipolar** and assume various sizes and shapes. Perikarya range from 15 to 60 μm in diameter; the nucleus is large, round, and often eccentrically placed in the cell; and binucleate cells are not uncommon. Lipofuscin granules are more frequent in neurons of autonomic ganglia than in craniospinal ganglia. The ganglia of larger sympathetic chains are encapsulated by satellite cells, but a capsule may be absent around ganglia in the walls of the viscera. Unlike craniospinal ganglia, neurons of autonomic ganglia receive numerous synapses and are influenced by other neurons.

Nerve Fibers

KEY WORDS: myelin, Schwann cell, neurilemma (sheath of Schwann), node of Ranvier, internode, internodal segment, neurokeratin, major dense line, intraperiod line, mesaxon, action potential, saltatory conduction

A nerve fiber consists of an axon (axis cylinder) and a surrounding sheath of cells. Large nerve fibers are enclosed by a lipoprotein material called **myelin;** smaller nerve fibers may or may not be surrounded by myelin. Thus nerve fibers may be classed as myelinated or unmyelinated (Fig. 8-5).

Axons of peripheral nerves are enclosed by a sheath of flattened cells called **Schwann cells.** These cells are thin, attenuated cells with flattened, elongate nuclei located near the center of the cells. Their cytoplasm contains a small Golgi complex, a few scattered profiles of granular endoplasmic reticulum, scattered mitochondria, microtubules, microfilaments, and at times, numerous lysosomes. The outer cell membrane is covered by a basal lamina. A continuous basal lamina covers successive Schwann cells and the axonal surfaces of the nodes. The integrity of the basal lamina is vital to regenerating axons and Schwann cells after injury. Schwann cells invest nerve fibers from near their beginnings almost to their terminations. The resulting **neurilemma (sheath of Schwann)** is continuous with the capsule of satellite cells that surrounds the perikarya of neurons in craniospinal ganglia. Schwann cells produce and maintain the myelin on nerve fibers of the peripheral nervous system.

The neurilemmal and myelin sheaths are interrupted by small gaps at regular intervals along the extent of the nerve fiber. These interruptions represent regions of discontinuity between individual Schwann cells and are called **nodes of Ranvier** (Fig. 8-6). The neurilemmal and myelin sheaths are made up of a series of small individual segments, each of which consists of the region between two consecutive nodes of Ranvier and makes up one **internode** or **internodal segment.** An internode represents the area occupied by a single Schwann cell and its myelin and is about 1 to 2 mm in length. Myelin is not a secretory product of Schwann cells but a mass of lipoprotein and glycolipid that

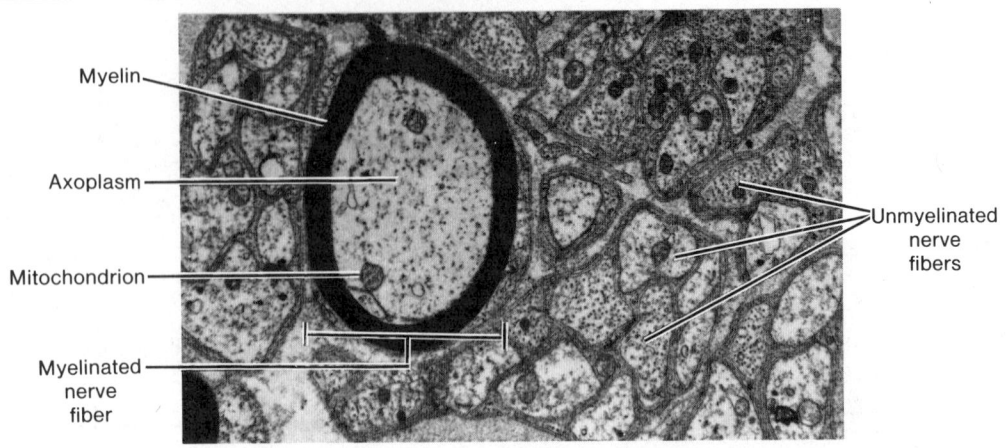

Fig. 8-5. A cross section of a mixed peripheral nerve illustrating a myelinated and several unmyelinated nerve fibers (TEM, × 8,000).

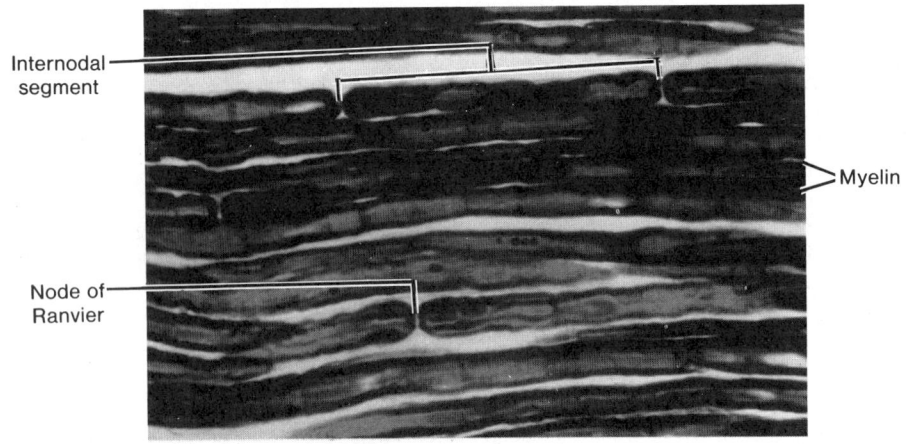

Fig. 8-6. A longitudinal section of human peripheral nerve stained with osmium illustrates several nodes of Ranvier (LM, × 400).

includes galactocerebroside, which is abundant in myelin and that results from successive layering of the Schwann cell plasmalemma as it wraps around the axon. Myelin may appear homogeneous (Fig. 8-7), but after some methods of preparation, it may be represented by a network of residual protein called **neurokeratin.**

Ultrastructurally, myelin appears as a series of regular, repeating light and dark lines. The dark lines, called **major dense lines,** result from apposition of the inner surfaces of the Schwann cell plasmalemma (Figs. 8-8 and 8-9). The

less dense **intraperiod lines** represent the fusion of the outer surfaces of the Schwann cell plasmalemma and lie between the repeating major dense lines. Oblique discontinuities often are observed in the myelin of peripheral nerves and form the Schmidt-Lantermann incisures or clefts. The clefts represent local separations of the myelin lamellae.

Unmyelinated fibers (axons) also are surrounded by Schwann cells. In electron micrographs, 15 or more unmyelinated axons often are found in recesses in the plas-

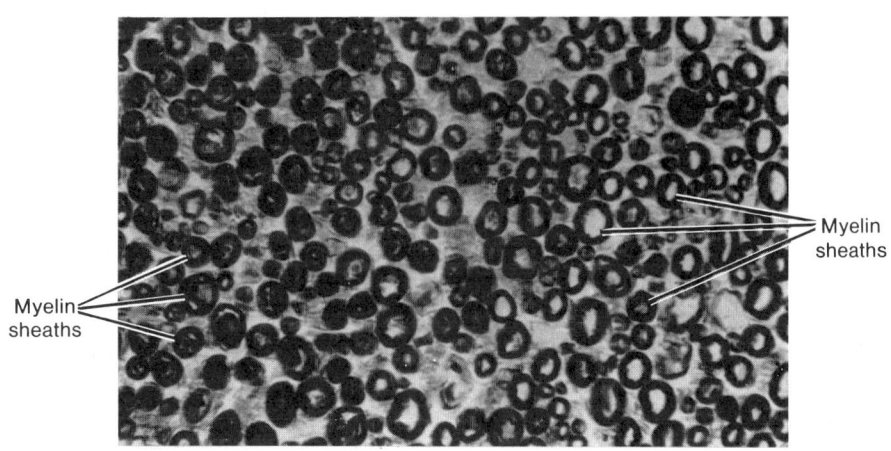

Fig. 8-7. A cross section of human peripheral nerve stained with osmium to demonstrate myelin sheaths (LM, × 400).

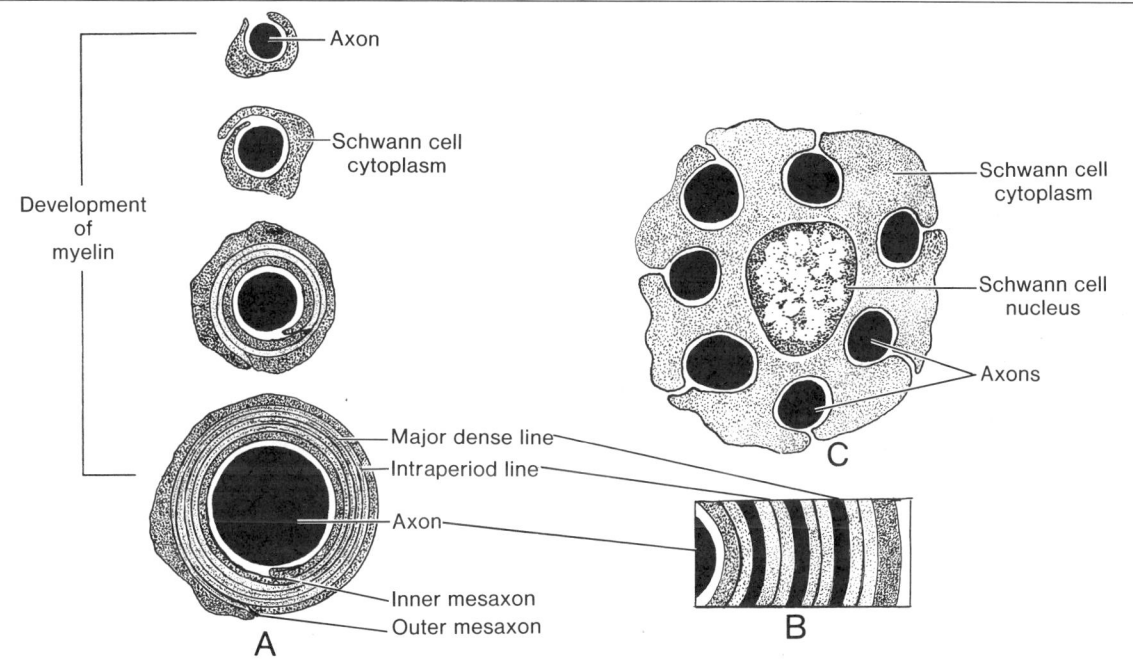

Fig. 8-8. Myelinated fiber (A, B) and nonmyelinated nerve (C).

malemma of a single Schwann cell (see Figs. 8-5 and 8-8). The cell membrane of the Schwann cell closely surrounds each axon and is intimately associated with it. As the cell membrane of the Schwann cell encircles the axon, it courses back to come in contact with itself. This point of contact is called the **mesaxon.** In myelinated nerves, the primary infoldings of the Schwann cell membrane at the axon-myelin junction is called the *internal mesaxon.* The junction between the superficial lamellae of the myelin sheath and the Schwann cell plasmalemma is the *external mesaxon.*

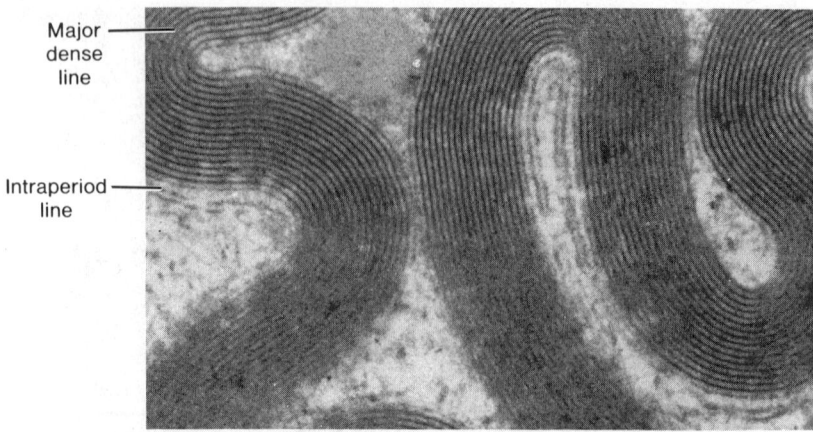

Major
dense
line

Intraperiod
line

Fig. 8-9. A region of myelin showing major dense and intraperiod lines (TEM, × 54,000).

The speed with which an impulse is transmitted along a nerve fiber is proportional to the diameter of the fiber. The diameter of heavily myelinated nerve fibers is much greater than that of unmyelinated fibers, and therefore, conduction is faster in myelinated fibers. Nerve fibers in peripheral nerves can be classed according to the diameter and speed of conduction. Type A nerve fibers are large, myelinated motor and sensory fibers, 3 to 20 μm in diameter, that conduct impulses at 15 to 120 m/second. More finely myelinated fibers with diameters up to 3 μm conduct impulses at 3 to 15 m/second and are type B fibers. They represent many of the visceral sensory fibers. Type C fibers are small, unmyelinated nerve fibers that conduct impulses at 0.5 to 2.0 m/second.

The first event in development of an **action potential** along a nerve fiber is a sudden increase in permeability to sodium ion, resulting in depolarization of the axon and development of a negative charge along the axon surface (Fig. 8-10A). The local current created between the depolarized and resting surface membranes causes an increased permeability of the membrane to sodium ion. A cycle of membrane activation is established that results in the transmission of the depolarization process (a nerve impulse) along the nerve fiber. Once depolarization occurs, it will pass along the entire cell membrane of the axon. The nerve impulse lasts for only a short time, and repolarization is brought about by a sodium pump mechanism in the cell membrane of the axon. The generation of a nerve impulse is an "all or nothing" phenomenon.

In the faster-conducting myelinated nerves, impulses travel from node to node; this method of impulse conduction is called **saltatory conduction** (Fig. 8-10B). Because depolarization occurs only at the nodes of Ranvier, nerve impulses jump from node to node across the intervening internodal segment. The axon at each node is slightly thicker than that associated with internodal segments, and the plasmalemma of the node region contains most of the sodium voltage-gated channel proteins held in place at this location by the link protein ankyrin, which attaches them to the cytoskeleton. This explains, in part, the higher velocity of nerve transmission in myelinated nerve fibers.

Peripheral Nerves

KEY WORDS: mixed nerve, epineurium, fascicle, perineurium, endoneurium

Peripheral nerves consist of several nerve fibers united by surrounding connective tissue. They are **mixed nerves** consisting of sensory (afferent) and motor (efferent) nerve fibers that may be myelinated or unmyelinated. The sheath of connective tissue surrounding a peripheral nerve makes up the **epineurium** and unites several bundles of nerve fibers into a single unit, a peripheral nerve. The epineurium consists of fibroblasts, longitudinal collagen fibers, and scattered fat cells. Each bundle or **fascicle** of nerve fibers is enclosed by concentric layers of flattened, fibroblast-like cells that form a dense sheath called the **perineurium.** The perineurium is limited externally and internally by a basal lamina. Tight junctions occur between cells, and the tightly adherent cells form a perineurial compartment for individ-

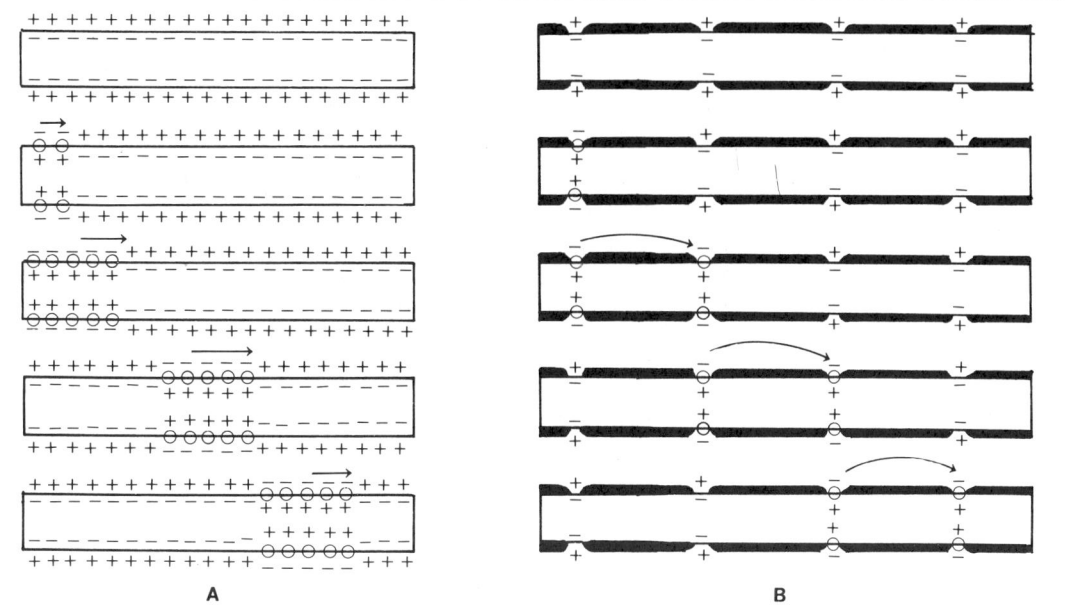

Fig. 8-10. Schematic representation of action potential in (A) nonmyelinated and (B) myelinated fibers.

ual fascicles. Delicate collagenous fibers, reticular fibers, fibroblasts, and macrophages lie between individual nerve fibers within each fascicle and constitute the **endoneurium.** Small blood vessels are found primarily in the epineurium, but in thicker regions of the endoneurium they may occur as delicate capillary networks. The perineurium and epineurium of peripheral nerves become continuous with the meninges of the central nervous system.

Peripheral Nerve Endings

KEY WORDS: somatic efferent, motor end-plate, motor unit, visceral efferent, receptor, free (naked) nerve ending, encapsulated nerve ending

Nerve endings increase the surface contact area between nervous and nonnervous structures and allow stimuli to be transmitted from nerve endings to muscles, causing them to contract, or to epithelial cells, causing them to secrete. When stimulated, dendrites at the periphery generate impulses that are transferred along the nerve fiber to sensory ganglia and ultimately to the central nervous system.

Peripheral efferent (motor) nerve fibers can be divided into somatic and visceral efferent groups. **Somatic efferent** fibers end in skeletal muscle as small, oval expansions

called **motor end-plates** (Fig. 8-11). The skeletal muscle cells supplied by a single motor neuron constitute a **motor unit.** Muscles with a large number of motor units can make more precise movements than those with fewer units for the same number of muscle cells. **Visceral efferent** nerve fibers stimulate smooth muscle, cardiac muscle, and glandular epithelium. Visceral motor endings of smooth muscle terminate as two or more swellings that pass between individual muscle cells. In cardiac muscle, numerous thin nerve fibers end near the surface of individual muscle cells but form no specialized contacts with them. Endings in glands are from unmyelinated nerves and form elaborate, delicate nets along the external surface of the basal lamina of the gland cells: fine branches penetrate the basal lamina and pass between individual epithelial cells.

Terminal peripheral nerve fibers that are excitable to stimuli are **receptors** and can transform chemical and physical stimuli into nerve impulses. Receptors vary in morphology, may be quite complex, and often are grouped into free (naked), diffuse, and encapsulated nerve endings.

Free (naked) nerve endings arise from myelinated and unmyelinated fibers of relatively small diameter. They are widely distributed and, although most numerous in the skin, are present in the connective tissue of visceral

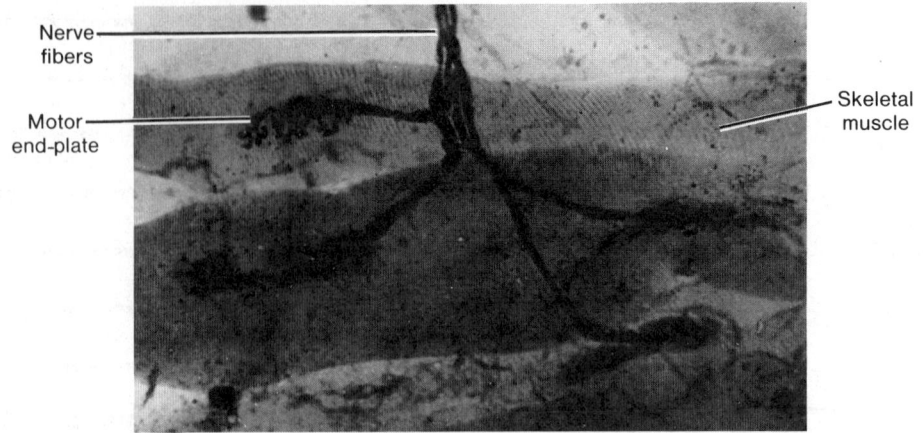

Fig. 8-11. Motor end-plates of human skeletal muscle (LM, × 400).

organs, in deep fascia, in muscles, and in mucous membranes. As myelinated fibers near their terminations, they lose their myelin and form unmyelinated terminal arborizations. Unmyelinated nerve fibers end in numerous fibrils that terminate as small, knoblike thickenings. Many naked nerve endings act as pain receptors. A diffuse type of naked receptor encircles the base of hair follicles to form peritrichal nerve endings that are sensitive to hair movement. In addition to the naked interepithelial nerve endings that terminate among epithelial cells, there are some endings that form concave neurofibrillar discs applied to a single modified epithelial cell of stratified squamous epithelium. These more specialized nerve endings are called the *tactile discs of Merkel.*

In **encapsulated nerve endings,** the terminals are enclosed in a capsule of connective tissue (Fig. 8-12). They vary considerably in size, shape, location, and type of stimuli perceived. As the nerve fibers near their respective capsules, they generally lose their myelin and neurolemmal sheaths and enter the capsule as naked nerve endings. A number of encapsulated nerve endings, their locations, and proposed functions are summarized in Table 8-1.

Synapse

KEY WORDS: terminal bouton, calyx (basket ending), bouton en passage, synaptic vesicle, presynaptic membrane, synaptic cleft, postsynaptic membrane, subsynaptic web, threshold of excitation, gap junction

The site where impulses are transmitted from one neuron to another is called a *synapse* and may occur between axon and dendrite (axodendritic), axon and cell body (axosomatic),

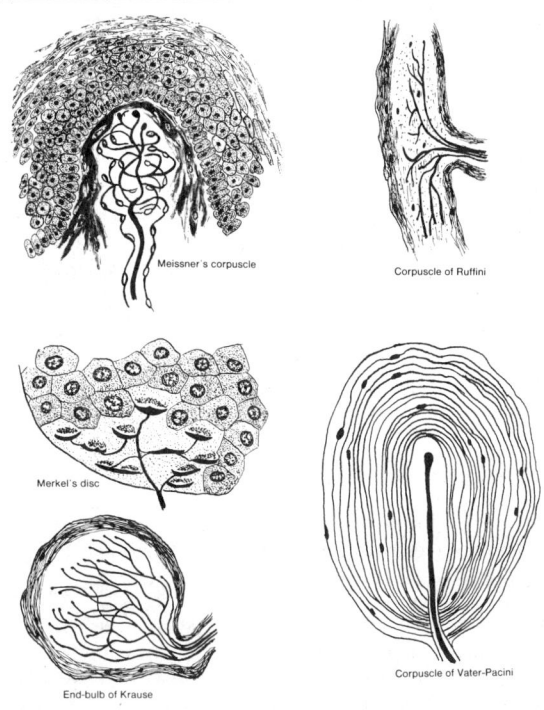

Fig. 8-12. Diagram of types of sensory nerve endings.

axon and axon (axoaxonic), dendrite and dendrite (dendrodendritic), dendrite and perikaryon (somatodendritic), or cell bodies of adjacent neurons (somatosomatic). The number of synapses associated with a single neuron varies accord-

Table 8-1. Encapsulated Sensory Nerve Endings

Name of Receptor	Locations	Functions
Lamellated corpuscles of Vater-Pacini	Dermis of skin, periosteum, mesenteries, pleura, nipples, pancreas, tendons, joint capsules, ligaments, walls of viscera, penis, clitoris	Proprioception, pressure, vibration
Corpuscles of Meissner	Dermal papillae of digits, lips, genitalia, and nipples	Touch
End-bulbs of Krause	Dermis, conjunctiva, oral cavity	Cold, pressure
Genital corpuscles	Penis, clitoris, nipples	Touch, pressure
Corpuscles of Ruffini	Dermis, joint capsules	Heat
Neurotendinous endings (of Golgi)	Tendons	Proprioception
Neuromuscular spindles	Skeletal muscle	Proprioception

ing to the type of neuron. Small Golgi type II neurons (granule cells) of the cerebellum may have only a few synaptic points, whereas other neurons, such as Golgi type I (Purkinje cells of the cerebellum), may have several hundred thousand. Physiologically, excitatory and inhibitory synapses occur.

Synaptic endings of axons usually occur as small swellings at the tips of axon branches and are called **terminal boutons** (Fig. 8-13). The axon may end in thin branches that surround the perikaryon or dendrites of another neuron; these endings are called **calyces (basket endings).** Synaptic contacts also may occur at intervals along the terminal segment of an axon, forming **boutons en passage.**

Synaptic endings vary considerably in form from one type of neuron to another but generally show some common features. Each neuron is a distinct cellular unit, and there is no cytoplasmic continuity between cells. In some

ways, synaptic points resemble desmosomes and may aid in maintaining contact between nerve cells. Ultrastructurally, the presynaptic area (terminal bouton) contains clusters of mitochondria and numerous membrane-bound, electron-lucent vesicles with diameters of 40 to 60 nm (Figs. 8-14 and 8-15). These **synaptic vesicles** contain neurotransmitter substances such as dopamine, norepinephrine, acetylcholine, and serotonin; amino acids; and a number of peptides. The vesicles congregate near the **presynaptic membrane,** a slightly thickened area of the axon plasmalemma at the synaptic contact. Pre- and postsynaptic membranes are separated by a narrow space, 20 to 30 nm wide, called the **synaptic cleft,** which may contain fine filaments and glycosaminoglycan. The **postsynaptic membrane** shows increased electron density and is slightly thickened, and its internal surface is associated with a network of dense filamentous material called the

Fig. 8-13. A perikaryon of a human spinal neuron, the external surface of which is associated with numerous terminal boutons (LM, × 1,000).

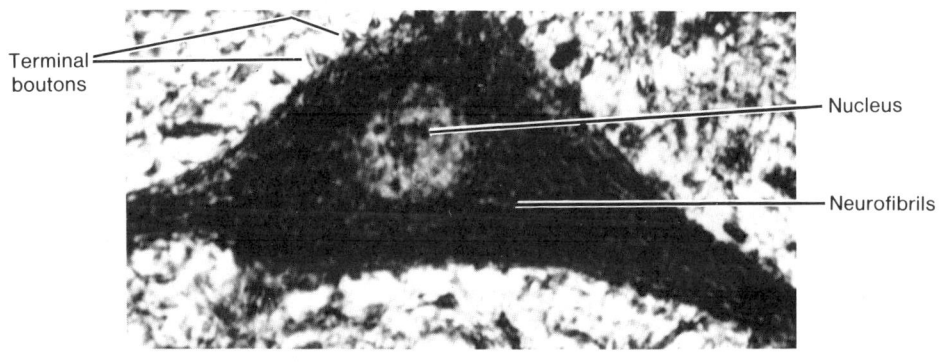

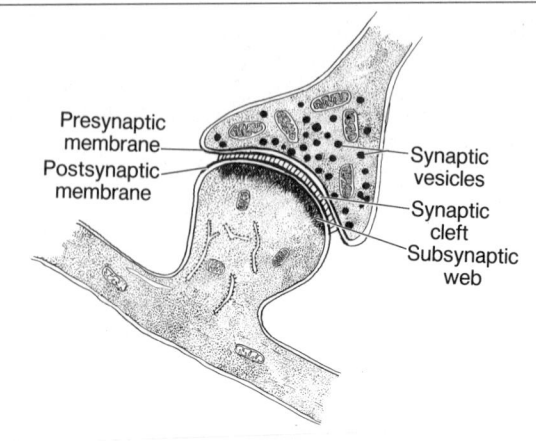

Presynaptic membrane
Postsynaptic membrane
Synaptic vesicles
Synaptic cleft
Subsynaptic web

Fig. 8-14. Diagram of a synapse.

subsynaptic web. The cytoplasm of the postsynaptic region lacks synaptic vesicles and shows fewer mitochondria than are found in the presynaptic area.

As an action potential reaches the presynaptic area of an axon, the synaptic vesicles fuse at sites along the presynaptic membrane and release transmitter substance into the synaptic cleft. The transmitter substance reacts with special receptors in the postsynaptic membrane, causing an immediate increase in the permeability of the postneuronal membrane to sodium ion, thus changing the resting membrane potential. If the potential of the postsynaptic membrane rises above a certain level, the **threshold of excitation,** an action potential is transmitted by the postsynaptic neuron. The first parts of the postsynaptic neuron to initiate an action potential are the axon hillock and initial segment of the axon. The initial segment of an axon is a few microns long and lacks a myelin sheath. In most neurons, the cell membrane of this region has a lower threshold of excitation than does that of the perikaryon or dendrites.

If transmitter substances decrease the permeability of the postsynaptic membrane, the threshold of excitation, rises and the net effect is inhibitory.

Acetylcholine is the transmitter substance found most commonly in the electron-lucent synaptic vesicles of the central nervous system and motor end-plates. Synaptic vesicles of most sympathetic postganglionic axons are characterized by electron-dense cores and contain catecholamines.

As a wave of depolarization reaches the presynaptic membrane, increased levels of calcium ion result, triggering a release by exocytosis of transmitter substance into the synaptic cleft. The transmitter released interacts with receptors within the postsynaptic membrane to elicit one of three possible effects (depolarization, hyperpolarization, neuromodulation) in the target neuron. If the transmitter substance (acetylcholine, glutamate) binds to specific ion-channel receptor proteins, the ion (sodium) channels in the postsynaptic membrane open, resulting in depolarization and excitation of the target neuron. If, on the other hand, the transmitter substance released is gamma-amino butyric acid (GABA), or glycine, which binds to ion-channel

Fig. 8-15. A region of dendritic spines illustrating both presynaptic and postsynaptic membranes (TEM, × 39,000).

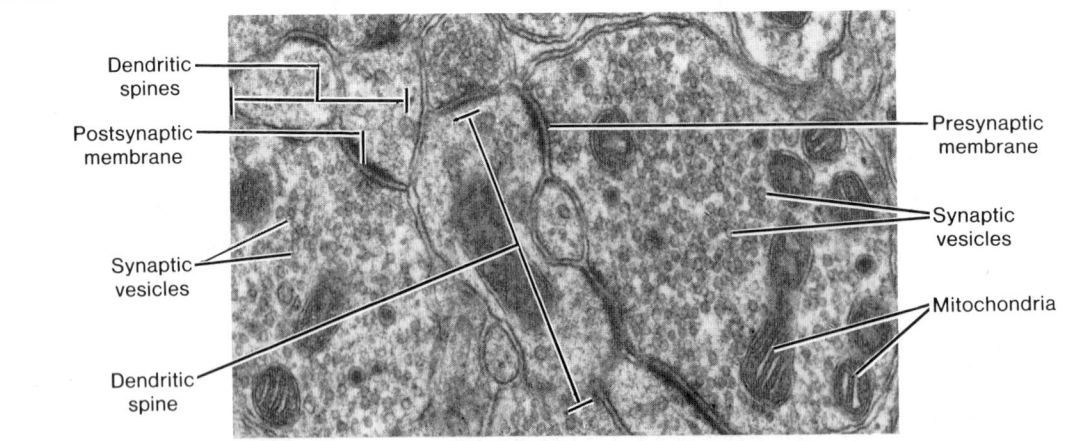

Dendritic spines
Postsynaptic membrane
Synaptic vesicles
Dendritic spine
Presynaptic membrane
Synaptic vesicles
Mitochondria

receptor proteins that allow entry of negatively charged ions (primarily chloride ion) into the cell, then hyperpolarization occurs, resulting in inhibition. In some instances transmitters (5-HT, dopamine, small neuropeptides) bind to second-messenger–coupled receptors and not intrinsic ion channels. These receptors are linked to G-proteins and second-messenger molecule systems that through a cascade of intracellular events modify the sensitivity or modulate the target neuron to depolarization.

Two groups of synaptic vesicle are thought to exist within the axon terminal: a small group that is released with the arrival of an action potential and a larger reserve group that is linked to the cytoskeleton by actin filaments. Synaptic vesicles of the reserve group move to a releasable group as needed. Synapsin I, a phosphoprotein associated with the cytoplasmic surface of synaptic vesicles, links actin to the vesicle surface, which results in the accumulation of reserve vesicles. Phosphorylation of synapsin I, due to the influx of calcium ion triggered by the action potential, disassociates this link, allowing the vesicles to join the releasable group. An additional phosphoprotein, synapsin II, is thought to mediate the interaction of synaptic vesicles with the cytoskeleton. The proteins synaptotagmin and synaptophysin associated with synaptic vesicles are thought to regulate the binding and fusion of the vesicles with the presynaptic membrane.

Following vesicle fusion with the presynaptic membrane and release of transmitter, equal amounts of cell membrane are taken up by the formation of clathrin-coated vesicles via endocytosis. These vesicles may fuse with cisternae of endoplasmic reticulum containing transmitter substances. New vesicles form by budding from these cisternae to return to the pool of reserve synaptic vesicles.

Acetylcholine released from the axon terminal is rapidly hydrolyzed by acetylcholine esterase into acetate and choline in the synaptic cleft. Choline is then transported back through the presynaptic membrane by a choline transporter, converted to acetyl choline by the enzyme choline acetyltransferase, and repackaged into newly formed synaptic vesicles. Inhibitory transmitters such as GABA, on the other hand, are taken up by specific GABA transporters located in the plasma membrane of surrounding astrocyte processes. In the astrocyte cytoplasm GABA is broken down to glutamine, which reenters the nerve terminal. Here, glutamine is enzymatically converted back into GABA and packaged into synaptic vesicles.

Neuropeptides, unlike many of the transmitters, must be synthesized in the perikaryon and transported to the axon terminal. Electrical coupling is another means of transfer-

ring impulses in some types of synapses and occurs at **gap junctions.** Gap junctions act as ion channels.

Neuroglia

KEY WORDS: astrocyte, protoplasmic astrocyte, fibrous astrocyte, glial fibrillary acid protein, end foot (foot process), oligodendrocyte, interfascicular oligodendrocyte, perineuronal satellite oligodendrocyte, microglia, ependyma, tela choroidea, choroid plexus, blood-brain barrier

Neurons form a relatively small proportion of the cells in the central nervous system. Most of the cells are nonneuronal supporting cells called *neuroglia*. Depending on the location in the central nervous system, neuroglia may outnumber neurons by as much as 50:1. Together, neuroglia account for over half the weight of the brain. Neuroglia generally are smaller than neurons and in light microscopic preparations can be identified by their small, round nuclei, which are scattered among neurons and their processes. Neuroglia include ependymal cells, astrocytes, oligodendrocytes, and microglia. Astrocytes and oligodendrocytes often are collectively referred to as *macroglia.*

Morphologically, **astrocytes** are divided into protoplasmic and fibrous types. **Protoplasmic astrocytes** are found mainly in the gray matter and are characterized by numerous, thick branched processes. **Fibrous astrocytes** occur mainly in white matter and are distinguished by long, thin, generally unbranched processes (Fig. 8-16). The two forms of astrocyte may represent a single type that varies in morphology according to its location and metabolic state. In electron micrographs, astrocytes usually show an electron-lucent, largely organelle-free cytoplasm and euchromatic nuclei. The cytoplasm of the cell body and its processes contains bundles of intermediate filaments called glial filaments, or **glial fibrillary acid protein** (GFAP). These filaments are particularly abundant in protoplasmic astrocytes. Astrocytes may provide a framework to support the neurons and contribute to their nutrition and metabolic activity, but their precise function remains unknown.

The astrocyte processes vary considerably in appearance. Some are thick and long; others are thin and sheetlike. The processes often expand to form **end feet (foot processes)** that are aligned along the internal surface of the pia mater and around the walls of blood vessels (see Fig. 8-16). The largest concentration of end feet occurs beneath the pial surface, where they surround the brain and spinal cord to form a membrane-like structure called the *glia limitans.*

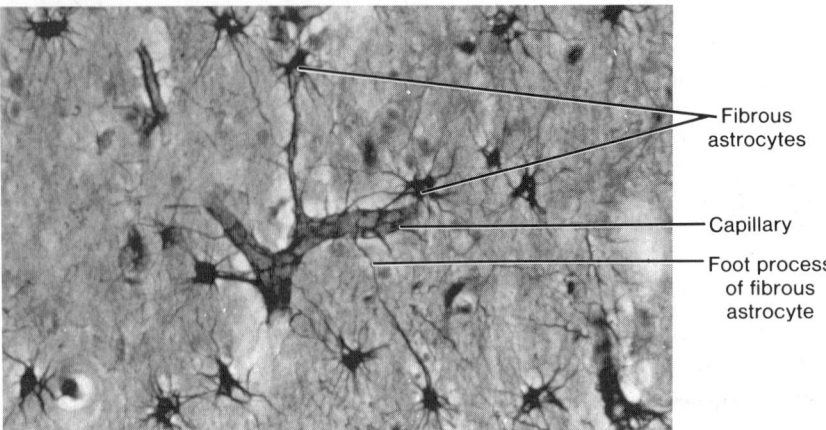

Fig. 8-16. Fibrous astrocytes from white matter of a human spinal cord (LM, × 300).

Smaller neuroglial cells with fewer processes and more deeply stained nuclei are **oligodendrocytes,** of which there are two types. **Interfascicular oligodendrocytes** occur mainly in the fiber tracts that form the white matter of the brain and spinal cord. **Perineuronal satellite oligodendrocytes** are restricted to the gray matter and are closely associated with the cell bodies of neurons. The cytoplasm of both types of oligodendrocytes contains free ribosomes, short cisternae of granular endoplasmic reticulum, an extensive Golgi complex, and numerous mitochondria. Large numbers of microtubules form parallel arrays that run throughout the cytoplasm of the cell body and into its processes.

Schwann cells are not present in the central nervous system; oligodendrocytes serve as the myelin-forming cells of this region. Myelinated nerve fibers of the central nervous system have nodes of Ranvier, but unlike peripheral myelinated fibers, they do not show incisures. Nodes in the central nervous system are bare, while those of the peripheral nervous system are partly covered by extensions of cytoplasm from adjacent Schwann cells into the paranodal area. The myelin sheath begins at the end of the initial segment of the axon, a few microns from the axon hillock, and ensheathes the rest of the axon to near its end. Each internodal segment of a nerve fiber in the central nervous system is formed by a single cytoplasmic process from a nearby oligodendrocyte, which wraps around the axon. Unlike Schwann cells of the peripheral nervous system, a single oligodendrocyte provides the internodal segments of the myelin sheath for several separate, but adjacent axons

(Fig. 8-17). In some regions, as in the optic nerve, a single oligodendrocyte may be responsible for the internodal segments on 40 to 50 axons. Because perineuronal satellite oligodendrocytes lie close to the perikarya of neurons, it has been proposed that oligodendrocytes influence the metabolism and nutrition of neurons. In addition, glia appear to be involved in calcium homeostasis and in the recycling of certain types of neurotransmitters.

In the central nervous system, perineuronal satellite and interfascicular oligodendrocytes have the same relationship to perikarya and their axons as that which exists between satellite cells and Schwann cells of peripheral nerve fibers. Schwann cells and satellite cells could be considered as neuroglial elements of the peripheral nervous system.

Fig. 8-17. Myelination in the central nervous system.

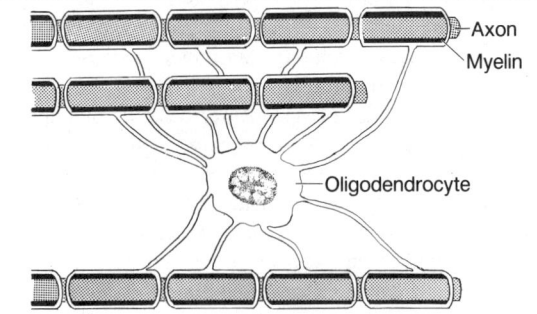

Small cells called **microglia** also are present in the central nervous system. Microglia have extensive ramifying processes, a characteristic phenotype that suggests they may be dendritic antigen-presenting cells. They also express class II MCH molecules. In areas of damage or disease, they are thought to proliferate and become phagocytic and are supplemented by monocytes from the blood that transform into additional macrophages. As material is phagocytosed, the cells enlarge and then are called *Gitterzellen,* or compound granular corpuscles. Monocytes appear to be the precursors of microglial cells, which thus are part of the mononuclear system of phagocytes with ultimate derivation from the bone marrow.

The **ependyma** lines the central canal of the spinal cord and ventricles of the brain. It consists of a simple epithelium in which the closely packed cells vary from cuboidal to columnar. The luminal surfaces of the cells show large numbers of microvilli and, depending on the location, may show cilia. Adjacent cells are united by desmosomes and zonula adherens; zonula occludens generally are not seen. Hence cerebrospinal fluid in the central canal and ventricles can pass between ependymal cells to enter the parenchyma of the central nervous system. The bases of ependymal cells have long, threadlike processes that branch, enter the substance of the brain and spinal cord, and may extend to the external surface, where they contribute end feet to the glia limitans.

The ependyma forms a secretory epithelium in the ventricles of the brain, where it is in direct contact with a highly vascular region of the pia mater, the **tela choroidea.** The modified ependyma and tela choroidea form the **choroid plexus,** which secretes cerebrospinal fluid. Ependymal cells of the choroid plexus are columnar and closely packed and bear numerous microvilli. Unlike those in other regions of the ependyma, cell apices here are joined by zonula occludens, which prevent passage of material between cells. Ependymal cells lie on a continuous basal lamina that separates them from a connective tissue that contains small bundles of collagenous fibers, pia-arachnoid cells, and numerous blood vessels. Capillary endothelial cells in this region, unlike those elsewhere in the brain, have numerous fenestrations. Fluid readily moves through the capillary wall but is prevented from entering the ventricles by the zonula occludens. Ependymal cells secrete sodium ions into the ventricles, and chloride ions, water, and other substances follow passively. Protein and glucose concentrations are relatively low. Cerebrospinal fluid is formed continuously, moves slowly through the ventricles of the brain, and enters the subarachnoid space. It surrounds and protects the central nervous system from mechanical injury and is important in the metabolic activities of the central nervous system.

Certain substances in the blood are prevented from entering the central nervous system, although they readily gain access to other tissues. Capillaries deep in the spinal cord and brain are sheathed by the end feet of astrocytes, and the nonfenestrated endothelial cells are united by occluding tight junctions. The endothelial cells do not exhibit transendothelial transport vesicles typical of endothelial cells found in capillaries elsewhere. Additionally, the internal plasmalemma of the endothelial cells is thought to have special properties that prevent passage of some substances. Together, the tight junctions and internal plasmalemma of the endothelial cells form the **blood-brain barrier.**

Highly vascularized areas — the circumventricular organs — are present in specific regions along the midline of the walls of the ventricles. These organs include the organum vasculosum, lamina terminalis, subfornical organ, subcommissural organ, and area postrema. Except for the subcommissural organ, the ependymal epithelium of these areas lacks cilia and contains stellate cells called tanycytes. Such regions are devoid of an internal (subependymal) layer and the external (subpial) layer of glial processes. The underlying capillaries are lined by a fenestrated endothelium. These regions lack the morphologic elements of a true blood-brain barrier and, except for the subcommissural organ, are able to accumulate vital dyes. Large molecules are thought to traverse these regions, which provide areas of exchange between the central nervous system and blood.

Spinal Cord

KEY WORDS: gray matter, internuncial neuron, central canal, ependyma, white matter

The spinal cord is subdivided into a central H-shaped region of gray matter and a surrounding layer of white matter. **Gray matter** consists mainly of perikarya of neurons, their dendrites, and surrounding neuroglial cells and is arranged into two dorsal and two ventral horns. The dorsal horns contain perikarya of multipolar neurons receiving sensory impulses that enter the spinal cord from the peripheral nervous system. Neurons of the dorsal horns transmit the impulses to other neurons in this and other areas of gray matter and are referred to as **internuncial neurons.** The multipolar neurons in the ventral horns are the largest in the spinal cord and transmit motor impulses from the spinal cord to the periphery. In the thoracic and upper lumbar regions of the spinal cord, small multipolar neurons form an intermediolateral horn that provides preganglionic sympathetic fibers for the autonomic nervous system. The **central canal** of the spinal cord lies in the center of the crossbar of the H-shaped gray matter and is lined by **ependyma.**

The **white matter** consists mainly of myelinated axons and lacks the perikarya and dendrites of neurons. It is subdivided into anterior, lateral, and posterior funiculi by the dorsal and ventral horns of the gray matter. A funiculus consists of several tracts, each of which in turn contains several bundles of nerve fibers. Nerve fibers in each tract carry similar impulses, motor or sensory, that either ascend or descend along the long axis of the spinal cord.

Cerebellar and Cerebral Hemispheres

The cerebellar and cerebral hemispheres differ from the spinal cord in that the gray matter is located at the periphery and the white matter lies centrally. Both regions of the brain consist of an outer cortex of gray matter and a subcortical region of white matter.

Cerebral Cortex

KEY WORDS: **laminated appearance, pyramidal cell, apical dendrite, Betz cell, stellate (nonpyramidal) cell**

The cerebral cortex is 1.5 to 4.0 mm thick and contains about 14 billion neurons plus nerve processes and supporting glial cells. In all but a few regions it is characterized by a **laminated appearance.** Perikarya generally are organized into five layers. Starting at the periphery of the cerebral cortex, the general organization of neurons is molecular layer (I), external granular layer (II), external pyramidal layer (III), internal granular layer (IV), internal pyramidal layer (V), and multiform layer (VI).

The molecular layer (I) is a largely cell-free zone just below the surface of the cortex. Neurons of similar type tend to occupy the same layer in the cerebral cortex, although each cellular layer is composed of several different cell types. For convenience of description, these neurons often are placed in two major groups: pyramidal cells and stellate or nonpyramidal cells. The perikarya of **pyramidal cells** are pyramidal in shape and have large **apical dendrites** that usually are oriented toward the surface of the cerebral cortex and enter the overlying layers; the single axons enter the subcortical white matter. They are found in

layers II, III, V, and, to a lesser extent, layer VI. Very large pyramid-shaped neurons (**Betz cells**) are present in the internal pyramidal layer (V) of the frontal lobe. **Stellate (nonpyramidal) cells** lack the pyramid-shaped perikarya and the large apical dendrite. They occur in all layers of the cerebral cortex but are concentrated in the internal granular layer (IV). Impulses entering the cortex are relayed primarily to stellate cells and then transmitted to pyramidal cells in the various layers by the vertical axons of the stellate cells. Axons of pyramidal cells generally leave the cortex and extend to other regions of the brain and spinal cord.

The cerebral cortex functions in vision, hearing, speech, voluntary motor activities, and learning.

Cerebellar Cortex

KEY WORDS: **molecular layer, stellate cell, basket cell, Purkinje cell layer, granule cell layer, mossy fiber, climbing fiber**

The cerebellar cortex is characterized by three layers: an outer molecular layer, a middle Purkinje cell layer, and an inner granule cell layer.

The **molecular layer** is mainly a synaptic area with relatively few nerve cells. It consists primarily of unmyelinated axons from granule cells, the axons running parallel to the cortical surface. It also contains large dendrites of the underlying Purkinje cells and in its superficial portion contains small scattered neurons called **stellate cells.** Other small neurons located deep in this layer and adjacent to Purkinje cells are called **basket cells.**

The **Purkinje cell layer** is formed by the cell bodies of Purkinje cells — large, pear-shaped neurons aligned in a single row and characterized by large branching dendrites that lie in the molecular layer. They represent Golgi type I neurons and number about 15 million. Three-dimensionally, the large dendritic trees occupy a narrow plane, reminiscent of fan coral, and are so arranged that each dendritic tree is parallel to its neighbor. A single small axon from the Purkinje cell passes through the granule cell layer and synapses with neurons in the central cerebellar area.

The **granule cell layer** consists of numerous closely packed, small neurons whose axons enter the molecular layer to synapse with dendrites of Purkinje cells. The granule cells represent Golgi type II neurons. They have small, round nuclei with coarse chromatin patterns and only scant cytoplasm; dendrites are short and clawlike. The axons enter the molecular layer, bifurcate, and run parallel to the surface but perpendicular to the wide plane of the Purkinje dendritic tree. The axons of granule cells synapse with about 450 Purkinje cells in a relationship similar to that of wires coursing along telephone poles. Axons of granule cells also synapse with stellate and basket neurons in the molecular layer. Another type of small neuron, the Golgi cell, is found in the outer zone of the granule cell layer.

Two types of afferent nerve fibers enter the cerebellar cortex from outer regions of the central nervous system. These are the **mossy fibers,** which synapse with granule cells, and the **climbing fibers,** which enter the molecular layer and wind about the dendrites of Purkinje cells.

The cerebellum primarily modulates and coordinates the activities of skeletal muscle.

The gray matter of the spinal cord, cerebrum, and cerebellum consists of a complex, highly ordered meshwork of dendritic, axonal, and glial processes that envelop the perikarya of associated neurons. This network, called the *neuropil,* provides a vast area for synaptic contact and interaction between nerve processes and forms an organizing framework. It is important for coordinating activities in the central nervous system.

Meninges

KEY WORDS: dura mater, epidural space, subdural space, arachnoid, trabeculae, subarachnoid space, pia mater

In addition to the covering of bone supplied by the skull and vertebral column, the central nervous system is contained within three connective tissue membranes called *meninges.* The outermost, the **dura mater,** consists primarily of dense collagenous connective tissue. Around the brain it forms two layers and serves as a periosteum for the cranium and as the covering of the brain. The periosteal layer is rich in blood vessels, nerves, and cells. The inner layer is less vascular, and its interior surface is lined by a simple squamous epithelium. Where it covers the spinal cord, the dura is a single layer of dense irregular connective tissue that contains scattered elastic fibers. The outer surface is covered by a simple squamous epithelium, separated from the periosteum of the vertebrae by an **epidural space** filled by loose connective tissue that is rich in fat

Fig. 8-18. Diagram of the pia-arachnoid.

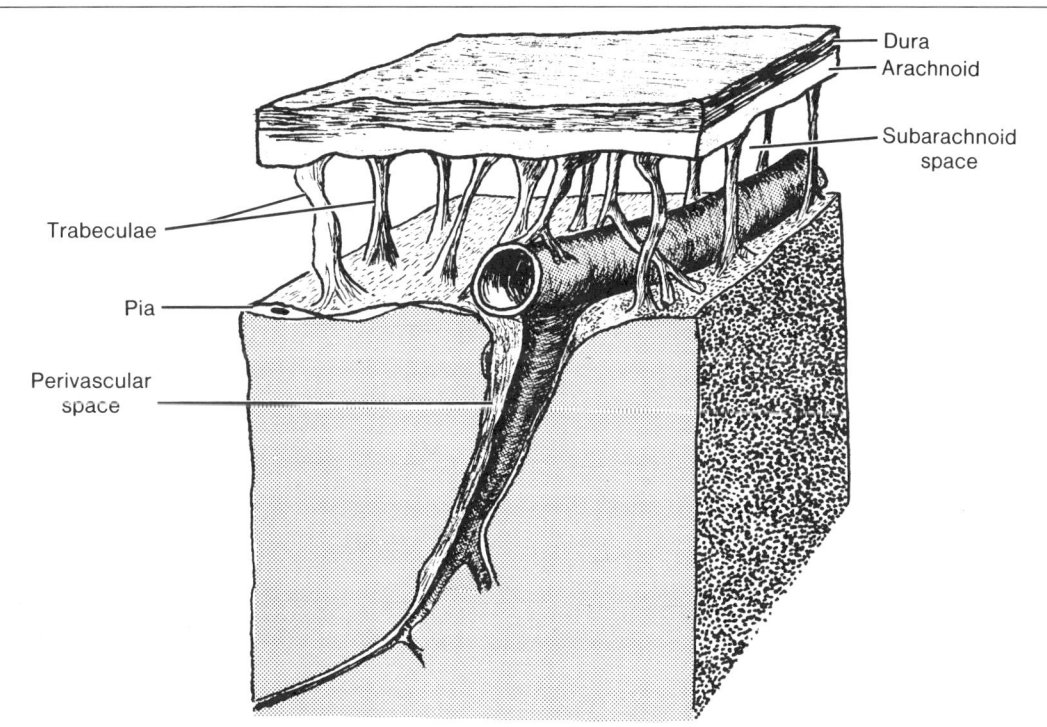

cells and veins. The inner surface of the dura of the spinal cord is lined by a simple squamous epithelium. Between it and the next meninx, the arachnoid, is a narrow, fluid-filled **subdural space.**

The **arachnoid** is a thin, weblike, avascular membrane made up of fine collagenous fibers and scattered elastic fibers. The outer region forms a smooth sheet, while the inner surface gives rise to numerous strands (**trabeculae**) that extend into and blend with the underlying pia mater (Fig. 8-18 on page 155). The inner surface of the arachnoid, the trabeculae, and the external surface of the pia are covered by a simple squamous epithelium that also lines a large space, the **subarachnoid space,** which contains cerebrospinal fluid.

The **pia mater** is a thin, vascular membrane that closely invests the brain and spinal cord. Because the pia and arachnoid are closely related, they often are considered together as the pia arachnoid. The pia consists of delicate collagenous and elastic fibers, fibroblasts, macrophages, and scattered mesenchymal cells. Blood vessels entering and leaving the brain substance are invested by a sheath of pia mater. A perivascular space surrounds the vessels as they extend into the substance of the central nervous system and is continuous with the subarachnoid space.

Review Table 8-1. Key Features of Primary Cell Types of Nervous Tissue

Location of Perikarya	Morphology	Other
Neurons of the Peripheral Nervous System		
Cranial and spinal ganglia	Large, round, pseudounipolar; distinct central nucleus with nucleolus	Each perikaryon surrounded by satellite cells
Sympathetic chain ganglia, parasympathetic ganglia in walls of organs	Size variable, stellate; multipolar; nucleus often eccentric; lipofuscin granules common	Axons usually unmyelinated
Neurons of the Central Nervous System		
Gray matter of spinal cord	Multipolar, stellate; size variable; distinct nuclei and nucleoli	Perikarya occur in groups or columns
Gray matter of cerebral cortex	Multipolar; pyramidal with large apical dendrite; stellate cells with round nuclei; lack apical dendrite	Perikarya organized into 5 layers
Gray matter of cerebellar cortex	Multipolar; Purkinje cell and small neurons of granule layer characteristic	Perikarya organized into 3 layers
Glia of the Peripheral Nervous System		
Schwann cells	Flattened, elongated; heterochromatic nucleus in center of cell	Envelop axon to form neurilemmal sheath; form myelin in peripheral nervous system
Satellite cells (amphicytes)	Single layer of low cuboidal cells; nuclei heterochromatic, cytoplasm clear and indistinct	Surround neuronal perikarya; continuous with neurilemma sheath
Glia of the Central Nervous System		
Fibrous astrocytes	Pale-staining ovoid or spherical nucleus, sparse chromatin; long thin, unbranched processes	Primarily in white matter
Protoplasmic astrocytes	Pale-staining ovoid or spherical nucleus, sparse chromatin; short, thick, branching processes	Primarily in gray matter
Interfascicular oligodendrocytes	Small cell (6–8 μm); ovoid or spherical, deeply staining nuclei; few processes	Envelop axons of white matter; form myelin in central nervous system
Perineuronal oligodendrocytes	Same as above	Gray matter; closely associated with perikarya of neurons
Ependyma	Simple cuboidal to columnar	Central canal of spinal cord; ventricles of brain

Functional Review: Nervous Tissue

An essential function of nervous tissue is communication. Nervous tissue is able to fulfill this role because of its ability to react to various stimuli and transmit the impulse from one region to another. Thus the organism is able to react to the external environment and internal events to integrate and coordinate body functions. In cortical regions of the brain, complex relationships between different types of neurons provide intellect, memory, and conscious experience and serve for the interpretation of special impulses from the eye and ear into the sensations of sight and sound, respectively. All contribute to the personality and behavior of the individual.

Neurons are the structural and functional units of nervous tissue. They generally are complex in shape and usually have several cytoplasmic processes. Dendrites conduct nerve impulses to the perikaryon, and a single axon conducts impulses from the perikaryon to other neurons or effector organs. Receptor organs of nerve tissue convert mechanical, chemical, or other stimuli into nerve impulses that are transmitted along a physiologic dendrite to its cell body located in spinal or cranial ganglia. From ganglia the impulses are relayed to other neurons in the central nervous system. Nerve fibers of most somatic sensory neurons are large and myelinated, and impulses are relayed quickly to the central nervous system to elicit a response. Most effector and internuncial neurons are multipolar, and their numerous dendrites summate the excitatory and inhibitory impulses that influence a specific neuron. Dendrites of this type of neuron usually are unmyelinated. When the threshold of activity is attained, an action potential is generated in the region of the axon hillock and/or initial segment of the axon. These regions of the neuron have a lower threshold of excitation than the dendrites or perikaryon. The impulse is conducted by the axon to other neurons or an effector organ to elicit a response. The speed of conduction depends on fiber diameter and usually reflects the degree of myelinization, both in peripheral nerve fibers and those of the central nervous system. Myelin is formed by consecutive layerings of the plasmalemma of adjacent supporting cells. In the peripheral nervous system, myelin is formed by Schwann cells scattered at intervals along a single axon. Oligodendrocytes are responsible for myelinization in the central nervous system and form internodal segments on more than one axon. Myelin is thought to act as an insulating material to prevent the wave of depolarization (action potential) from traveling the length of the axolemma as occurs in unmyelinated nerve fibers. Depolarization occurs only at the nodes of Ranvier in myelinated nerves, and the impulses jump the internodal segment to adjacent nodes. This saltatory conduction partly explains the greater speed of nerve impulses transmitted by myelinated nerve fibers.

Communication between neurons and effector organs takes place at specific contact points called synapses, where nerve impulses are transferred directly to other cells by chemical transmitters or electrical coupling. The terminal axons of some neurons, such as those in the hypothalamic and paraventricular regions, may secrete peptides or other agents directly into the bloodstream. These neurons influence cells that are not in direct contact with them and thus act as endocrine cells.

Neurons are surrounded by supporting cells. In the peripheral nervous system, neurons are intimately associated with Schwann and satellite cells that, like astrocytes and oligodendrocytes of the central nervous system, provide structural and metabolic support to their associated neurons. The ependyma represents glial cells that form a simple epithelial lining for the ventricles of the brain and the central canal of the spinal cord. In some areas of the ventricles, ependyma comes in direct contact with vascularized regions of the pia mater to form the choroid plexus. The choroid plexus actively and continuously produces cerebrospinal fluid, which surrounds and cushions the brain and spinal cord and plays an important role in their metabolism. The meninges form protective coverings for the brain and spinal cord and contain numerous blood vessels that supply the structures of the central nervous system.

————— *Review Questions and Answers* —————

Questions

46. Define a nerve fiber as found in the peripheral nervous system.
47. Discuss the role of Nissl substance in a neuron.
48. Compare myelin in the peripheral and central nervous systems.
49. What role does myelin play in the transmission of a nerve impulse?
50. Briefly discuss the synapse and its role in nervous tissue.

Answers

46. A peripheral nerve fiber consists of an axon (axis cylinder) and a surrounding neurilemmal sheath (sheath of Schwann), derived from ectoderm. Nerve fibers may be classified as myelinated or unmyelinated depending on whether a myelin sheath is present.

47. Nissl substance is found throughout the cytoplasm of the perikaryon and dendrites of most neurons but is absent from the axon hillock and axon. It consists of cisternae of granular endoplasmic reticulum arranged in parallel. Nissl bodies represent the sites of protein synthesis in neurons. The protein is transported down the length of the axon for use in metabolism or repair in this region or in some instances is secreted. Protein synthesis does not occur in the axoplasm.

48. Myelin of the peripheral nervous system does not form a continuous sleeve but is in the form of internodal segments, each of which is produced and maintained by a single Schwann cell. It is not a secretory product but represents a mass of lipid-protein that results from the successive layering of the Schwann cell plasmalemma wrapped around the axon. Oblique discontinuities (clefts of Schmidt-Lantermann) are often observed in the myelin sheaths of peripheral nerves and represent local separations of myelin lamellae.

    Myelin of the central nervous system also is arranged in numerous internodal segments around axons and, as in peripheral nerves, forms as the result of successive layering of a plasmalemma around an axon. Unlike myelin of the peripheral nerves, however, myelin in the central nervous system is formed and maintained by interfascicular oligodendrocytes. Whereas each internodal segment of myelin in the peripheral nervous system is formed by a single Schwann cell, in the central nervous system a single oligodendrocyte is responsible for the internodal segments on several separate but adjacent axons. Myelinated nerve fibers of the central nervous system lack the clefts of Schmidt-Lantermann.

49. The speed with which an impulse is transmitted along a nerve fiber is proportional to the diameter of the nerve fiber. The diameter of heavily myelinated nerve fibers is much larger than that of unmyelinated fibers, and therefore, the speed of impulse conduction is faster in myelinated nerves. In the unmyelinated nerve fibers, the action potential travels along the length of the axolemma, whereas in the faster-conducting myelinated nerves, the impulses are conducted from node to node, jumping the internodal segments (saltatory conduction). The myelin sheath is thought to act as an insulating material, and depolarization occurs only at the nodes of Ranvier. This explains, in part, the speed of nerve transmission by myelinated nerve fibers.

50. A synapse represents a specialized site of attachment between neurons or between nerve cells and effector cells where impulses are transferred from one cell to another. Impulses may be transferred either by chemical transmitters or electrical coupling. Synapses may occur between axon and dendrite, axon and cell body, axon and axon, dendrite and dendrite, dendrite and perikaryon, or cell bodies of adjacent neurons. Both excitatory and inhibitory synapses occur. Although synapses vary in morphology in different types of neurons, they exhibit some common features. The presynaptic area generally contains numerous mitochondria and clusters of synaptic vesicles and is separated from the postsynaptic membrane by a narrow synaptic cleft that measures 20 to 30 nm in diameter. The postsynaptic cell membrane shows an increase in electron density and is slightly thickened, and its internal surface is associated with a subsynaptic web. The postsynaptic cytoplasm is further characterized by a lack of synaptic vesicles and contains only a few scattered mitochondria. In this type of synapse, as an action potential reaches the presynaptic area, the synaptic vesicles fuse at specific sites along the presynaptic membrane and release their contained transmitter substance into the synaptic cleft. If the transmitter substance acts to decrease the electrical

polarization of the postsynaptic membrane, an action potential (impulse) in the postsynaptic neuron is likely to be generated. If, on the other hand, the effect of the transmitter substance on the postsynaptic membrane increases the electrical polarization, it acts to elevate the threshold of firing and is inhibitory. Some neuro-transmitters stimulate second-messenger molecule systems, which, through a cascade of intracellular events, serve to modulate neuronal activity. Electrical coupling is another method by which impulses can be transferred between cells and usually occurs at gap junctions.

9 The Eye

Eyes are photoreceptors containing systems of convex surfaces that transmit and focus light on a surface sensitive to the intensity and wavelength (color) of light. Images formed there are transmitted to the brain, where they are interpreted, correlated, and translated into the sensation of sight. Eyes can regulate the amount of light admitted and change their focal lengths.

In the human, the eyes are almost spherical and are located in and protected by bony sockets provided with a padding of fat. Because the eyes are offset from each other, the images formed are correspondingly offset and provide for binocular vision. Integration of the binocular images results in a three-dimensional quality of sight through which depth and spacial orientation are recognized.

General Structure

The wall of the eyeball consists of an outer corneoscleral coat, a central coat called the *uvea*, and an inner coat called the *retina*.

The corneoscleral coat is divided into a smaller, transparent, anterior cornea that allows light to enter the eye and the sclera, a larger, white segment. The sclera is a tough fibrous layer that protects the delicate internal structures of the eye and helps maintain the shape of the eyeball.

The uveal coat consists of the choroid, ciliary body, and iris. The choroid is the vascular part of the uvea, immediately beneath the retina. It is continuous with the ciliary body, which forms a beltlike structure around the interior of the eyeball, just forward of the anterior margin of the retina. The ciliary body controls the diameter and shape of the lens, which focuses light on the retina. The iris, the last portion of the uvea, is continuous with the ciliary body and has a central opening, the pupil, whose diameter can be changed to regulate the amount of light passing into the eyeball.

The retina consists of pigment-containing photoreceptors that, on exposure to light, produce chemical energy that is converted into the electric energy of nerve impulses. A chain of conducting neurons in the retina transmits the impulses to the optic nerve. The optic disc is the point where nerve fibers from the retina gather to form the optic nerve and leave the eyeball.

The interior of the eyeball is subdivided into an anterior chamber, a posterior chamber, and a vitreal cavity. Each is filled with a transparent medium that helps maintain the shape and turgor of the eye. The anterior and posterior chambers are continuous through the pupil and contain a watery aqueous humor, which provides nutrients for the anterior structures of the eye. The large vitreal cavity is filled with a viscous, transparent gel called the *vitreous humor* or *vitreous body*. The general structure of the eye is shown in Figure 9-1.

Corneoscleral Coat

The corneoscleral coat forms the outermost layer of the eyeball and consists of the cornea and sclera.

Sclera

KEY WORDS: fibrous outer tunic, lamina cribrosa

The sclera forms the tough, opaque, **fibrous outer tunic** of the posterior five-sixths of the eyeball. It varies in thickness in different regions but averages about 0.5 mm. It is composed of flat bundles of collagen that run in various directions parallel to the scleral surface. Delicate networks of elastic fibers and elongated, flattened fibroblasts extend between the collagen bundles. Melanocytes occur in the deeper layers. The sclera thins and forms a fenestrated membrane, the **lamina cribrosa,** at the point where fibers of the optic nerve penetrate the sclera to exit the eye. The sclera protects the interior of the eye, aids in maintaining its shape, and serves as the site of attachment for the extrinsic muscles of the eye.

Cornea

KEY WORDS: corneal epithelium, Bowman's membrane, substantia propria, keratocyte, Descemet's membrane, corneal endothelium

The transparent cornea is the most anterior part of the eye. It is about 11 mm in diameter, approximately 0.8 mm thick

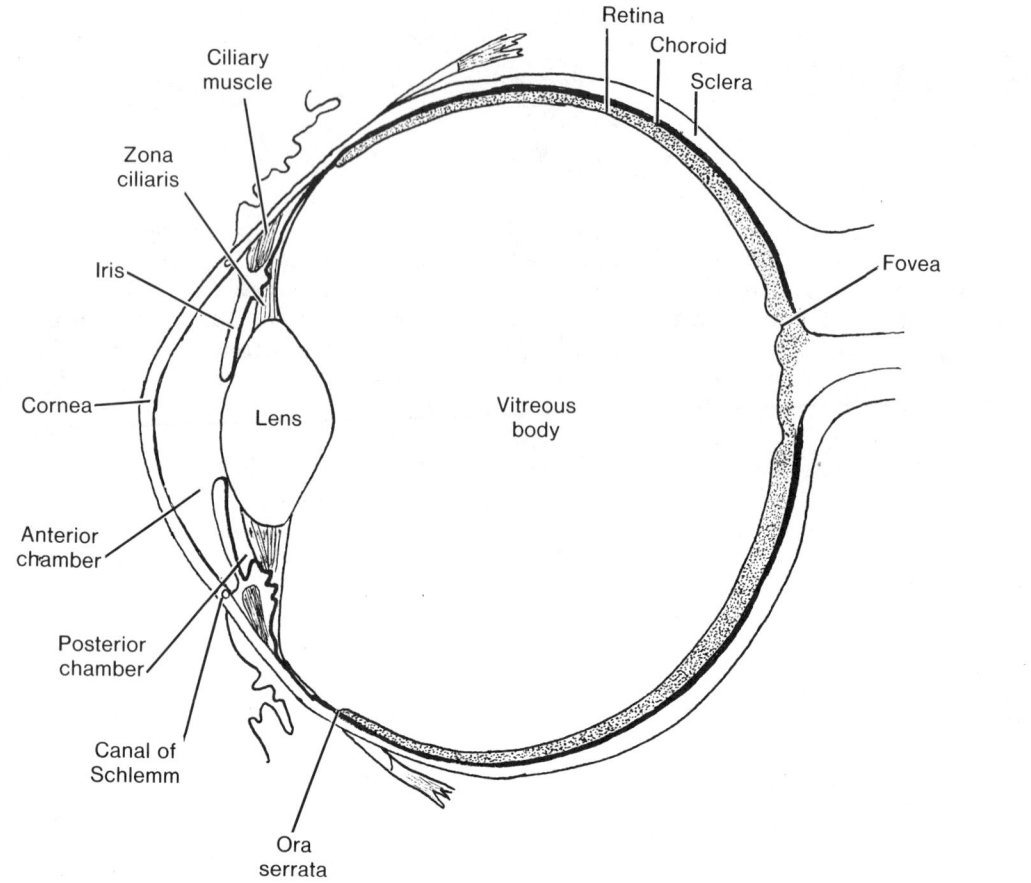

Fig. 9-1. General structure of the eye.

near the center, and 1.00 mm at the periphery. Its curvature is considerably greater than that of the posterior sclera. The cornea is uniform in structure throughout and consists of corneal epithelium, Bowman's membrane, substantia propria, Descemet's membrane, and corneal endothelium.

The **corneal epithelium** is stratified squamous with a smooth outer surface and averages about 50 μm in depth (Fig. 9-2). It usually consists of five layers of large squamous cells that have few organelles but often contain glycogen. The superficial cells retain their nuclei, and their external surfaces form numerous fine ridges (microplicae) that help retain moisture on the corneal surface. The epithelium lies on a distinct basal lamina. The lateral membranes of adjacent cells are extensively interdigitated and joined by numerous desmosomes. The corneal epithelium

contains many free nerve endings and is very sensitive to a number of stimuli, especially pain.

The basal lamina of the corneal epithelium lies on the outer layer of the substantia propria and is called **Bowman's membrane;** it ends abruptly at the peripheral margin of the cornea. Bowman's membrane has a homogeneous appearance and consists of a feltwork of small collagen fibrils (type I) and lacks elastin.

The **substantia propria,** or stroma, forms the bulk of the cornea (see Fig. 9-2). It consists of numerous bundles of collagen fibers (type I and type V) arranged in thin lamellae that run parallel to the corneal surface. The collagen bundles in each successive lamina run in different directions and cross at various angles. Adjacent lamellae are knit tightly together by interlacing collagenous fibers. The

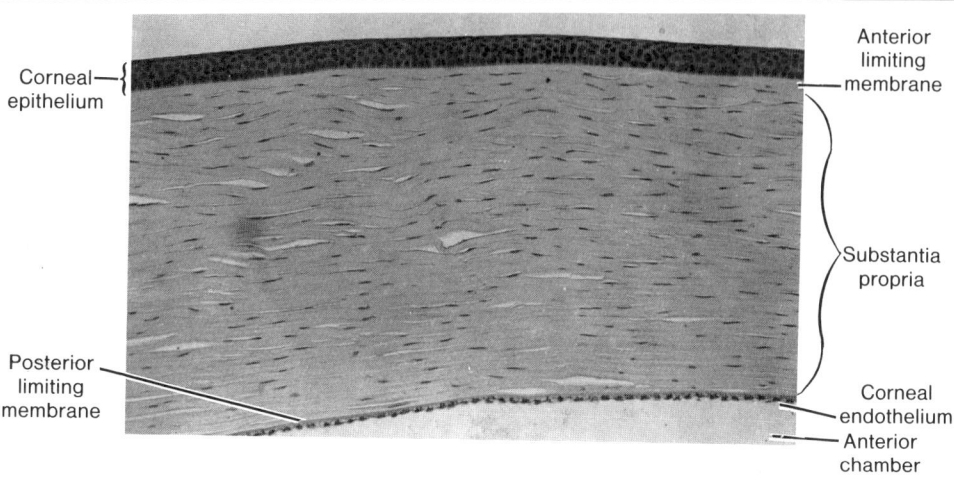

Corneal epithelium

Anterior limiting membrane

Substantia propria

Posterior limiting membrane

Corneal endothelium

Anterior chamber

Fig. 9-2. A photomicrograph of cornea showing its subcomponents (LM, × 100).

fibers, bundles, and lamellae of collagen are embedded in a matrix of sulfated proteoglycans rich in chondroitin sulfate and keratosulfate. These proteoglycans are not present in the sclera. Flattened fibroblast-like cells, the **keratocytes,** are present between the bundles of collagen fibers. Blood vessels and lymphatics normally are absent in the substantia propria, although occasional lymphoid wandering cells are seen.

Descemet's membrane is a homogeneous membrane, 6 to 8 μm thick, lying between the posterior surface of the substantia propria and the corneal endothelium. It corresponds to a thick basal lamina secreted by the corneal endothelium. It is resilient and elastic, although elastic fibers are not present, and appears to consist mainly of the type VIII form of collagen.

The **corneal endothelium** lines the inner surface of the cornea and consists of a layer of large, hexagonal squamous cells (see Fig. 9-2). The apices of the cells are joined by tight junctions, and the cytoplasm contains numerous mitochondria and vesicles. The cells are involved in transporting materials from the anterior chamber.

The transparency of the cornea is due to the uniform diameter and orderly arrangement of collagen fibers and to the properties of the ground substance. The cornea is avascular and depends on the aqueous humor and blood vessels of the surrounding limbus to supply its nutrients. Ion pumps of the corneal endothelial cells maintain a critical fluid level within the substantia propria. If excess fluid accumulates, the cornea becomes opaque. Corneal endothelial cells do not divide after birth and therefore are not

replaced. Each person is born with a complete complement of corneal endothelial cells.

Limbus

KEY WORDS: external scleral sulcus, internal scleral sulcus, scleral spur, trabecular meshwork, Schlemm's canal

The limbus represents a zone of transition 1.5 to 2.0 mm wide between the transparent cornea and the opaque sclera. The outer surface bears a shallow groove, the **external scleral sulcus,** where the more convex cornea joins the sclera. A similar structure, the **internal scleral sulcus,** lies on the inner surface of the limbus. The **scleral spur** is a small ridge of tissue projecting from the posterior lip of the internal scleral sulcus and attached anteriorly to tissue that comprises the outflow system for fluid in the anterior chamber. The limbus is further characterized by the terminations of Bowman's and Descemet's membranes. The collagen fibers and bundles of the cornea become larger here and their arrangement more irregular as they blend with those of the sclera. Numerous small blood vessels form arcades in this region and nourish the peripheral portions of the avascular cornea.

Descemet's membrane is replaced by the spongy **trabecular meshwork** attached to the anterior part of the scleral spur. It contains numerous flattened, anastomosing trabeculae that consist of collagenous fibers and ground substance covered by an attenuated endothelium that is continuous with the corneal endothelium. The trabeculae form a labyrinth of spaces that communicate with the anterior chamber. Between the bulk of the limbal stroma and the trabecular meshwork, a flattened, endothelial-lined canal, **Schlemm's canal,** courses around the circumference of the cornea. Aqueous humor enters the trabecular spaces from the anterior chamber and crosses the endothelial lining of the trabeculae, the juxtacanalicular connective tissue, and finally, the endothelium of the canal to enter its lumen. Several channels arise from the peripheral wall of the canal to join veins in the limbus and eventually drain to episcleral veins. Obstruction to the drainage of aqueous humor causes a rise in intraocular pressure, a characteristic of the condition called *glaucoma.*

Uveal Layer

The uvea, the middle vascular coat of the eye, is divided into choroid, ciliary body, and iris.

Choroid

KEY WORDS: **suprachoroid lamina, perichoroidal space, vessel layer, choriocapillary layer, glassy (Bruch's) membrane**

The choroid is a thin, brown, highly vascular membrane that lines the inner surface of the posterior sclera. Its outer surface is connected to the sclera by thin avascular lamellae that form a delicate, pigmented layer called the **suprachoroid lamina.** The lamellae mainly consist of fine elastic fibers between which are numerous large, flat melanocytes and scattered macrophages. The lamellae cross a potential cleft, the **perichoroidal space,** between the sclera and choroid to enter the choroid proper.

The choroid proper consists of three regions. The outer **vessel layer** is made up of loose connective tissue with numerous melanocytes and contains the larger branches of the ciliary arteries and veins. The central **choriocapillary layer** consists of a net of capillaries lined by a fenestrated endothelium; these vessels provide for the nutritional needs of the outer layers of the retina. The choriocapillary layer ends near the ora serrata. The innermost layer of the choroid is the **glassy (Bruch's) membrane** that lies between the remainder of the choroid and the pigment epithelium of the retina. With the light microscope it appears as a homogeneous layer, 1 to 4 μm thick. Ultrastructurally it shows five separate strata made up of basal laminae of capillaries in the choriocapillary layer, the pigment epithelium of the retina, and between them, two thin layers of collagen separated by a delicate elastic network.

Ciliary Body

KEY WORDS: **ciliary epithelium, inner nonpigmented cell, outer pigmented cell, blood-aqueous barrier, aqueous humor, ciliary process, zonule fiber, ciliaris muscle**

The ciliary body is located between the ora serrata of the neural retina and the outer edge of the iris, where it attaches to the posterior aspect of the scleral spur at the corneoscleral junction (Fig. 9-3). It forms a thin triangle when

Fig. 9-3. A segment of human eye illustrating the relationship between iris, lens, and ciliary body (LM, × 250).

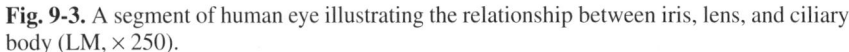
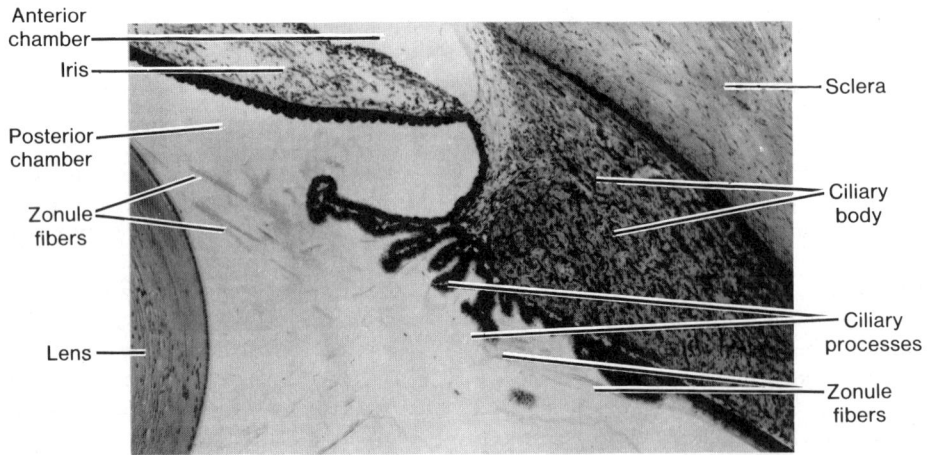

seen with the light microscope and consists of an inner vascular tunic and a mass of smooth muscle immediately adjacent to the sclera. The internal surface is covered by **ciliary epithelium,** a continuation of the pigment epithelium of the retina that lacks photosensitive cells.

Ciliary epithelium consists of an **inner** layer of **nonpigmented cells** and an outer layer of pigmented cells, each resting on a separate basal lamina; it is unusual in that the cell apices of both layers are closely apposed (Fig. 9-4). The **outer pigmented cell** layer is separated from the stroma of the ciliary body by a thin basal lamina continuous with that underlying the pigment epithelium in the remainder of the retina. The basal lamina of the nonpigmented layer lies adjacent to the posterior chamber of the eye and is continuous with the inner limiting membrane of the retina. The basal plasmalemma of the nonpigmented cells shows numerous infoldings and is involved in ion transport. The cells contain numerous mitochondria and a well-developed, supranuclear Golgi complex. The adjacent pigmented cells of the outer layer also show prominent basal infoldings, and the cytoplasm is filled with melanin granules. The apices of cells of the inner, nonpigmented epithelium are united by well-formed tight junctions that form the anatomic portion of the **blood-aqueous barrier,** which selectively limits passage of materials between the blood and the interior of the eye.

The ciliary epithelium elaborates **aqueous humor,** which differs from blood plasma in its electrolyte composition and lower protein content. Aqueous humor fills the posterior chamber, provides nutrients for the lens, and posteriorly, enters the vitreous humor. Anteriorly, it flows from the posterior chamber through the pupil into the anterior chamber and aids in nourishing the cornea. It leaves the anterior chamber via the trabecular meshwork and Schlemm's canal to the episcleral veins.

In its anterior part, the inner surface of the ciliary body is formed by 60 to 80 radially arranged, elongated ridges called **ciliary processes,** which are lined by ciliary epithelium and contain a highly vascularized stroma and scattered melanocytes. **Zonule fibers,** which hold the lens in place, are produced mainly by the nonpigmented cell layer of the ciliary epithelium and are attached to its basal lamina. They extend from the ciliary processes to the equator of the lens.

Most of the ciliary body consists of smooth muscle, the **ciliaris muscle,** which controls the shape and therefore the focal power of the lens. The muscle cells are organized into regions with circular, radial, and meridional orientations. Numerous elastic fibers and melanocytes form a sparse connective tissue between the muscle bundles. The ciliaris is important in eye accommodation, and when it contracts, it draws the ciliary processes forward, thus relaxing the suspensory ligament (zonule fibers) of the lens, allowing the lens to become more convex and to focus on objects near the retina.

Iris

KEY WORDS: pupil, anterior chamber, posterior chamber, myoepithelial cell, sphincter muscle

The iris is a thin disc suspended in the aqueous humor between the cornea and lens and has a central, circular aperture called the **pupil.** The iris is continuous with the ciliary body at the periphery and divides the space between the cornea and lens into anterior and posterior chambers (see Fig. 9-3). The **anterior chamber** is bounded anteriorly by

Fig. 9-4. Details of the ciliary epithelium from a human eye (LM, × 400).

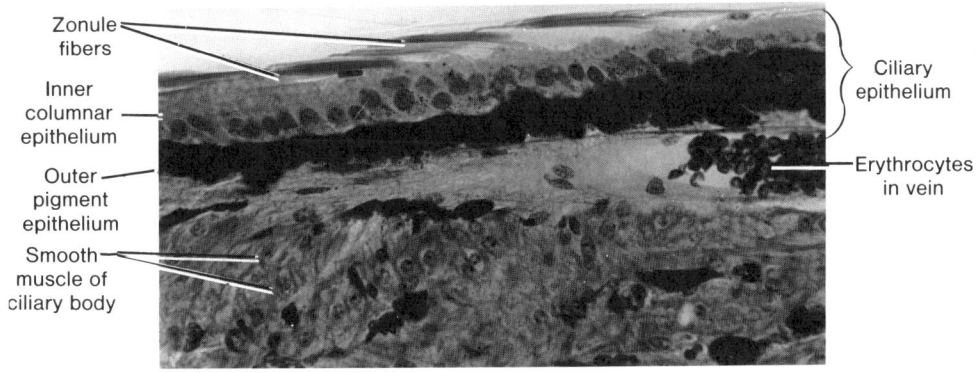

the cornea and posteriorly by the iris and central part of the lens. The **posterior chamber** is a narrow space between the peripheral part of the iris in front and the peripheral portion of the lens, ciliary zonule, and ciliary processes. The two chambers communicate through the pupil. The anterior surface of the iris is irregular due to numerous fissures or crypts. The margin attached to the ciliary body forms the ciliary margin; that surrounding the pupil is the pupillary margin.

The stroma of the iris consists primarily of loose, vascular connective tissue with scattered collagenous fibers, melanocytes, and fibroblasts embedded in a homogeneous ground substance. The anterior surface lacks a definite endothelial or mesothelial covering but is lined in part by a discontinuous layer of melanocytes and fibroblasts. Tissue spaces of the stroma often appear to communicate with the anterior chamber. The posterior surface is covered by two rows of cuboidal, pigmented cells that are continuous with the ciliary epithelium. Where it passes onto the posterior surface of the iris, the inner nonpigmented layer of the ciliary epithelium becomes heavily pigmented. Cells of the outer layer have less pigment and are modified into **myoepithelial cells** to form the dilator of the iris. The dilator of the pupil consists only of a single layer of radially arranged myoepithelial cells whose contraction increases the diameter of the pupil.

The **sphincter muscle** consists of a circularly arranged, compact bundle of smooth muscle cells near the pupillary margin. Contraction of the sphincter muscle reduces the diameter of the pupil. The iris acts as a diaphragm, modifying the amount of light that enters the eye, thus permitting a range of vision under a variety of lighting conditions.

The color of the iris is determined by the amount and distribution of pigment. In the various shades of blue eyes, melanin is restricted to the posterior surface of the iris, whereas in gray and brown eyes melanin is found in increasing concentration within melanocytes present throughout the stroma of the iris. In albinos, melanin pigment is absent and the iris takes on a pink color due to the vasculature of the iridial stroma.

Retina

KEY WORDS: **pigment epithelium, neural retina (retina proper), ora serrata, macula lutea, fovea centralis, optic disc**

The retina, the innermost of the three tunics, is a delicate sheet of nervous tissue that forms the photoreceptor of the eye. The outer surface is in contact with the choroid; the inner surface is adjacent to the vitreous body. The posterior retina consists of an outer **pigment epithelium** and an inner **neural retina (retina proper).** The retina decreases in thickness anteriorly, and the nervous component ends at a ragged margin called the **ora serrata.** A thin prolongation of the retina covers the posterior aspect of the iris, where it forms the iridia retinae. It also extends anteriorly to cover the ciliary processes as the ciliary epithelium. The forward extension of the retina consists only of the pigmented layer and an inner layer of columnar epithelial cells; the nervous component is lacking.

The neural retina is anchored only at the optic disc, where nerve fibers congregate before passing through the sclera to form the optic nerve, and at the ora serrata. Although cells of the pigment epithelium interdigitate with photoreceptor cells of the neural retina, there is no anatomic connection between the two components of the retina, and after trauma or disease, the neural retina may detach from the pigment epithelium.

The exact center of the posterior retina corresponds to the axis of the eye, and at this point, vision is most perfect. The region appears as a small, yellow, oval area called the **macula lutea.** In its center is a depression about 1.5 mm in diameter called the **fovea centralis,** where the sensory elements are most numerous and most precisely organized. Nearer the periphery of the retina, neural elements are larger, fewer, and less evenly distributed. About 4.0 mm to the nasal side of the macula lutea is the **optic disc,** the site of formation and exit of the optic nerve. Lacking photoreceptors, this area is insensitive to light and forms the "blind spot" of the retina.

Neural Retina

KEY WORDS: **three-neuron conducting chain, photoreceptor cell, bipolar neuron, ganglion cell, association neuron, glial cell**

Except for the extreme periphery and the fovea centralis, the neural retina is made up of the following layers, listed in order from the choroid, as seen with the light microscope: (1) layer of rods and cones, (2) external limiting membrane, (3) outer nuclear layer, (4) outer plexiform layer, (5) inner nuclear layer, (6) inner plexiform layer, (7) ganglion cell layer, (8) layer of nerve fibers, and (9) internal limiting membrane (Fig. 9-5). The retina consists essentially of a **three-neuron conducting chain** that ultimately forms the nerve fibers in the optic nerve (Fig. 9-6). The neural elements of the conducting chain are **photore-**

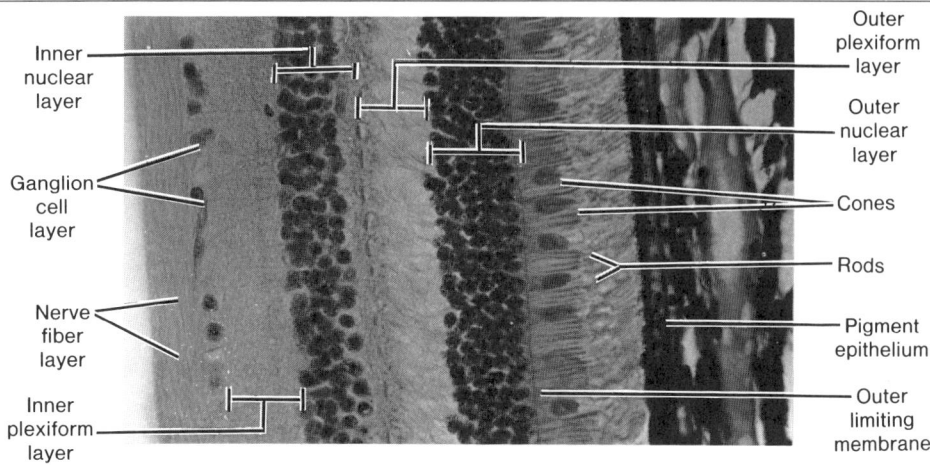

Fig. 9-5. A section through a human neural retina illustrating its subcomponents (LM, × 250).

ceptors (rod and cone cells), **bipolar neurons,** and **ganglion cells.**

When stimulated by light, photoreceptor cells transmit an action potential to the bipolar neurons, which in turn synapse with ganglion cells. Unmyelinated axons from the ganglion cells enter the layer of retinal nerve fibers and unite at the optic disc to form the optic nerve, which transmits the impulse to the brain. Other intraretinal cells are either **association neurons** (horizontal cells, amacrine cells) or **glial cells.** These elements are organized into the various layers of the retina and can be summarized as shown in Table 9-1.

Glial Cells

KEY WORDS: Müller's cells, inner limiting membrane, outer limiting membrane

The largest and most prominent glial elements in the neural retina are **Müller's cells,** which extend between the outer and inner limiting membranes of the retina. Their cytoplasmic processes run between the cell bodies and processes of neurons in the retina and provide physical support for the neural elements. The basal lamina of Müller's cells lies adjacent to the vitreal body and forms the **inner limiting membrane** of the retina. The apices of Müller's cells are joined to adjacent photoreceptor cells by junctional complexes and form the **outer limiting membrane.** The cytoplasm is rich in smooth endoplasmic reticulum and contains abundant glycogen. In addition to

providing structural support, Müller's cells are thought to aid in the nutritional maintenance of other retinal elements and have been equated with astrocytes of the central nervous system.

A small number of spindle-shaped glial cells also are present around ganglion cells and between axons that form the nerve fiber layer of the retina.

Photoreceptor Cells

KEY WORDS: rod cell, rod proper, outer segment, rhodopsin, inner segment, ellipsoid, vitreal portion, myoid, outer fiber, cell body, inner fiber, spherule, synaptic ribbon, cone cell, iodopsin, cone pedicle

The two types of photoreceptors are rod cells and cone cells (Fig. 9-7). **Rod cells** are long, slender cells that lie perpendicular to the layers of the retina. They are 40 to 60 μm long and 1.5 to 3.0 μm wide, depending on their location in the retina, and number between 75 million and 170 million in each retina (see Fig. 9-5). The scleral (outer) third of each rod, called the *rod proper,* lies between the pigment epithelium and the outer limiting membrane. The scleral end of the rods is surrounded by processes that extend from the underlying pigment epithelial cells; the inner or vitreal end extends into the outer plexiform layer.

Each **rod proper** consists of outer and inner segments connected by a slender stalk containing nine peripheral doublets of microtubules. The doublets originate from a basal body in the vitreal end of the inner segment, but the

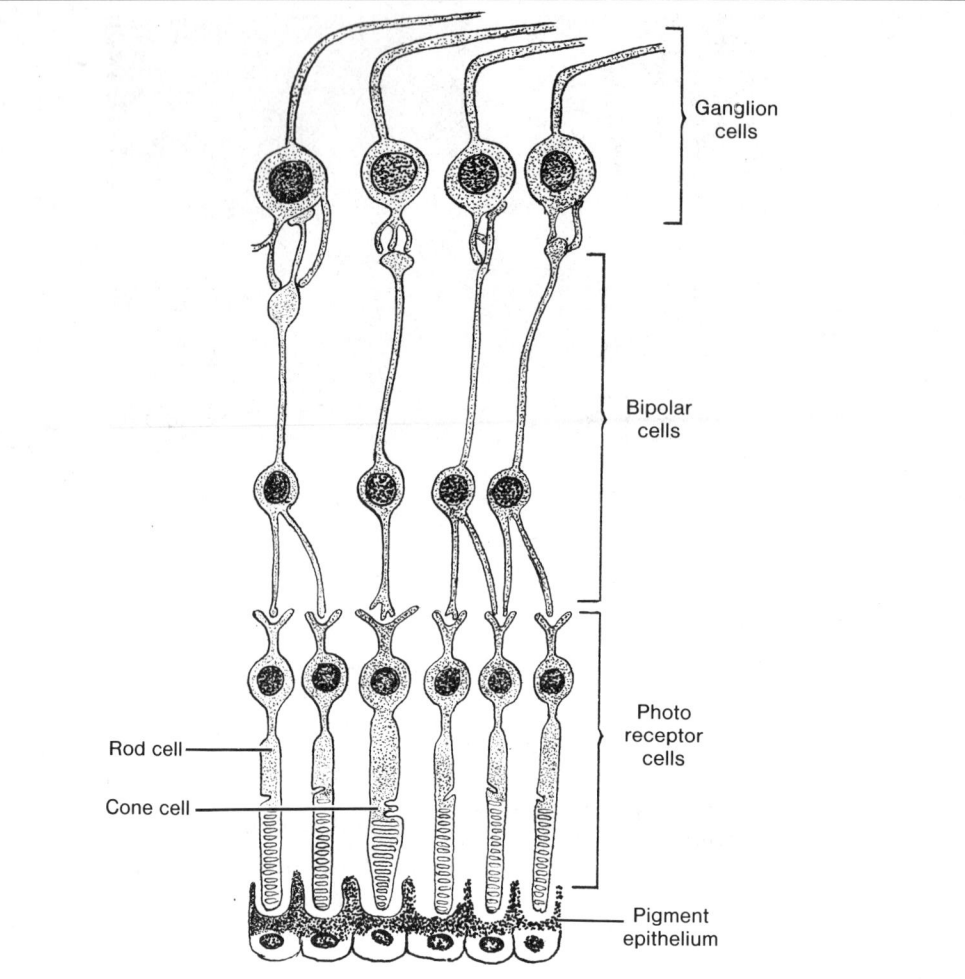

Fig. 9-6. Three-neuron chain system.

connecting stalk differs from a typical cilium in that it lacks a central pair of microtubules (Fig. 9-8).

The **outer segment** contains hundreds of flattened membranous sacs or discs of uniform diameter. These sacs contain **rhodopsin,** the substance responsible for absorption of light. After exposure to light, rhodopsin changes its molecular configuration from cis to trans and breaks down, resulting in hyperpolarization of the plasmalemma of the rod cell, with formation of an electrical potential that is transferred to dendrites of associated bipolar cells. After stimulation, rhodopsin is rapidly reconstituted.

The **inner segment** consists of an outer scleral region called the **ellipsoid** (which contains numerous mitochon-

dria) and a **vitreal portion** that houses the Golgi complex, free ribosomes, and myoid. Microtubules are numerous in both regions. The **myoid** region contains elements of granular and smooth endoplasmic reticulum and synthesizes and packages proteins that are transported down the connecting stalk to the scleral end of the outer rod segment to be used in the assembly of new membranous discs. Older sacs are shed from the tips of the rods during the morning hours and are replaced by new discs. The discarded discs are phagocytized and destroyed by cells of the pigment epithelium.

The remainder of the rod cell consists of an outer fiber, a cell body, and an inner fiber. The **outer fiber** is a thin

Table 9-1. Contents of Layers of Retina

Layers of Retina	Principal Contents
1. Layer of rods and cones	Inner and outer segments of rods and cones
2. External limiting membrane	Formed by junctional complexes between the scleral tips of Müller's cells and adjacent photoreceptor cells
3. Outer nuclear layer	Cell bodies and nuclei of rods and cones
4. Outer plexiform layer	Rod spherules, cone pedicles, dendrites of bipolar neurons, processes of horizontal cells
5. Inner nuclear layer	Nuclei of bipolar, horizontal, amacrine, and Müller's cells
6. Inner plexiform layer	Axons of bipolar neurons, dendrites of ganglion cells, processes of amacrine cells
7. Ganglion cell layer	Ganglion cells, scattered neuroglial cells
8. Nerve fiber layer	Nonmyelinated nerve fibers from ganglion cells, small neuroglial cells, processes of Müller's cells
9. Internal limiting membrane	Vitreal processes of Müller's cells and their basal lamina

process that extends from the inner segment of the rod proper to the **cell body,** which contains the nucleus. The **inner fiber** joins the cell body to the **spherule,** a pear-shaped synaptic ending that contains numerous synaptic vesicles and a **synaptic ribbon.** The latter consists of a dense proteinaceous plaque that lies perpendicular to the presynaptic surface, often bounded by numerous vesicles. The cell body and nucleus of the rods are located in the outer nuclear layer of the retina; the spherule lies within the outer plexiform layer.

Cone cells are 75 μm or more in length at the fovea, decreasing to 40 μm at the edge of the retina. Each retina contains between 6 million and 7 million cones. In general, cone cells resemble the rods but are flask-shaped with short, conical outer segments and relatively broad inner segments united by a modified stalk similar to that of rod cells (see Figs. 9-5 and 9-7). The membranous sacs of the outer cone segments differ in that they may remain attached to the surrounding cell membrane and decrease in diameter as they approach the tip of the cone.

The inner segment, also called the *ellipsoid portion,* shows a region with numerous mitochondria and a myoid part that contains the Golgi complex and elements of smooth and granular endoplasmic reticulum. As in rods, the cones synthesize proteins that pass to the outer segments, where they are used in the formation of new membranous sacs. Older sacs appear to be shed in the evening and are phagocytized in the pigment epithelial cells. Unlike those in rod cells, the sacs decrease in size as they approach the tip of the cone. The visual pigment of cone cells, **iodopsin,** is associated with the outer segments. Absorption of light and generation of an electrical impulse are similar to that oc-

curring in rods. Cones function in color perception and visual acuity, responding to light of relatively high intensity. Detection of color is believed to depend on the presence of several pigments in the cones, whereas rods are thought to contain only one form of pigment.

Except for those in the outer fovea, cones lack an outer fiber, and the inner cone segment blends with the cell body. The nuclei of cone cells are larger and paler than those of the rods and form a single row in the outer nuclear layer, adjacent to the outer limiting membrane. Each cone has a thick inner fiber that runs to the outer plexiform layer, where it forms a club-shaped synaptic ending, the **cone pedicle,** which synapses with processes from bipolar and horizontal neurons.

Bipolar Cells

KEY WORDS: rod bipolar cell, flat cone bipolar cell, invaginating midget bipolar cell, flat midget bipolar cell

Like photoreceptor cells, bipolar cells lie perpendicular to the retinal layers with their cell bodies and nuclei in the inner nuclear layer of the retina. They give rise to one or more dendrites that extend into the outer plexiform layer, where they synapse with terminals of photoreceptor cells. A single axon extends into the inner plexiform layer and synapses with processes from ganglion cells. Several types of bipolar neurons have been described: **Rod bipolar cells** make contact with several rod cells; **flat cone bipolar cells** form synapses with several cone pedicles; **invaginating midget** and **flat midget bipolar cells** synapse with a single cone pedicle. Bipolar cells relay impulses from the rod and cone cells to the ganglion cells of the next layer.

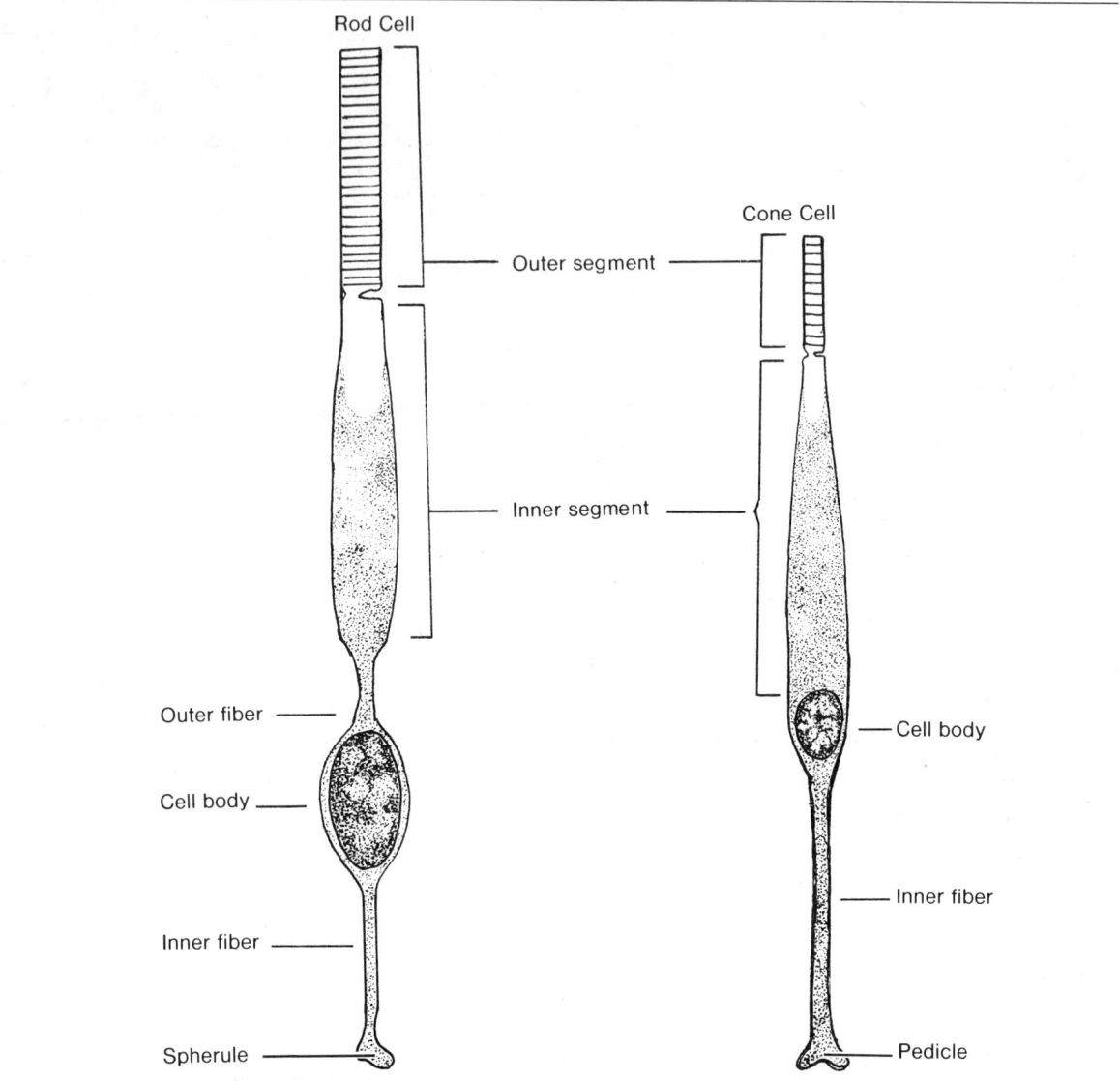

Fig. 9-7. Diagram of rod and cone cells.

Ganglion Cells

KEY WORDS: diffuse ganglion cell, midget ganglion cell

Dendrites of ganglion cells synapse with axons of bipolar neurons and represent the third and terminal link in the neuron chain. Axons of the various ganglion cells pass along the vitreal surface in the nerve fiber layer of the retina and join other axons at the optic disc to form the optic nerve. Although several morphologic varieties of ganglion cells have been described, two forms have been identified by their synaptic relations with bipolar neurons. **Diffuse ganglion cells** synapse with several types of bipolar cells, and **midget ganglion cells,** a monosynaptic type, synapse with a single midget bipolar cell.

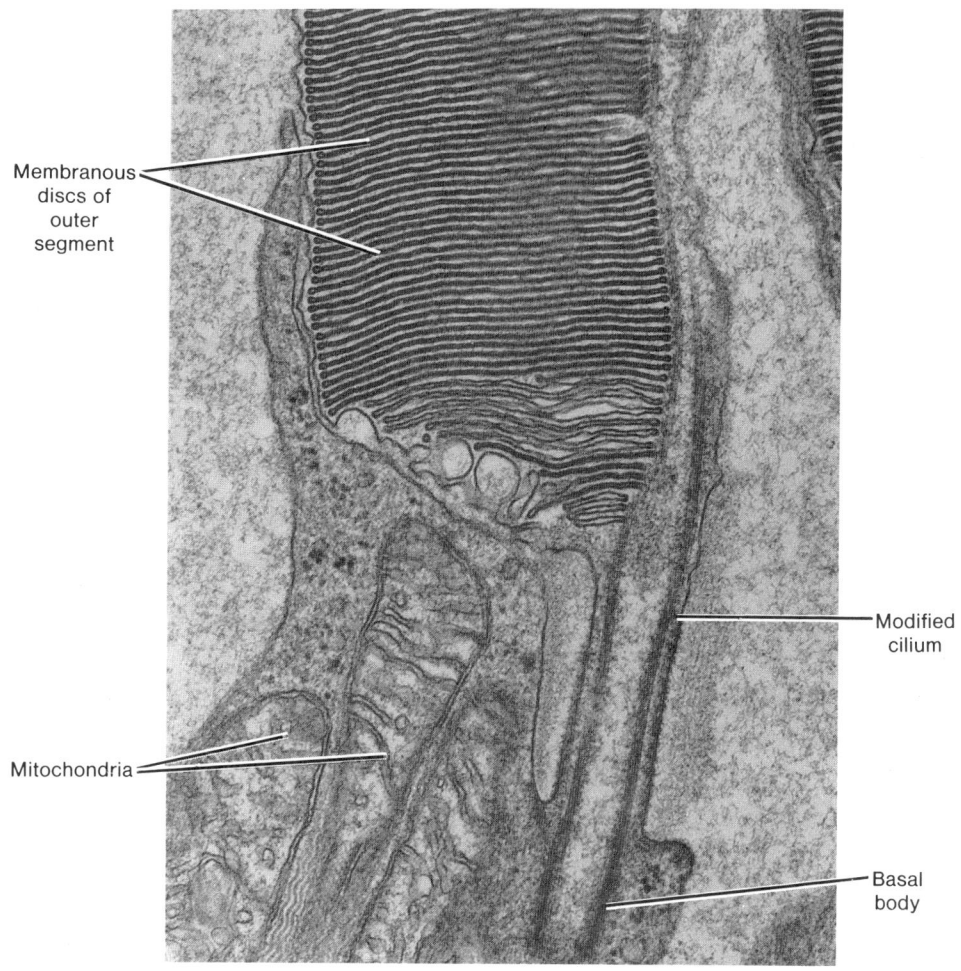

Membranous
discs of
outer
segment

Modified
cilium

Mitochondria

Basal
body

Fig. 9-8. The junction between the outer and inner segment of a rod (TEM, × 45,000).

Association Neurons

KEY WORDS: horizontal cell, amacrine cell, interplexiform cell

The cell bodies and nuclei of **horizontal cells** lie in the inner nuclear layer. Their dendrites enter the outer plexiform layer and synapse with a single cone pedicle. The axons also enter the outer plexiform layer and run parallel to the adjacent retinal layers to end on terminal twigs that synapse with several rod spherules. Although the functional significance of horizontal cells is not clear, because they synapse with cones of one area, with rods of another area, and with bipolar cells, it has been suggested that they may raise or lower the functional threshold of these cells.

Amacrine cells are pear-shaped neurons that lack axons but have several dendrites. The cell bodies lie in the inner nuclear layer, and their dendrites extend into the inner plexiform layer.

Perikarya of **interplexiform cells** also lie in the inner nuclear layer and send processes to both plexiform layers. They receive input from amocrine cells, and their output is with both horizontal and bipolar neurons.

All three types of association neurons are thought to act to modulate the passage of impulses from the photoreceptors to ganglion cells.

Fovea Centralis

The fovea centralis is a funnel-like depression on the posterior surface of the retina (Fig. 9-9), in direct line with the visual axis. At this point, those vitreal layers of the retina which are beyond the outer nuclear layer are displaced laterally, giving light an almost free pathway to the photoreceptors. The central region of the fovea, about 1.5 mm in diameter, consists only of cones that are longer and thinner

than cones elsewhere in the retina. They are closely packed and number about 25,000 to 35,000. Vision is most acute in this part of the retina.

Pigment Epithelium

KEY WORDS: cylindrical sheath, elongated microvillus, lipofuscin granule, melanin granule, residual body, vitamin A

The pigment epithelium of the retina consists of a simple layer of hexagonal cells that tend to increase in diameter near the ora serrata (Fig. 9-10). Their basal cell mem-

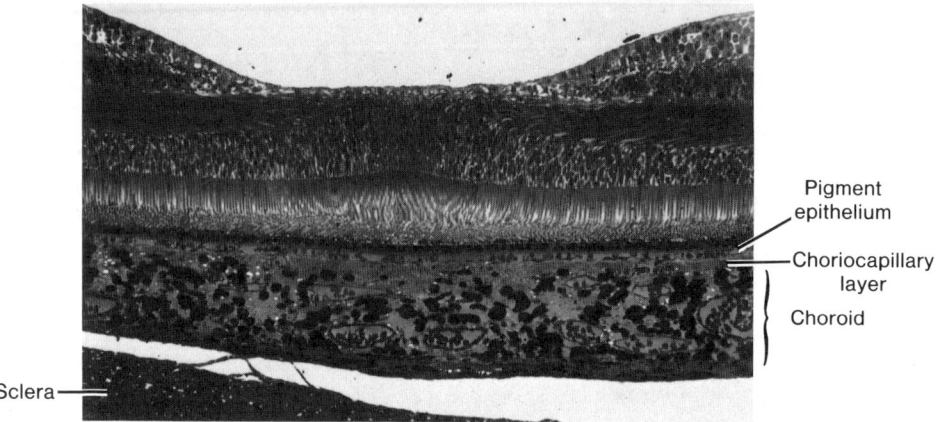

Fig. 9-9. The fovea centralis of a neural retina (LM, × 100).

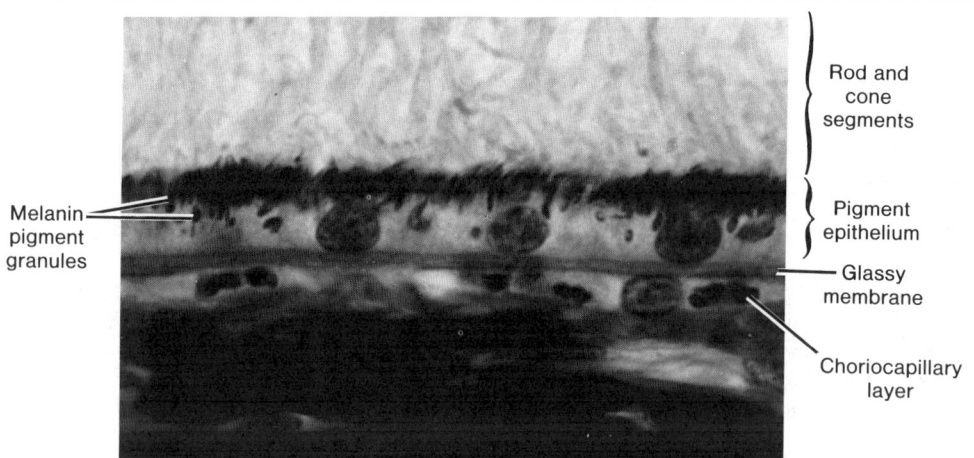

Fig. 9-10. Pigment epithelium of a human retina. Note the underlying choriocapillary layer (LM, × 1,000).

branes show numerous infoldings with associated mitochondria and are thought to be active in transport. The basal lamina contributes to the glassy (Bruch's) membrane of the choroid as seen by the light microscope. The lateral cell membranes of adjacent cells show some interdigitations and, near the apex, are united by tight junctions. Two types of cytoplasmic processes arise from the apices of the cells: **Cylindrical sheaths** invest the tips of the rod and cone outer segments, and **elongated microvilli** extend toward photoreceptor cells. The cytoplasm is characterized by abundant mitochondria near the base of the cell and by many **lipofuscin** and **melanin granules** (Fig. 9-11). Pigment epithelial cells may show numerous **residual bodies** that contain remnants of phagocytized membrane material shed by rod outer segments.

Pigment epithelium absorbs light after it has passed through the neural retina, thereby preventing reflection within the eye, and the apical tight junctions prevent undesirable substances from entering the intercellular spaces of the neural retina. The pigment epithelium also stores **vitamin A,** a precursor of rhodopsin, needed by the outer rod membranes.

Vascular Supply of the Retina

The outer nuclear and plexiform layers and the layer of rod and cone inner segments lack blood vessels. These parts of the retina are nourished by capillaries of the choriocapillary layer of the choroid. Nutrients cross the pigmented epithelium to enter the intercellular spaces of the outer neural retina. The inner layers of the retina are supplied by retinal vessels arising from the central retinal artery, which enters the eye in the optic nerve. Capillary networks from this source lie in the nerve fiber layer and in the inner plexiform layer.

Lens

KEY WORDS: capsule, anterior lens cell, lens fiber, lens substance, cortex, nucleus, zonule fiber

The lens is a transparent, biconvex epithelial body placed immediately behind the pupil between the iris and vitreous body. It is about 10 mm diameter and 3.7 to 4.5 mm thick. The posterior surface is more convex than the anterior

Fig. 9-11. Melanin and lipofuscin granules within a human pigment epithelial cell (TEM, × 6,000).

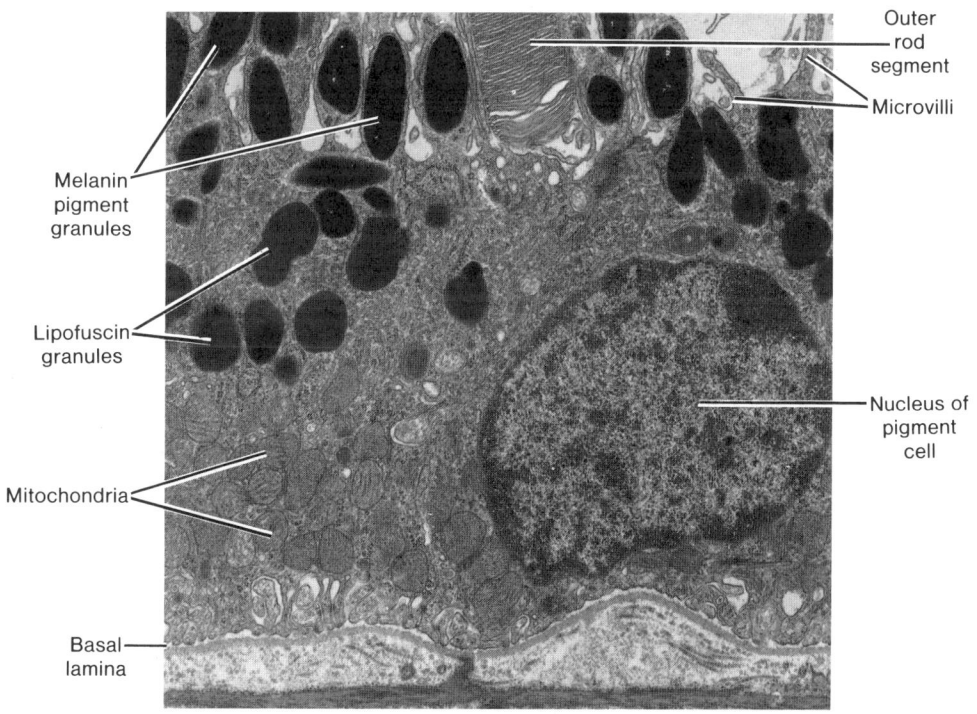

surface. The lens consists of a capsule, anterior lens cells, and lens substance.

On its anterior and posterior surfaces, the lens is covered by a homogeneous, proteoglycan-rich **capsule** 10 to 18 μm thick (Figs. 9-12 and 9-13). A single layer of cuboidal cells, the **anterior lens cells,** lies immediately beneath the capsule and forms the epithelium of the lens, which is restricted to the anterior surface. The posterior surface has no epithelium. The apices of the anterior lens cells face inward, toward the lens; the basal surfaces rest on a basal lamina. Anterior lens cells contain ion pumps that maintain the proper fluidity of the lens. Near the equator of the lens, the cells increase in height and gradually differentiate into **lens fibers** that make up the bulk of the lens, referred to as the **lens substance.** The lens grows throughout life by addition of new fibers to the periphery of the lens substance. The outer layers of the lens substance form the **cortex.** Nearer the center, the lens substance consists of condensed, concentrically arranged fibers that give it a more homogeneous appearance; this area is called the **nucleus** of the lens.

At the equator of the lens, the anterior lens cells elongate and push into the lens to lie beneath the epithelium anteriorly and the capsule posteriorly (see Fig. 9-13). As the cells increase in length, they lose their nuclei and their basal attachment to the capsule and become lens fibers that show a homogeneous, finely granular cytoplasm with few organelles. They appear hexagonal in cross section and measure 7 to 10 μm in length. The lateral membranes of the lens fibers (cells) in the cortex show large protrusions that interdigitate with concavities in adjacent fibers, in a ball and socket manner. Nexus junctions are common at these points. The ball and socket joints maintain the relative positions of the fibers as the lens changes shape during focusing. The lens is avascular and derives nourishment totally from diffusion of material from the aqueous humor and vitreous body.

The lens is held in place by a system of fibers called **zonule fibers** that make up the suspensory ligament of the lens. Zonule fibers arise from the ciliary epithelium that covers the ciliary processes and attach to the lens capsule just anterior and posterior to the equator of the lens. Zonule fibers consist of small bundles of fine filaments, approximately 12 nm in diameter, which may correspond to the microfibril component of elastic fibers (see Fig. 9-13).

The thickness and convexity of the lens are controlled by the ciliary muscle. If the ciliary muscle contracts, the ciliary body and choroid are pulled forward and centrally, releasing tension on the zonule fibers, and the lens "sags" to become thicker and more convex. This permits focusing on near objects. When the ciliary muscle relaxes, the zonule fibers are placed under tension, and the lens becomes thinner and less convex to focus on far objects.

Fig. 9-12. A region of lens at its equator showing the transformation of anterior lens (epithelial) cells into fiber cells and their subsequent migration into the lens interior (LM, × 125).

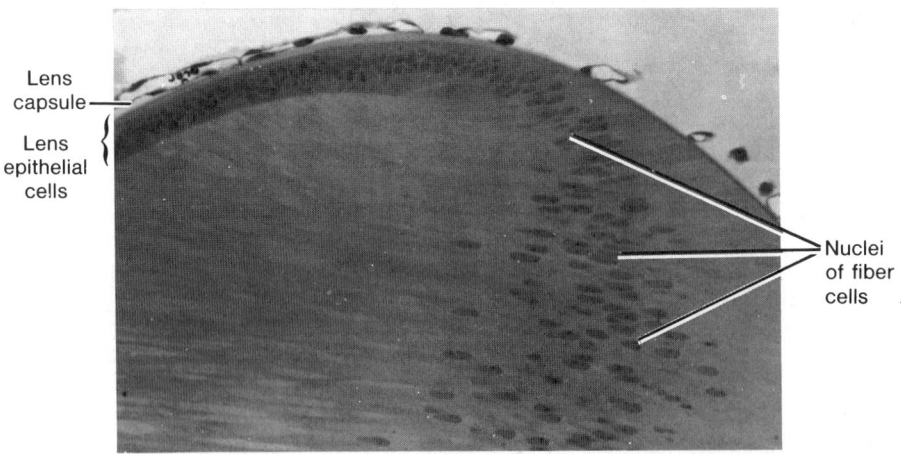

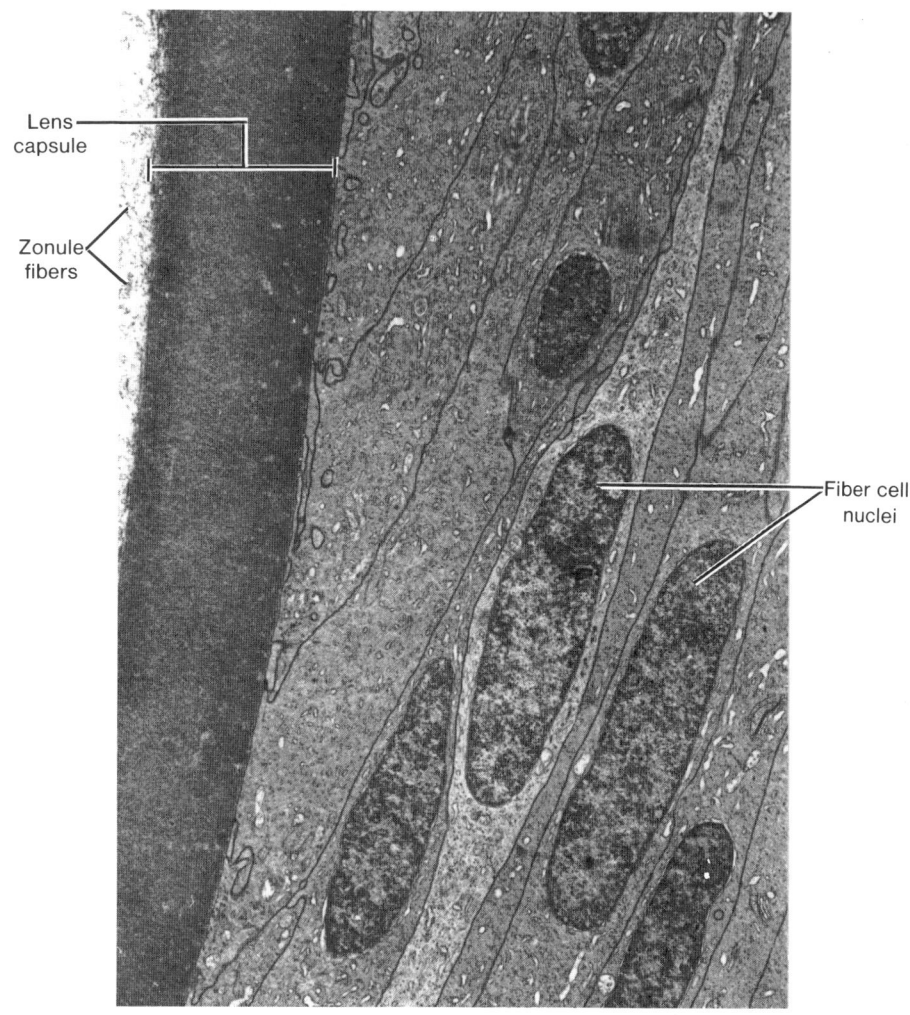

Fig. 9-13. Ultrastructural features of fiber cells near the lens equator. The lens capsule and zonule fibers also are shown (TEM, × 4,500).

Vitreous Body

KEY WORDS: hyaluronic acid, hyalocyte, hyaloid canal

The vitreous body is a colorless, transparent gelatinous mass that fills the vitreal cavity. It is 99 percent water and contains **hyaluronic acid** and other hydrophilic polysaccharides, as well as thin fibrils of collagen (type II and type XI) arranged randomly in a network. The fibrils are most prominent near the periphery. A few cells, the **hyalocytes,** and occasional macrophages are present in the outermost parts of the vitreous body. Hyalocytes may be responsible for the formation and maintenance of the vitreous body. A thin, cylindrical network of fibrils, the **hyaloid canal,** extends from the optic disc to the posterior surface of the lens. It represents the site of the embryonic hyaloid artery. The vitreous body provides nutrients to the lens and adjacent structures and maintains the correct consistency and shape of the eye.

Accessory Structures

Eyelids

KEY WORDS: conjunctiva, tarsal plate, tarsal (meibomian) gland, eyelash, gland of Zeis, Moll's glands

The eyelids protect the anterior aspects of the eyes. They consist of a connective tissue core covered externally by thin skin. The interior surface of each lid and the anterior surface of each eye (except for the cornea) are covered by a mucous membrane called the **conjunctiva.** The palpebral conjunctiva lines the interior of the lid; the bulbar conjunctiva covers the anterior surface of the eye. Both consist of a stratified columnar epithelium interspersed with goblet cells, resting on a dense connective tissue that is rich in elastic fibers and may contain numerous lymphocytes. The goblet cells provide mucus, which lubricates and prevents abrasion of the corneal epithelium. At the margin of the lid, the palpebral conjunctiva is continuous with the keratinized stratified epithelium of the skin.

The external surface of the eyelid consists of skin — an external layer of keratinized stratified squamous epithelium and an inner layer of loose connective tissue with many elastic fibers. Fine hairs are present, associated with many small sebaceous and sweat glands. Skeletal muscle from the orbicularis oculi and superior levator palpebrae muscles is present in the substance of the lid. The curved **tarsal plate,** composed of dense fibrous connective tissue, is a major structure in the lid. It conforms to the shape of the eyeball and maintains the shape of the lid. Large sebaceous glands, the **tarsal (meibomian) glands,** are embedded in the tarsal plate, arranged in a single row with their ducts opening at the margin of the lid. Each gland consists of a long central duct surrounded by numerous secretory alveoli. They elaborate a lipid secretion that lubricates the lid margins.

The **eyelashes** are thick, short, curved hairs arranged in two or three irregular rows along the margin of the lid. Their follicles extend into the tarsal plate. Sebaceous glands associated with eyelashes are called the **glands of Zeis;** the sweat glands located between the follicles are called **Moll's glands.**

Lacrimal Glands

KEY WORDS: compound tubuloalveolar, serous, myoepithelial cell, lacrimal duct, lacrimal sac, nasolacrimal duct

Lacrimal glands are well-developed, **compound tubuloalveolar** glands of the **serous** type, located in the superior temporal region of the orbit. Each gland consists of several separate glandular units that empty into the conjunctival sac via 6 to 12 ducts. The secretory units consist of tall columnar cells with large, pale secretory granules. Numerous **myoepithelial cells** lie between the bases of the secretory cells and their basal lamina. The ducts are lined at first by simple cuboidal epithelium, which becomes stratified columnar in the larger ducts. Small accessory lacrimal glands can occur in the inner surface of the eyelid.

Secretions of the lacrimal gland moisten, lubricate, and flush the anterior surface of the eye and the interior of the eyelids. Excess tears collect at a medial expansion of the conjunctival sac, which is drained by the **lacrimal duct** to a lacrimal sac at the medial corner of each eye. The lacrimal duct is lined by stratified squamous epithelium. The **lacrimal sac** and the **nasolacrimal duct,** which drains the sac into the inferior meatus of the nasal cavity, are lined by pseudostratified columnar epithelium. The remainder of their walls consists of dense connective tissue.

Review Table 9-1. Key Features of the Eye

Region	Epithelium	Major Components	Other Features
Sclera	None	Fibrous tunic	Lamina cribrosa
Cornea	Nonkeratinized stratified squamous	Bowman's membrane Substantia propria Descemet's membrane	Corneal endothelium
Choroid	None	Suprachoroid layer Choriocapillary layer Glassy membrane	Vessel layer Melanocytes
Ciliary body	Inner nonpigmented cells Outer pigmented cells	Ciliaris muscle Ciliary processes	Zonule fibers
Iris	Two layers of pigmented cells on posterior surface	Iridial stroma Sphincter muscle	Melanocytes
Neural retina	Pigment epithelium	Rods, cones, bipolar, and ganglion cells	Association neurons, glial cells
Lens	Anterior lens cells	Lens fibers	Lens capsule, zonule fibers
Vitreous body	None	Hyaluronic acid Thin collagen fibers	Hyalocytes

---

Functional Review: The Eye

The corneoscleral coat, together with the intraocular pressure of the fluid contents within the eye, maintains the proper shape and size of the eyeball. Light entering the eye must cross several transparent media (cornea, aqueous humor, lens, and vitreous body) before reaching receptors in the retina. There are no blood vessels in the transparent elements, which rely on diffusion of materials for their nutrition. Peripheral regions of the cornea receive nutrients from adjacent vessels in the limbus; the remainder of the cornea depends on diffusion of nutrients from the aqueous humor. The lens receives all its nutrition from the aqueous humor, and it is thought to be secreted continuously into the posterior chamber by the ciliary epithelium. The aqueous humor enters the anterior chamber through the pupil and diffuses posteriorly into the vitreous chamber of the eye. From the anterior chamber the aqueous humor passes through the trabecular network into Schlemm's canal and then into adjacent episcleral veins. The aqueous humor supplies nutrients to the transparent media of the eye and is responsible for maintaining the correct intraocular pressure.

Stationary refraction occurs through the transparent cornea, in contrast to variable refraction that occurs in the lens as it changes shape to focus near or far objects on the retina during eye accommodation. The other media have negligible refraction. The lens is held in place by zonule fibers that extend from surrounding ciliary processes and focuses an inverted, real image on the retina. The convexity and thickness of the lens are controlled by the ciliary muscle acting through the ciliary processes and zonule fibers. If the ciliary muscle contracts, the ciliary body and choroid are pulled centrally and forward, releasing tension on the zonule fibers; this allows the lens to become thicker and more convex, enabling the eye to focus on near objects. When the ciliary muscle relaxes, the ciliary body slides posteriorly and peripherally, increasing the tension on the zonule fibers and making the lens thinner and less convex to focus on far objects.

The iris is continuous with the ciliary body at the periphery and divides the space between the cornea and lens into anterior and posterior chambers. The chambers communicate via the pupil, an aperture in the iris through which light passes into the lens and vitreous chamber. The dilator of the pupil consists of a single layer of radially arranged myoepithelial cells along the posterior surface of the iris. Contraction of these cells increases the diameter of the pupil. The sphincter of the iris consists of smooth muscle arranged around the margin of the iris that acts as a diaphragm to modify the amount of light entering the eye and permits vision under a variety of light conditions.

The rods and cones of the retina are the photoreceptors that collect visual impressions (light patterns) and translate them into nerve impulses. The membranous sacs of the outer rod segments contain rhodopsin, which, on exposure to light, changes from cis to trans and breaks down, producing hyperpolarization of the rod cell plasmalemma and formation of an electrical potential that is transferred to dendrites of associated bipolar cells. Rods have a lower threshold to light intensity than cones and are important in dark and light discrimination and in night vision.

The visual pigment associated with cone outer segments is iodopsin. Light absorption and generation of an electrical impulse in cones follow a sequence similar to that in rods. Cones serve for color perception and visual acuity, responding to light of relatively high intensity. Detection of color depends on the presence of different pigments in the cones. The central region of the fovea centralis consists only of cones and is in direct line with the visual axis of the eye. Here the inner layers of the retina that are beyond the outer nuclear layer are displaced laterally, allowing light almost a free pathway to the photoreceptors; in this portion of the retina vision is most acute.

The pigment epithelium of the retina absorbs light after it has passed through the neural retina and prevents reflection within the eye; melanocytes in the choroid, iris, and other regions of the eye serve a similar purpose. The cells of the pigment epithelium phagocytize the membranous sacs as they are shed from the tips of the outer cone and rod segments. Some components of the digested membranes are carried back to the photoreceptor cells to be reused. The pigment epithelium is a storage site for vitamin A, a precursor of rhodopsin, which is recycled to the membranes of the outer rod segments. Apical tight junctions between cells of the pigment epithelium form a barrier to prevent unwanted materials from entering the neural retina. The neurons are nourished by diffusion from capillaries in the choriocapillary layer of the choroid.

Basically, the retina represents a three-neuron chain of receptors (rods and cones), bipolar neurons, and ganglion cells equivalent to three-neuron sensory chains of the peripheral nervous system. Rods and cones can be equated with any other sensory receptors, bipolar neurons with craniospinal ganglia, and retinal ganglion cells with internuncial neurons in the spinal cord and brainstem. Thus the arrangement of nervous elements in the sensory chain of the retina is identical to that in other sensory pathways. The retina represents an extension of the brain, modified to form a special receptor. Other neurons, horizontal cells, amacrine cells, and interplexiform cells in the retina serve as association neurons.

The eyelids are mobile folds of skin that protect the anterior of the eye from injury, desiccation, and excessive light. Lacrimal glands secrete tears that moisten and lubricate the anterior surface of the eye. Tears prevent the cornea from drying and flush the exposed surface of the eye and inner surface of the eyelid, washing away potentially dangerous agents.

Review Questions and Answers

Questions

51. Trace the flow of the aqueous humor and briefly discuss its function.
52. Briefly discuss the structure and function of the iris.
53. Describe the three-neuron conducting chain of the retina.
54. What is the function of the pigment epithelium of the retina?
55. Briefly describe the function of the accessory structures of the eye.

Answers

51. The ciliary epithelium is thought to elaborate the aqueous humor, which fills the posterior chamber of the eye and provides nutrients for the lens. Posteriorly, the aqueous humor enters, to a limited extent, the vitreous body, while anteriorly it flows through the pupil into the anterior chamber, where it aids in nourishing the cornea. The aqueous humor flows into the trabecular spaces from the anterior chamber, crosses the en-

dothelium that lines the trabeculae, passes through the juxtacanalicular connective tissue, and finally traverses the endothelium of the Schlemm's canal to enter its lumen. Several channels arise from the peripheral wall of Schlemm's canal to join with veins in the limbus and eventually drain to episcleral veins. In addition to its nutritive role, the aqueous humor also is important for maintaining the proper intraocular pressure and shape of the cornea and eye, features that are essential for good vision.

52. The iris is a thin, circular disc suspended between the cornea and lens that divides the space between these structures into anterior and posterior chambers. It has a central, circular aperture, the pupil, through which the two chambers communicate, and at its margin the iris is attached to the ciliary body. The iris consists primarily of a loose, vascular connective tissue containing much ground substance and is covered on the posterior surface by two rows of pigmented cells. The outer cells of this pigmented layer constitute the dilator of the eye and are modified myoepithelial cells arranged radially in a single layer; contraction of the dilator increases the diameter of the pupil. A compact, circularly arranged bundle of smooth muscle cells is present near the pupillary margin and forms the sphincter muscle of the iris. Contraction of the sphincter muscle reduces the diameter of the pupil; thus the iris serves as a diaphragm to modify the amount of light entering the eye.

53. The conducting chain of neurons in the retina basically consists of three neuronal elements: photoreceptors (rods and cones), bipolar neurons, and ganglion cells. It is equivalent to other chains of sensory neurons found in the peripheral nervous system. The rods and cones, bipolar neurons, and ganglion cells are perpendicular to the retinal layers. Bipolar neurons make synaptic contact with rod and cone cells and relay impulses generated by these photoreceptors to the dendrites of the overlying ganglion cells, the third link in the neuron chain of the retina. Axons from the ganglion cells run along the vitreal surface in the nerve fiber layer of the retina before they join with other axons at the optic disc to form the optic nerve.

54. The pigment epithelium of the retina absorbs light after it has passed through the neural retina and prevents reflection within the eye. Its component cells actively phagocytize and break down the membranous sacs that are shed at the tips of the outer rod and cone segments. The pigment epithelium serves as a storage site for vitamin A (a precursor for rhodopsin), which, together with amino acids and proteins, is recycled back to the rods. Apical tight junctions between the cells of the pigment epithelium act as a barrier to prevent undesirable substances from entering the neural retina, which is nourished by diffusion from the capillaries of the choriocapillary layer of the choroid.

55. The primary accessory structures of the eye are the eyelids and the lacrimal glands, both of which protect the anterior surface of the eye. The eyelids are mobile folds of skin that provide protection from physical injury, excessive light, and desiccation, while the lacrimal glands secrete tears that moisten and lubricate the anterior surface of the eye. Together with the eyelids, tears prevent the cornea from drying out. Tears also flush the interior of the eyelid and the anterior surface of the eyeball, washing away potentially dangerous materials from this surface of the eye.

10 The Ear

The ear contains receptors specialized for hearing and for awareness of head position and movement. The ear is subdivided into the external ear, which receives and directs sound waves from the external environment; the middle ear, which transforms sound waves into mechanical vibrations; and the inner ear, where these mechanical vibrations are converted to nerve impulses and relayed to the brain to be interpreted as sound. The inner ear also contains the vestibular organs that function in balance. The general structure of the ear is shown in Figure 10-1.

External Ear

KEY WORDS: aurical (pinna), elastic cartilage, external auditory meatus, ceruminous gland, cerumen

The **auricle (pinna)** of the external ear consists of an irregular plate of **elastic cartilage** surrounded by a thick perichondrium rich in elastic fibers. In humans the covering skin contains a few small hairs, their associated sebaceous glands, and an occasional sweat gland. The skin adheres tightly to the perichondrium except on the posterior surface, where a subcutaneous layer is present. The auricle is associated with sheets of skeletal muscle.

The **external auditory meatus** is about 2.5 cm long and follows an open S-shaped course. It consists of an outer part whose cartilaginous walls are continuous with the auricular cartilage and an inner portion whose walls are formed by the temporal bone. The external auditory meatus is lined by keratinized stratified squamous epithelium that is continuous with the epidermis of the auricle. The epithelium is firmly anchored to the surrounding perichondrium or periosteum by dense collagenous connective tissue. Numerous small hairs are present in the epithelium of the outer part of the meatus and are associated with large sebaceous glands in the underlying connective tissue. A special form of coiled apocrine sweat gland, the **ceruminous gland,** occurs in the skin that lines the meatus. Depending on the state of activity, the secretory cells may vary in shape from cuboidal to columnar. Each secretory tubule is surrounded by a network of myoepithelial cells that lies between the basal lamina and the bases of the epithelial cells. The ducts may open directly on the surface or, together with adjacent sebaceous glands, into hair follicles. Their secretory product is **cerumen,** a waxy material that prevents drying of the skin that lines the external auditory meatus. Hairs and glands occur only along the roof of the inner (boney) part of the external auditory meatus.

Middle Ear

KEY WORDS: tympanic cavity, auditory ossicle, tensor tympani, stapedius, auditory tube, tympanic membrane

The **tympanic cavity** is an irregularly shaped space in the petrous portion of the temporal bone. It is continuous anteriorly with the auditory (eustachian) tube and posteriorly with the mastoid air cells (cavities) of the temporal bone. The lateral wall consists chiefly of the tympanic membrane (eardrum), which forms a partition separating the tympanic cavity from the external auditory meatus. The inner, boney wall of the middle ear makes contact with the inner ear via two small, membrane-covered openings called the *oval* and *round windows.* The membrane of the oval window contains the base of an auditory ossicle, the stapes. The membrane that covers the round window often is referred to as the *secondary tympanic membrane.*

The tympanic cavity contains the **auditory ossicles,** a chain of three bones (malleus, incus, stapes) that unites the tympanic membrane of the middle ear with the oval window of the inner ear. The ossicles consist of compact bone and are united by small synovial joints. The bones are suspended in the air-filled tympanic cavity by thin strands of connective tissue called *ligaments.* Small skeletal muscles, the **tensor tympani** and **stapedius,** are associated with two of the ossicles; the tendon of the tensor tympani attaches to the malleus, and the tendon from the stapedius attaches to the stapes. The bulk of the muscles themselves is contained within small canals in the temporal bone.

The auditory ossicles, their suspending ligaments, and the inner walls of the tympanic cavity are covered by a thin mucous membrane that consists of a simple squamous

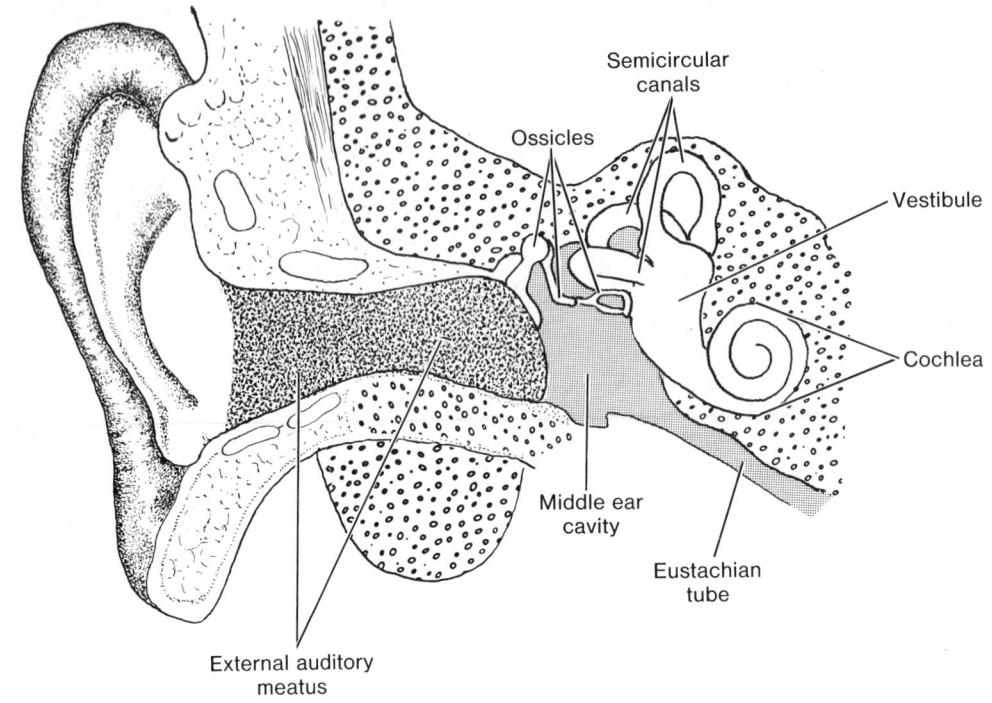

Fig. 10-1. Structure of the ear.

epithelium lying on a thin layer of connective tissue. The mucous membrane is firmly attached to the periosteum of the temporal bone, lines the interior of the mastoid air cells, and covers the inner surface of the tympanic membrane. Where the tympanic cavity joins the auditory tube, the lining epithelium becomes ciliated columnar, interspersed with secretory cells.

The **auditory tube** (eustachian tube) is about 4 cm long and connects the anterior part of the tympanic cavity with the nasopharynx. The auditory tube acts as a passageway to ventilate the tympanic cavity and allows equalization of pressure between the middle ear and pharynx. Near the tympanic cavity the supporting wall consists of compact bone, whereas in its medial two-thirds, the auditory tube is supported by a J-shaped elastic cartilage. The boney portion is lined by a simple columnar epithelium that becomes ciliated pseudostratified columnar epithelium in the cartilaginous part. Cilia beat toward the pharynx. Goblet cells are present near the pharyngeal opening, and mixed, compound, tubuloalveolar glands often are found in the underlying connective tissue. A mass of lymphoid tissue, the

tubal tonsil, fills the connective tissue and infiltrates the epithelium.

The **tympanic membrane** forms most of the lateral wall of the tympanic cavity and is made up of three layers. The outer layer consists of stratified squamous epithelium, which reflects onto the tympanic membrane from the external auditory meatus. Two layers of collagenous fibers and fibroblasts form the middle layer. In the external layer, the fibers are arranged radially, whereas those of the inner layer run a circular pattern. A simple squamous epithelium and its supporting connective tissue form the third layer and are continuous with the lining of the tympanic cavity. The fibrous layers of the tympanic membrane enter an encircling ring of fibrocartilage that unites the eardrum to the surrounding bone. The inner surface of the tympanic membrane attaches to the malleus.

Sound waves received by the tympanic membrane cause it to vibrate slightly. The vibrations then are transmitted to the fluid-filled chambers of the inner ear via the ossicles. The tympanic membrane is about 18 times as large as the oval window of the inner ear, which contains the foot plate of the stapes. Because of this arrangement, the eardrum and auditory ossicles act as a piston that exerts pressure on the confined fluid of the inner ear. Thus the tympanic membrane and auditory ossicles not only transmit vibrations but also are able to amplify weak forces of sound waves without expending energy.

The tensor tympani and stapedius muscles protect delicate structures in the inner ear by dampening ossicle movement resulting from loud or sudden noises. They also are thought to regulate the degree of tension in the tympanic membrane so that sounds of moderate intensity can be picked up in a noisy environment.

Inner Ear

KEY WORDS: boney labyrinth, perilymph, membranous labyrinth, endolymph

The inner ear consists of a system of canals and cavities within the petrous portion of the temporal bone. The compact bone surrounding the canals and cavities forms the **boney labyrinth** and is filled with a fluid called **perilymph**. A series of fluid-filled membranous structures, collectively called the **membranous labyrinth,** lie suspended in the perilymph. The membranous labyrinth is filled with **endolymph,** a fluid with an ionic composition similar to that of intracellular fluid, being rich in potassium ions and low in sodium ions. Perilymph resembles extracellular fluid, having a high content of sodium ions and a low concentration of potassium ions. Endolymph is produced by an area of the cochlear duct called the *stria vascularis;* the exact site of perilymph formation is unknown.

The boney and membranous labyrinths consist of two major components: the vestibular labyrinth, which contains sensory elements for equilibrium, and the cochlea, which contains sensory structures for hearing.

Vestibular Labyrinth

KEY WORDS: semicircular canal, utricle, saccule, crista ampullaris, macula utriculi, macula sacculi, type I hair cell, kinocilium, type II hair cell, supporting cell, planum semilunatum, cupula cristae ampullaris, otolithic membrane, otoliths (otoconia)

The vestibular labyrinth consists of three **semicircular canals** that are continuous with an elliptical structure called the **utricle.** The utricle in turn unites with an anteromedial spherical part of the vestibular system called the **saccule** by a thin duct that joins with a similar duct from

the saccule. The two ducts unite to form the slender endolymphatic duct, which terminates as a small expansion called the *endolymphatic sac.*

Most of the membranous labyrinth that forms the vestibular portion of the inner ear is lined by a simple squamous epithelium. The remainder of this wall consists of fine connective tissue fibers and stellate fibroblasts. Thin trabeculae of connective tissue extend from the wall of the membranous labyrinth and cross the perilymphatic space to blend with the periosteum of the surrounding bone. The trabeculae suspend the membranous component of the semicircular canals, utricle, and saccule in the perilymph contained within the osseous labyrinth. The perilymphatic connective tissue is rich in blood capillaries that supply the various segments of the membranous labyrinth.

In specific regions of each subdivision of the membranous labyrinth, the epithelium assumes a stratified appearance and serves in sensory reception (Figs. 10-2 and 10-3). The sensory epithelium of each semicircular canal is restricted to the dilated ampullary portion and, together with an underlying core of connective tissue, forms a transverse ridge that projects into the lumen of the ampulla. The connective tissue contains many nerve fibers. These sensory neuroepithelial regions of the semicircular canals form the **cristae ampullaris** (Figs. 10-2 and 10-4). Similar raised regions of neuroepithelium are found in the utricle and saccule and form the **macula utriculi** and the **macula sacculi,** respectively (Figs. 10-2 and 10-5). The macula utriculi covers an area approximately 2 mm square along the superoanterior wall. It lies on a plane perpendicular to the macula sacculi, which occupies an area measuring 2×3 mm on the anterior wall of the saccule. The sensory epithelium of the cristae and maculae consists of type I and type II sensory hair cells (see Figs. 10-4 and 10-5) and supporting (sustentacular) cells.

The **type I hair cell** is flask-shaped with a narrow apical region and a rounded base that contains the nucleus (Fig. 10-6). A large part of the cell is enveloped by a cuplike afferent nerve ending (calyx). Nearby efferent nerve endings may synapse with the nerve calyx but do not directly contact the type I hair cell; a narrow intercellular cleft, 30 nm wide, separates the hair cell from the nerve ending. The space narrows to approximately 5 nm at gap junctions that are scattered between the two compartments. Synaptic ribbons often are found in the cytoplasm of type I hair cells, immediately adjacent to the plasmalemma. Mitochondria are concentrated around the nucleus and at the cell apex, and numerous microtubules are present, especially in the apical region. The cytoplasm of the nerve calyx shows

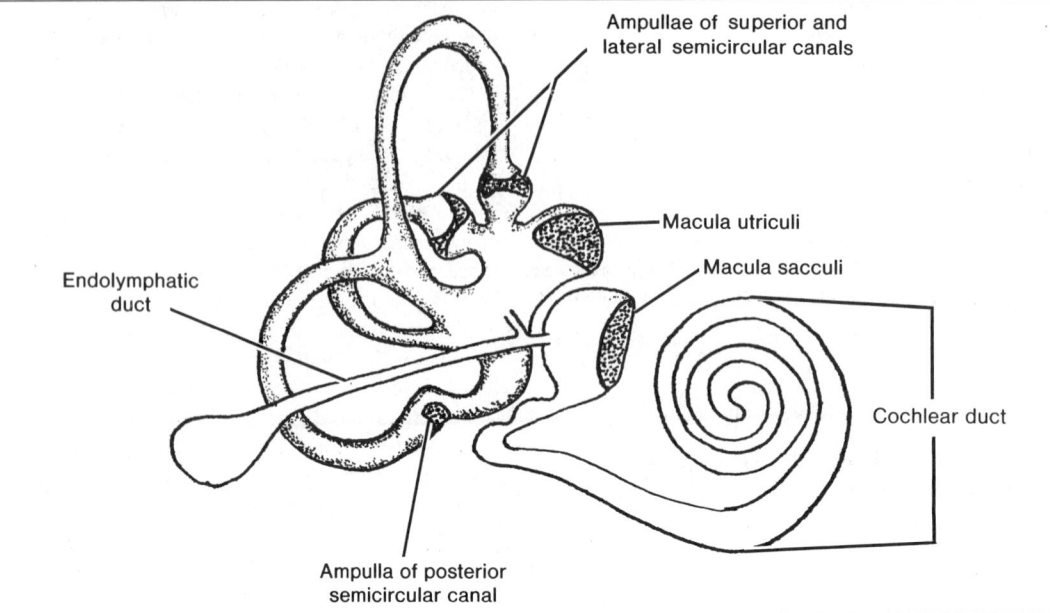

Fig. 10-2. Location of sensory regions in the membranous labyrinth (stippled areas).

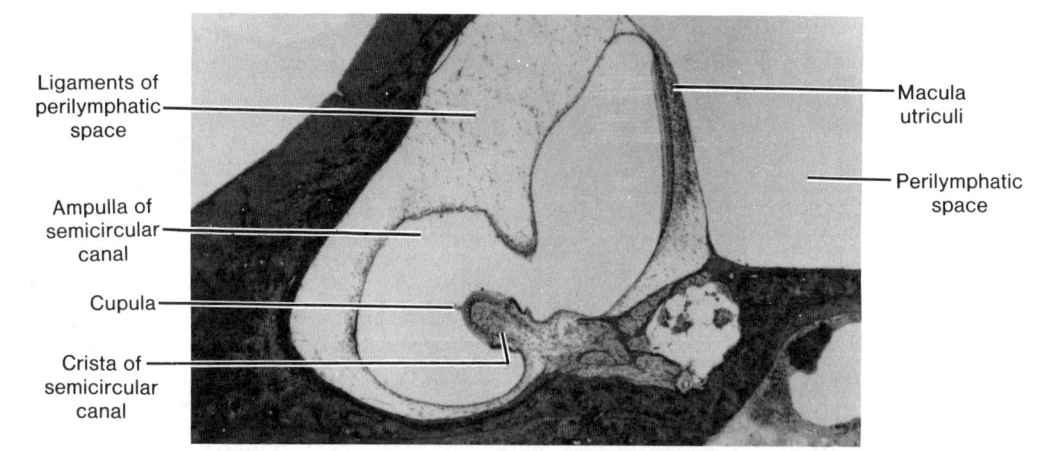

Fig. 10-3. A section through the vestibular portion of the inner ear illustrates both a macula utriculi and a crista within the ampulla of a semicircular canal (LM, × 40).

scattered mitochondria and many vesicles that range from 50 to 200 nm in diameter.

The apical membrane of the type I hair cell bears 50 to 100 large, specialized microvilli, known as "hairs" (Figs. 10-6 and 10-7). These nonmotile elements are limited by a plasmalemma, have cytoplasmic cores that contain numer-

ous fine filaments, and are constricted at their base just before they join the rest of the cell. The longitudinal filaments leave the microvilli and enter into a thick mat of filaments that forms a terminal web in the apical cytoplasm of the cell. The microvilli progressively increase in height from about 1 μm on one side to about 100 μm on

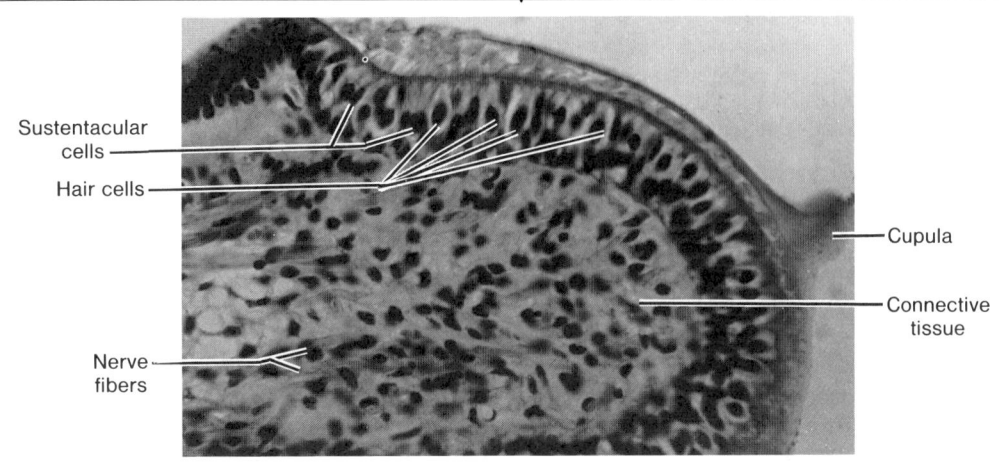

Fig. 10-4. A region of a crista ampullaris showing both hair and sustentacular cells as well as the cupula (LM, × 250).

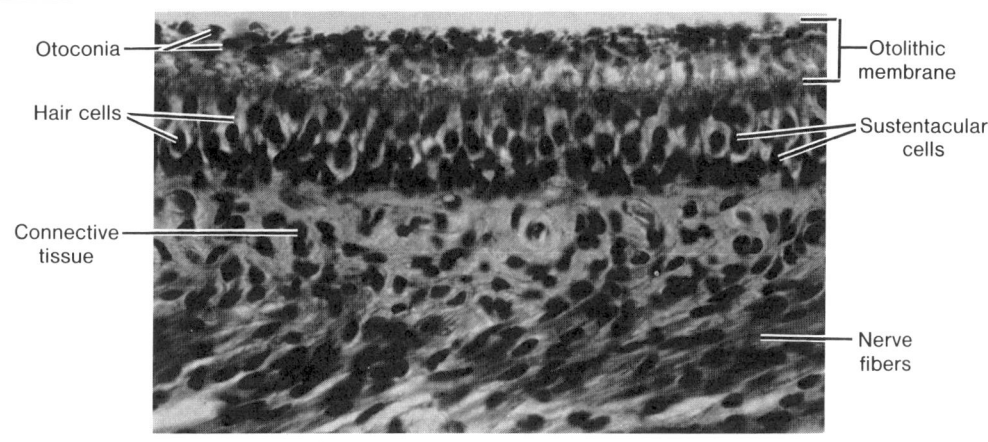

Fig. 10-5. Hair cells and sustentacular cells of a macula utriculus. Note the underlying nerve fibers and the position of the otolithic membrane (LM, × 250).

the opposite side of the cell. A single, eccentrically placed cilium is present on the apical surface and is peculiar in that the two central microtubules end shortly after originating from the basal body. The cilium is believed to be nonmotile and often is called a **kinocilium.**

Type II hair cells are simple columnar cells surrounded by numerous separate afferent and efferent nerve endings, rather than by a single afferent nerve ending as seen surrounding type I hair cells (see Fig. 10-6). The type II hair cell also bears a single, eccentrically placed, nonmotile

cilium and 50 to 100 large microvilli that are arranged identically to those of type I cells. The cytoplasm contains scattered profiles of granular endoplasmic reticulum, abundant mitochondria, smooth-surfaced tubules, and numerous vesicles 20 nm in diameter. A well-developed Golgi complex occupies a supranuclear position. Synaptic ribbons are present in the cytoplasm of type II hair cells, immediately opposite the surrounding nerve terminals.

Adjacent columnar-shaped **supporting cells** (sustentacular cells) extend from the basal lamina to the free surface

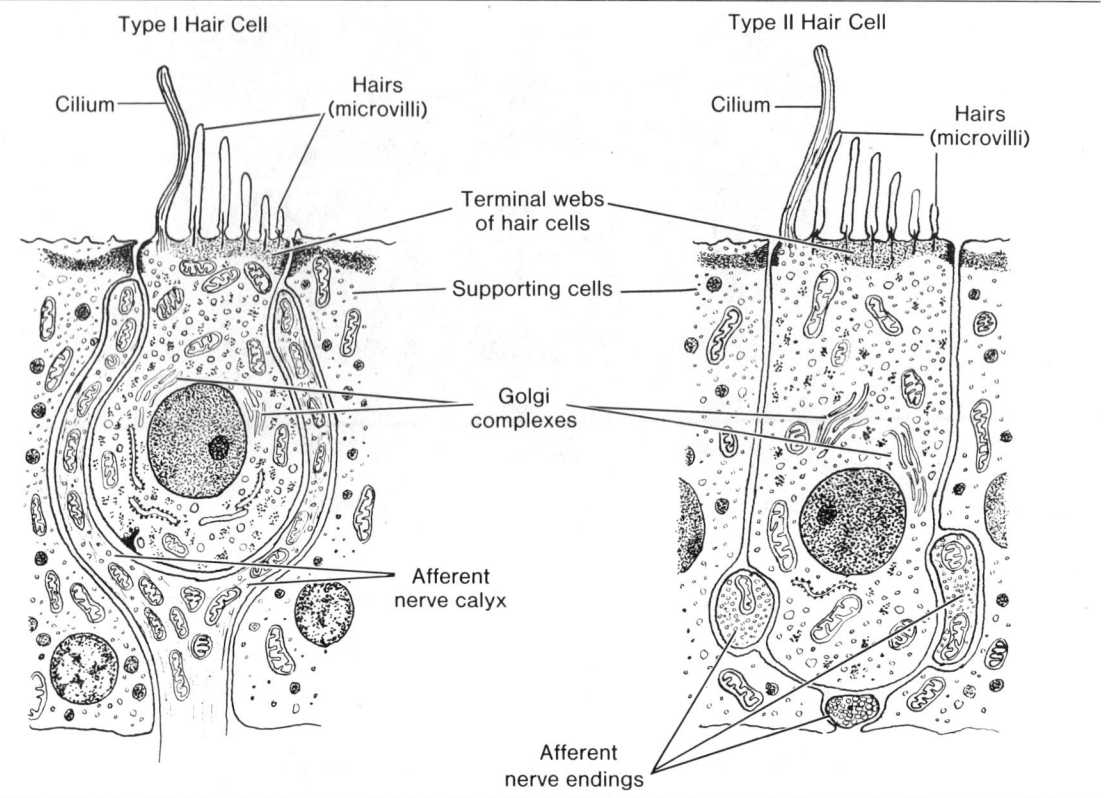

Fig. 10-6. Diagram of type I and type II hair cells.

of the sensory epithelium. They follow a very irregular course throughout the epithelium and in electron micrographs show a well-developed terminal web at the cell apex, a prominent Golgi complex, numerous secretory granules, and bundles of microtubules that extend from the basal cytoplasm to the terminal web. The microtubules form an integral part of the cytoskeleton and provide rigidity to the supporting cells. Although the function of the supporting cells is uncertain, it has been suggested that they are concerned with the metabolism of endolymph or that they contribute to the nutrition of hair cells. The supporting cells at the periphery of the sensory epithelium form a simple columnar layer, the **planum semilunatum,** which lacks hair cells.

The microvilli of the hair cells in the cristae are embedded in a gelatinous structure called the **cupula cristae ampullaris,** which is composed of viscous proteoglycans that project from the surface of the crista into the ampullary lumen of each semicircular canal (see Fig. 10-4). Support-

ing cells in the sensory epithelium also may contribute to the cupula. The sulfated proteoglycans may be secreted by the planum semilunatum.

Microvilli of hair cells from the maculae of the utricle and saccule also are embedded in a viscous proteoglycan that forms the **otolithic membrane** (see Fig. 10-5). Additionally, numerous crystalline bodies called **otoliths (otoconia)** are suspended in this layer. These bodies consist of protein and calcium carbonate.

The gelatinous cupula of the cristae projects across the lumen of the ampullary region of the semicircular canal like a swinging door. During angular movement of the head, the cupula is displaced by the motion of endolymph contained within this part of the membranous labyrinth. Displacement of the cupula excites the sensory hair cells, which, in turn, generate an action potential that is received by surrounding nerve terminals. Similarly, gravitational forces on the otolithic membrane and the otoconia embedded in it cause a shearing action on the microvilli of the

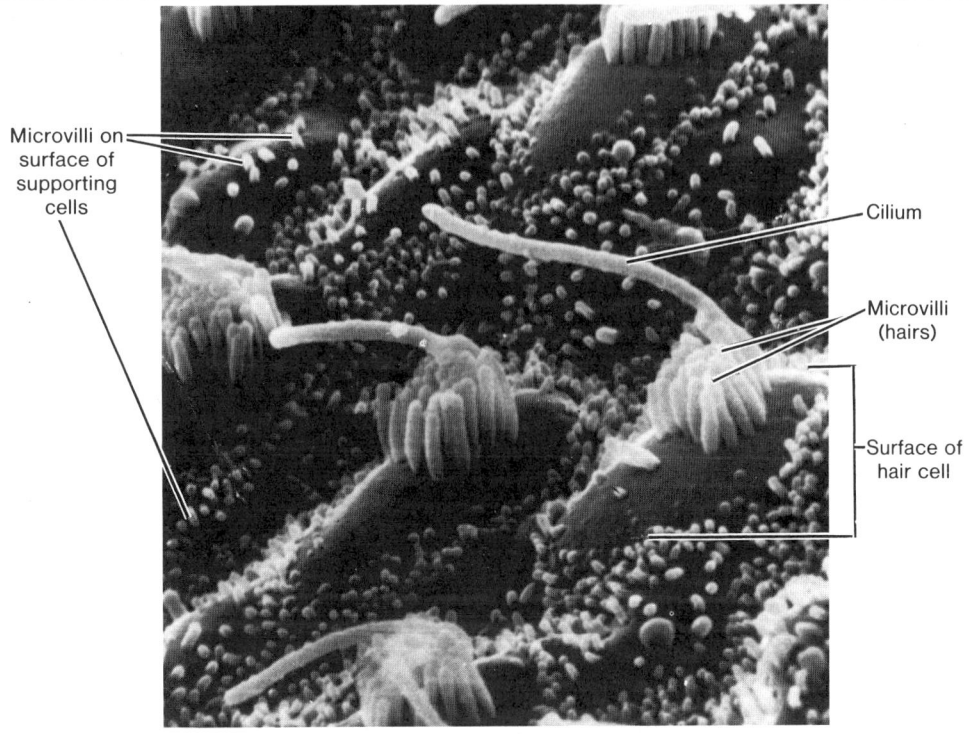

Microvilli on surface of supporting cells

Cilium

Microvilli (hairs)

Surface of hair cell

Fig. 10-7. The apices of vestibular hair cells. Note the single cilium and adjacent microvilli (SEM, × 6,000).

underlying hair cells in the macula. Linear acceleration also results in stimulation of hair cells in the macula.

Although the exact mechanism is unknown, the sensory epithelium in the vestibular organs transforms the mechanical energy of endolymph movement into the electrical energy of a nerve impulse. The bending or displacement of microvilli is thought to result in depolarization of hair cells, the impulse being transferred to surrounding nerve endings to result in the generation of a nerve impulse. Efferent nerve endings probably have inhibitory functions to control the threshold of activity of hair cells.

Cochlea

KEY WORDS: modiolus, spiral ganglion, spiral lamina, basilar membrane, spiral ligament, vestibular membrane, scala vestibuli, cochlear duct, scala tympani, ductus reuniens, cecum cupulare, helicotrema

Like the vestibular portion of the inner ear, the cochlea consists of an outer portion of compact bone and a central membranous part contained in perilymph. The osseous

part of the cochlea spirals for two and three-fourths turns around a cone-shaped axis of spongy bone called the **modiolus** (Fig. 10-8). Blood vessels, nerve fibers, and the perikarya of afferent bipolar neurons, called the **spiral ganglion,** lie within the boney substance of the modiolus. A projection of bone, the **spiral lamina,** extends from the modiolus into the lumen of the cochlear canal along its entire course. A fibrous structure, the **basilar membrane,** extends from the spiral lamina to the **spiral ligament,** a thickening of periosteum on the outer boney wall of the cochlear canal. The thin **vestibular membrane** extends obliquely across the cochlear canal from the spiral lamina to the outer wall of the cochlea.

The basilar and vestibular membranes divide the cochlear canal into an upper **scala vestibuli,** an intermediate **cochlear duct,** and a lower **scala tympani.** A cross section of the cochlea is shown in Figure 10-9.

The cochlear duct is part of the endolymphatic system and is connected to the saccule of the membranous labyrinth by the small **ductus reuniens.** The opposite end of the cochlear duct ends at the apex of the cochlea as the

Fig. 10-8. Cochlea (SEM, × 100).

blindly ending **cecum cupulare.** The scala vestibuli and the scala tympani contain perilymph and communicate at the apex of the cochlea through a small opening called the **helicotrema.** At the base of the cochlea, the scala tympani is closed by the secondary tympanic membrane, which fills the round window. This membrane separates the perilymph in the scala tympani from the middle ear. The scala vestibuli extends through the perilymphatic channels of the vestibule to end at the oval window, which is closed by the foot of the stapes. Movement of the stapes in the oval window exerts pressure on the perilymph in the scala vestibuli. Because fluid cannot be compressed, waves of pressure either pass through the cochlear duct, displacing it to

enter the scala tympani, or enter the scala tympani directly through the helicotrema. The pressure is released from the confined perilymphatic spaces of the cochlea by the elasticity of the secondary tympanic membrane, which bulges into the tympanic cavity of the middle ear (Fig. 10-10).

Cochlear Duct

KEY WORDS: basilar membrane, spiral crest, vestibular membrane, stria vascularis, marginal cell, basal cell, intraepithelial capillary, spiral prominence

The cochlear duct is a triangular space that follows the spiral course of the cochlea (Fig. 10-11). Its floor is formed

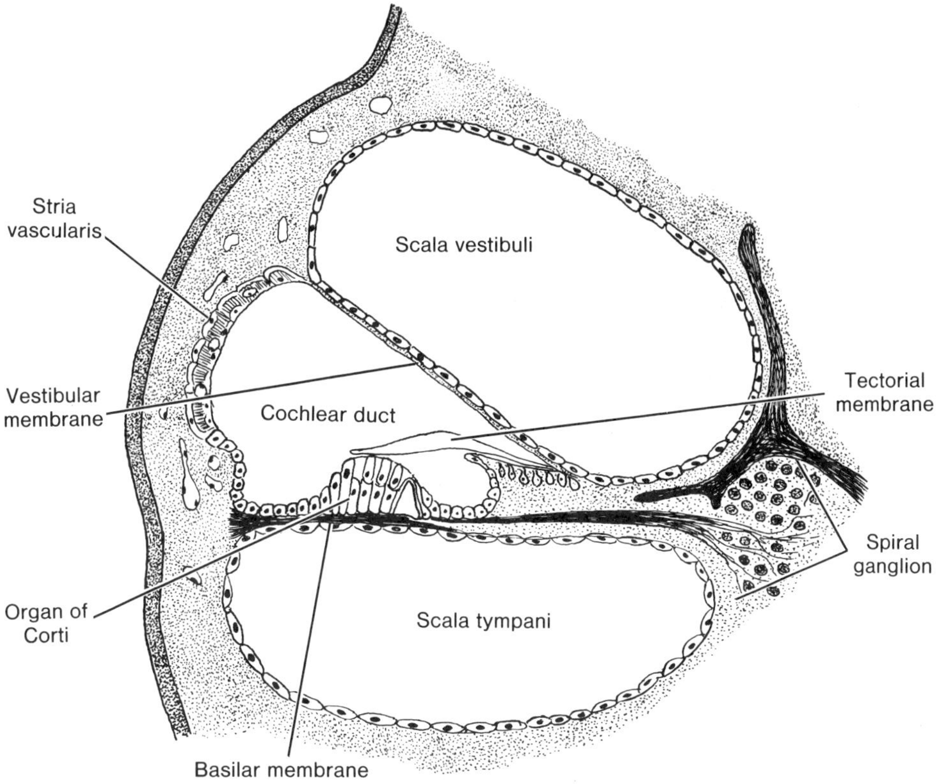

Stria
vascularis

Scala vestibuli

Vestibular
membrane

Cochlear duct

Tectorial
membrane

Organ of
Corti

Spiral
ganglion

Scala tympani

Basilar membrane

Fig. 10-9. Cross section of the cochlea.

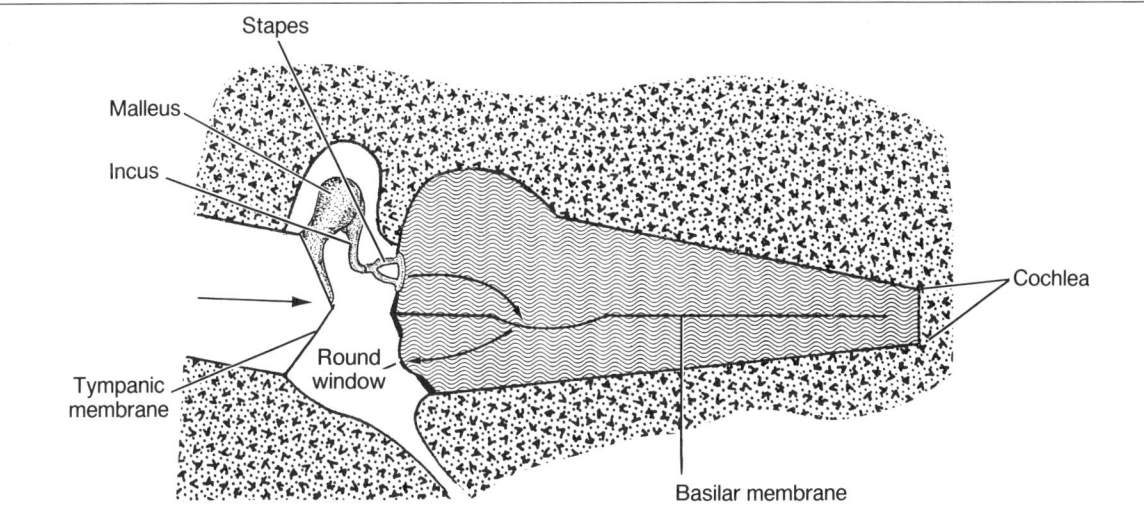

Stapes

Malleus

Incus

Cochlea

Tympanic
membrane

Round
window

Basilar membrane

Fig. 10-10. Mechanics of hearing.

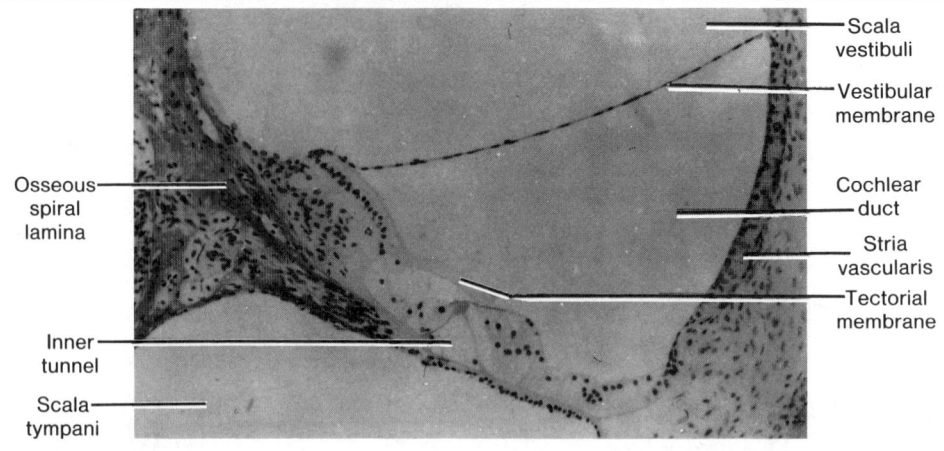

Osseous spiral lamina

Inner tunnel

Scala tympani

Scala vestibuli

Vestibular membrane

Cochlear duct

Stria vascularis

Tectorial membrane

Fig. 10-11. A section through the cochlear duct illustrates its relationship to the scala vestibuli and scala tympani (LM, × 100).

by the basilar membrane and its roof by the vestibular membrane. The **basilar membrane** consists of a layer of collagen-like fibers embedded in an amorphous matrix and extends from the osseous spiral lamina of the modiolus to the **spiral crest,** a well-vascularized periosteal region along the outer wall of the cochlea. The **vestibular membrane** consists of two layers of simple squamous cells separated mainly by their basal laminae.

The outer wall of the cochlear duct is formed by a vascular area called the **stria vascularis.** It occurs along the entire length of the cochlear duct and consists of a pseudostratified columnar epithelium that rests on a vascular connective tissue and contains basal and marginal cells. The epithelium is continuous with the simple squamous epithelium that lines the interior of the vestibular membrane.

Marginal cells show deep infoldings of the basal and lateral cell membranes and are associated with numerous mitochondria, suggesting that these cells are involved in fluid transport. They may participate in the elaboration of endolymph. **Basal cells** show few mitochondria or basal infoldings. The epithelium of the stria vascularis differs from that found elsewhere in the body in that it contains **intraepithelial capillaries.**

The epithelium of the stria vascularis is continuous with a simple layer of attenuated cells that overlies the **spiral prominence,** a highly vascularized thickening of the periosteum. The spiral prominence lies beneath the stria vascularis and extends the length of the cochlear duct. The

cells become cuboidal where the epithelium reflects onto the basilar membrane from the outer wall of the cochlear duct.

Organ of Corti

KEY WORDS: **supporting cell, inner pillar cell, outer pillar cell, inner phalangeal cell, outer phalangeal cell, border cell, inner spiral tunnel, spiral limbus, interdental cell, tectorial membrane, cells of Hensen, cells of Claudius, inner hair cell, outer hair cell**

The cochlear duct contains a region of specialized cells, the organ of Corti (Figs. 10-12 and 10-13) that transforms vibrations of the basilar membrane into nerve impulses. The avascular organ of Corti extends along the length of the cochlear duct and lies on the basilar membrane. It consists mainly of supporting and hair cells. **Supporting cells** are tall and columnar and consist of inner and outer pillar cells, inner and outer phalangeal cells, border cells, cells of Hensen, and cells of Claudius.

Inner and **outer pillar cells** form the boundaries of the inner tunnel, a space that lies within and extends the length of the organ of Corti. In both types of pillar cells, the nucleus is located in the broad base. The elongated cell bodies contain numerous microtubules that form a cytoskeleton. Inner pillar cells are slightly expanded at the apex and extend over the outer pillar cells. The apices of the outer pillar cells also expand slightly to fit into the concave apical undersurface of the inner pillar cells. In addition to

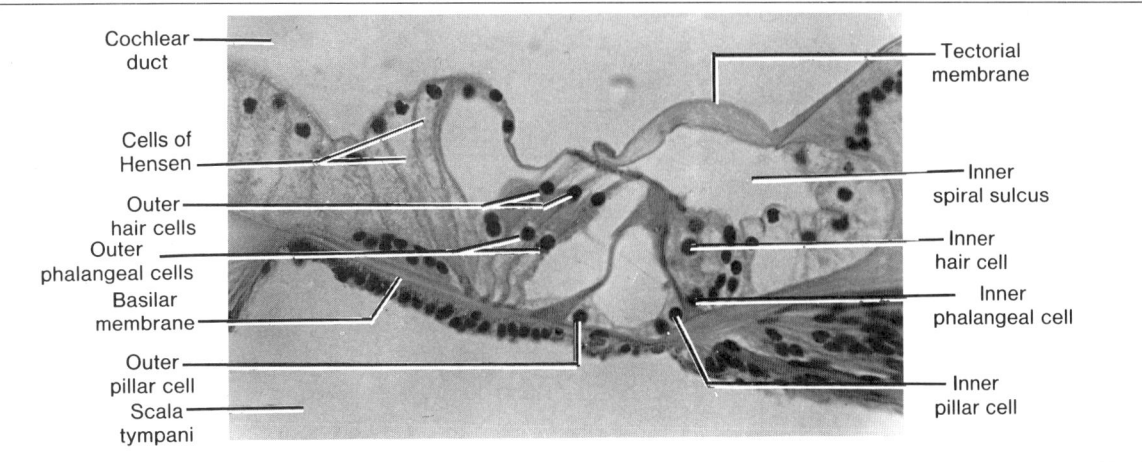

Fig. 10-12. Cells and other components of the organ of Corti (LM, × 250).

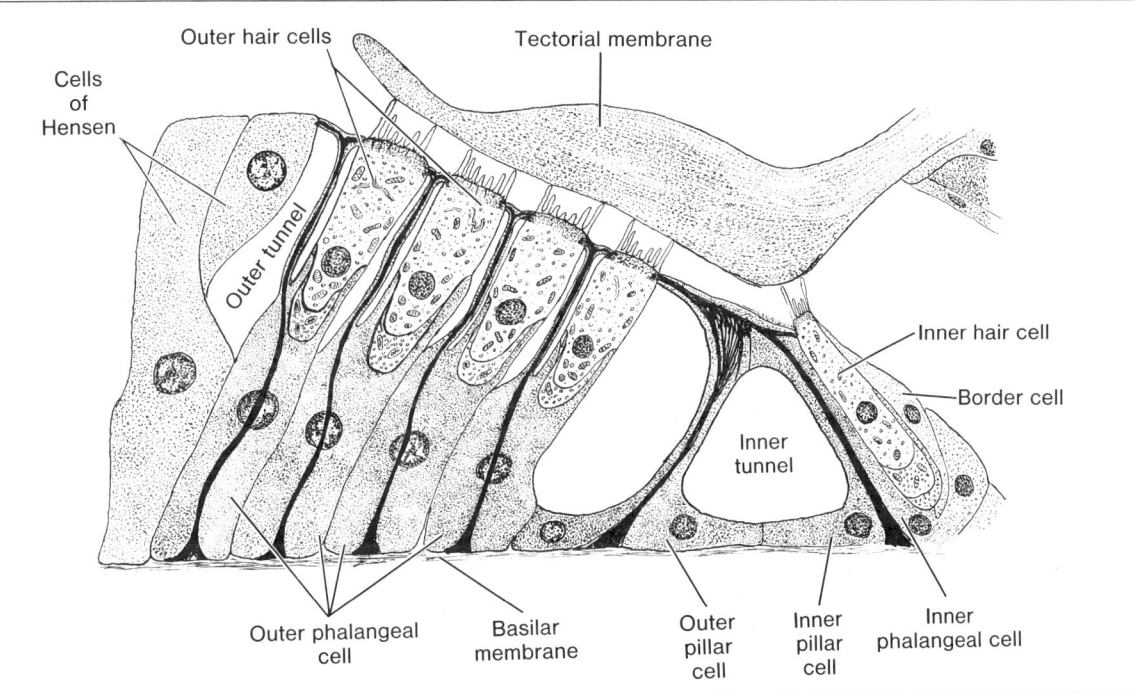

Fig. 10-13. Diagram of the organ of Corti.

forming the boundaries of the inner tunnel, the pillar cells provide structural support for adjacent cells.

The phalangeal cells serve as supporting elements for the sensory hair cells (see Fig. 10-13). The **inner phalangeal cells** form a single row immediately adjacent to the inner pillar cells and surround the sensory inner hair cells except at their apical regions. In contrast, the columnar **outer phalangeal cells** form three or four rows and support outer hair cells, which also are arranged in rows. The apex of each outer phalangeal cell forms a cuplike structure that surrounds the basal one-third of an outer hair cell. Afferent and efferent nerves are located at the base.

Each outer phalangeal cell gives rise to a slender cytoplasmic process filled with microtubules. The process extends to the surface, where it expands into a flat plate that attaches to the apical edges of the outer hair cell and is supported laterally by outer phalangeal cells and the outer hair cells in the adjacent row. The apical plates of the outer phalangeal cells provide additional support for the outer hair cells, the upper two-thirds of which are not supported by adjacent cells and are surrounded by large, fluid-filled intercellular spaces. The fluid contained in these spaces is believed to be similar to that within the inner tunnel.

Border cells are supporting cells associated with the inner edge of the organ of Corti. These slender cells undergo a transition to the squamous cells that line the **inner spiral tunnel.** This space is formed by the **spiral limbus,** which consists of periosteal connective tissue that extends from the osseous spiral lamina and bulges into the cochlear duct. Vertically arranged collagenous fibers, often called *auditory teeth,* are contained within the substance of the limbus. **Interdental cells** are specialized cells along the upper surface of the spiral limbus. The cells extend between collagen fibers to reach the lumen of the cochlear duct. On the upper surface of the limbus, they form a continuous sheet and are united by tight junctions.

The interdental cells secrete a sheet of material that forms the **tectorial membrane,** which consists of a gelatinous matrix rich in proteoglycans and a filamentous protein thought to be similar to epidermal keratin. The tectorial membrane extends over the interdental cells and beyond the substance of the spiral limbus to overlie the hair cells of the organ of Corti (see Fig. 10-13). The tips of microvilli extending from sensory hair cells are firmly embedded in the undersurface of the tectorial membrane (Fig. 10-14).

Near the outer edge of the organ of Corti, immediately adjacent to the outer phalangeal cells, is another group of columnar supporting cells called **cells of Hensen.** They decrease in height and transform into the adjacent **cells of Claudius,** which form the outer edge of the organ of Corti.

Fig. 10-14. The external border of the tectorial membrane is attached to cells of Hensen. Note the exposed outer hair cells where the membrane is lifted up (SEM, × 380).

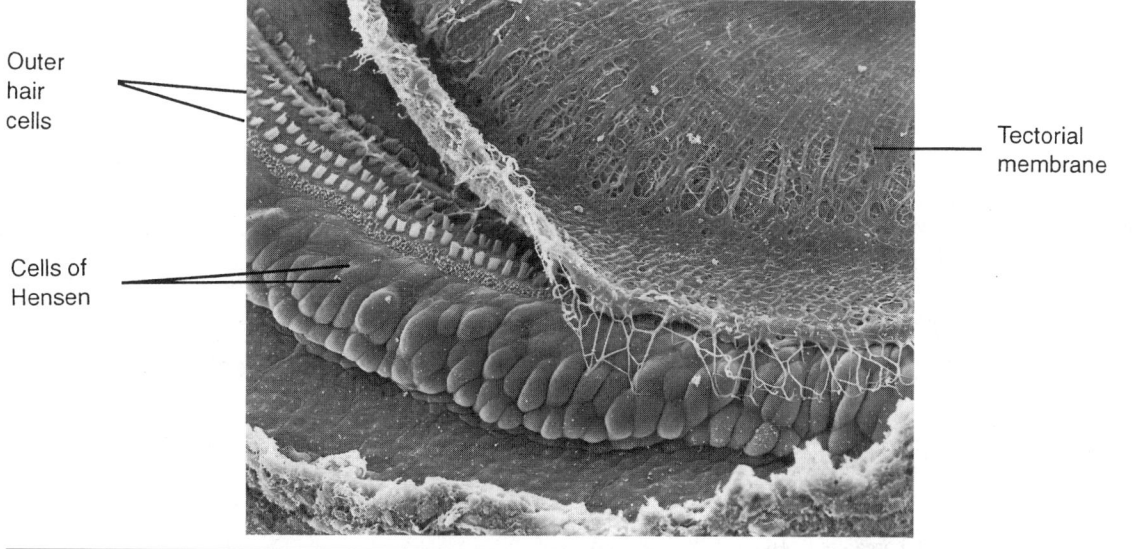

Outer hair cells

Cells of Hensen

Tectorial membrane

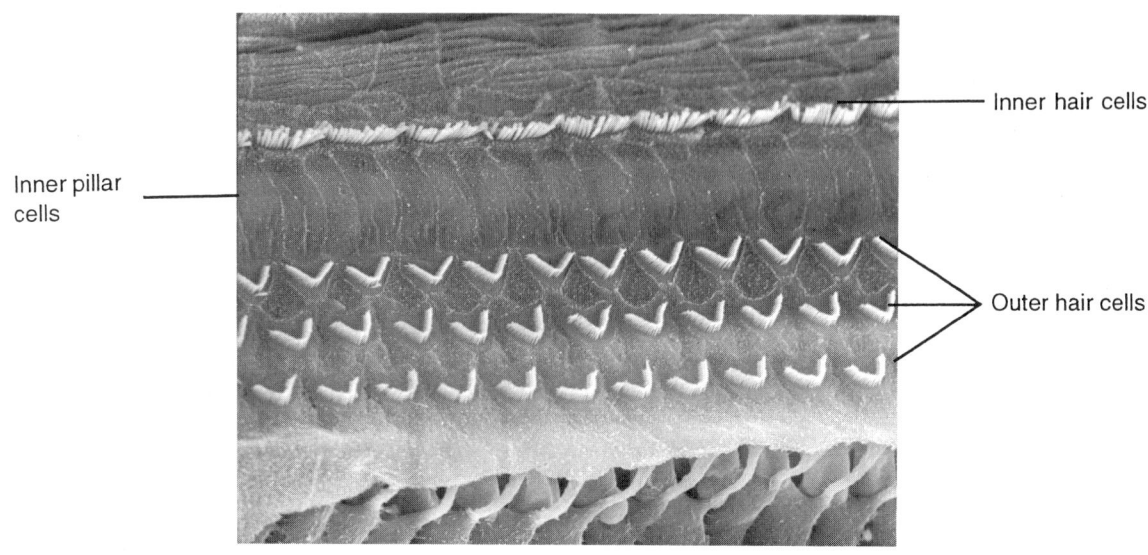

Fig. 10-15. The surface of the organ of Corti showing the arrangement of microvilli on inner and outer hair cells. They are separated by inner pillar cells (SEM, × 1,350).

The sensory hair cells are divided into inner and outer hair cells. **Inner hair cells** form a single row along the inner aspect of the organ of Corti and extend the length of the cochlea (Fig. 10-15). Like type I hair cells of the vestibular labyrinth, the inner hair cells of the cochlea are short, flask-shaped cells with narrow necks. Well-developed microvilli extend from the apical surface of the inner hair cells, but unlike type I hair cells, they lack a kinocilium. The microvilli contain numerous filaments that extend into a dense terminal web in the apical cytoplasm. Mitochondria aggregate just beneath the terminal web and also are scattered throughout the remainder of the cytoplasm, which contains scattered ribosomes and profiles of smooth endoplasmic reticulum. Numerous nerve endings synapse with the bases of the inner hair cells on a plane below the level of the nucleus.

The columnar **outer hair cells** usually form three or four rows and are supported by the apices of the outer phalangeal cells (see Fig. 10-15). Microvilli are arranged in the shape of a V. There is no cilium. Mitochondria aggregate at the base of the outer hair cell, which makes contact with afferent and efferent nerve fibers. In all other respects, the outer hair cells are similar to the inner hair cells, and both are receptors of sound in the organ of Corti.

Review Table 10-1. Key Features of the Vestibular Labyrinth

Subdivision	Receptor Region	Neuroepithelium	Other Features
Semicircular canals	Cristae ampullaris	Type I and II hair cells	Cupula cristae ampullaris
Utricle (1)	Macula (of the utricle)	Type I and II hair cells	Otolithic membrane
Saccule (1)	Macula (of the saccule)	Type I and II hair cells	Otolithic membrane
Cochlear duct (1)	Organ of Corti	Inner hair cells	Tectorial membrane
		Outer hair cells	Basilar membrane

Functional Review: The Ear

The pinna of the external ear collects sound and directs it into the external auditory meatus. Sound waves cause the tympanic membrane to vibrate slightly at the same frequency as the sound waves. Auditory ossicles suspended in the tympanic cavity of the middle ear conduct the vibrations to the perilymph channels of the inner ear. The force of the vibrations created by the sound waves is amplified with no expenditure of energy, because the tympanic membrane is much larger than the foot plate of the stapes. Together, the intervening ossicles act as a piston. With each movement, the foot plate of the stapes exerts pressure on the perilymph, which is confined to the scala vestibuli and scala tympani of the inner ear. The pressure waves pass through the perilymph from the scala vestibuli to the scala tympani and are released from the perilymphatic space by bulging of the secondary tympanic membrane into the air-filled tympanic cavity. As the pressure waves move from scala vestibuli to scala tympani, they must first pass through the intervening cochlear duct or travel to the helicotrema at the apex of the cochlea. When a pressure wave passes through the cochlear duct, the vestibular and basilar membranes are displaced slightly, creating strong shearing forces between microvilli at the apex of inner and outer hair cells and in the overlying tectorial membrane. The shearing action is thought to stimulate the hair cells, which in turn elicit a nerve impulse in the surrounding afferent nerve endings. Supporting cells, especially the pillar and phalangeal cells, provide strong structural support to the cell bodies of their hair cells so that the latter are not displaced by the shearing action on the microvilli.

Large areas of the basilar membrane vibrate at several frequencies, but sound waves of a given frequency produce a maximum displacement along specific regions of the basilar membrane. At low frequencies, maximum displacement occurs farther from the oval window. Because both efferent and afferent endings terminate on hair cells in the organ of Corti, both sensory reception and inhibition are said to occur. The inhibitory mechanism stems from the central nervous system and may aid in discrimination of loudness and pitch.

The two skeletal muscles of the middle ear, the tensor tympani and stapedius, dampen the movement of the ossicles and protect the delicate structures of the inner ear from loud and sudden noises. Contraction of these muscles may play a role in regulating the degree of tension placed on the tympanic membrane so that sounds of moderate intensity can be transmitted in a noisy environment.

Nerve impulses from the organ of Corti are transmitted via the bipolar neurons of the spiral ganglion to the cochlear division of the eighth cranial nerve. From here, impulses are relayed to appropriate regions of the brain to be interpreted as sounds.

The vestibular portion of the inner ear is important for coordinating and regulating locomotion and equilibrium. Types I and II hair cells of the cristae in the ampullary region of each semicircular canal are stimulated during angular movement of the head. Sensory stimulation of the receptors results from movement of endolymph and displacement of the cupula that overlies the cristae. Displacement of the cupula, in which microvilli of underlying hair cells are embedded, causes a shearing force at the apices of the hair cells. Similarly, forces created during linear acceleration act on the otolithic membrane and its contained otoconia, again resulting in shearing forces being applied to the apices of hair cells. Hence linear acceleration results in the stimulation of hair cells in the macula, while angular motion results in stimulation of hair cells in the cristae.

Hair cells of the vestibular organs, as well as those of the cochlea, transform mechanical movements of the endolymph (whether produced by displacement of the cupula, otolithic membrane, or basilar membrane) into the electrical energy of a nerve impulse. Bending or displacement of microvilli on the various types of hair cells is thought to result in depolarization of the hair cell, with the stimulus then being transferred to surrounding afferent nerve endings to generate a nerve impulse. Associated efferent nerve endings are thought to have an inhibitory action and may elevate the threshold of activity of hair cells.

Review Questions and Answers

Questions

56. Briefly discuss the primary function of structures comprising the middle ear.
57. Describe the fluid dynamics of the cochlear division of the inner ear.
58. Compare the mechanisms of stimulation for sensory regions in the vestibular and cochlear regions of the membranous labyrinth.

Answers

56. The middle ear consists of an air-filled space called the tympanic cavity that is continuous anteriorly with the eustachian tube and posteriorly with mastoid air cells. The lateral wall of the tympanic cavity is formed by the tympanic membrane, and its inner boney wall makes contact with the inner ear via two small, membrane-covered apertures called the oval and round windows. The tympanic cavity contains a chain of auditory ossicles (malleus, incus, and stapes) and two small skeletal muscles, the tensor tympani and stapedius.

    The auditory tube connects the anterior portion of the tympanic cavity with the nasopharynx and serves as a passage to ventilate the tympanic cavity and allow pressure equilibration between the middle ear and throat.

    The tympanic membrane receives sound waves and transmits the resulting vibrations to the fluid-filled chambers of the inner ear via the auditory ossicles. The tympanic membrane is about 18 times as large as the oval window, which contains the foot plate of the stapes, and because of this relationship, the tympanic membrane and the auditory ossicles function as a hydraulic piston that exerts pressure on the confined fluid (perilymph) in the chambers of the inner ear. These structures not only transmit the vibrations created by the sound waves to the inner ear but also amplify the weak forces of the sound waves without an expenditure of energy. The tensor tympani and stapedius muscles protect delicate structures in the inner ear by dampening ossicle movement that results from loud or sudden noise. They also may regulate the degree of tension in the tympanic membrane so that sound waves of moderate intensity can be transmitted in a noisy environment.

57. Movement of the foot plate of the stapes in the oval window (fenestra ovalis) exerts pressure on the perilymph in the scala vestibule. Because fluid in a confined space cannot be compressed, the waves of pressure either pass through the cochlear duct and displace it to enter the scala tympani, or enter the scala tympani directly by passing through the helicotrema. Pressure is released from the scala tympani of the cochlea by the elasticity of the secondary tympanic membrane, which bulges into the air-filled tympanic cavity of the middle ear. Pressure waves passing through the cochlear duct produce a slight displacement of the basilar membrane and the organ of Corti, creating a strong shearing force between the microvilli of the inner and outer hair cells and the overlying tectorial membrane. This shearing action is thought to depolarize the hair cells, which in turn generate nerve impulses in the surrounding afferent nerve endings.

58. Type I and type II hair cells of the cristae in the ampullary region of the semicircular canals are stimulated during angular movement of the head. Sensory stimulation of these receptors results from the movement of endolymph that displaces the gelatinous cupula cristae ampullaris, causing a shearing force between the microvilli at the apices of the hair cells, which underlie the cupula. Forces created during linear acceleration act on the otolithic membrane and the contained otoconia and result in a shearing force on the apices of the type I and type II hair cells in the maculae. In the cochlear region, a similar shearing force occurs between the microvilli on the inner and outer hair cells and the overlying tectorial membrane in which their tips are embedded. The shearing forces are generated by displacement of the basilar membrane and organ of Corti.

11 Cardiovascular and Lymph Vascular Systems

Unicellular animals acquire their oxygen and nutrients directly from the external environment by simple diffusion and other activities of their cell membranes. However, in multicellular animals, most of the cells lie deep in the body, with no access to the external environment, and materials are carried to them through a closed system of branching tubes. Together with a muscular pump, the tubular system makes up the cardiovascular (blood vascular) system. The entire structure consists of a heart that acts as a pump, arteries that carry blood to organs and tissues, capillaries through which exchange of materials occurs, and veins that return the blood to the heart.

A second system of vessels, the lymph vascular system, drains interstitial fluid from organs and tissues and returns it to the blood. It lacks a pumping unit and is unidirectional, carrying lymph toward the heart only. There is no lymphatic equivalent of blood arteries.

Heart

The heart is a modified blood vessel that serves as a double pump and consists of four chambers. On the right side, the atrium receives blood from the body and the ventricle propels it to the lungs. The left atrium receives blood from the lungs and passes it to the left ventricle, from which it is distributed throughout the body. The wall of the heart consists of an inner lining layer, a middle muscular layer, and an external layer of connective tissue.

Endocardium

KEY WORDS: squamous (endothelial) cell, tight (occluding) junction, gap junction, subendothelial layer, subendocardial layer

The endocardium forms the inner lining of the atria and ventricles and is continuous with and comparable to the inner lining of blood vessels. It consists of a single layer of polygonal **squamous (endothelial) cells** with oval or rounded nuclei. **Tight (occluding) junctions** unite the closely apposed cells, and **gap junctions** permit the cells to communicate with one another. Electron micrographs show a few short microvilli, a thin proteoglycan layer over the luminal surface, and the usual organelles in the cytoplasm. The endothelial cells rest on a continuous layer of fine collagen fibers, separated from it by a basement membrane. The fibrous layer is called the **subendothelial layer.** Deep to it is a thick layer of denser connective tissue that forms the bulk of the endocardium and contains elastic fibers and some smooth muscle. Loose connective tissue constituting the **subendocardial layer** binds the endocardium to the underlying heart muscle and contains collagen fibers, elastic fibers, and blood vessels. In the ventricles it also contains the specialized cardiac muscle fibers of the conducting system.

Myocardium

KEY WORDS: cardiac muscle, pectinate muscle, trabeculae carnea, atrial natriuretic factor (ANF)

The myocardium is the middle layer of the heart and consists mainly of **cardiac muscle.** It is the thinnest in the atria and thickest in the left ventricle. The myocardium is arranged in layers that form complex spirals about the atria and ventricles. In the atria, bundles of cardiac muscle form a latticework and locally are prominent as the **pectinate muscles,** while in the ventricles, isolated bundles of cardiac muscle form the **trabeculae carnea.** In the atria, the muscle cells are smaller and contain a number of dense granules not seen elsewhere in the heart. These cells have properties associated with endocrine cells. They are most numerous in the right atrium and release the granules when stretched. The granules contain **atrial natriuretic factor (ANF),** which increases the glomerular filtration rate and excretion of sodium ions and water by acting on the medullary collecting tubules of the kidney.

Elastic fibers are scarce in the ventricular myocardium but are plentiful in the atria, where they form an interlacing network between muscle fibers. The elastic fibers of the myocardium become continuous with those of the endocardium and outer layer of the heart (epicardium).

Epicardium

KEY WORDS: visceral layer of pericardium, mesothelial cell, subepicardial layer

The epicardium is the **visceral layer of the pericardium,** the fibrous sac that encloses the heart. The free surface of the epicardium is covered by a single layer of flat to cuboidal **mesothelial cells,** beneath which is a layer of connective tissue that contains numerous elastic fibers. Where it lies on the cardiac muscle, the epicardium contains blood vessels, nerves, and a variable amount of fat. This portion is called the **subepicardial layer.**

The parietal layer of the epicardium consists of connective tissue lined by mesothelial cells that face those covering the epicardium. The two layers are separated only by a thin film of fluid, produced by the mesothelial cells, that allows the layers to slide over each other during contraction and relaxation of the heart.

Cardiac Skeleton

KEY WORDS: annuli fibrosi, dense connective tissue, trigona fibrosi, septum membranaceum

The so-called cardiac skeleton consists of several connective tissue structures to which the cardiac muscle is attached. The main part of the cardiac skeleton is formed by the **annuli fibrosi,** rings of **dense connective tissue** that surround the openings to the aorta and pulmonary artery. Also contributing to the skeleton are masses of fibrous tissue, the **trigona fibrosi,** that occur between the atrioventricular and atrial openings and the **septum membranaceum,** which is the upper, fibrous part of the interventricular septum. The fibrous rings contain elastic fibers and some fat cells and, with aging, may calcify.

Valves

KEY WORDS: chordae tendineae, papillary muscles

Valves are present between the atria and ventricles and at the openings to the aorta and pulmonary vessels. Regardless of their location, the valves are similar in structure. The atrioventricular valves are attached to the annuli fibrosi, the connective tissue of which extends into each valve to form its core. The valves are covered on both sides by endocardium that is thicker on the ventricular side. Scattered smooth muscle cells are present on the atrial side of the valves, while on the ventricular side elastic fibers are prominent. Thin tendinous cords called **chordae tendineae** attach the ventricular sides of the valves to projec-tions of cardiac muscle called **papillary muscles.** Valves of the pulmonary arteries and aorta are thinner but show the same general structure as the atrioventricular valves.

Conducting System

KEY WORDS: specialized cardiac muscle, sinoatrial node, atrioventricular node, atrioventricular bundle, Purkinje fiber (cell), nodal cell

The conducting system of the heart consists of **specialized cardiac muscle** fibers and is responsible for initiating and maintaining cardiac rhythm and for ensuring coordination of the atrial and ventricular contractions. The system consists of the sinoatrial node and the atrioventricular bundle.

The **sinoatrial node** is located in the epicardium at the junction of the superior vena cava and right atrium. Impulses initiated in the node spread throughout the ordinary cardiac muscle of the atria to reach the **atrioventricular node** located on the right side of the interatrial septum. From here, impulses travel rapidly along the **atrioventricular bundle** in the membranous part of the interventricular septum. The bundle divides into two trunks that pass into the ventricles, where they break up into numerous twigs that connect with the ordinary cardiac muscle fibers. Thus the impulse is carried to all parts of the ventricular myocardium.

The specialized fibers of the atrioventricular bundle and its branches are called **Purkinje fibers (cells)** and differ from ordinary cardiac muscle in several respects (Fig. 11-1). Purkinje fibers are larger and contain more sarcoplasm, but myofibrils are less numerous and usually have a peripheral location. The fibers are rich in glycogen and mitochondria and often have two (or more) nuclei. Intercalated discs are uncommon, but numerous desmosomes are scattered along the cell boundaries. **Nodal cells** are smaller than ordinary cardiac muscle cells and contain fewer and more poorly organized myofibrils. Intercalated discs are lacking between cells of the sinoatrial nodes.

Blood Vessels

The blood vessels are variously sized tubes arranged in a circuit, through which blood is delivered from the heart to the tissues and back to the heart from all parts of the body. By this circulatory system, oxygen from the lungs, nutrients from the intestines and liver, and regulatory substances such as hormones are distributed to all the organs and tissues. Waste products also are emptied into the circu-

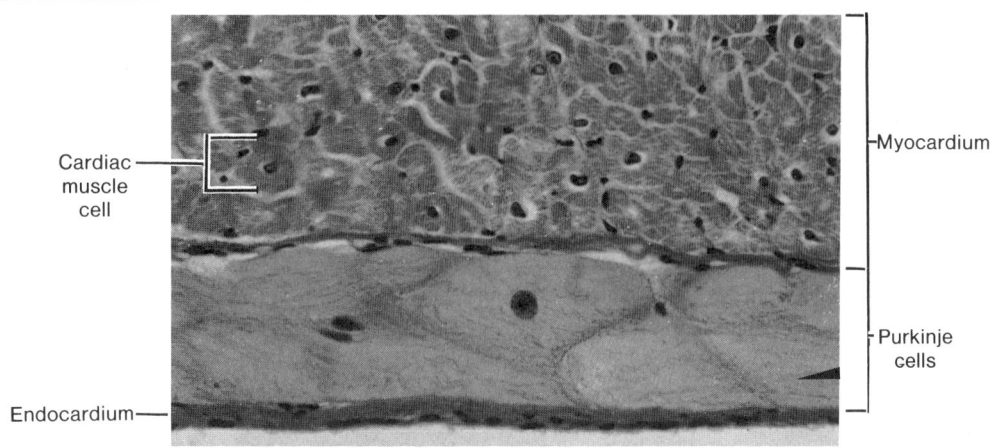

Fig. 11-1. A region of human endocardium that contains Purkinje cells. Compare the size and staining intensity of these cells with cardiac muscle cells forming the myocardium (LM, × 250).

lation and carried to organs such as the lungs and kidneys for elimination.

Structural Organization

KEY WORDS: tunica intima, tunica media, tunica adventitia

Although blood vessels differ in size, distribution, and function, structurally they share many common features. As in the heart, the walls of blood vessels consist of three major coats or tunics. Differences in the appearance and functions of the various parts of the circulatory system are reflected by structural changes in these tunics or by reduction and even omission of some of the layers. From the lumen outward, the wall of a blood vessel consists of a tunica intima, tunica media, and tunica adventitia.

The **tunica intima** corresponds to and is continuous with the endocardium of the heart. It consists of an endothelium of flattened squamous cells resting on a basal lamina and is supported by a subendothelial connective tissue.

The **tunica media** is the equivalent of the myocardium of the heart and is the layer most variable both in size and structure. Depending on the function of the vessel, this layer contains variable amounts of smooth muscle and elastic tissue.

The **tunica adventitia** also varies in thickness in different parts of the vascular circuit. It consists mainly of col-

lagenous connective tissue and corresponds to the epicardium of the heart, but it lacks mesothelial cells.

Arteries

As the arteries course away from the heart they undergo successive divisions to provide numerous branches whose calibers progressively decrease. The changes in size and the corresponding changes in structure of the vessel wall are continuous, but three classes of arteries can be distinguished: large elastic or conducting arteries, medium-sized muscular or distributing arteries, and small arteries and arterioles.

Elastic or Conducting Arteries

KEY WORDS: tunica intima, endothelial coat, caveolae, subendothelial layer, tunica media, elastic lamina, tunica adventitia, vasa vasorum

The aorta is the main conducting artery, but others in this class are the common iliacs and pulmonary, brachiocephalic, subclavian, and common carotid arteries. A major feature of this type of artery is the width of the lumen, which, by comparison, makes the wall of the vessel appear thin.

The **tunica intima** is relatively thick and is lined by a single layer of flattened, polygonal endothelial cells that rest on a complete basal lamina. Adjacent endothelial cells may be interdigitated or overlap and are extensively linked by occluding and communicating junctions. The cells contain the usual organelles, but these are few in number. Peculiar membrane-bound, rod-shaped granules, consisting of several tubules embedded in an amorphous matrix, are present and have been shown to contain a factor (VIII) associated with blood coagulation that is synthesized by the cells and released into the blood. A proteoglycan layer, the **endothelial coat,** covers the luminal surface of the cell membrane. On the basal surface, the endothelial cells are separated from the basal lamina by an amorphous matrix. Numerous invaginations or **caveolae** are present at the luminal and basal surfaces of the plasmalemma, and the cytoplasm contains many small vesicles 50 to 70 nm in diameter. These are thought to form at one surface, detach, and cross to the opposite side, where they fuse with the plasmalemma to discharge their contents. In humans, about one-fourth of the total thickness of the intima is formed by the **subendothelial layer,** a layer of loose connective tissue that contains elastic fibers and a few smooth muscle cells.

The **tunica media** is the thickest layer and consists largely of elastic tissue that forms 50 to 70 concentric, fenestrated sheets or **elastic laminae,** each about 2 or 3 μm thick and spaced 5 to 20 μm apart (Fig. 11-2). Successive laminae are connected by elastic fibers. In the spaces between the laminae are thin layers of connective tissue that contain collagen fibers and smooth muscle cells arranged circumferentially. The smooth muscle cells are flattened, irregular, and branched and are bound to adjacent elastic laminae by elastic microfibrils and collagen fibers. An appreciable amount of amorphous ground substance also is present. Smooth muscle cells are the only cells present in the media of elastic arteries and synthesize and maintain the elastic fibers.

Elastic arteries are those nearest the heart and, because of the great content of elastic tissue, are expandable. As blood is pumped from the heart during its contraction, the walls of elastic arteries expand; when the heart relaxes, the elastic recoil of these vessels serves as an auxiliary pumping mechanism to force blood onward between beats.

The **tunica adventitia** is relatively thin and contains bundles of collagen fibers and a few elastic fibers, both of which have a loose, helical arrangement. Fibroblasts, mast cells, and rare longitudinally oriented smooth muscle cells also are present. Tunica adventitia gradually blends into surrounding loose connective tissue.

The walls of large arteries are too thick to be nourished only by diffusion from the lumen. Consequently, these vessels have their own small arteries, the **vasa vasorum,** that may arise as branches of the main vessel itself or derive from neighboring vessels. They form a plexus in the tunica adventitia and generally do not penetrate deeply into the media (see Fig. 11-2).

Muscular or Distributing Arteries

KEY WORDS: tunica intima, endothelium, subendothelial layer, internal elastic lamina, tunica media, smooth muscle, external elastic lamina, tunica adventitia

Fig. 11-2. A section through the wall of human aorta stained with orcein to selectively demonstrate elastic laminae (LM, × 40).

The muscular or distributing arteries are the largest class, and except for the elastic arteries, the named arteries of gross anatomy fall into this category. All the small branches that originate from elastic trunks are muscular arteries, and the transition is relatively abrupt, occurring at the openings of the branching arteries. Compared with the luminal diameter, the walls of these arteries are thick and make up over one-fourth of the cross-sectional diameter. The thickness of the wall is due mainly to the large amount of smooth muscle in the media, which, by contraction and relaxation, helps to regulate the blood supply to organs and tissues. The general organization of these vessels is similar to that of elastic arteries, but the proportion of cells and fibers differs.

The **tunica intima** consists of an endothelium, a subendothelial layer, and an internal elastic lamina. The **endothelium** and **subendothelial layers** are similar to those of elastic arteries, but as the size of the vessel decreases, the subendothelial layer becomes thinner. It contains fine collagen fibers, a few elastic fibers, and scattered smooth muscle cells that are longitudinal in orientation. The **internal elastic lamina** is a fenestrated band of elastin that forms a prominent, scalloped boundary between the tunica media and the tunica intima (Fig. 11-3). The basal surfaces of the endothelial cells send slender processes through discontinuities in the elastic lamina to make contact with cells in the tunica media.

The **tunica media** is the thickest coat and consists mainly of **smooth muscle** cells arranged in concentric, helical layers. The number of layers varies from 3 to 4 in smaller arteries to 10 to 40 in the large muscular arteries. The muscle cells are surrounded by basal laminae and communicate with each other by gap junctions. Reticular fibers and small bundles of collagen fibers interspersed with elastic fibers are present between the layers of smooth muscle. The number and distribution of elastic fibers correlate with the caliber of the artery. In the smaller vessels elastic fibers are scattered between the muscle cells, whereas in larger arteries the elastic tissue forms circular, loose networks. At the junction of the tunica media and the tunica adventitia, the elastic tissue forms a second prominent fenestrated membrane called the **external elastic lamina.**

The **tunica adventitia** is prominent in muscular arteries and in some vessels may be as thick as the media. It consists of collagen and elastic fibers that are longitudinal in orientation. This coat continues into surrounding loose connective tissue with no clear line of demarcation.

Small Arteries and Arterioles

Small arteries and arterioles differ from large muscular arteries only in size and the thickness of their walls. The distinction between small arteries and arterioles is mainly one of definition: Arterioles are those small arteries with a diameter less than 250 to 300 μm and in which the media contains only one or two layers of smooth muscle cells (Figs. 11-4 and 11-5).

Fig. 11-3. A section through the wall of a human muscular artery stained with orcein to selectively demonstrate elastic laminae (LM, × 200).

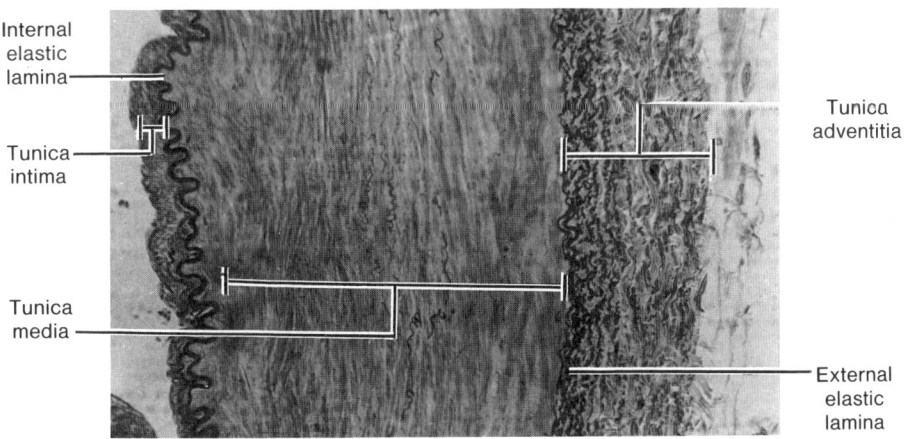

Internal elastic lamina

Tunica intima

Tunica media

Tunica adventitia

External elastic lamina

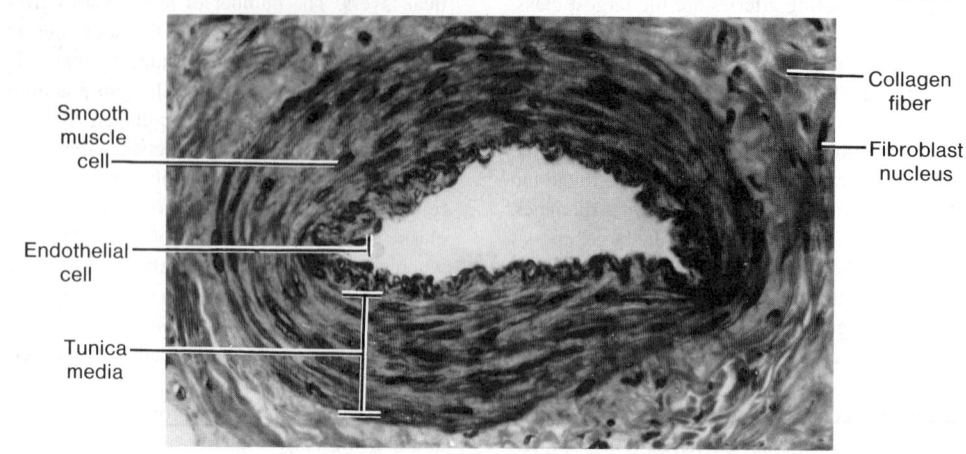

Fig. 11-4. A section through the wall of a small human artery (LM, × 250).

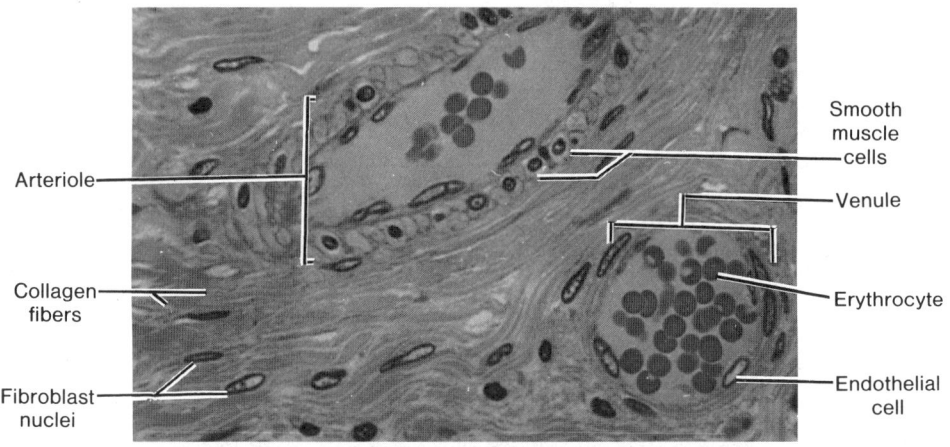

Fig. 11-5. A region of loose connective tissue containing an arteriole and venule (LM, × 250).

In the progression from distributing artery to arteriole, the subendothelial connective tissue progressively decreases until in arterioles, a subendothelial layer is lacking and the tunica intima consists only of endothelium and a fenestrated internal elastic lamina. The endothelial cells are joined by occluding junctions. Basal processes make contact with smooth muscle cells through the fenestrations in the elastic lamina. The thin basal lamina becomes indistinct in the smallest arterioles. The layers of smooth muscle of the media progressively decrease in number, and at a diameter of about 30 μm, the muscle coat of arterioles consists of a single layer of circumferentially oriented cells. The internal elastic membrane is thin and disappears in terminal arterioles; there is no definite external elastic lamina. Tunica adventitia also decreases in thickness, becoming extremely thin in the smallest arterioles.

Compared with the luminal size, the walls of arterioles are thick. Because of the smooth muscle in the media, arterioles control the blood to capillary beds they feed and are the main regulators of blood pressure.

Metarterioles

Metarterioles are intermediate between capillaries and arterioles and control the flow of blood through capillary beds. They also are called *capillary sphincter areas* or *precapillary arterioles*. Although poorly defined morphologically, their lumina generally are wider than those of the capillaries they serve, and circularly arranged smooth muscle cells are scattered in their walls. Here individual smooth muscle cells spaced a short distance apart completely encircle the endothelial tube.

Capillaries

Capillaries are the smallest functional units of the blood vascular system and are inserted between arterial and venous limbs of the circulation. They branch extensively to form elaborate networks, the extent of which reflects the activity of an organ or tissue. Highly branched, closely packed networks of capillaries are present in the lungs, liver, kidneys, glands, and mucous membranes. Based on the appearance of the endothelium and basal lamina in electron micrographs, capillaries are classed as continuous or fenestrated. Regardless of the type, the basic structure of capillaries is the same and represents an extreme simplification of the vessel wall. The tunica intima consists of endothelium and a basal lamina; the tunica media is absent and the tunica adventitia is greatly reduced.

Continuous Capillaries

KEY WORDS: **continuous endothelium, continuous basal lamina, tunica adventitia, pericytes**

Continuous capillaries are the common type of capillary found throughout the connective tissues, muscle, and central nervous system. Their lumina range from 5 to 10 μm in diameter and in the smaller vessels may be encompassed by a single endothelial cell. In larger capillaries, three or four cells may enclose the lumen. The wall consists of a **continuous endothelium** and a thin tunica adventitia. The endothelial cells rest on a **continuous basal lamina** and show the same features as endothelial cells elsewhere in the vascular tree (Fig. 11-6). They contain the usual organelles, caveolae and vesicles, a proteoglycan layer, a few short microvilli on the luminal surfaces, and a variable number of membrane-bound granules. Multivesicular bodies also can be found. Adjacent cells may abut each other or overlap obliquely or show sinuous interdigitations. At intervals, adjacent cell membranes unite in

occluding junctions, but it is not certain that these form complete belts as in other epithelia.

The **tunica adventitia** is thin and contains some collagen and elastic fibers embedded in a small amount of ground substance. Fibroblasts, macrophages, and mast cells are present also.

Pericytes are irregular, branched, isolated cells that occur at intervals along capillaries, enclosed by the basal lamina of the endothelium. The cells resemble fibroblasts and characteristically contain a few dense bodies and numerous cytoplasmic filaments. Actin and myosin have been demonstrated in the cytoplasm by immunohistochemistry, indicating a contractile function. However, their exact function is unknown.

Fenestrated Capillaries

KEY WORDS: **fenestra, diaphragm, continuous basal lamina**

Fenestrated capillaries have the same structure as the continuous type but differ in that the endothelial cells contain numerous **fenestrae** (pores) that appear as circular openings 70 to 100 nm in diameter (Fig. 11-7). Fenestrae may be distributed at random or in groups and usually are closed by thin **diaphragms** that show central, knoblike thickenings. Each diaphragm is a single-layered structure — thinner than a single-unit membrane — so it would appear unlikely that it is formed by apposition of two cell membranes. The **basal lamina** is **continuous** across the fenestrae on the basal side of the endothelium (see Fig. 11-7). The fenestrated capillaries of the renal glomerulus differ in that the pores lack diaphragms and the basal lamina is much thicker than in other capillaries.

Sinusoids

KEY WORDS: **gaps, discontinuous basal lamina, phagocyte**

Sinusoids are thin-walled vessels that are much wider and have lumina that are more irregular in outline than those of capillaries. Sinusoids are found in the liver, spleen, bone marrow, and some endocrine glands. Their endothelia are attenuated and may be continuous as in the marrow, or the cells may be separated by **gaps** and rest on a **discontinuous basal lamina.** Sinusoids often are associated with **phagocytes** either as a component of the lining, as in the liver, or closely applied to the exterior of the wall, as in the spleen. The endothelial cells themselves show no greater capacity for phagocytosis than do endothelial cells in other

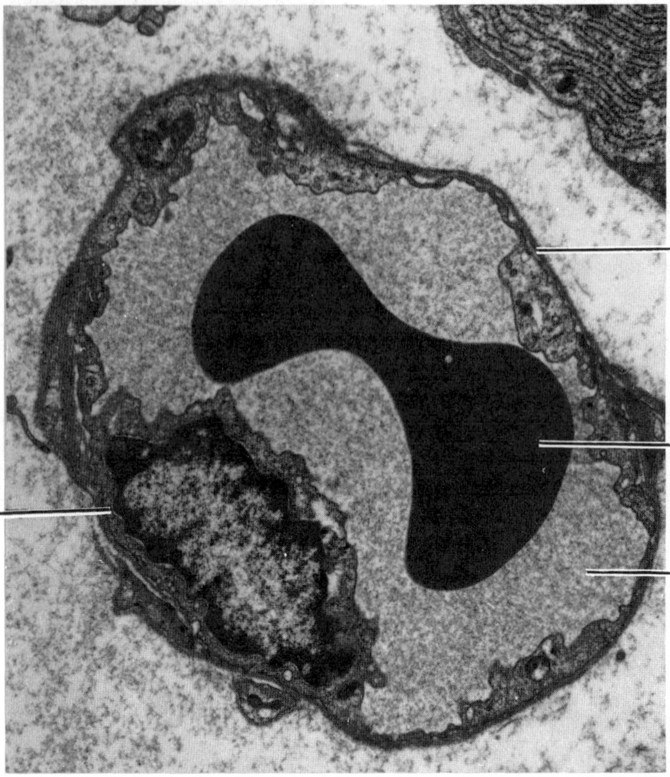

Fig. 11-6. A capillary from the lamina propria of a human stomach. Note the erythrocyte within its lumen (TEM, × 5,000).

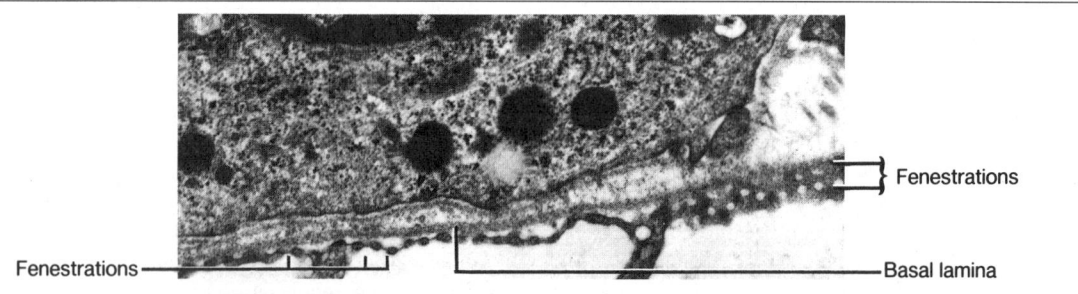

Fig. 11-7. Fenestrated endothelium from a capillary within a parathyroid gland (TEM, × 15,000).

vessels. They do show more active endocytosis and more numerous lysosomes than are found in endothelia elsewhere.

Diffusion

Exchange of nutrients and wastes between tissues and blood occurs across the thin endothelium of capillaries. It has been estimated that no active cell lies more than 30 or 40 μm away from a capillary. The most important way materials cross the endothelium is by diffusion, and the total capillary surface available for exchange has been calculated to be over 100 square meters. Lipid-soluble materials diffuse directly through the endothelial cells; water and water-soluble materials are transmitted through water-filled, physiologic "pores" that may be the equivalent of the cytologic pores of the fenestrated capillaries. In the continuous capillaries, caveolae and cytoplasmic vesicles have been implicated in transepithelial transport of water-soluble materials; pinocytosis also may contribute to transfer of substances of high molecular weight. Fluid exchange also occurs through intercellular clefts between endothelial cells.

Endothelial Cell Secretion

KEY WORDS: nitric oxide, endothelin I, intercellular adhesion molecules, platelet-activating factor, P-selectin, prostacyclin

Endothelial cells are now known to secrete several substances of the underlying extracellular matrix (types II, IV, and V collagens, fibronectin, laminin), as well as several factors involved with maintenance of vascular tone, blood clotting, and emigration of leukocytes. Endothelial cells synthesize and release **nitric oxide,** which causes relaxation of adjacent smooth muscle, thereby decreasing the vascular tone of blood vessels. Endothelial cells also can respond to anoxia by secreting a peptide called **endothelin I,** a potent vasoconstrictor of long duration that binds to smooth muscle cells of arteries, thereby elevating blood pressure. Endothelial cells at sites of inflammation are capable of synthesizing and inserting **intercellular adhesion molecules, platelet-activating factor,** and **P-selectin** into their luminal plasmalemma. These factors bind to circulating leukocytes, allowing them to migrate through the endothelium to the site of inflammation. Endothelial cells also contain **prostacyclin,** an agent that inhibits platelet adhesion, as well as thromboplastic substances and antihemophilic factor VIII of the blood coagulation system. Thus

endothelial cells are equipped to initiate clotting and repair minor defects in the lining of vessels to prevent the leakage of blood.

Veins

The venous side of the circulatory system carries blood from the capillary beds to the heart, and in their progression, the veins gradually increase in size and their walls thicken. Their structure basically is the same as that of arteries, and the three coats — tunica intima, media, and adventitia — can be distinguished but are not as clearly defined. In general, veins are more numerous and larger than the arteries they accompany, but their walls are thinner because of a reduction of muscular and elastic elements. Since their walls are less sturdy, veins tend to collapse when empty and in sections may appear flattened, with irregular, slitlike lumina.

Structurally, veins vary much more than arteries. The thickness of the wall does not always relate to the size of the vein, and the same vein may differ structurally in different areas. Histologic classification of veins is less satisfactory than for arteries, but several subdivisions usually are made — namely, venules and small, medium, and large veins.

Venules

KEY WORDS: continuous endothelium, pericyte, thin adventitia, incomplete tunica media

Venules arise from the union of several capillaries to form vessels 10 to 50 μm in diameter. The junctions between venules and capillaries are important sites of fluid exchange between tissues and blood.

The tunica intima consists of a thin, **continuous endothelium,** the cells of which are loosely joined by poorly developed intercellular junctions (see Fig. 11-5). The thin basal lamina is pierced by **pericytes** that appear to make contact with the endothelial cells.

The tunica media is missing in the smallest venules, and the relatively **thin adventitia** contains a few collagen fibers, scattered fibroblasts, mast cells, macrophages, and plasma cells. As the vessels increase in size to reach diameters of 50 μm, circularly oriented, scattered smooth muscle cells begin to appear and form a somewhat discontinuous and **incomplete tunica media.** The adventitia increases in thickness and consists of longitudinally arranged collagen fibers that form an open spiral around the vessel. Fibroblasts often are irregular in shape and bear thin processes.

Small Veins

KEY WORDS: **continuous endothelium, tunica media, tunica adventitia**

Small veins vary from about 0.2 to 1.0 mm in diameter. The tunica intima consists only of a **continuous endothelium** that rests on a thin basal lamina. Smooth muscle cells make up a **tunica media,** which contains one to four layers of cells. Between the smooth muscle cells is a thin network of elastic and collagen fibers. The **tunica adventitia** forms a relatively thick coat and contains longitudinally oriented collagen fibers and some thin elastic fibers.

Medium Veins

KEY WORDS: **thin tunica intima, narrow subendothelial layer, tunica media, thick tunica adventitia, vasa vasorum, valve**

The medium class of veins includes most of the named veins of gross anatomy except for major trunks. They vary in size from 1 to 10 mm in diameter. The **thin tunica intima** consists of endothelial cells resting on a basal lamina, but a **narrow subendothelial layer** may be present and contains fine collagen fibers and scattered thin elastic fibers. The elastic fibers may form a network at the junction of tunica intima and media, but a poorly defined inter-

nal elastic lamina is formed only in the larger vessels. In most medium veins, the **tunica media,** though well developed, is thinner than in corresponding arteries (Fig. 11-8). The **thick tunica adventitia** forms the bulk of the wall and is larger than the tunica media. It consists of collagen and elastic fibers and frequently contains longitudinal smooth muscle cells. **Vasa vasorum** are present in the larger vessels of this class.

Most medium veins are equipped with **valves,** pocketlike flaps of tunica intima that project into the lumen, their free edges oriented in the direction of flow. They consist of a core of connective tissue covered on both sides by endothelium. A rich network of elastic fibers is present in the connective tissue beneath the endothelium on the downstream side of the valves. As blood flows toward the heart, the valves are forced against the vessel wall, but with backflow, the valves are forced outward against each other to occlude the vessel and prevent reversal of blood flow.

Large Veins

KEY WORDS: **thick tunica adventitia, vasa vasorum, thin tunica media, tunica intima**

In large venous trunks (venae cavae; renal, external iliac, splenic, portal, and mesenteric veins), the **thick tunica adventitia** forms the greater part of the wall (Fig. 11-9). It consists of a loosely knit connective tissue with thick, longitudinal bundles of collagen and elastic fibers. Smooth muscle layers, also longitudinal in orientation, are present

Fig. 11-8. A segment of wall from a human medium-sized vein stained with orcein to selectively demonstrate elastic laminae (LM, × 100).

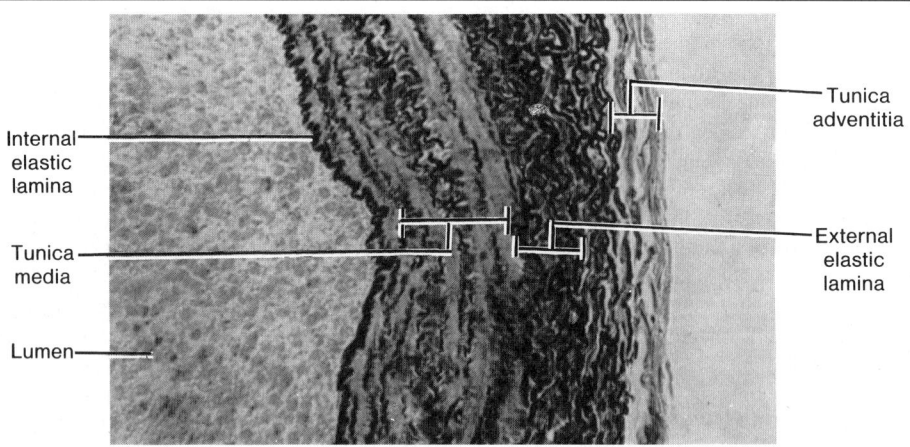

Internal elastic lamina

Tunica media

Lumen

Tunica adventitia

External elastic lamina

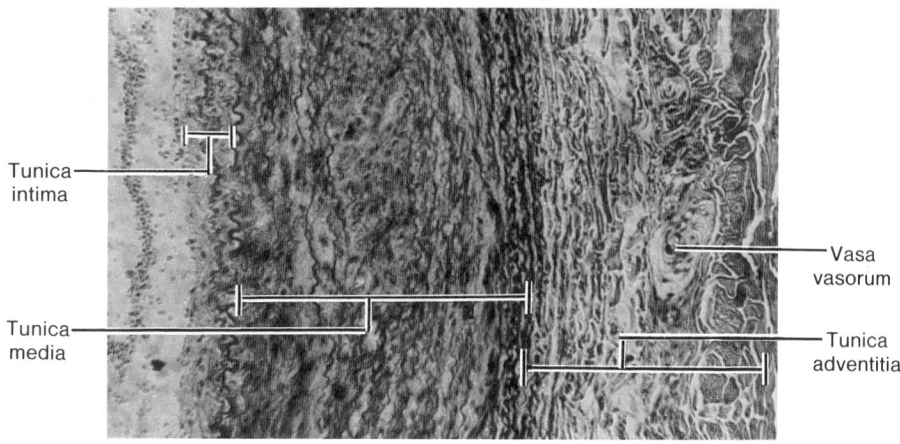

Fig. 11-9. A portion of the wall of human inferior vena cava stained with orcein selectively demonstrates scattered elastic laminae (LM, × 40).

and are especially well developed in the inferior vena cava. **Vasa vasorum** are present and may extend into the media. A **thin tunica media** is poorly developed and may even be absent; otherwise it has the same organization as that in medium veins. The **tunica intima** is supported by a subendothelial layer that may become prominent in larger trunks.

Veins with Special Features

Some veins lack a tunica media, such as the trabecular veins of the spleen; veins of the retina, bone, and maternal placenta: most meningeal and cerebral veins; and those of the nail bed. Veins in the pregnant uterus contain smooth muscle in all three coats; in the intima the fibers run longitudinally rather than circularly, as they do also in the tunica intima of the saphenous, popliteal, femoral, umbilical, and internal jugular veins. At their junctions with the heart, the adventitia of the pulmonary veins and venae cavae are provided with a coat of cardiac muscle; the fibers run longitudinally and circularly about the vessels for a short distance.

Arteriovenous Anastomoses

In addition to their capillary connections, arteries and veins may unite by shunts called *arteriovenous anastomoses.* Generally these arise from side branches of arterioles that pass directly to venules. They are thick-walled, muscular vessels of small caliber that usually are coiled and surrounded by a connective tissue sheath. They are plentiful in the plantar and palmar surfaces, fingertips, toes, lips, and nose and also occur in the thyroid. When open, the anastomoses shunt blood around the capillary bed and thus regulate the blood supply to many tissues.

Carotid and Aortic Bodies

KEY WORDS: epithelioid (glomus) cells, chemoreceptor

Carotid bodies are encapsulated structures that occur on each side of the neck at the bifurcations of the common carotid arteries. They consist of masses of large, polyhedral **epithelioid (glomus) cells** that are closely related to a rich network of sinusoidal vessels. The cells have large, pale nuclei and light, finely granular cytoplasm. In electron micrographs, the granules show dense cores and contain catecholamines and 5-hydroxytryptamine. A second cell type, the type II glomus cell, appears much the same as type I but lacks granules. Many nerves ramify throughout the structures. Carotid bodies act as **chemoreceptors** and monitor blood for changes in oxygen and carbon dioxide content, pH, and pressure.

Aortic bodies are similar structures that lie close to the aorta between the angle of the subclavian and carotid arteries on the right and near the origin of the subclavian artery on the left. They are thought to have functions similar to those of carotid bodies. Both these structures are derived from the neural crest.

Lymph Vascular System

The lymph vascular system consists of endothelial-lined tubes that recover intercellular fluid not picked up by the blood vascular system and return it to the blood. The fluid (lymph) carried by the lymphatics is a blood filtrate formed as fluid crosses the blood capillaries into the tissues. Unlike the blood vascular system, lymph flow is unidirectional — from tissues to the union of the lymphatic and blood vascular systems at the base of the neck.

The lymphatic vascular system begins in the tissues as blindly ending capillaries that drain into larger collecting vessels and then into two main lymphatic trunks. Lymph nodes occur along the course of the vessels and filter the lymph. Lymphatics are present in most tissues but are absent from bone marrow, the central nervous system, coats of the eye, internal ear, and fetal placenta.

Lymph Capillaries

KEY WORDS: thin continuous endothelium, discontinuous basal lamina, anchoring filament

Lymph capillaries are thin-walled, blind tubes that branch and anastomose freely to form a rich network in organs and tissues. They are wider and more irregular than blood capillaries. The wall of a lymph capillary consists only of a **thin continuous endothelium** and a **discontinuous basal lamina** that is present only in patches or may even be absent. Adjacent endothelial cells may overlap, but junctional complexes are few and clefts occur between the cells. Externally, the endothelium is surrounded by a small amount of collagenous connective tissue. Fine filaments run perpendicularly from the collagen bundles and attach to the outer surfaces of the endothelium as **anchoring filaments** that maintain the patency of the vessel.

Collecting Lymph Vessels

KEY WORDS: tunica intima, tunica media, tunica adventitia, valve

Collecting lymph vessels differ from lymph capillaries in size and the thickness of their walls. Although three coats — intima, media, and adventitia — are described as in blood vessels, they are not clearly delineated.

The **tunica intima** consists of an endothelium supported by a thin network of longitudinal elastic fibers. The **tunica media** is composed of smooth muscle with a predominantly circular arrangement, though some fibers run longitudinally. Between the muscle fibers are a few fine elastic fibers. The **tunica adventitia** is the thickest coat and consists of bundles of collagen fibers, elastic fibers, and some smooth muscle cells, all of which have a longitudinal orientation.

Valves are numerous along the course of lymphatic vessels and occur at closer intervals than in veins. They arise in pairs as folds of the intima, as in veins.

Lymphatic Trunks

KEY WORDS: tunica intima, subendothelial layer, internal elastic lamina, tunica media, tunica adventitia

Collecting lymphatic ducts ultimately gather into two main trunks, the thoracic and right lymphatic ducts. The thoracic duct is the longer and has a greater field of drainage. It begins in the abdomen, passes along the vertebral column, and opens into the venous system at the junction of the left jugular and subclavian veins. It receives lymph from the lower limbs, abdomen, left upper limb, and left side of the thorax, head, and neck. The right lymphatic duct receives lymph only from the upper right part of the body and empties into the brachiocephalic vein.

The structure of the trunks is the same, generally resembling that of a large vein. The **tunica intima** consists of a continuous endothelium supported by a **subendothelial layer** of fibroelastic tissue with some smooth muscle. Near the junction with the tunica media, the elastic fibers condense into a thin **internal elastic lamina.** The thickest layer is the **tunica media,** which contains more smooth muscle than does the media of large veins. The smooth muscle cells have a predominantly circular arrangement and are separated by abundant collagenous tissue and some elastic fibers. The **tunica adventitia** is poorly defined and merges with the surrounding connective tissue. It contains bundles of longitudinal collagen fibers, elastic fibers, and occasional smooth muscle cells. The wall of the thoracic duct contains nutrient blood vessels similar to the vasa vasorum of large blood vessels.

Review Table 11-1 Key Features of Arteries and Veins

Arteries	Tunica Intima	Tunica Media	Tunica Adventitia	Other
Large	Thick subendothelial layer with elastic fibers and smooth muscle; internal elastic lamina prominent	Major coat; 40–60 elastic laminae; smooth muscle, fibrous connective tissue between laminae	Thin; no external elastic lamina	Compared to lumen, wall is thin; vasa vasorum extend through adventitia and about one-third of the way through media
Small and medium	Thin subendothelial layer present; prominent internal elastic lamina	Major coat; up to 40 layers of smooth muscle, some elastic tissue	May equal thickness of media; external elastic lamina present	Muscular (distributing) arteries, including most of named arteries
Arteriole	Thin, no recognizable subendothelial layer; internal elastic lamina	Forms the main coat; 1–5 layers of smooth muscle	Thin; no external elastic lamina; fibroelastic	Wall (relative to lumen) is thicker than in any other vessel
Capillaries				
Continuous	Endothelium plus basement membranes; pericytes, tight junctions, desmosomes	Absent	Scant reticular fibers	5–10 μm in diameter
Fenestrated	Endothelial cells bear openings in cytoplasm, usually closed by diaphragm	Absent	Scant reticular fibers	In glomeruli, fenestrae not closed by diaphragm
Veins				
Venule	Endothelium only; no internal elastic lamina	Very thin; 1–3 layers of smooth muscle cells	Thick compared with total wall; wholly collagenous	
Small and medium	Thin; subendothelial layer lacking in smaller veins	Thin; layer of smooth muscle, elastic fibers, and collagen fibers	Well developed; thick fibroelastic layer; no elastic lamina	
Large	Thicker than in small veins; a delicate internal elastic lamina may be present	Thin or even lacking	Thickest coat; fibromuscular; no elastic lamina	

Functional Review: Cardiovascular and Lymph Vascular Systems

The myocardium of the heart provides the propelling force that moves blood through the systemic and pulmonary circulations. Cardiac muscle is characterized by its ability to contract spontaneously without external stimulus. The heart has a system of specialized fibers that regulate the contractions into rhythmic beats coordinated throughout the chambers of the heart. The contraction impulse is initiated in the sinoatrial (SA) node, the pacemaker of the heart. The cells of the SA node are unique in that they spontaneously reach the threshold at which an action potential is automatically triggered. The impulse passes through ordinary cardiac muscle fibers of the atria to reach the atrioventricular node, where it is delayed, allowing the atria to complete their contractions before the impulse is passed to the ventricles. The impulse then is conducted rapidly along the atrioventricular bundle and by its branches is spread throughout the ventricular myocardium. The granular cells of the atria secrete the peptide hormone ANF, which plays a role in regulating blood pressure and volume and in controlling the excretion of water and sodium. Thus the heart functions, to a limited extent, as an endocrine organ.

Flow of blood from the heart is intermittent, and the arteries immediately joining the heart are subjected to pulses of high pressure. Because of the large amounts of elastic tissue in their walls, the large conducting vessels can expand and absorb the forces without damage. During the intervals between contractions, the elastic recoil of the vessel wall serves as an additional propelling mechanism, helps smooth out the blood flow, and protects delicate capillaries and venules from severe changes in pressure.

Distribution of blood to organs is controlled largely by muscular arteries. The caliber of these vessels is regulated by the degree of contraction of the tunica media in response to changes in blood pressure and the demands of organs. Pressure in a capillary bed is controlled by arterioles: The amount of blood flow is controlled by the opening and closing of shunts formed by arteriovenous anastomoses.

Exchange of materials between blood and tissues occurs at the level of capillaries and venules. Passage of fluid across capillary walls depends partly on blood pressure and partly on colloidal osmotic pressure in the capillaries. Blood pressure promotes passage of fluid into tissues; osmotic pressure promotes resorption. At the arterial end of capillary beds, blood pressure is greater than osmotic pressure and fluid is driven out of the vessel; at the venous end, osmotic pressure is the greater and fluid is driven into the vessel. Venules, with their thin walls and loosely joined endothelial cells, also are important sites of fluid exchange. Lipid-soluble materials readily diffuse through endothelial cells, whereas water-soluble materials pass through pores and intercellular clefts. The so-called physiologic pores may correspond to the fenestrae of fenestrated endothelium or to caveolae and cytoplasmic vesicles of continuous capillaries. Bulky material, including whole cells, readily crosses the walls of sinusoids.

Blood is returned to the heart via the veins, which are able to constrict or enlarge and store large quantities of blood. From 70 to 75 percent of the circulating blood is contained in the veins, and much of it can be made available to organs as needed. Blood is moved along the veins by the massaging action of muscles and fascial tissues and by special arrangements of vessels in the soles of the feet that act as pumps when pressure is exerted on them. Blood flows away from the area of compression but is prevented from flowing backward by the valves.

Although most of the fluid filtered from the blood capillaries is resorbed by them, about 10 percent is not recovered by the blood vascular system, and it is this fluid that is recovered by the lymphatics. Substances of high molecular weight such as proteins pass into the blood capillaries with difficulty but are recovered by lymphatic capillaries almost unhindered. Lymphatic channels are one of the main routes by which nutrients, especially fats, are absorbed from the gastrointestinal tract.

Blood vascular endothelium serves as a permeability barrier — a moderator of selective transport — and has important functions in maintaining the fluidity of blood. The endothelium provides a physical surface to which platelets do not normally stick, so it prevents their aggregation; it also synthesizes prostacyclin, an agent that inhibits platelet adhesion. However, endothelium also contains thromboplastic substances and antihemophilic factor VIII of the blood coagulation system. Thus the endothelium is also equipped to initiate clotting and repair minor defects to prevent leakage of blood. The luminal surface bears a negative charge, as do red cells, thus preventing their aggregation on the wall of the blood vessel and resulting eddying and turbulent flow. Endothelial cells secrete several other substances involved with the maintenance of vascular tone (nitric oxide, endothelin I), as well as factors concerned with the emigration of leukocytes at the sites of inflammation.

Review Questions and Answers

Questions

59. List the layers that comprise the walls of the heart, and name the corresponding structures in the blood vessels.
60. What is the most prominent coat in the large conducting arteries? What tissue component characterizes this coat? What functional advantage does this give to the arteries?
61. Of what does the conducting system of the heart consist? What is its function?
62. How do veins differ from arteries?
63. Name two types of capillaries, and describe the structural changes that distinguish them. How do sinusoids differ from capillaries?
64. How is blood flow through a capillary bed controlled?

Answers

59. Layers comprising the wall of the heart are the endocardium, myocardium, and epicardium and correspond to the tunica intima, tunica media, and tunica adventitia of blood vessels.
60. Conducting arteries show a prominent tunica media, characterized by the abundance of elastic tissue that is arranged in sheets or laminae. Because of the large amount of elastic tissue, the conducting arteries are expansile and able to absorb the force of the blood as it is ejected from the heart. The recoil of the elastic wall serves as a propelling mechanism for blood during the periods between the heart beats, thus helping to smooth out the blood flow and protect the delicate capillaries and venules from violent changes in pressure.
61. The conducting system of the heart consists of the sinoatrial node, the atrioventricular node, and the atrioventricular bundle and its branches. Each of these components consists of modified cardiac muscle fibers. The conducting system is responsible for the coordination of the intrinsic contractions of the cardiac muscle, ensuring the maintenance of cardiac rhythm and coordination of the atrial and ventricular contractions.
62. Veins are more numerous and generally of greater size, and their walls are thinner than those of arteries mainly because of a reduction in the muscular and elastic elements. Veins also are much more variable in their structure — the thickness of their walls does not necessarily relate to the overall size of the vein, even in the same vein. Usually, but not invariably, the tunica adventitia forms the prominent layer in the wall of the veins.
63. Capillaries are classed as continuous or fenestrated based on the appearances of the endothelium and basal lamina. Continuous capillaries are characterized by an uninterrupted, continuous lining of endothelial cells resting on a complete basal lamina. Fenestrated capillaries are characterized by fenestrae or pores in the endothelial cells and a continuous basal lamina. The fenestrae usually are closed by a thin diaphragm. Sinusoids are more irregular in outline and have wider bores than other capillaries. These vessels show a loose arrangement of the endothelium, and gaps are present between adjacent endothelial cells. The basal lamina is discontinuous. Phagocytes usually are closely associated with the walls of sinusoids. In all types of capillaries, the walls consist of tunica intima and an attenuated tunica adventitia; the media is lacking.
64. Blood flow through a capillary bed is controlled by the activity of the arterioles and by the operation of the arteriovenous anastomoses. The distribution of blood to the organs or tissues in which the capillary bed lies is controlled by the muscular distributing arteries.

12 Lymphatic Organs

Lymphatic organs are divided into primary (central) and secondary (peripheral) organs. Primary lymphatic organs are the first to develop and include the thymus in mammals and the thymus and bursa of Fabricius in birds. The mammalian equivalent of the bursa is unknown but probably is the bone marrow, although the spleen and the lymphatic tissue associated with the gastrointestinal tract (Peyer's patches, lamina propria of small intestine, appendix) have been considered also. The secondary lymphatic organs are the lymph nodes, spleen, tonsils, and lymphatic tissues associated with the gastrointestinal tract. Lymphatic organs consist of accumulations of diffuse and nodular lymphatic tissues that are organized into discrete structures. The lymphatic tissues that compose the organs are isolated from the surrounding tissues by a well-defined capsule.

Tonsils

Tonsils are aggregates of lymphatic nodules associated with the pharynx and oropharynx. A ring of lymphatic tissue, in varying degrees of completeness, is present at the entrance to the esophagus and respiratory tract. The structures are spread through different areas — oropharynx, nasopharynx, and tongue — and form the palatine, pharyngeal, and lingual tonsils, respectively. The general structure of a tonsil is shown in Figure 12-1.

Tonsils do not filter lymph; they have no afferent vessels leading to them, nor do they have an internal system of sinuses for the filtering of lymph. However, at the inner surfaces of the tonsils, plexuses of blindly ending lymph capillaries form the beginnings of efferent lymphatic

Fig. 12-1. General structure of a tonsil.

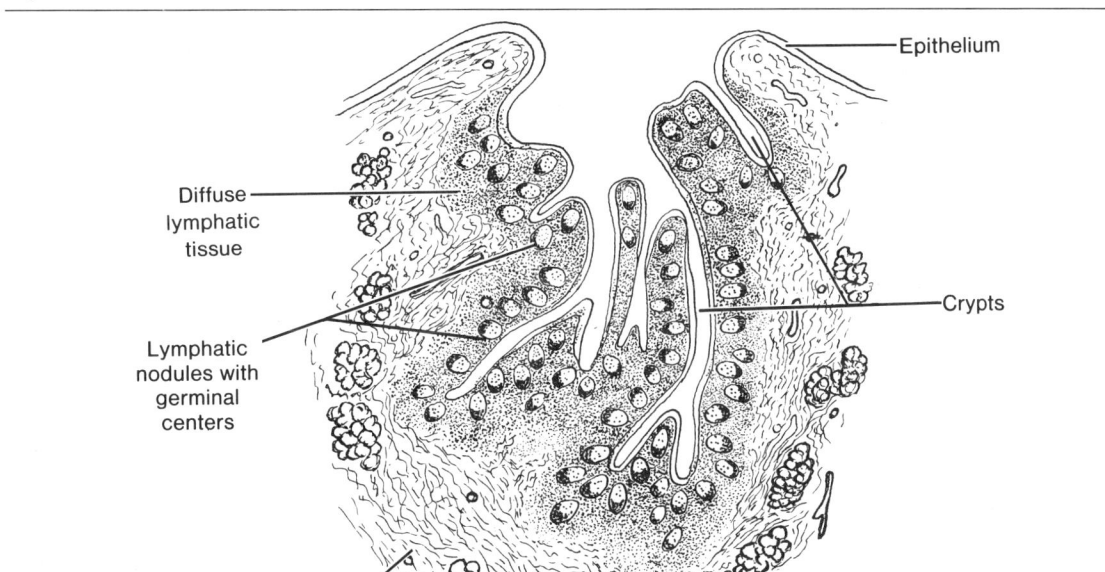

vessels. Tonsils contribute to the formation of lymphocytes, many of which migrate through the covering epithelium and appear in the sputum as salivary corpuscles. Bacteria that penetrate the lymphoid tissue of the tonsils act as antigens to stimulate the production of antibodies.

Palatine Tonsils

KEY WORDS: stratified squamous epithelium, crypt, partial capsule

The palatine tonsils are paired, oval lymphatic organs located laterally at the junction of the oral cavity and oropharynx. A **stratified squamous epithelium** continuous with that of the oral cavity and oropharynx covers the free surface of the tonsil and is very closely associated with the lymphatic tissue. Deep invaginations of the epithelium form the tonsillar **crypts** that reach almost to the base of the tonsil. Secondary crypts may branch from these deep pockets and also are lined by stratified squamous epithelium that tends to thin out in the deeper parts of the crypts. Lymphatic nodules, many with prominent germinal centers, usually are arranged in a single layer beneath the epithelium, embedded in a mass of diffuse lymphatic tissue. Large numbers of lymphocytes infiltrate the epithelium, especially that lining the lower ends of the crypts, often to such an extent that the demarcation between epithelium and lymphatic tissue is obscure.

A **partial capsule** beneath the basal surface of the tonsil separates it from surrounding structures to which the capsule adheres. Septa of loosely arranged collagen fibers extend from the capsule into the tonsillar tissue and partially divide the crypts and their associated lymphatic tissue from one another. Fibers from the septa spread out and become continuous with the reticular fibers of the lymphatic tissue. The connective tissue is infiltrated by lymphocytes of various sizes, plasma cells, and mast cells. Neutrophil granulocytes may be present and are numerous during inflammation of the tonsils. Small mucous glands lie outside the capsule, and their ducts drain to the free surface of the tonsil or, rarely, empty into a tonsillar crypt.

Lingual Tonsils

KEY WORDS: root of the tongue, crypt, stratified squamous epithelium, mucous gland

The lingual tonsils form nodular bulges in the **root of the tongue,** and their general structure is similar to that of the palatine tonsil. **Crypts** are deep, may be branched, and are lined by **stratified squamous epithelium** that invaginates from the surface. The associated lymphatic tissue consists of diffuse and nodular types. **Mucous glands** embedded in the underlying muscle of the tongue drain by ducts, most of which open into the bases of the crypts.

Pharyngeal Tonsil

KEY WORDS: nasopharynx, pseudostratified ciliated columnar epithelium

The pharyngeal tonsil is located on the posterior wall of the **nasopharynx.** Its surface epithelium is a continuation of that lining the respiratory passages — namely, **pseudostratified ciliated columnar epithelium** that contains goblet cells. Patches of squamous epithelium may be present, however, and tend to become more common with aging and smoking. The crypts are not as deep as in the palatine tonsils, and the epithelium forms numerous shallow folds. A thin capsule separates the pharyngeal tonsil from underlying tissues and provides fine septa that extend into the substance of the tonsil. Small mucoserous glands lie beneath the capsule and empty onto the surface of the folds.

Lymph Nodes

Lymph nodes are small encapsulated lymphatic organs set in the course of lymphatic vessels. They are prominent in the neck, axilla, groin, and mesenteries and along the course of large blood vessels in the thorax and abdomen.

Structure

KEY WORDS: hilus, afferent lymphatic, efferent lymphatic, diffuse lymphatic tissue, nodular lymphatic tissue, capsule, trabeculae, reticular network, cortex, outer cortex, deep (inner) cortex, thymus-dependent area, follicular dendritic cell, medulla, medullary cord

Grossly, lymph nodes appear as flattened, ovoid or bean-shaped structures with a slight indentation at one side, the **hilus,** through which blood and lymphatic vessels enter or leave. Lymph nodes are the only lymphatic structures that are set into the lymphatic circulation and thus are the only lymphatic organs to have afferent and efferent lymphatics. **Afferent lymphatics** enter the node at multiple sites, anywhere over the convex surface; **efferent lymphatics** leave the node at the hilus. Both sets of vessels have valves that allow unidirectional flow of lymph through a node. The valves of afferent vessels open toward the node; those of efferent vessels open away from the node.

Essentially, lymph nodes consist of accumulations of

diffuse and **nodular lymphatic tissue** enclosed in a capsule that is greatly thickened at the hilus. The **capsule** consists of closely packed collagen fibers, scattered elastic fibers, and a few smooth muscle cells that are concentrated about the entrance and exit of lymphatic vessels. From the inner surface of the capsule, branching **trabeculae** of dense irregular connective tissue extend into the node and provide a kind of skeleton for the lymph node (Fig. 12-2). The spaces enclosed by the capsule and trabeculae are filled by an intricate, three-dimensional **reticular network** of reticular fibers and their associated reticular cells. The meshes of the network are crowded with lymphatic cells, so disposed as to form an outer cortex and an inner medulla.

The **cortex** forms a layer beneath the capsule and extends for a variable distance toward the center of the node. It consists of lymphatic nodules, many with germinal centers, set in a bed of diffuse lymphatic tissue. Trabeculae are arranged fairly regularly and run perpendicular to the capsule, subdividing the cortex into several irregular "compartments." However, trabeculae do not form complete walls but rather represent bars or rods of connective tissue, and compartmentalization of the cortex is incomplete. All the cortical areas communicate with each other laterally and above and below the plane of section where trabeculae are deficient.

The cortex usually is divided into an **outer cortex** that lies immediately beneath the capsule and contains nodular and diffuse lymphatic tissue and a **deep (inner) cortex** that consists of diffuse lymphatic tissue only. There is no sharp boundary between the two zones, and their proportions differ from node to node and with the functional status of the node. In general, B-lymphocytes are concentrated in the nodular lymphatic tissue, while T-cells are present in the diffuse lymphatic tissue. The deep cortex becomes depleted of cells after thymectomy and has been called the **thymus-dependent area.** This area also contains **follicular dendritic cells,** a cell type with a pale-staining nucleus, few cytoplasmic organelles, and numerous, long processes that interdigitate with similar projections from nearby lymphocytes. They represent a form of antigen-presenting cell. The deep cortex continues into the medulla without interruption or clear demarcation.

The **medulla** appears as a paler area of variable width, generally surrounding the hilus of the node. It consists of diffuse lymphatic tissue arranged as irregular **medullary cords** that branch and anastomose freely. Medullary cords contain abundant plasma cells, macrophages, and lymphocytes. The trabeculae of the medulla are more irregularly arranged than those of the cortex.

Fig. 12-2. General structure of a lymph node.

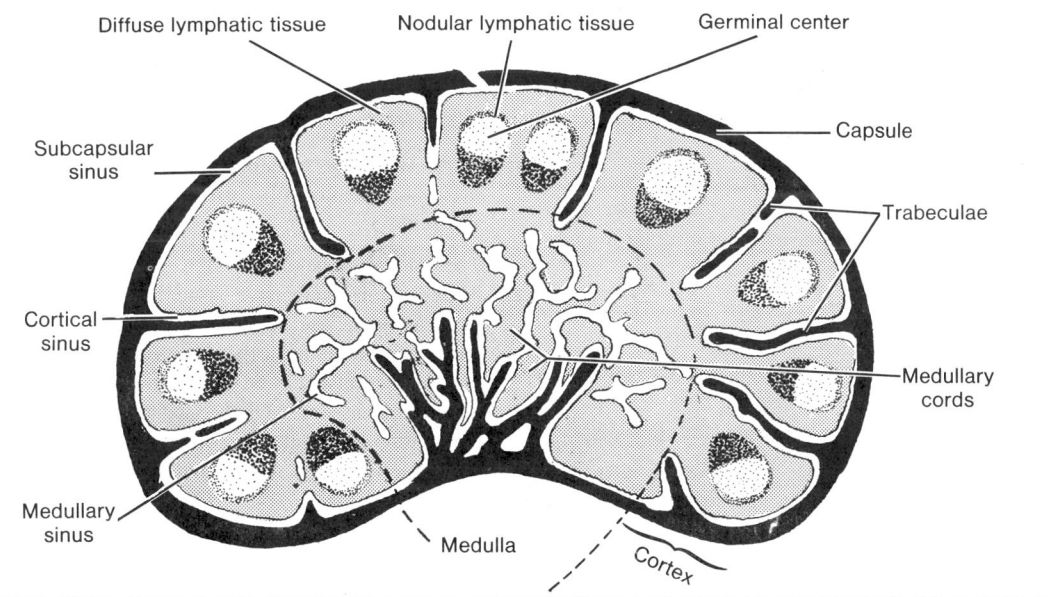

Lymph Sinuses

KEY WORDS: subcapsular (marginal) sinus, cortical (trabecular, intermediate) sinus, medullary sinus

Within the lymph node is a system of channel-like spaces, the lymph sinuses, through which lymph percolates. Lymph enters the node through afferent lymphatic vessels that pierce the capsule anywhere along the concave border and empty into the **subcapsular (marginal) sinus,** which separates the cortex from the capsule (see Fig. 12-2). The sinus does not form a tubular structure but is present as a wide space extending beneath the capsule, interrupted at intervals by the trabeculae. The subcapsular sinus is continuous with similar structures, the **cortical (trabecular, intermediate) sinuses,** which extend radially into the cortex, usually along the trabeculae. These in turn become continuous with **medullary sinuses** that run between the medullary cords and trabeculae of the medulla (see Fig. 12-2). At the hilus, the medullary and subcapsular sinuses unite, penetrate the capsule, and become continuous with the efferent lymphatics.

In sections, sinuses appear as regions in which the reticular net is coarser and more open. Stellate cells supported by reticular fibers crisscross the lumen and are joined to each other and to the cells bordering the lumen by slender processes. Numerous macrophages are present in the luminal network and also project from the boundaries of the sinus. The cells that form the margins of the sinuses and extend through the sinus space generally are regarded as flattened reticular cells. However, they also have been considered to be attenuated endothelial cells akin to those which line the lymphatic vessels with which they are continuous.

Sinuses in the cortex are less numerous than in the medulla and are relatively narrow. Those in the medulla are large and irregular and show repeated branchings and anastomoses. They run a tortuous course in the medullary parenchyma and account for the irregular cordlike arrangement of the lymphatic tissue in this area.

Blood Vessels

KEY WORDS: postcapillary venule, high endothelium

The major blood supply reaches the node through the hilus, and only occasionally do smaller vessels enter from the convex surface. The arteries (arterioles) at first run within the trabeculae in the medulla but soon leave these and enter the medullary cords to pass to the cortex. Here they break up into a rich capillary plexus that distributes to the diffuse lymphatic tissue and also forms a network around the nodules.

The capillaries regroup to form venules that run from the cortex and enter the medullary cords as small veins. These in turn are tributaries of larger veins that pass out of the node at the hilus. In the deep cortex, the venules take on a special appearance and have been called **postcapillary venules.** These vessels are characterized by a **high endothelium** that varies from cuboidal to columnar and at times appears to occlude the lumen. The walls of these vessels often are infiltrated with small lymphocytes that generally are presumed to be passing into the lymph node. The cells pass between adjacent endothelial cells, indenting their lateral walls as they cross through the endothelium. The significance of the tall endothelium is not known, but it has been suggested that as the lymphocytes sink deeper between them, the adjacent endothelial cells resume their original relationship to each other above the lymphocytes and seal off the interendothelial cleft, thereby limiting the loss of plasma from the venule.

Spleen

The spleen embodies the basic structure of a lymph node and can be regarded as a modified, enlarged lymph node inserted into the blood flow. Unlike lymph nodes, the spleen has no afferent lymphatics and no lymphatic sinus system, and the lymphatic tissue of the spleen is not arranged into cortex and medulla. It does have a distinctive pattern of blood circulation and specialized vascular channels that facilitate the filtering of blood.

Structure

KEY WORDS: capsule, hilus, trabeculae, reticular network, splenic pulp, white pulp, red pulp, splenic sinusoid

The spleen is enclosed in a well-developed **capsule** of dense irregular connective tissue (Fig. 12-3). Elastic fibers are present between bundles of collagen fibers and are most abundant in the deeper layers of the capsule. Smooth muscle fibers also may be present in small groups or cords, but the amount varies. On the medial surface of the spleen, the capsule is indented to form a cleftlike **hilus** through which blood vessels, nerves, and lymphatics enter or leave the spleen.

Broad bands of connective tissue, the **trabeculae,** extend from the inner surface of the capsule and pass deeply into the substance of the spleen to form a rich branching

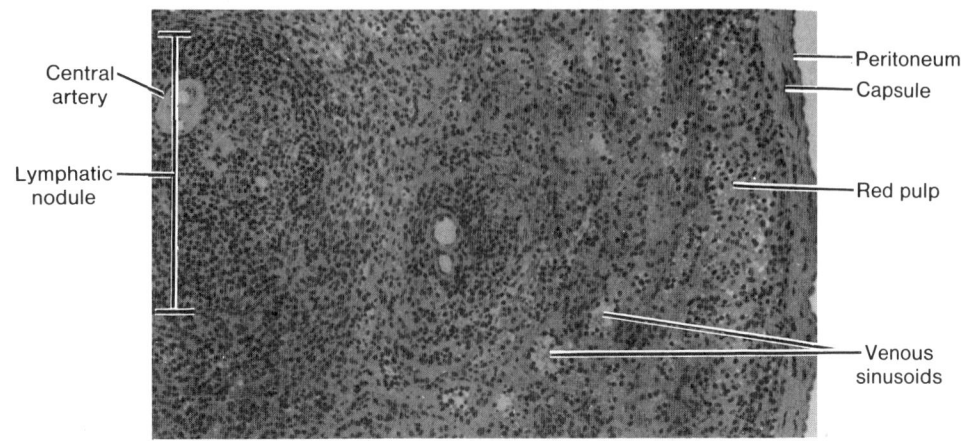

Fig. 12-3. A segment of spleen near the capsule illustrates a region of red pulp as well as a lymphatic nodule (white pulp) (LM, × 100).

and anastomosing framework. As in lymph nodes, the trabeculae subdivide the organ into communicating compartments. The spaces between trabeculae are filled by a **reticular network** of fibers and associated reticular cells. The meshes vary in size and tend to be smaller around blood vessels and aggregates of lymphatic tissue.

The substance of the spleen is called the **splenic pulp,** and sections from a fresh spleen show a clear separation of the tissue into rounded or elongated grayish areas set in a greater mass of dark red tissue. Collectively, the scattered gray areas form the **white pulp** and consist of diffuse and nodular lymphatic tissue. The dark red tissue is the **red pulp** and consists of diffuse lymphatic tissue that is suffused with blood. The large number of red cells present imparts the color to the red pulp. The red pulp contains large, branching, thin-walled blood vessels called **splenic sinusoids** (sinuses), and such spleens are said to be sinusal.

White Pulp

KEY WORDS: **periarterial lymphatic sheath, lymphatic nodule, splenic follicle, marginal zone, interdigitating cell**

The white pulp generally is associated with the arterial supply of the spleen and forms the **periarterial lymphatic sheaths** that extend about the arteries, beginning as soon as the vessel leaves the trabeculae to enter the splenic pulp (see Fig. 12-8). The sheaths have the structure of diffuse lymphatic tissue and contain the usual cellular elements of lymphatic tissue. Small lymphocytes, mostly T-cells, make up the bulk of the cells. Here and there along the course of the sheath, the lymphatic tissue expands to incorporate **lymphatic nodules** that resemble the cortical nodules of lymph nodes (Figs. 12-3 and 12-8). They represent sites of B-lymphocytes, and many contain germinal centers. The lymphatic nodules of the spleen are called **splenic follicles.** At the periphery of the lymphatic sheath, the reticular net is more closely meshed than elsewhere, and the reticular fibers and cells form concentric layers that tend to delimit the lymphatic tissue from the red pulp. This forms the **marginal zone.** Cells similar to the follicular dendritic cells of lymph nodes are present in the white pulp of the spleen. These dendritic-like cells are called **interdigitating cells** and also represent antigen-presenting cells.

Red Pulp

KEY WORDS: **splenic sinusoid, splenic cord**

The red pulp is a diffuse lymphatic tissue associated with the venous system of the spleen. The reticular meshwork is continuous throughout the red pulp and is filled with large numbers of free cells, including all those usually found in the blood. Thus, in addition to the cells of lymphatic tissue, the red pulp is suffused with red cells, granular leukocytes, and platelets (see Fig. 12-3). Occasionally, macrophages can be seen that contain ingested red cells or granulocytes or that are laden with a yellowish brown pigment, hemosiderin, derived from the breakdown of hemoglobin.

The red pulp is riddled with large, irregular blood vessels, the **splenic sinusoids,** between which the red pulp assumes a branching, cordlike arrangement to form the **splenic cords** (Fig. 12-4). The sinusoids have wide lumina (20 to 40 μm diameter), and their walls have a unique structure (Fig. 12-5). Almost the entire wall is made up of elongated, fusiform endothelial cells that lie parallel to the long axis of the vessel (Figs. 12-6 and 12-7). The cells lie side by side around the lumen but are not in contact and are separated by slitlike spaces. Outside the endothelium, the wall is supported by a basement membrane that is not continuous but forms widely spaced, thick bars that encircle the sinusoid. The bars of basement membrane are joined to each other by thinner strands of the same material and are continuous with the reticular network of the splenic cords.

Fig. 12-4. A region of splenic red pulp clearly demonstrating a longitudinal profile of a sinusoid and an adjacent splenic (Billroth's) cord (LM, × 250).

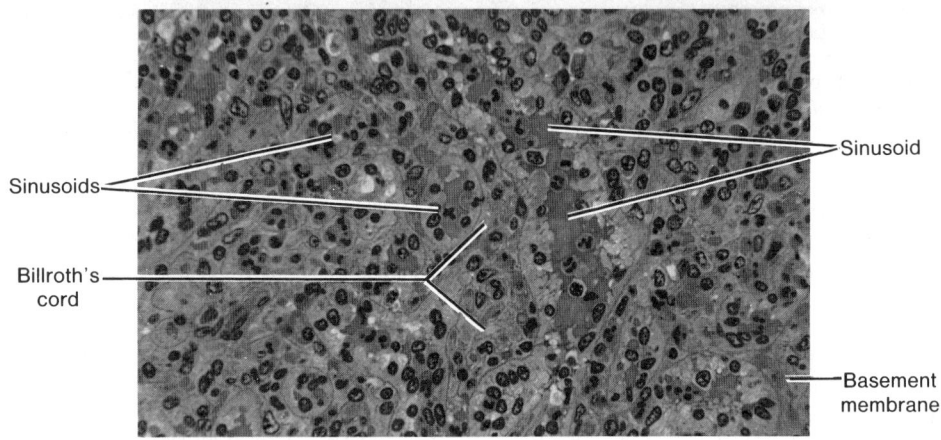

Fig. 12-5. Structure of a splenic sinusoid.

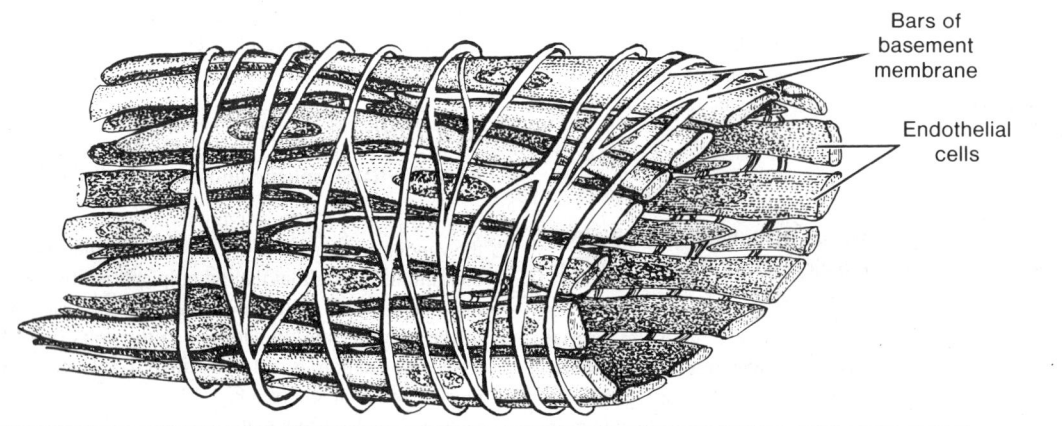

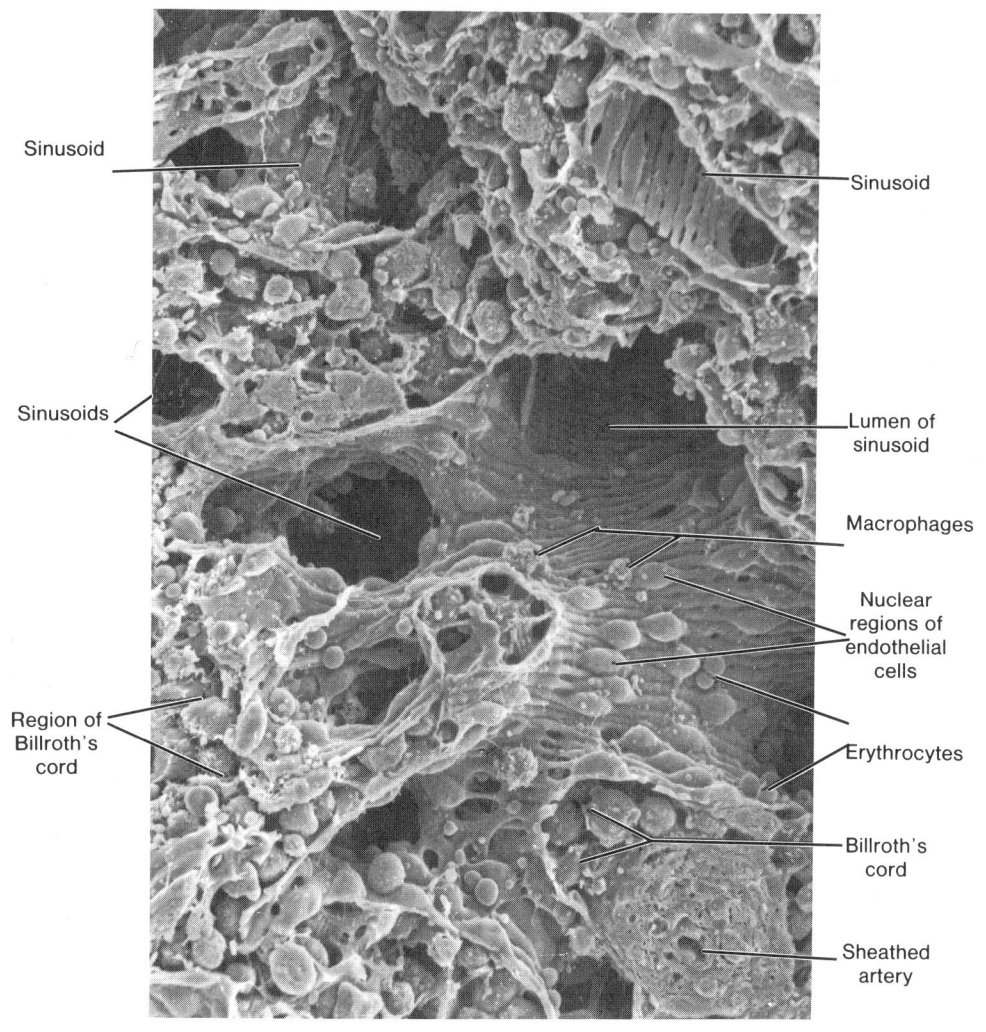

Sinusoid

Sinusoids

Region of
Billroth's
cord

Sinusoid

Lumen of
sinusoid

Macrophages

Nuclear
regions of
endothelial
cells

Erythrocytes

Billroth's
cord

Sheathed
artery

Fig. 12-6. Several profiles of sinusoids within the red pulp of a spleen (SEM, × 700).

Blood Supply

KEY WORDS: trabecular artery, central artery, penicillar artery, sheathed capillary, open circulation, closed circulation, pulp vein, trabecular vein

The architecture of the spleen is best understood in relation to the blood circulation, which shows some special features (Fig. 12-8). Branches of the splenic artery enter the spleen at the hilus, divide, and pass within trabeculae into the interior of the organ. These branches have an overlapping, segmental arrangement, with each main branch serving a defined area of the spleen. The **trabecular arteries,** as they now are called, branch repeatedly, finally leaving the trabeculae as **central arteries** that immediately become surrounded by the lymphatic tissue of the periarterial lymphatic sheath. Where the sheath expands to form nodules, the central artery (more appropriately, the follicular arteriole) is displaced to one side and assumes an

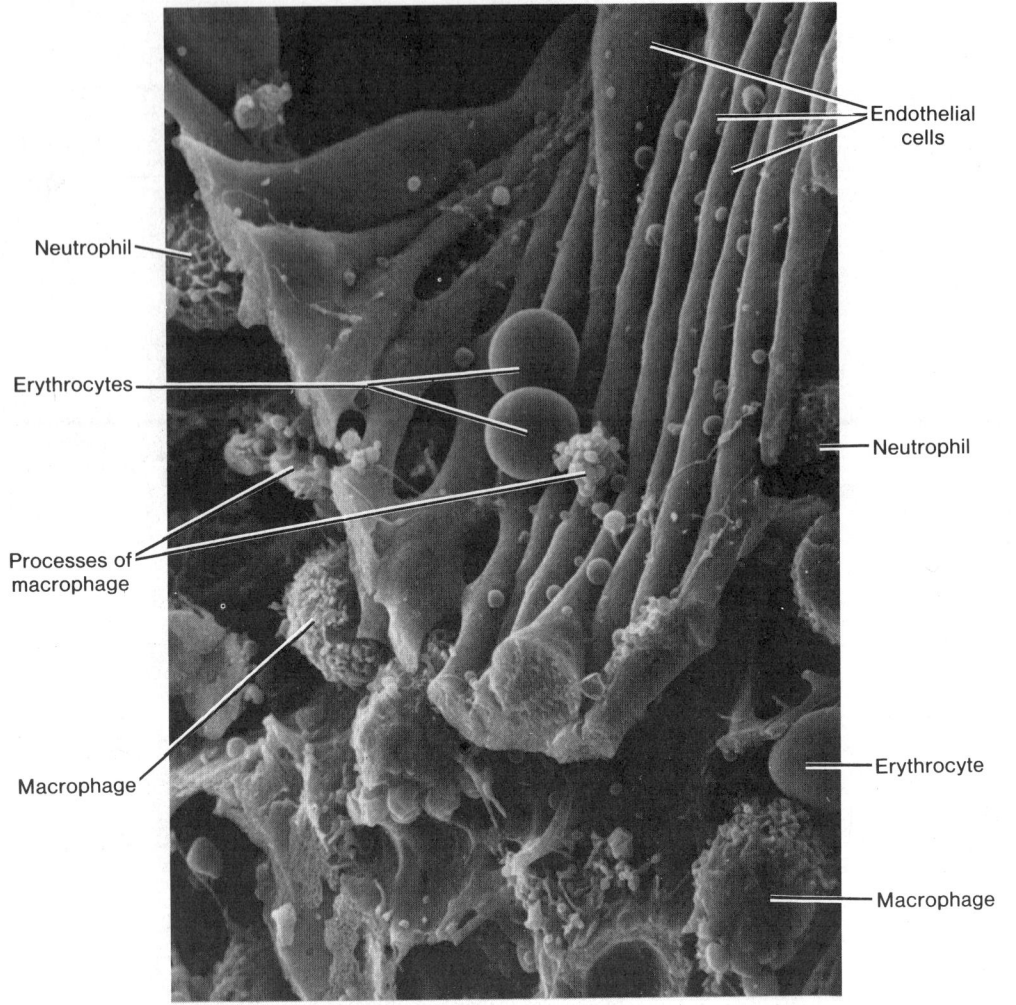

Neutrophil

Erythrocytes

Processes of
macrophage

Macrophage

Endothelial
cells

Neutrophil

Erythrocyte

Macrophage

Fig. 12-7. Endothelial cells comprising a sinusoidal wall in the red pulp of a spleen (SEM, × 3,000).

eccentric position in the nodule. Only rarely does the vessel retain a central position in the nodular tissue. Throughout its course in the white pulp, the central artery provides numerous capillaries that supply the sheath and then pass into the marginal zone surrounding the white pulp. How these capillaries end is not certain. Many arterial branches appear to open directly into the marginal zone, often ending in a funnel-shaped terminal part. There is no direct venous return, and the white pulp is associated only with the arterial supply. Although the term *artery* commonly is used for the different levels of the splenic vasculature, once the vessel has reached lymphatic tissue, it is actually an arteriole.

The central artery continues to branch, and its attenuated stem passes into a splenic cord in the red pulp, where it divides into several short, straight **penicillar arteries,** some of which show a thickening of their wall and now form **sheathed capillaries.** The sheath consists of compact masses of concentrically arranged cells and fibers that become continuous with the reticular network of the red

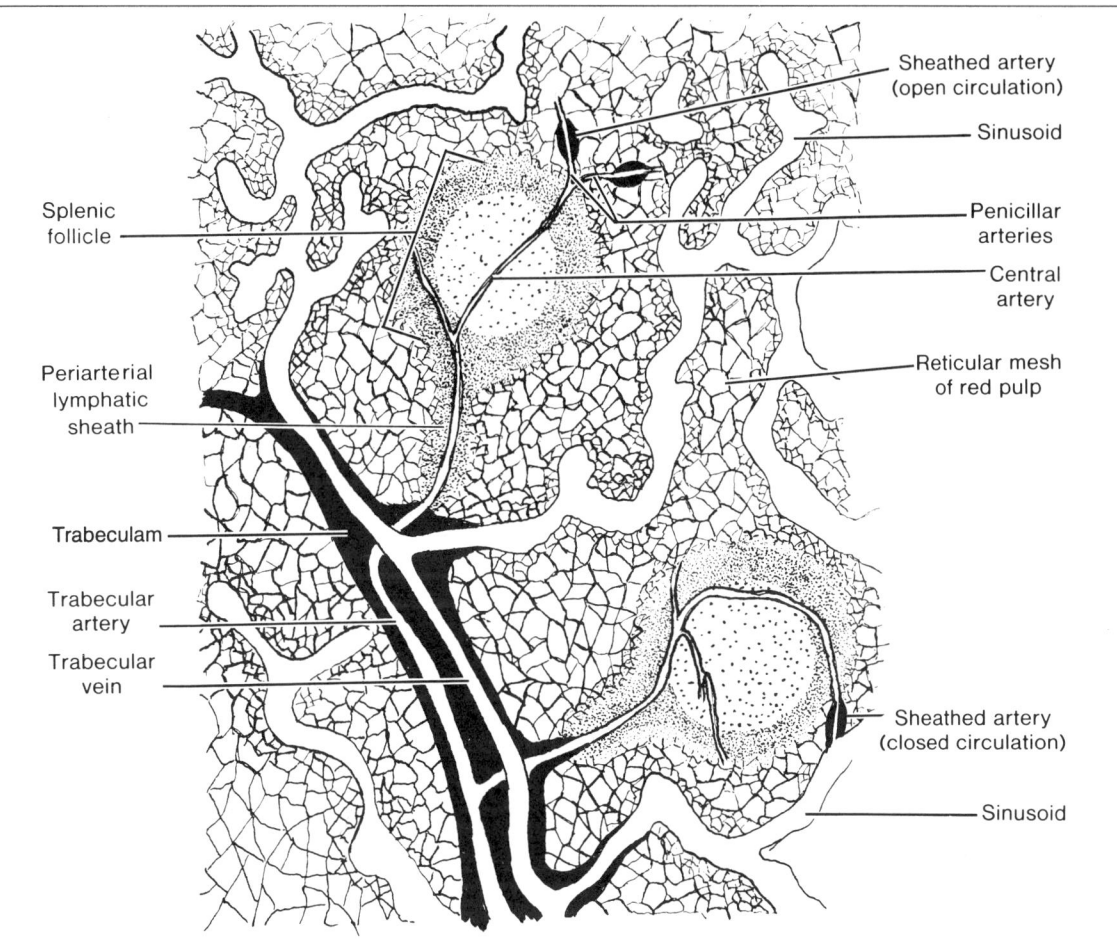

Fig. 12-8. Blood supply of the spleen.

pulp. Close to the capillary the cells of the sheath are rounded, while at the periphery of the sheath the cells become stellate. The cells of the sheath are avidly phagocytic.

Not all the capillaries are sheathed, and occasionally, a single sheath may enclose more than one capillary. Sheathed capillaries may continue as simple capillaries or may divide to produce two to four nonsheathed, terminal capillaries. How these end still is a matter of some debate and has led to the "open" and "closed" theories of circulation. According to the **open circulation** theory, the capillaries empty into the meshwork of the red pulp in the splenic cords. The blood then slowly seeps through the red pulp and finds its way into the venous system through the

walls of the splenic sinusoids. The **closed circulation** theory holds that the capillaries open directly into the lumina of the venous sinusoids; blood leaves the sinusoid at the capillary end and reenters at the venous end because of a decreasing pressure gradient between the two ends of the sinusoid. A compromise between the two views suggests that a closed circulation in a contracted spleen becomes open when the spleen is distended.

Splenic sinusoids drain into **pulp veins,** which are supported by a thin muscle coat and, more externally, are surrounded by reticular and elastic fibers. A sphincter-like activity of the smooth muscle has been described at the junction of the pulp veins and sinusoids. Pulp veins enter trabeculae, where, as **trabecular veins,** they pass in company with the artery. Trabecular veins unite and leave at the hilus as the splenic vein.

The spleen has no afferent lymphatics, but efferents arise deep in the white pulp and converge on the central arteries. The lymphatics enter trabeculae and form large vessels that leave at the hilus. Areas of the spleen supplied by sinusoids (i.e., the red pulp) lack lymphatics.

Thymus

The thymus is a bilobed, encapsulated lymphatic organ situated in the superior mediastinum dorsal to the sternum and anterior to the great vessels that emerge from the heart. The thymus is the only primary lymphatic organ and is the first organ of the embryo to become lymphoid. Unlike the spleen and lymph nodes, it is well developed and relatively large at birth. The greatest weight (30 to 40 g) is achieved at puberty, after which the organ undergoes progressive involution and is partially replaced by fat and connective tissue.

Structure

KEY WORDS: lobe, capsule, septum, lobule, cortex, medulla, reticular network, epithelial reticular cell, cytoreticulum

The thymus consists of two **lobes** closely applied and joined by connective tissue. Each lobe arises from a separate primordium, and there is no continuity of thymic tissue from one lobe to the other. A thin **capsule** of loosely woven connective tissue surrounds each lobe and provides **septa** that extend into the thymus, subdividing each lobe into a number of irregular **lobules.** Each lobule consists of a **cortex** and **medulla** (Fig. 12-9). In the usual sections, the medulla may appear as isolated, pale areas completely surrounded by denser cortical tissue. However, serial sections show that the lobules are not isolated by the septa; the medullary areas are continuous with one another throughout each lobule.

The free cells of the thymus are contained within the meshes of a **reticular network,** which, however, differs from that of other lymphatic tissues. In the thymus the reticular cells take origin in endoderm rather than mesenchyme and are not associated with reticular fibers. These **epithelial reticular cells** are stellate in shape with greatly branched cytoplasm that contains few organelles. The large, round or oval euchromatic nuclei show one or more small but prominent nucleoli. The cytoplasmic processes are in contact with the processes of other reticular cells and at their points of contact are united by desmosomes. Thus the stroma of the thymus consists of a **cytoreticulum** composed of epithelial cells. Reticular fibers are found only in relation to blood vessels.

Cortex

KEY WORDS: lymphocyte, no lymphatic nodules, reticular cell, macrophage

The cortex consists of a thick, deeply stained layer extending beneath the capsule and along the septa. Most of the cells of the cortex are **lymphocytes** that are closely packed with little intervening material between them. Because of mutual compression, the cells appear polyhedral in shape.

Fig. 12-9. A section of human thymus showing the central medulla and surrounding cortex (LM, × 10).

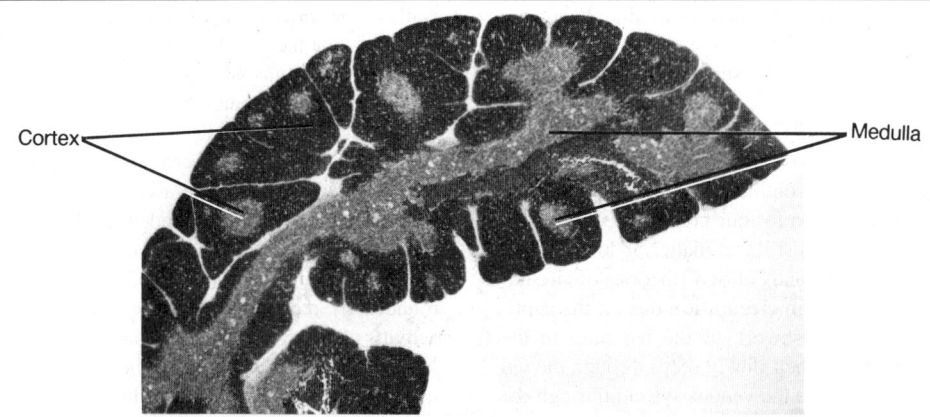

Large, medium, and small lymphocytes are present, the latter being the most abundant; they are indistinguishable from small lymphocytes found elsewhere. Large lymphocytes tend to concentrate in the outer cortex beneath the capsule and represent stem cells that have newly emigrated from the bone marrow. They form only a small part of the cell population. Small lymphocytes become increasingly more numerous toward the deeper cortex, where degenerating cells with pyknotic nuclei also are found. Unlike lymph nodes, there are **no lymphatic nodules** in the cortex of the thymus, nor is there an internal sinus system.

Reticular cells in the cortex are highly branched, but their processes are obscured by the mass of lymphocytes. They form a continuous layer at the periphery of the cortex, separating it from the capsule and septa. These epithelial reticular cells contain tonofilaments and membrane-bound structures that appear to be secretion granules. **Macrophages** are consistently present in small numbers, scattered throughout the cortex. They are difficult to distinguish from reticular cells by light microscopy unless phagocytosed material can be seen in their cytoplasm. In electron micrographs they are distinguished from epithelial reticular cells by the lack of desmosomes. Macrophages that have engulfed degenerating cells can be found scattered throughout the thymus and tend to increase in number toward the junction of the cortex and medulla.

Medulla

KEY WORDS: **pleomorphic reticular cell, dendritic interdigitating cell, small lymphocyte, thymic corpuscle**

The medulla occupies the central region of the thymus, where it forms a broad, pale band of tissue that is continuous throughout each lobule. Often, however, it appears to be isolated within a lobule, surrounded by a complete layer of cortex. Lymphocytes are less numerous than in the cortex, and the epithelial reticular cells are not as widely dispersed or as highly branched. The **reticular cells** tend to be **pleomorphic** and vary from stellate cells with long processes to rounded or flattened cells with many desmosomes and abundant tonofilaments. **Dendritic interdigitating cells** are found in the medulla and corticomedullary region. They act as antigen-presenting cells for more mature T-lymphocytes.

The free cells of the medulla are mainly **small lymphocytes,** but a small and variable number of macrophages are present. Plasma cells, mast cells, and eosinophil granulocytes can be found. Mitotic figures are rare. Collagen and reticular fibers extend for short distances from blood vessels and wind between the epithelial cells.

Rounded or ovoid epithelial structures, the **thymic corpuscles,** are a prominent feature of the medulla (Fig. 12-10). These bodies vary in size from 10 to 100 μm or more in diameter and consist of flattened epithelial reticular cells, wrapped about one another in concentric lamellations, that are joined by numerous desmosomes. Many contain granules of keratohyalin. The cells at the center of the structure undergo hyalinization or necrosis and may become lysed to leave a cystic structure. Some of the cells at the periphery of the corpuscle retain their connections with the surrounding cytoreticulum. The function of the thymic corpuscles is unknown; they have been regarded

Fig. 12-10. Thymic corpuscles from the medulla of a human thymus (LM, × 250).

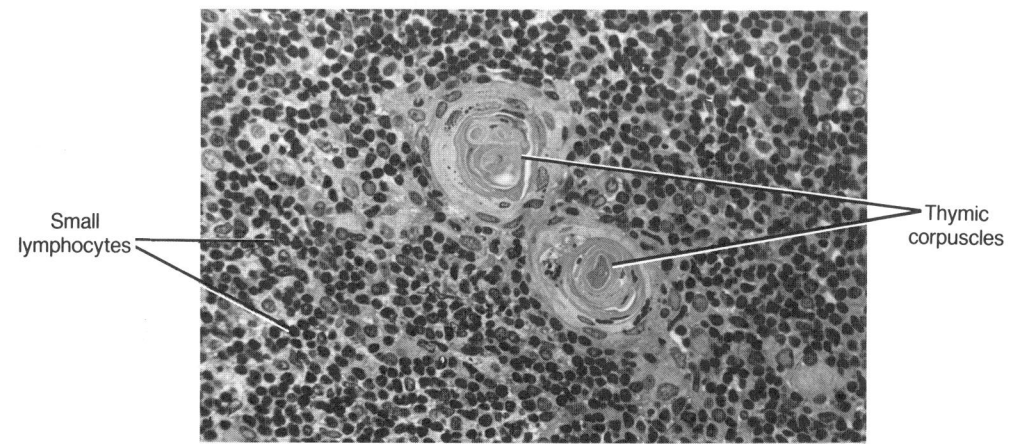

Small lymphocytes

Thymic corpuscles

purely as degenerated structures, but there is some evidence that they may be secretory bodies.

Blood Supply

KEY WORDS: corticomedullary junction, cortical capillary, blood-thymic barrier

The arteries to the thymus penetrate the organ within the connective tissue of the septa. Arteriolar branches from these vessels run along the **corticomedullary junction** and provide arterioles and capillaries to the medulla and capillaries to the cortex; vessels larger than capillaries are not found in the cortex. Within the cortex, the capillaries run toward the capsule, where they form branching arcades before passing back through the cortex to drain into venules and thence into veins that accompany the arterioles in the corticomedullary region and medulla. The veins leave via the septa and ultimately unite to form a single thymic vein.

The **cortical capillaries** are enveloped by a collar of connective tissue that forms part of the **blood-thymic barrier.** This envelope in turn is surrounded by a continuous layer of epithelial reticular cells. The perivascular connective tissue space varies in width and is traversed by reticular fibers that accompany the vessel. Within the perivascular space are granular leukocytes, plasma cells, macrophages, and lymphocytes. The blood-thymic barrier in the cortex thus consists of the capillary endothelium and its basal lamina, the perivascular connective tissue sheath, and the layer of epithelial reticular cells and their associated basal lamina. There is little movement of macromolecules across this barrier, and cortical lymphocytes develop in relative isolation from antigens in a privileged environment. Vessels of the medulla and corticomedullary junction are permeable to circulating macromolecules. As T-lymphocytes differentiate in the cortex of the thymus, they acquire surface marker molecules of major histocompatibility complex (MHC_I and MHC_{II}) and synthesize receptors for recognition of foreign antigens. Those lymphocytes which recognize self MHC are destroyed in the thymus, whereas those T-lymphocytes capable of reacting to nonself go on to form clones of cells that mature and are released as immunocompetent T-lymphocytes. Lymphocytes leave the thymus by entering postcapillary venules located in the medulla and near the corticomedullary junction.

Involution

Growth of the thymus during fetal life is very rapid, and the organ attains its greatest relative size by the time of birth. It continues to grow, at a lesser rate, until puberty, after which it undergoes progressive involution. The organ decreases in weight, loses cortical lymphocytes, and shows an increase in the size and number of thymic corpuscles. The septa also increase in width. The cortical areas become infiltrated by fat cells, and the replacement may become extensive. By the time involution commences, T-cells have disseminated into the secondary lymphatic tissue throughout the body. The thymic parenchyma does not disappear completely even in old age, and the thymus maintains some activity in the adult.

Other Constituents

The thymus frequently shows a number of peculiar structural elements, the significance of which is not known. Among these structures are cystlike spaces lined by cells with brush borders, cilia, or mucus-producing cells or by reticular cells that contain microvillus-lined vacuoles. Most peculiar are the "myoid" cells, which have an imperfect resemblance to striated muscle. These cells may resemble embryonal or adult muscle fibers complete with typical banding patterns: Z lines and A, I, and M lines have been described. It is not known whether these various inclusions have functional significance or represent aberrant differentiation of embryonal elements.

Dendritic (Antigen-Presenting) Cells

Dendritic (antigen-presenting) cells arise in the bone marrow but, unlike monocytes (macrophages), have low levels of lysosomal enzymes. In contrast, they have high levels of class II major histocompatibility complex (MHC) molecules essential for presenting new antigens to T-lymphocytes. Dendritic cells have the capacity to ingest foreign proteins, process (partially digest) the antigens, and insert selected portions of an antigen into their cell membrane, thereby retaining the product for long periods of time. Dendritic cells are found in lymph nodes, spleen, thymus, mesenteries, and skin (Langerhans' cells) and under mucosal surfaces. Despite the morphologic similarities of exhibiting fine ramifying cytoplasmic processes and containing relatively few lysosomes, each dendritic antigen-presenting cell appears to have developed specialized surface receptors for a given specific microenvironment at the site in which it is found. After a period of time, they leave the specific environment (skin, for example), make their way to lymphoid tissue, and gradually present the antigens collected in a specific environment to T-lymphocytes. Thus the dendritic antigen-presenting cells play a very important role in the immune system by monitoring a number of different environments.

Review Table 12-1. Key Features of Lymphatic Organs

	Tonsil	Lymph Node	Spleen	Thymus
Capsule	Partial, at base	Yes	Yes	Thin
Trabeculae	No	Yes	Yes	Forms thin septae
Lymphatic tissue arranged as cortex and medulla	No	Yes	No	Yes
Nodular lymphatic tissue	Yes	Yes	Yes	No
Special features	Crypts; closely associated with overlying epithelium Palatine and lingual: stratified squamous, nonkeratinized Pharyngeal: pseudostratified, ciliated columnar	Subcapsular, trabecular, and medullary sinuses; only lymphatic organ with afferent and efferent lymphatics	Lymphatic tissue arranged as red and white pulp; splenic follicles (nodules) associated with central arteries; sinusoids in red pulp	Thymic corpuscles in medulla; stroma is a cytoreticulum of endodermal origin

Functional Review: Lymphatic Organs

The assembly of lymphatic tissue — lymphatic organs spread throughout the connective tissue — constitutes the immune system that protects the body against invading foreign macromolecules and constantly monitors the circulating fluids (blood and lymph) for abnormal components. Antigenic material evokes a specific defense reaction, the immune response, that gives rise either to B-cells that produce specific antibodies or to a variety of T-cells that attack foreign cells directly, release nonspecific toxic agents, assist B-cells in the production of antibody, or direct macrophage response and maturation. Antibodies may act by binding to the antigen to neutralize it, inhibit the entrance of the antigen into a cell, enhance its phagocytosis, or initiate its lysis. Germinal centers are associated with production of humoral antibody, especially during a secondary response, and may represent a clone of B-cells geared to the production of a single antibody. Stimulation by antigen results in proliferation and differentiation of lymphocytes, some of which go on to react against the antigen, while others remain as committed memory cells. On a second exposure to the same antigen, memory cells react swiftly and with great efficiency.

Tonsils are the major bulwark against oral and pharyngeal routes of infection. They contribute to the formation of lymphocytes, are able to mount immunologic responses, and take part in the immunologic defenses of the body. Tonsils appear to be especially valuable sources of interferon, an antiviral factor, and may be of greater importance in fighting infection than has been recognized previously.

Lymph nodes contribute to the production of lymphocytes, as indicated by the mitotic activity, especially in germinal centers. The lymphocytopoietic activity of quiescent nodes does not seem to be great, and the bulk of the cells in an unstimulated node are recirculating lymphocytes. B- and T-cells enter the node through the postcapillary venules, but some also enter through afferent lymphatics. Once in the parenchyma, B-cells home in on the lymphatic nodules of the outer cortex; T-cells make their way to the deep cortex and the region of the thymus-dependent area. The cells remain in the lymph node for variable periods of time and then leave to recirculate in the blood and other lymphatic organs. Only small lymphocytes recirculate.

Lymph nodes form an extensive filtration system that removes foreign particles from the lymph, preventing their spread through the body. A single lymph node can remove 99 percent of the particulate matter presented to it, and lymph usually passes through several nodes before entering the bloodstream. Sinuses form a settling chamber through which lymph flows slowly, while baffles of crisscrossing reticular fibers form a mechanical filter and produce an eddying of the lymph. Most of the lymph that enters a node flows centrally through the sinuses, where it comes into contact with numerous phagocytes. Filtration by a lymph node may be impaired if the number of particles is excessive or if the organism is exceptionally virulent. Agents not destroyed in a lymph node may disseminate throughout the body, and the node itself may become a focus of chronic infection.

Lymph nodes are immunologic organs and participate in humoral and cellular immune responses. Antigen accumulates about and within primary nodules and at the junction of the deep and outer cortex. Some of the antigen appears to be held on the surfaces of macrophages and follicular dendritic cells and thus is available to react with T- and B-cells as these move into the node and sort out within the parenchyma.

Antigens that evoke antibodies react with uncommitted B-cells to elicit their activation and proliferation. The first antibody-producing cells appear in the cortex and then migrate into medullary cords, where they accumulate as plasma cells. After this cortical response, intense antibody formation occurs in newly formed germinal centers. This is the primary response, elicited on first exposure to an antigen, and takes several days to develop. Relatively few cells respond, and a low titer of antibody results. On a second exposure to the same antigen, a rapid and greatly enhanced production of antibody occurs. This is the secondary response, marked by an explosive development of germinal centers that expand rapidly because of the presence of conditioned memory cells produced during the primary response.

Cell-mediated responses (rejection of foreign grafts, reaction to molds, fungi, and some bacteria) involve the T-cells. The deep cortex of the draining nodes become very thick, and lymphoblasts in this area divide rapidly. After a short lag, small lymphocytes leave the nodes, infiltrate the affected region, and destroy the antigen. Committed T-cells also are disseminated throughout the body as memory cells and form part of the recirculating pool of small lymphocytes. Since an associated humoral response also develops, newly formed germinal centers appear in the cortex.

Normal humans have no significant splenic reservoir of erythrocytes (only 30 to 40 ml), as in some species, but there are indications that even this small volume can be mobilized. About one-third of the platelets in the body are sequestered in the spleen, where they form a reserve available on demand.

The spleen serves as a filter for blood, removing particulate matter that is taken up by macrophages in the marginal zone, splenic cords, and sheathed capillaries. The sinusoidal endothelium and the reticular cells of the reticular net have no special phagocytic powers and contribute little to the clearing of foreign materials from the blood.

The spleen also forms the graveyard for worn out red cells and platelets and possibly for granular leukocytes as well. As the blood filters through the splenic cords, it comes under constant scrutiny and monitoring by macrophages. Viable cells are allowed to pass through the spleen, but damaged or aged cells are retained and phagocytosed. Several factors are probably involved in the recognition of old cells. As the cell ages, changes in the surface may permit antigenic reaction with opsonizing antibodies that enhance phagocytosis. The sinusoidal wall is a barrier to the reentrance of cells into the circulation, since the cells must insinuate themselves through the narrow slits between sinusoidal endothelial cells. Normal cells are pliant and able to squeeze through the interendothelial clefts, but cells such as spherocytes or sickled red cells are unable to pass through the sinusoidal barrier. As red cells age, their membranes become more permeable to water, and the relatively slow passage through the red pulp may allow the cells to imbibe fluid, swell, and become too rigid to pass into the sinusoids. Red cells that have been excluded are phagocytosed, and the cells appear to be taken up intact without prior lysis or break down. Components of the red cells that can be reused in production of new blood cells are recovered by the spleen; it is especially efficient in conserving iron freed from hemoglobin and returning it to accessible stores.

A "pitting function" has been described for the spleen. Red cells that contain rigid inclusions (malarial parasites or the iron-containing granules of siderocytes, for example) but that are otherwise normal are not destroyed, but the inclusions are removed at the wall of the sinusoid. The flexible part of the erythrocyte passes through the sinusoidal wall, but the rigid inclusion is held back by the narrow intercellular clefts and is stripped from the cell. The rigid portion remains in the splenic pulp and is phagocytosed; the rest of the cell enters the lumen of the sinusoid.

The spleen has great importance in the immune system, mounting a large scale production of antibody against blood-borne antigens. However, antigen introduced by other routes also evokes a response in the spleen, since an antigen soon finds its way into the bloodstream. The reactions in the spleen are the same as those occurring in lymph nodes and include primary and secondary responses. In the primary response, clusters of antibody-forming cells first appear in the periarterial lymphatic sheaths, then increase in number and concentrate at the edge of the sheath. Immature and mature plasma cells appear, and germinal centers develop in the splenic nodules. Ultimately, plasma cells become numerous in the marginal zone between white and red pulp and in the red pulp cords, either by direct emigration from the white pulp or indirectly via the circulation. During a secondary response, germinal center activity dominates, occurs almost immediately, and is of large scale.

The spleen has important hemopoietic functions in all vertebrates and in lower forms is the primary blood-forming organ, producing all types of blood cells. The spleen produces red cells, platelets, and granulocytes only during embryonic life, but production of lymphoid cells continues throughout life. In some conditions (leukemia and some anemias) the red pulp contains islets of hemopoietic tissue, even in the adult. This probably results from sequestration of circulating stem cells from the blood rather than from activation of an indigenous population of potential stem cells in the spleen.

Removal of the spleen (splenectomy) emphasizes its primary functions. The peripheral blood shows an increase in the number of platelets and abnormal erythrocytes. Older erythrocytes often contain Howell-Jolly bodies that normally would have been removed by the splenic cord-sinus system. Following splenectomy, individuals are at risk of developing bacterial septicemia. Under normal circumstances, blood-borne vectors would stimulate the immune component of the spleen and prevent infection via the blood.

The thymus is essential for production of T-cells, lymphocytes that are involved in cell-mediated immune responses such as rejection of foreign grafts and immunologic responses to fungi, viruses, and certain bacteria. Although not directly involved in elaborating conventional antibodies, T-cells do cooperate with B-cells in producing antibodies against antigens such as foreign red cells. All these responses are impaired in animals that have been thymectomized at birth. There is a marked decrease in the number of circulating small lymphocytes, and the deep cortex of lymph nodes and the periarterial lymphatic sheaths of the spleen fail to develop.

Most of the lymphocytes in the thymus seem to be inert and acquire immunologic capabilities after passage through the spleen. T-cells have a long life span and recirculate between lymphatic organs (except for the thymus), lymph, and blood. Thymic lymphocytes show a high rate of mitotic activity, but many die within the thymus or emigrate to other organs.

Thymic tissue contained in diffusion chambers and transplanted into thymectomized newborn animals partly prevents or reverses the effects of thymectomy. Permeable molecules but not whole cells diffuse from the chamber, indicating the presence of a thymic-produced humoral factor. Several peptides have been isolated from thymic extracts and appear to have some regulatory and stimulating effects on the thymus. The best characterized of these is thymosin, an agent that restores T-cell deficiencies in thymectomized mice; it is regarded as a hormone that induces T-cell differentiation. Other agents have been described: thymopoietin, which induces T-cell maturation; thymic humoral factor, which enhances T-cell mediated graft rejection; and serum thymic factor, which induces development of surface markers on T-cells. The agents appear to be produced locally in the thymus, possibly by the reticular epithelial cells, and to have local effects on the lymphocyte population in the thymus.

In addition to encapsulated lymphatic tissue, an equally large amount of nonencapsulated lymphoid tissue occurs in the walls of the respiratory, urogenital, and gastrointestinal tracts that provides immunologic protection along these mucosal surfaces. Here it is generally referred to as mucosa-associated lymphoid tissue (MALT); if specifically associated with the gastrointestinal tract it is termed gut-associated lymphoid tissue (GALT).

Review Questions and Answers

Questions

65. Given a section of a lymphatic organ, how would you identify it as a lymph node, a tonsil, or the thymus?
66. What constitutes the thymus-blood barrier?
67. What is the orientation of germinal centers in tonsils, spleen, lymph nodes, and thymus?
68. Describe a splenic sinusoid. In what part of the spleen are these structures found?
69. What is the distribution of T-lymphocytes in the lymphatic organs? What is their function?

Answers

65. The presence of a complete capsule and compartmentalization of the tissue by septa or trabeculae would eliminate the possibility of the tissue being a tonsil. Although both the thymus and lymph node show a cortex and medulla, the cortex of the thymus lacks lymphatic nodules, which usually are prominent in a lymph node and often contain germinal centers. Also lacking in the thymus is an internal system of lymphatic sinuses, which are prominent in a lymph node and consist of marginal (subcapsular), cortical (trabecular), and medullary sinuses. The medulla of the thymus appears rather homogeneous and contains thymic (Hassall's) bodies. That of the lymph node is arranged in irregular cords separated by the medullary sinuses. In the absence of a complete capsule and trabeculae, the tonsil could be identified by the close association of lymphatic tissue to an overlying epithelium that is deeply invaginated to form crypts. The

lymphatic tissue contains lymphatic nodules, many of which contain germinal centers and usually are arranged in a single row immediately beneath the epithelium.

66. The thymus-blood barrier exists in the cortex of the thymus, which is supplied only by capillaries. The barrier consists of the capillary endothelium and its basal lamina, the perivascular connective tissue sheath, and a complete investing layer of epithelial reticular cells and their associated basal laminae.

67. In the tonsil the germinal centers are oriented so that the light pole is directed toward the covering epithelium. In the lymph nodes, the light poles are directed toward the capsule and the subcapsular sinus. The light pole in the spleen is oriented toward the red pulp, and since the thymus lacks lymphatic nodules there are no germinal centers (see Chap. 6).

68. The splenic sinusoids are irregular, tortuous blood vessels whose walls consist only of fusiform, longitudinally oriented endothelial cells and a discontinuous basement membrane. The endothelial cells are not united by any type of junction and so have large gaps between them. The basal lamina forms widely spaced bars that encircle the endothelial cells. All other elements of a blood vessel wall are lacking. Sinusoids are found in the red pulp of the spleen.

69. T-lymphocytes are found in the deep or inner cortex of the lymph nodes (the thymic-dependent area), in the periarterial lymphatic sheaths of the spleen, and generally in the thymus. They function in the cell-mediated immune responses and cooperate with B-cells in the production of conventional antibodies.

13 Skin

The skin covers the exterior of the body and constitutes a large organ with several functions. In humans, the skin accounts for about 7 percent of the total body weight. It protects the body from mechanical injury and loss of fluid, acts as a barrier against noxious agents, aids in temperature regulation, excretes various waste products, and through its receptors for sensations of heat, cold, touch, and pain, provides information about the external environment.

Structure

KEY WORDS: epidermis, dermis, appendage, integument, thick skin, thin skin, epidermal ridge, dermal papilla

The skin consists of a surface layer of epithelium called the **epidermis** and an underlying layer of connective tissue, the **dermis** (corium). A loose layer of connective tissue, the *hypodermis,* attaches the skin to underlying structures but is not considered a component of skin. The **appendages** of the skin — hair, nails, and sweat and sebaceous glands — are local specializations of the epidermis. Together, the skin and its appendages form the **integument.**

The thickness of the skin varies in different parts of the body, and the proportions of dermis and epidermis also differ. Between the scapula, the dermis is especially thick, whereas on the palms of the hands and soles of the feet, the epidermis is thickened. The terms **thick skin** and **thin skin** refer to the thickness of the epidermis and not to the thickness of the skin as a whole. In thick skin the epidermis is especially well developed, whereas in thin skin it forms a relatively narrow layer.

At its junction with the dermis, the epidermis forms numerous ridgelike extensions, the **epidermal ridges,** that project into the underlying dermis. Complementary projections of the dermis fit between the epidermal ridges and form the **dermal papillae.** The surface patterns of the skin, such as the fingerprints, reflect the pattern of the dermal papillae.

Epidermis

KEY WORDS: stratified squamous epithelium, keratinocyte, keratinization, melanocyte, pigmentation, layers (strata)

The epidermis forms the surface layer of the skin and consists of **stratified squamous epithelium.** The cells, **keratinocytes,** undergo an orderly progression of maturation and **keratinization** to produce a superficial layer of dense, flattened, dead cells at the surface. A smaller population of cells, the **melanocytes,** is associated with **pigmentation** of the skin. The epidermis lacks blood vessels and is nourished by diffusion of material from vessels in the underlying dermis.

The epidermis can be divided into several **layers (strata)** that reflect the sequential differentiation of keratinocytes as they progress from the base of the epidermis to the surface, where they are exfoliated. In thick skin, the layers are stratum basale, stratum spinosum, stratum granulosum, stratum lucidum, and stratum corneum (Fig. 13-1). In thin skin the layers are narrower and less well defined and stratum lucidum is absent.

Stratum Basale

KEY WORDS: columnar epithelial cell, growing layer, stratum germinativum

Stratum basale abuts the underlying dermis, separated from it only by a basement membrane. It consists of a single row of **columnar epithelial cells** that follows the contours of the ridges and papillae. Stratum basale is the **growing layer** of the epidermis, and most, but not all, of the mitotic activity occurs in this layer. For this reason, stratum basale often is called the **stratum germinativum** (Fig. 13-2).

Each cell of stratum basale is limited by a typical trilaminar cell membrane and is joined to adjacent cells by numerous desmosomes. At the basal surfaces, hemidesmosomes aid in attaching the cells to the underlying basement membrane. The basement membrane at the dermoepidermal junction, as elsewhere, consists of three strata: a superficial lamina lucida, a lamina densa, and an underlying fibroreticular lamina. On the dermal side of the basement membrane, anchoring filaments of type VII collagen link the undersurface on the lamina densa to collagen fibers of the papillary dermis. Fibrillin microfibrils attach the lamina densa to elastic fibers. The cytoplasm of the basal keratinocytes is fairly dense and contains many scattered keratin intermediate filaments. Clusters of ribosomes are

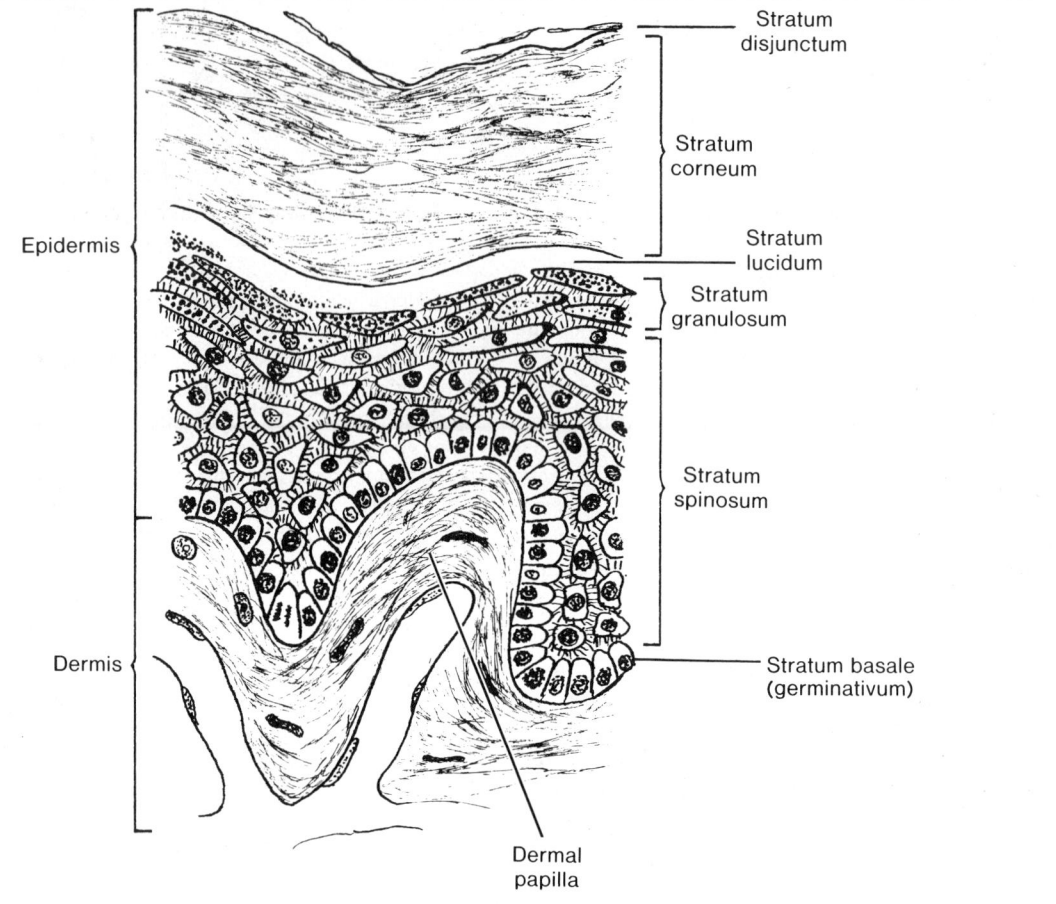

Stratum
disjunctum

Stratum
corneum

Stratum
lucidum

Stratum
granulosum

Stratum
spinosum

Stratum basale
(germinativum)

Epidermis

Dermis

Dermal
papilla

Fig. 13-1. Structure of thick skin.

prominent, and the cells contain a moderate number of mitochondria, some profiles of granular endoplasmic reticulum, and a few Golgi saccules.

Stratum Spinosum

KEY WORDS: prickle cell, membrane-coating granule, stratum malpighii

Stratum spinosum consists of several strata of irregular, polyhedral cells that become somewhat flattened in the outermost layers. The cells are closely applied to each other and joined at all surfaces by numerous desmosomes. During tissue preparation, the cells tend to shrink and pull apart except at these points of attachment. Thus the cells appear to have numerous short, spiny projections that ex-

tend between adjacent cells and commonly are called **prickle cells.** The projections are not areas of cytoplasmic continuity between cells but are sites of typical desmosomes.

In addition to the organelles seen in the basal cell, prickle cells in the upper layers of stratum spinosum contain ovoid granules, 0.1 to 0.5 μm in diameter, called **membrane-coating granules.** These consist of parallel laminae bounded by a double membrane and are rich in glycols and phospholipids that help maintain the barrier function of the skin. Keratin intermediate filaments are numerous and may form dense bundles that extend into the spinous processes, ending in the dense plaques of desmosomes. Stratum spinosum and stratum basale often are grouped together as **stratum malpighii.**

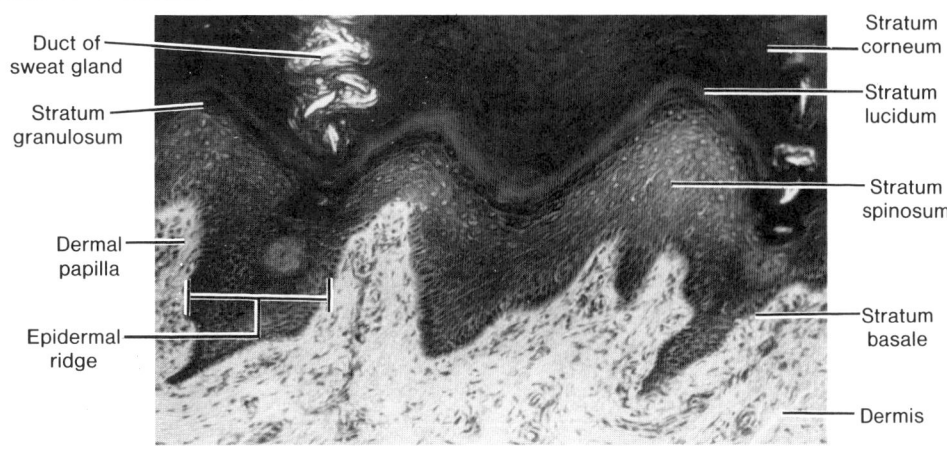

Fig. 13-2. The epidermis of human thick skin (palm) (LM, × 100).

Stratum Granulosum

KEY WORDS: keratohyalin granules

Stratum granulosum contains three to five layers of flattened cells whose long axes are parallel to the surface of the skin (see Fig. 13-2). The cells are distinguished by the presence of amorphous, densely packed particles, the **keratohyalin granules,** that vary in size and shape. The granules are not limited by membranes and are associated closely with bundles of keratin filaments. Chemically, they contain large amounts of sulfur-containing amino acids. The granules increase in number and size in the outermost layers of stratum granulosum, and the cells show evidence of degenerative changes. The nuclei stain more palely, and the contacts between adjacent cells become less distinct. The cells of the granular layer are viable but undergo programmed death as they pass into the succeeding horny layer. Membrane-coating granules increase in number in the cells of stratum granulosum. These rod-shaped granules fuse with the plasmalemma and empty their contents into the intercellular space. The lipid-rich contents act as a barrier between cells of this layer and those toward the surface and contribute to the sealing effect of skin, preventing water loss and entrance by foreign substances between cells.

Stratum Lucidum

KEY WORDS: loss of nuclei, thickened plasmalemma, thick skin

Stratum lucidum forms a narrow, undulating, lightly stained zone at the surface of stratum granulosum (see Fig. 13-2). It consists of several layers of cells so compacted together that outlines of individual cells cannot be made out. Traces of flattened nuclei may be seen, but generally this layer is characterized by the **loss of nuclei.** Only a few remnants of organelles are present, and the main constituent of the cytoplasm is aggregates of keratin filaments that now have a more regular arrangement, generally parallel to the skin surface. The **plasmalemma is thickened** and more convoluted, and the amount of intercellular material is increased. Stratum lucidum is prominent in the **thick skin** of the palms and soles but is absent from the epidermis in other parts of the body (Fig. 13-3).

Stratum Corneum

KEY WORDS: squame, keratin, soft keratin, desquamation, stratum disjunctum

Stratum corneum forms the outermost layer of the epidermis and is composed of scalelike cells, often called **squames,** that become increasingly flattened as they approach the surface. Squames are enclosed by a thickened, modified cell envelope and represent the remains of cells that have lost their nuclei, all their organelles, and their desmosomal attachments to adjacent cells. The cells are filled with **keratin,** which consists of tightly packed bundles of filaments embedded in an opaque, structureless material. The keratin filaments of stratum corneum consist of "soft" keratin as distinct from the "hard" keratin of nails and hair. **Soft**

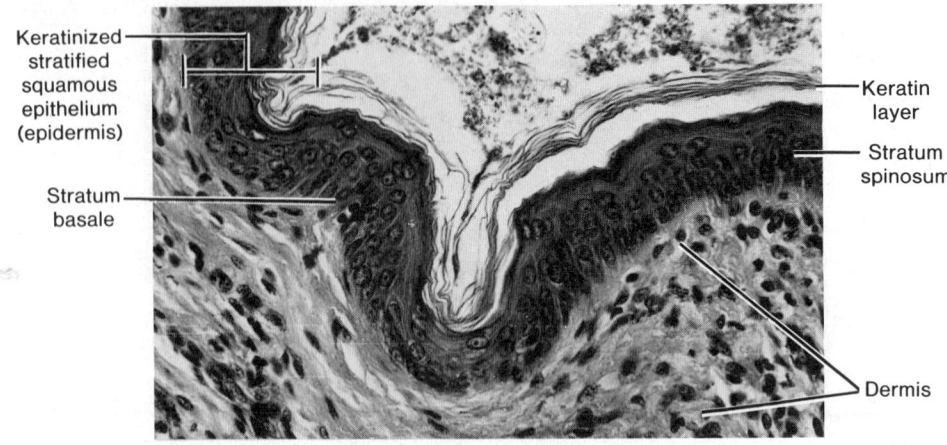

Keratinized
stratified
squamous
epithelium
(epidermis)

Keratin
layer

Stratum
spinosum

Stratum
basale

Dermis

Fig. 13-3. The epidermis of human thin skin (thigh) (LM, × 250).

keratin has a lower sulfur content and is somewhat more elastic than hard keratin. The outermost cells of stratum corneum are constantly shed or **desquamated;** this region often is referred to as **stratum disjunctum.**

The life span of keratinocytes in their passage from basal layer to desquamation is between 40 and 55 days. As the keratinocytes progress toward the surface, they become much broader and flatter so that, ultimately, only four surface squames are needed to cover the same area as 100 columnar basal cells. Thus a low rate of mitosis (less than seven per 1000 basal cells) can maintain the surface layer.

Pigmentation

KEY WORDS: melanin, melanocyte, epidermal-melanin unit, melanosome

Skin color is determined by the pigments carotene and melanin and by the blood in the capillaries of the dermis. Carotene is a yellowish pigment found in the stratum corneum and in fat cells of the dermis and hypodermis. Melanin, a brown pigment, is present mainly in the cells of stratum basale and the deeper layers of stratum spinosum.

Melanin is formed in specialized cells of the epidermis, the **melanocytes,** which originate from cells that emigrated from the neural crest during development. Melanocytes are scattered throughout the basal layer and send out numerous finger-like cytoplasmic processes that extend between the cells of stratum spinosum. Each melanocyte maintains a functional relationship with 36 keratinocytes in stratum spinosum. The combination of melanocytes and

keratinocytes forms an **epidermal-melanin unit.** In electron micrographs, epidermal melanocytes can be distinguished by the lack of keratin filaments and the content of melanin granules (**melanosomes**) in various stages of development. Melanocytes also contain the usual array of cytoplasmic organelles. While melanocytes lack desmosomal junctions with neighboring cells, they are affixed to the basement membrane by hemidesmosome-like junctions.

Mature melanosomes are transferred from melanocytes via their processes to the cells of the basal layer and stratum spinosum. In humans, melanosomes frequently aggregate above the nuclei of basal cells, but in the superficial layers of the epidermis the granules are more evenly dispersed and become progressively finer. Since epidermal cells constantly are shed from the surface, melanocytes must provide for continued renewal of melanin. Although melanocytes may die and be replaced by division of neighboring cells, this is a random event, and there does not appear to be a regular cycle of replacement of melanocytes.

The number of melanocytes and the size of the melanosomes is somewhat greater in Negroid and Mongolian skin than in Caucasian, but not sufficiently to account for the differences in skin color. Caucasoid melanosomes are less completely melanized and are packaged in groups, the melanosome complexes, surrounded by a membrane. The larger Negroid melanosomes are more widely dispersed within cells and do not form membrane-bound complexes. The distribution of pigment within the epidermis also differs, being concentrated in the lower zones of the malpighian layer in Caucasian skin and distributed through all cells of this layer in Negroid skin. The smaller melanosomes of Caucasians are degraded by lysosomes before they reach the upper strata.

The degree of pigmentation varies in different parts of the body. The axilla, circumanal region, scrotum, penis, labia majora, nipples, and areola are areas of increased pigmentation. In contrast, the palms and soles contain little or no pigment. Freckles represent local areas of pigmentation but paradoxically contain fewer melanocytes than adjacent paler skin. Pigmentation of freckles and the areolae and nipples may be intensified with pregnancy. A general increase in melanization occurs from exposure to sunlight, at first by darkening of existing melanin and later by increased formation of new melanin. The activated melanocytes produce larger melanosomes.

A deficiency of melanin also may occur. The white spotted patterns of piebaldism in humans result from abnormalities in the morphology of melanocytes in which the length and number of the cell processes is deficient. Complete albinism results from an inability of the melanocytes to synthesize melanin, not from an absence of melanocytes.

Other Cell Types

KEY WORDS: **Langerhans' cells, Merkel's cells**

Two other cell types present in the epidermis are Langerhans' and Merkel's cells. **Langerhans' cells** are present throughout the epidermis but are especially numerous in the upper layers of stratum malpighii. In routine sections, the cells have darkly stained, large, convoluted nuclei surrounded by a clear cytoplasm. With special stains they are seen to be stellate cells that extend long cytoplasmic processes into stratum spinosum. In electron micrographs, the nucleus shows a highly irregular outline, and the abundant cytoplasm lacks keratin filaments, desmosomes, and melanosomes but is characterized by membrane-bound, rod-shaped granules with a regular, granular interior. Some of these granules appear to be continuous with the cell membrane. Langerhans' cells trap antigens that penetrate the skin and transport them to regional lymph nodes. These antigen-presenting cells originate in bone marrow from a monocyte precursor.

Merkel's cells tend to lie close to naked sensory nerve endings in stratum basale and form Merkel cell–neurite complexes. Morphologically, Merkel's cells resemble cells of stratum spinosum, but in electron micrographs, they are distinguished by indented nuclei, prominent Golgi complexes, and numerous dense-cored vesicles. The vesicles tend to concentrate basally at the site where nerve endings

approach. The function of Merkel's cells is unknown, but they may have a mechanoreceptor function. They may play a role in sensory processes and have been compared with polypeptide- and hormone-containing cells, but a definite relationship with these has not been established.

Dermis

KEY WORDS: **reticular layer, papillary layer, dermal papilla, arrectores pilorum muscle**

The dermis (corium) varies in thickness in different regions of the body. It is especially thin and delicate in the eyelids, scrotum, and prepuce; very thick in the palms and soles; thicker on the posterior than anterior aspects of the body; and thicker in men than in women. The dermis is tough, flexible, and highly elastic and consists of a feltwork of collagen fibers with abundant elastic fibers in a glycosaminoglycan-containing matrix. The connective tissue is arranged into deep reticular and superficial papillary layers.

The **reticular layer** is the thicker, denser part of the dermis and consists of interwoven bundles of collagen fibers that mostly run parallel to the surface. Below, the reticular layer merges indistinctly into the subcutaneous tissue, which generally contains many fat cells. Collagen in the reticular layer is mostly type I. The **papillary layer** lies immediately below the epidermis and extends into it in the form of the dermal papillae. The papillary layer is not clearly demarcated from the reticular layer, but its fibers and bundles of collagen tend to be thinner and more loosely arranged. This layer also has a more cellular appearance. Most of the collagen is type III. Just below the epidermis, reticular fibers are present and have a vertical orientation. Elastic fibers are found deeper in the papillary layer.

Dermal papillae are small, conical projections with rounded or blunt apices that fit into corresponding pits on the undersurface of the epidermis. They are especially prominent on the palmar surface of the hands and fingers, where they are closely aggregated and arranged in parallel lines that correspond to the surface ridges of the epidermis. Capillary loops are present in the papillae (Fig. 13-4), and in some, especially those in the palms and fingers, nerve endings and tactile corpuscles are present (Fig. 13-5). The epidermis, like all epithelia, is avascular and must rely on diffusion through the basement membrane to meet its nutritional needs. Capillaries within the dermal papillae facilitate this exchange deep within the epidermis.

Smooth muscle is present in the deeper parts of the reticular layer in the penis, scrotum, perineum, and areola.

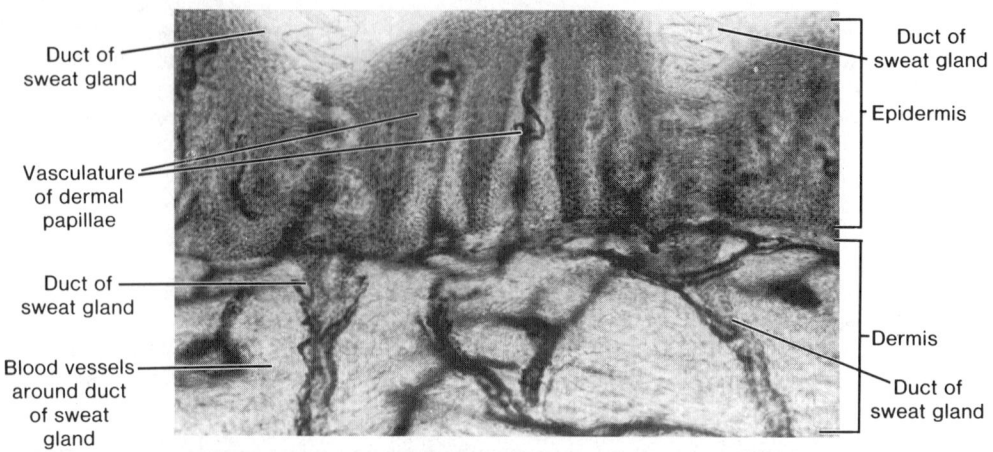

Duct of sweat gland

Vasculature of dermal papillae

Duct of sweat gland

Blood vessels around duct of sweat gland

Duct of sweat gland

Epidermis

Dermis

Duct of sweat gland

Fig. 13-4. A segment of human thick (palmar) skin injected with colored gelatin to illustrate the associated vasculature (LM, × 100).

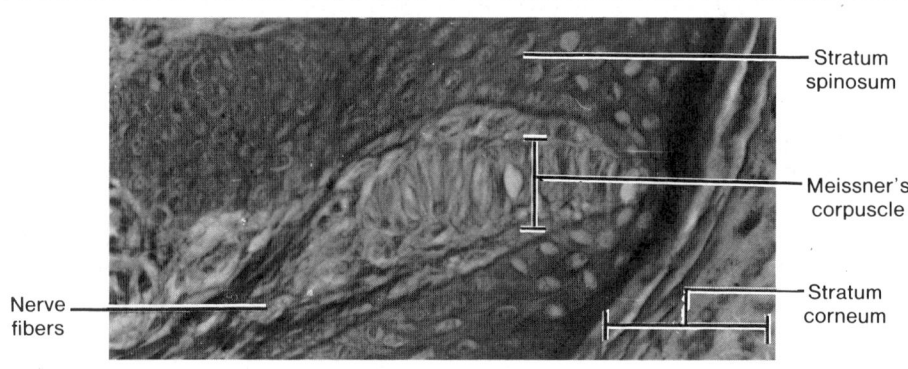

Stratum spinosum

Meissner's corpuscle

Nerve fibers

Stratum corneum

Fig. 13-5. A Meissner's corpuscle from a dermal papilla of human thick (palmar) skin (LM, × 400).

On contraction, the muscle produces wrinkling of the skin in these areas. The **arrectores pilorum muscles** are small bundles of smooth muscle associated with hairs.

Dermatoglyphics

Skin ridges are normal features of the fingers, hands, feet, and toes. They act as antislip devices and also are thought to improve the sense of touch. The ridges form three main patterns: arches, loops, and whorls. Loops are the most common pattern on the fingers and arches the rarest. The ridge patterns on the toes are similar to those on the fingers, but arches are more numerous and whorls are fewer. The ridge pattern is established early in life and is perma-

nent. In any individual the pattern varies from digit to digit, and it is unlikely that in a group of unrelated persons the sequence of patterns would be identical.

Appendages of the Skin

The appendages of the skin are derived from the epidermis and include hair, nails, and sweat and sebaceous glands.

Nails

KEY WORDS: body (nail plate), root, proximal nail fold, lateral nail fold, eponychium, hyponychium, lunule, nail bed, nail matrix, hard keratin

The nails (Fig. 13-6) are hard, elastic, keratinized structures that cover the dorsal surfaces of the tips of the fingers and toes. Each nail consists of a visible **body (nail plate)** and a proximal part, the **root,** which is implanted into a groove in the skin. The root is overlapped by the **proximal nail fold,** a fold of skin that continues along the lateral borders of the nail, where it forms the **lateral nail fold.** Stratum corneum of the proximal nail fold extends over the upper surface of the nail root and for a short distance onto the surface of the body of the nail, where it forms a thin cuticular fold called the **eponychium.** At the free border of the nail, the skin is attached to the underside of the nail, forming the **hyponychium.**

The nail is a modification of the cornified zone of the epidermis and consists of several layers of flattened cells with shrunken, degenerate nuclei. The cells are hard, tightly adherent, and throughout most of the body of the nail, clear and translucent. The pink color of the nails is due to transmission of color from the underlying capillary bed. Near the root, the nail is more opaque and forms a crescentic area, the **lunule,** which is most visible on the thumb, becoming smaller and more hidden by the proximal nail fold toward the little finger. The lunule represents the region from which nail formation occurs.

Beneath the nail lies the **nail bed,** which corresponds to the stratum malpighii of the skin. It consists of prickle cells and a stratum basale resting on a basement membrane. The underlying dermis is thrown into numerous longitudinal ridges that are very vascular. Near the root,

the ridges are smaller and less vascular. The nail bed beneath the root and lunule is thicker, actively proliferative, and concerned with growth of the nail; it is called the **nail matrix.** The nail bed beneath the rest of the nail is thinner and not involved with nail growth. Cells in the deepest layer of the matrix are cylindrical and show frequent mitoses, while above them are several layers of polyhedral cells and flattened squames that represent the differentiating cells of the nail. Nail keratin has a higher sulfur content than the keratin of the epidermis and is called **hard keratin.**

Hair

KEY WORDS: lanugo, vellus hair, terminal hair

Hairs are present on almost all surfaces of the skin except for the palmar surfaces of the hands, plantar surfaces of the feet, margin of the lip, prepuce, glans penis, clitoris, labia minora, and inner surfaces of the labia majora. They consist of elastic, keratinized threads that vary in length and thickness in different regions of the body and in different races. From the middle of fetal life, the skin is covered by fine hair called **lanugo;** this mostly is shed by birth and is replaced by downy vellus hair. **Vellus hairs** are retained in most regions, where they appear as short, soft, colorless hair such as that on the forehead. In the scalp and eyebrows, vellus hairs are replaced by coarser **terminal hair** that also forms the axillary and pubic hair and, in the male, the hair of the beard and chest.

Fig. 13-6. Longitudinal section of a nail.

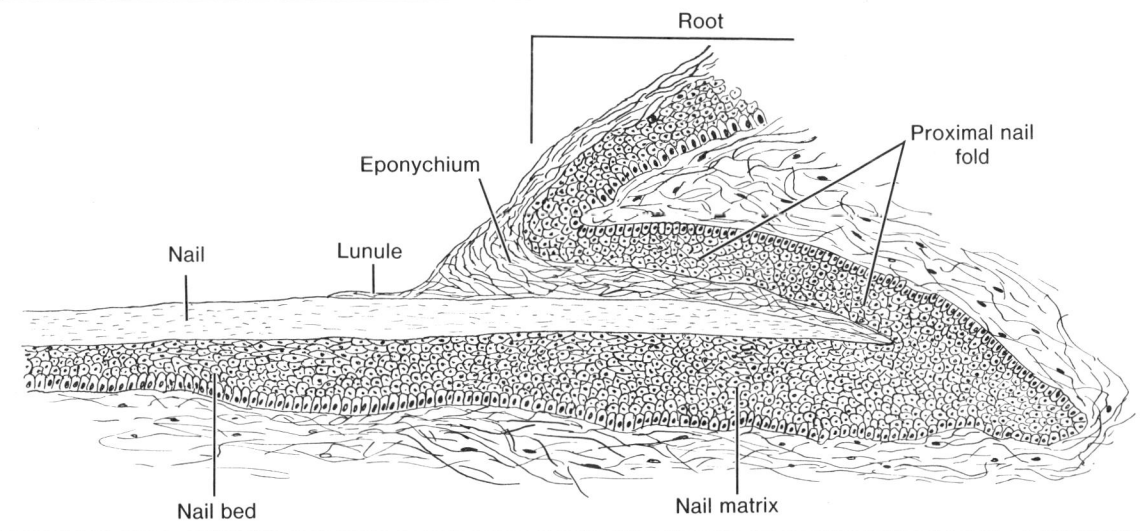

Structure of Hair

KEY WORDS: shaft, medulla, cortex, cuticle, root, hair bulb, papilla

Each hair consists of a root embedded in the skin and a hair **shaft** projecting for a variable distance above the surface of the epidermis. In cross section the shaft appears round or oval and is made up of three concentric layers. The **medulla** (core) is composed of flattened, cornified, polyhedral cells in which the nuclei are pyknotic or missing. There is no medulla in thin fine hair (lanugo), and a medulla may be absent in hairs of the scalp or extend only part of the way along the shaft. The bulk of the hair consists of the **cortex**, which is composed of several layers of intensely cornified, elongated cells tightly compacted together. Most of the pigment of colored hair is found in the cortex and is present in the cells and the intercellular spaces. Variable accumulations of air spaces are present between and within the cells of the cortex. Together with fading of pigment, increase in the number of air spaces is responsible for graying of the hair. The outermost layer is thin and forms the **cuticle.** It consists of a single layer of clear, flattened, squamous cells that overlap each other, shingle fashion, from below upwards.

The **root** of the hair is embedded in the skin. At the lower end, the root expands to form the **hair bulb,** which is indented at its deep surface by a conical projection of the dermis and called a **papilla.** Papillae contain blood vessels that provide nourishment for the growing and differentiating cells of the hair bulb.

The structure of the root differs somewhat from that of the shaft. In the lower part of the root, the cells of the medulla and cortex tend to be cuboidal in shape and contain nuclei of normal appearance. At higher levels in the root, nuclei become indistinct and finally are lost. The cells of the cortex become progressively flattened toward the surface of the skin.

Hair Follicle

KEY WORDS: inner epithelial root sheath, cuticle of root sheath, Huxley's layer, trichohyalin granules, Henle's layer, outer epithelial root sheath, glassy (vitreous) membrane

The root of each hair is enclosed within a tubular sheath called the *hair follicle,* which consists of an inner epithelial component and an outer connective tissue portion. The epithelial component is derived from the epidermis and consists of inner and outer root sheaths. The connective tissue sheath is derived from the dermis.

The **inner epithelial root sheath** corresponds to the superficial layers of the epidermis that have undergone specialization to produce three layers. The innermost layer is the **cuticle of the root sheath,** which abuts the cuticle of the hair shaft. The cells are thin and scalelike and are overlapped from above downward; the free edges of the cells interlock with the free edges of the cells of the hair cuticle. Immediately surrounding the cuticle of the root sheath are several layers of elongated cells that form **Huxley's layer.** The cells contain granules that are similar to keratohyalin granules but differ chemically; they are called **trichohyalin granules.** Huxley's layer is surrounded by **Henle's layer,** a single row of clear, flattened cells that contain keratin filaments. The cells of these three layers are nucleated in the distal parts, but as the sheath approaches the surface, the nuclei are lost.

The **outer epithelial root sheath** is a direct continuation of stratum malpighii. The cells of the outermost layer are columnar and arranged in a single row and at the surface become continuous with stratum basale of the epidermis. The inner layers of cells are identical to and continuous with the prickle cells of stratum spinosum.

The connective tissue portion of the follicle consists of three layers. A narrow clear band, the **glassy (vitreous) membrane,** is closely applied to the columnar cells of the outer epithelial root sheath and equates with the basement membrane. The middle layer consists of fine, circularly arranged collagen fibers. The outermost layer is poorly defined and contains collagen fibers arranged in loose, longitudinal bundles interspersed with some elastic fibers.

A bulbous expansion of the hair root surrounds the papilla. The cells of the hair bulb are not arranged in layers but form a matrix of growing cells. Matrix cells at the tip of the papilla differentiate into the medulla of a hair. Those on the slopes develop into the cortex and medulla. Laterally, cells of the bulb form the inner epithelial root sheath. Pigmentation occurs from the activity of melanocytes present in the matrix.

The structure of a hair and hair follicle are shown in Figure 13-7.

Hair Cycle

No hair grows forever; each cycles through a proliferative phase (anagen), a period of decreasing growth (catagen), and a resting phase (telogen). The cyclic activity continues throughout life, but the phases of the cycle change with age. At about the fifth month of gestation, all the hairs are in anagen, a uniformity of growth not seen again. Between 8 and 10 weeks before birth, some hair sites have reached catagen and telogen phases. The frontal and parietal scalp

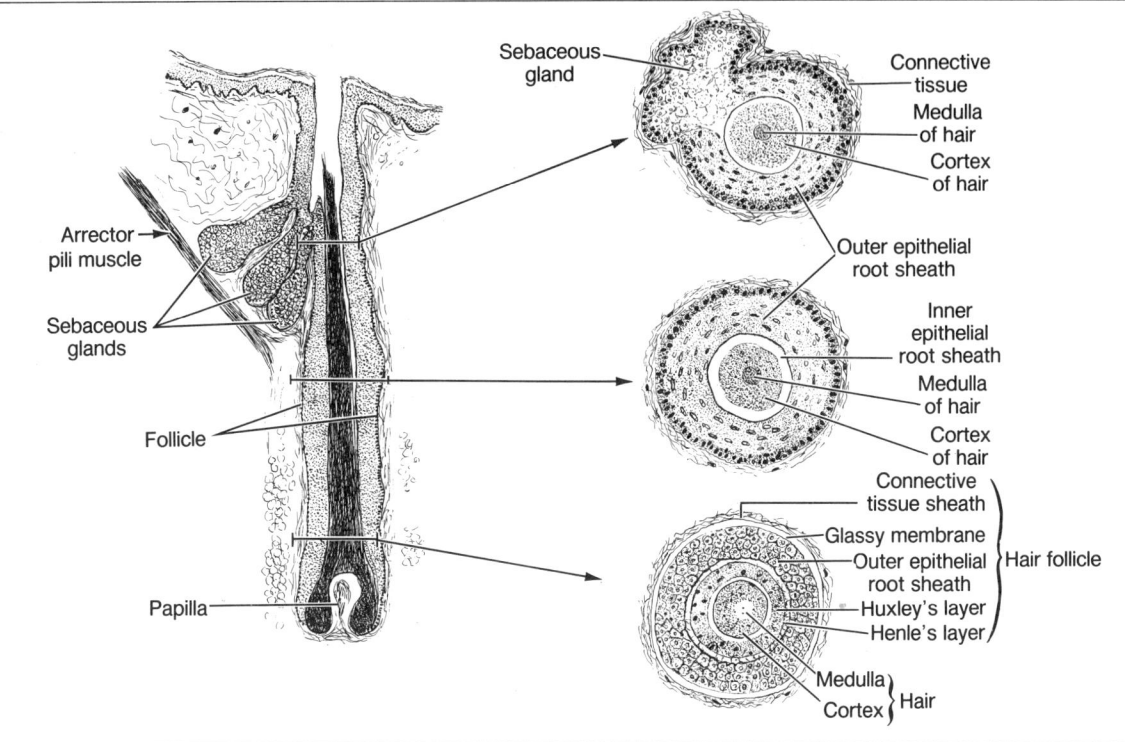

Fig. 13-7. Structure of a hair and hair follicle.

areas show the first shedding events; in the occipital region, hairs remain in anagen until after birth. From about 6 weeks before birth, telogen hairs again appear in the frontal and parietal scalp, indicating a second cycle of hair growth.

All hairs usually enter telogen immediately after birth, giving rise to a second period of shedding. After this, the phases are more irregular. About 18 weeks after birth, cycles are associated with individual hairs or groups of hairs. In adults, hair cycles vary with body region. The total hair cycle in the scalp extends over 300 weeks, with telogen occupying 18 to 19 weeks.

Sebaceous Glands

KEY WORDS: alveolus, holocrine, sebum

Sebaceous glands occur in most parts of the skin and are especially numerous in the scalp and face and around the anus, mouth, and nose. They are absent from the palms of the hands and soles of the feet. Generally, sebaceous glands are associated with hairs and drain into the upper part of the hair follicle, but on the lip, glans penis, inner surface of the prepuce, and labia minora, the glands open directly onto the surface of the skin, unrelated to hairs. The glands vary in size and consist of a cluster of two to five oval alveoli drained by a single duct.

The secretory **alveoli** lie within the dermis and are composed of epithelial cells enclosed in a well-defined basement membrane and supported by a thin connective tissue capsule. Cells abutting the basement membrane are small and cuboidal and contain round nuclei. The entire alveolus is filled with cells that, centrally, become larger and polyhedral and gradually accumulate fatty material in their cytoplasm. Nuclei become compressed and pyknotic and finally disappear (Fig. 13-8). Secretion is of the **holocrine** type, meaning the entire cell breaks down, and cellular debris, along with the secretory product, is released as **sebum.** Myoepithelial cells are absent from sebaceous glands, but the glands are closely related to the arrectores pilorum muscle. In the nipple, smooth muscle bundles are present in the connective tissue between alveoli.

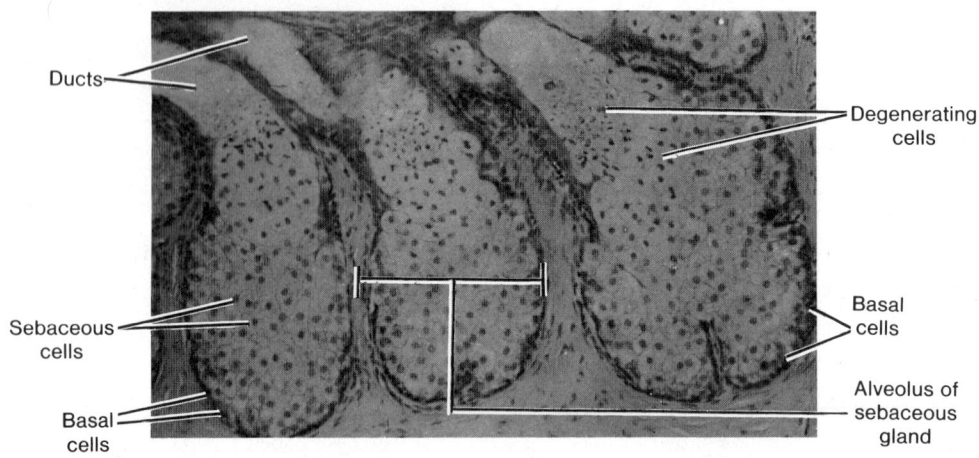

Ducts

Degenerating cells

Sebaceous cells

Basal cells

Basal cells

Alveolus of sebaceous gland

Fig. 13-8. A human sebaceous gland from the eyelid (LM, × 100).

The short duct of a sebaceous gland is lined by stratified squamous epithelium that is a continuation of the outer epithelial root sheath of the hair follicle. Replacement of secretory cells of the alveolus comes mostly from division of cells close to the walls of the ducts, near their junctions with the alveoli. Some replacement comes from cells at the periphery of the alveoli.

Collectively, the hair follicle, hair shaft, sebaceous gland, and erector pili muscle are referred to as the *pilosebaceous apparatus*. The pilosebaceous apparatus produces hair and sebum, the latter of which protects the hair and acts as a lubricant for the epidermis to protect it from the drying effects of the environment.

Sweat Glands

KEY WORDS: eccrine sweat gland, simple tubular, coiled, dark cell, clear cell, myoepithelial cell, apocrine sweat gland

Two classes of sweat glands are distinguished: ordinary or eccrine sweat glands and apocrine glands, such as those of the axilla and circumanal region.

Eccrine sweat glands are distributed throughout the skin except in the lip margins, glans penis, inner surface of the prepuce, clitoris, and labia minora. Elsewhere the numbers vary, being plentiful in the palms and soles and least numerous in the neck and back. Each gland is a **simple tubular** structure. The deep part is tightly **coiled** and forms the secretory part located in the dermis (Fig. 13-9). The secretory part consists of a simple epithelium resting on a thick basement membrane.

Two types of cells, clear cells and dark cells, are present. The **dark cells** are narrow at their bases and broad at the luminal surface; in electron micrographs they show numerous ribosomes, secretory vacuoles, and few mitochondria. These cells secrete glycoproteins, which have been identified in secretory vacuoles. **Clear cells** are broad at their bases and narrow at the apices. Intercellular canaliculi extend between adjacent clear cells, which contain glycogen, considerable smooth endoplasmic reticulum, and numerous mitochondria but few ribosomes. The plasmalemma at the base of the cell shows extensive, complex infoldings. Clear cells appear to secrete sodium chloride, urea, uric acid, ammonia, and water.

Eccrine glands empty via ducts onto the surface of the skin, on the summit of the skin ridges. The ducts consist of a stratified cuboidal epithelium.

Myoepithelial cells are present in the secretory portion, located between the basal lamina and the bases of the secretory cells. These stellate cells are contractile and are believed to aid in the discharge of secretions. The gland empties into a narrowed duct that at first is coiled and then straightens as it passes through the dermis to reach the epidermis. The lining of the duct consists of two layers of cuboidal cells. At their luminal surfaces, the cells of the inner layer show aggregations of filaments organized into a terminal web. The epithelium lies on a basal lamina, but myoepithelial cells are lacking. In the epidermis, the duct consists of a spiral channel that is simply a cleft between the epidermal cells; those cells immediately adjacent to the duct are circularly arranged.

Apocrine sweat glands are enlarged, modified eccrine

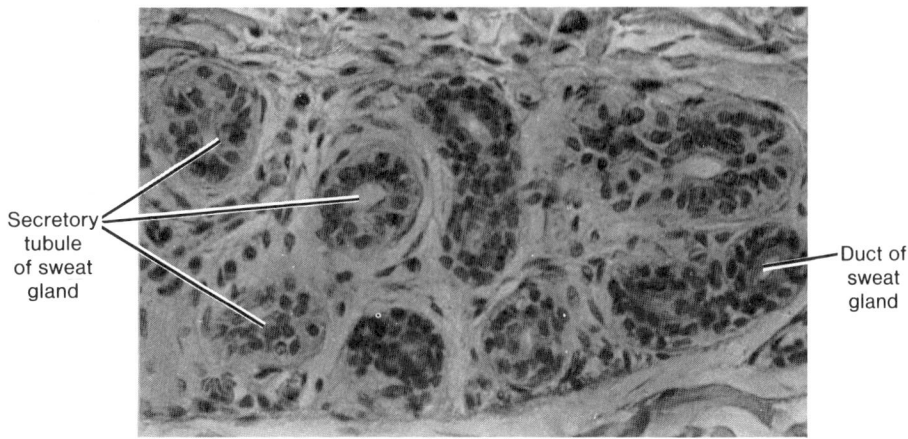

Fig. 13-9. An eccrine sweat gland from human thin skin (LM, × 250).

sweat glands. Their secretions are thicker than those of the ordinary sweat glands, and their histologic structure differs in several respects. Apocrine glands also are coiled, tubular glands lined by cuboidal cells, but the tubes are wider (Fig. 13-10). Their myoepithelial cells are larger and more numerous, and there is only one type of secreting cell, which resembles the dark cells of the eccrine glands. The ducts are similar to those of ordinary sweat glands but empty into a hair follicle rather than onto the surface of the epidermis. Secretion by apocrine glands is of the merocrine type and involves no loss of cellular structure. Although retained, the term *apocrine* is misapplied to these glands.

The ceruminous (wax) glands of the external auditory canal are apocrine glands in which the secretory portion and the duct may branch. Glands in the margins of the eyelids (Moll's glands) also are apocrine but differ in that the terminal portions show less coiling and have wider lumina.

Fig. 13-10. A human apocrine sweat gland from the axilla. Note that the secretory tubule is larger and the lumen more dilated when compared to that of an eccrine sweat gland (see 13-9) (LM, × 100).

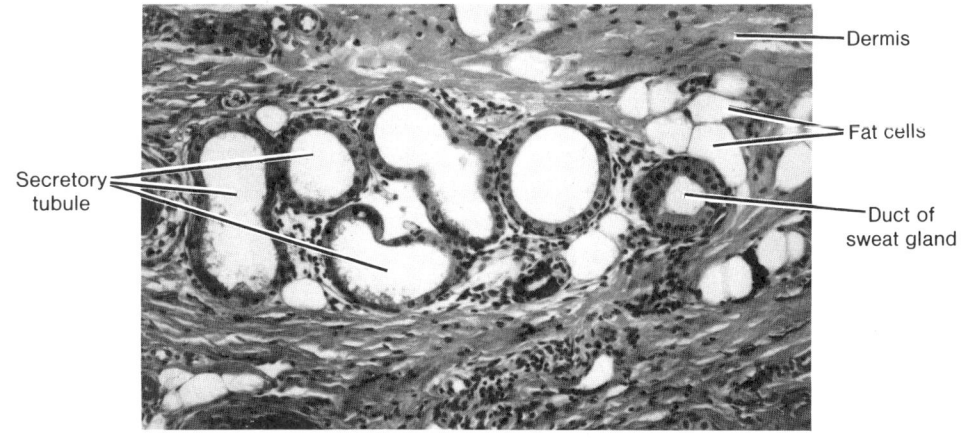

Review Table 13-1. Key Features of Thick and Thin Skin

	Epidermis	Dermis	Other
Thick skin	Shows five layers: stratum corneum, stratum lucidum, stratum granulosum, stratum spinosum, and stratum germinativum	Papillary and reticular layers very thick	Only on palms of hand, soles of feet; contains eccrine sweat glands, the only glands present in thick skin
Thin skin	Lacks stratum lucidum; stratum corneum, stratum granulosum, and stratum spinosum are thinner	As for thick skin, depth varies with area; thick dermis in skin between scapulae; delicate dermis over eyelids, scrotum, penis; thinner on anterior surface of body than on posterior surface	Contains hair follicles, eccrine sweat and sebaceous glands; apocrine sweat glands in axilla and groin

Functional Review: Skin

The outer, horny layer of the epidermis is highly impermeable to water and rather inert chemically. It is this portion of the epidermis that acts as the main barrier to mechanical damage, desiccation, and invasion by bacteria. Contents of membrane-coating granules contribute to or constitute the primary intercellular barrier to water. Epidermis has a high capacity for self-renewal, the new cells being derived mainly from stratum basale. The main function of melanin is to protect the germinal cells from the effects of ultraviolet irradiation. It has been suggested that melanin may capture harmful free radicals generated in the epidermis by ultraviolet light. Accumulation of melanin above the nuclei of basal cells affords maximum protection of the genome from the effects of irradiation. Poorly melanized skin is more subject to sunburn, degenerative changes, and skin cancer after chronic exposure to sunlight than is darker skin.

The dermis provides mechanical strength, due to the high content of collagen fibers; elastic fibers provide elasticity. The vascular supply is contained within the dermis, and the epidermis is nourished by diffusion from this vascular bed. The area of contact between dermis and epidermis is increased by the dermal papillae, which facilitate exchange of nutrients from the capillaries in the papillary layer of the dermis. They also provide an interlocking surface to more tightly secure the epidermis to the dermis.

Hairs in humans mainly are concerned with cutaneous sensations of touch. Each hair follicle is elaborately innervated, and some of the nerve endings are associated with tactile discs of the Merkel type. Regulation of body temperature occurs in part by the evaporation of sweat from eccrine sweat glands. Sebum from sebaceous glands acts as a lubricant for the skin, protects it from drying, and aids in waterproofing. It also has some slight antibacterial activity, but the effect appears to be minimal. Nails not only have a protective function, they also serve as rigid bases for the support of the pads at the ends of the digits and thus may have a role in tactile mechanisms.

Review Questions and Answers

Questions

70. Define the terms *thick skin* and *thin skin*. What are the main differences between them?
71. Briefly outline the structure of the hair follicle and relate its components to the epidermis and dermis.
72. What contributes to the color of the skin?
73. What functions are served by the integument?

Answers

70. The terms *thick skin* and *thin skin* refer to the thickness of the epidermis. In thick skin, the epidermis is well developed and its component layers are well defined. In thin skin, the epidermis is thinner, its layers are more poorly established, and a stratum lucidum is lacking. Thin skin also contains hair follicles and associated sebaceous glands.

71. The hair follicle surrounds the hair and is composed of an outer component, derived from the dermis, that consists of a poorly defined outer layer of collagenous and elastic fibers; a middle layer of circularly arranged connective tissue fibers; and an inner, homogeneous glassy membrane. The glassy membrane forms a boundary with the epithelial component of the hair follicle and is equivalent to the epithelial basement membrane. Immediately adjacent to the glassy membrane are layers that represent a direct extension of the stratum malpighii (stratum basale and stratum spinosum), which forms the outer epithelial root sheath. Interior to this is the inner epithelial root sheath, which represents modifications of the stratum granulosum and stratum corneum.

72. The color of the skin depends mainly on the presence and distribution of melanin, which gives a brown coloration. Melanin is produced by the melanocytes found scattered in the stratum basale and is distributed via their cytoplasmic processes to the cells of the basal layer and stratum spinosum. In the white race melanin tends to be confined to the lower regions of the malpighian layer, whereas in the black race it is distributed through all layers of the epidermis. Carotene, a pigment in the stratum corneum and the fat cells of the dermis and hypodermis, also contributes to the color of the skin, as does the blood in the capillaries of the dermis.

73. The integument consists of the skin and its appendages (hair, nails, and sweat and sebaceous glands). The skin forms a protective, waterproof covering that is chemically inert and acts as a barrier against penetration by noxious agents. It also prevents water loss from the body. Hairs in humans are largely concerned with cutaneous sensation. The skin, especially the epidermis, is a relatively poor conductor of heat, and in humans the paucity of hair is compensated for by an increase in the thickness of the epidermis. Temperature is regulated mainly by the evaporation of sweat produced by the sweat glands. Sebum elaborated by the sebaceous glands acts as a lubricant for skin, prevents drying, and aids in waterproofing. Nails protect the ends of the digits and may have a role in tactile mechanisms by providing a firm base for the pads of the fingertips.

14 Respiratory System

The respiratory tract consists of respiratory and conducting portions. Respiratory tissue, located in the lungs, is that portion of the tract where exchange of gases between air and blood occurs and is characterized by a close relationship between capillary blood and air chambers. The conducting portion delivers air to the respiratory tissue and is characterized by rigid walls that keep the airways open. This part of the tract consists of nasal cavities, pharynx, larynx, trachea, and various subdivisions of the bronchial tree. Parts of the conducting system are within the lungs (intrapulmonary) and parts are outside the lungs (extrapulmonary).

Nose

KEY WORDS: pseudostratified ciliated columnar epithelium, goblet cell, olfactory epithelium, lamina propria, venous sinus, mucoperiosteum, mucoperichondrium

The skin that covers the external surface of the nose extends for a short distance into the vestibule of the nasal cavity, where stiff hairs and their associated sebaceous glands aid in filtering the inspired air. More posteriorly, the vestibule is lined by nonkeratinized stratified squamous epithelium (Fig. 14-1). Throughout most of the remainder of the nasal cavity, the respiratory passage is lined by **pseudostratified ciliated columnar epithelium** that contains **goblet cells,** but a specialized **olfactory epithelium** is present in the roof of the nasal cavities (Figs. 14-2 and 14-3).

A layer of connective tissue, the **lamina propria,** underlies the epithelium and is separated from it by a basement membrane. Contained within the lamina propria are mucous glands, serous glands, and thin-walled **venous sinuses.** The venous sinuses warm incoming air and are especially prominent in the lamina propria that covers the middle and inferior conchae. Lymphatic tissue is present and becomes prominent near the nasopharynx. The deep layers of the lamina propria fuse with the periosteum or perichondrium of the nasal bones and cartilages, and at these sites, the nasal mucosa forms a **mucoperiosteum** or **mucoperichondrium,** respectively.

The surface of the epithelial lining is bathed by a film of mucus that constantly is moved toward the pharynx by the action of the ciliated epithelial cells. The mucus is derived from surface goblet cells and secretions from mucoserous

Fig. 14-1. Keratinized stratified squamous epithelium lining the vestibule of the human nose (LM, × 300).

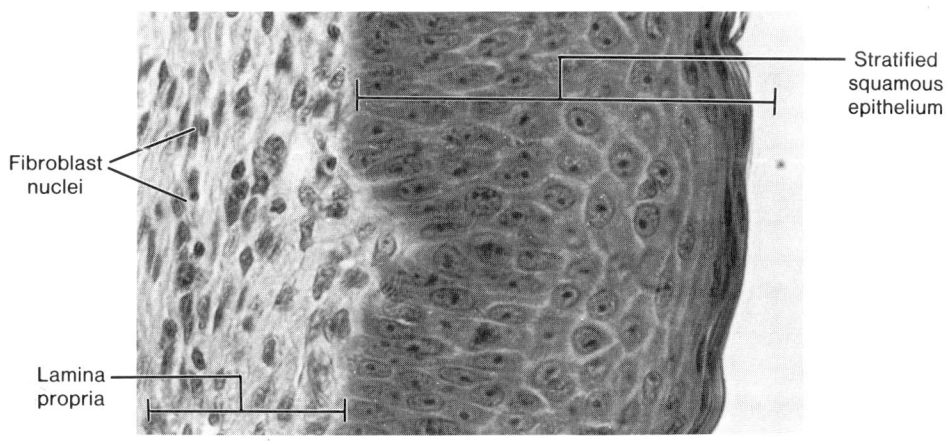

Stratified squamous epithelium

Fibroblast nuclei

Lamina propria

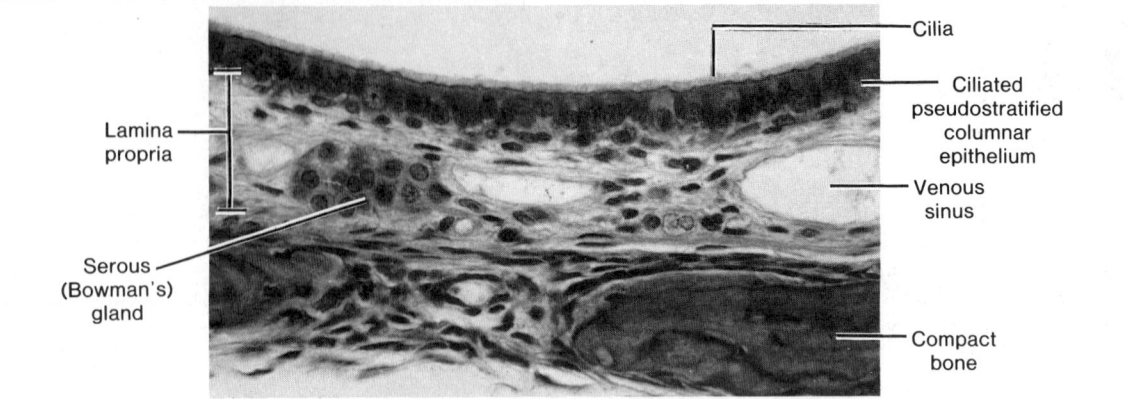

Cilia

Ciliated pseudostratified columnar epithelium

Venous sinus

Lamina propria

Serous (Bowman's) gland

Compact bone

Fig. 14-2. Ciliated pseudostratified columnar epithelium lining a nasal concha. Note the supporting bone. Compare the epithelium shown in this figure with 14-3 (LM, × 250).

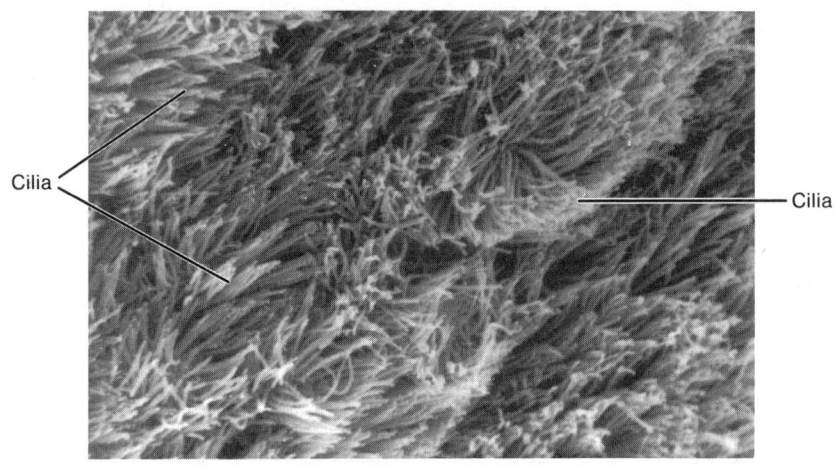

Cilia

Cilia

Fig. 14-3. Surface features of the epithelium covering the human inferior concha (SEM, × 400).

glands in the lamina propria. The layer of mucus helps to moisten the air and trap particulate matter. The mucous layer contains IgA and other immunoglobulins that protect against local infection. IgA is produced by plasma cells within the lamina propria and is taken up by secretory cells of adjacent glands. The IgA is then coupled to the secretory component of these cells, transported, and secreted onto the surface of the nasal mucosa. The same type of mucosa extends into the paranasal sinuses, but here the epithelium is thinner, there are fewer goblet cells, the lamina propria is thinner and contains fewer glands, and venous sinuses are absent. The paranasal sinuses (maxillary, sphenoid, frontal, ethmoid) lie within bones of the same name surrounding the nasal cavity and are continuous with it through small openings. Although cilia of the lining epithelium within the paranasal sinuses generally beat toward the nasal cavity, the ciliated epithelial cells are coordinated in such a way that the pathway ciliary motion follows is a large, open spiral, the pitch of which narrows at the opening to the nasal cavity.

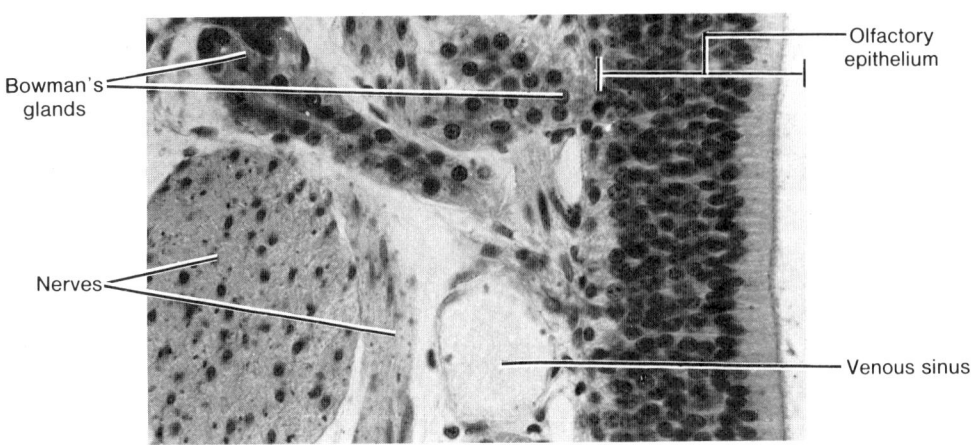

Fig. 14-4. A region of human superior concha showing olfactory epithelium and glands. Note the nerves and venous sinus in the lamina propria (LM, × 250).

Olfactory Epithelium

KEY WORDS: pseudostratified columnar epithelium, supporting (sustentacular) cell, olfactory cell, bipolar nerve cell, olfactory vesicle (knob), olfactory hair, fila olfactoria, basal cell, olfactory gland

Olfactory epithelium occurs in the roof of the nasal cavities, where it extends over the superior conchae and for a short distance on either side of the nasal septum. It also is a **pseudostratified columnar epithelium,** but it lacks goblet cells and is much thicker than the respiratory lining epithelium (Fig. 14-4).

Three primary cell types are present: supporting cells, basal cells, and sensory or olfactory cells. **Supporting (sustentacular) cells** are tall with narrow bases and broad apical surfaces that bear long, slender microvilli. An oval nucleus is located just above the center of the cell. Apically, well-developed junctional complexes join the supporting cells to adjacent olfactory cells. **Olfactory cells** are the sensory component and are **bipolar nerve cells** evenly distributed among the supporting cells (Fig. 14-5). They are spindle-shaped with rounded nuclei located centrally in an expanded area of cytoplasm. Apically, the cell tapers to a single slender process (a modified dendrite) that extends between the supporting cells to reach to the surface, where it expands into a bulblike **olfactory vesicle (knob).** Six to eight long **olfactory hairs** extend from the olfactory vesicle and pass parallel to the surface of the ep-

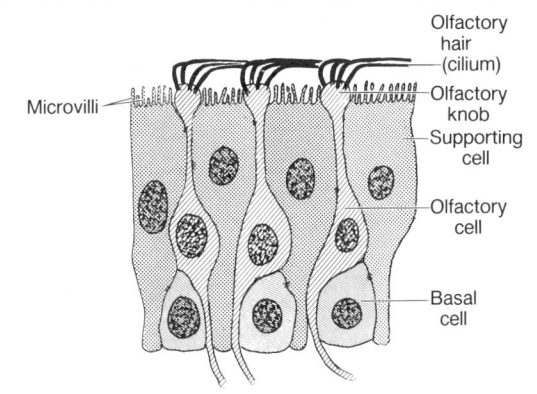

Fig. 14-5. Diagram of olfactory epithelium.

ithelium, embedded in a film of fluid (Fig. 14-6). The processes are modified, nonmotile cilia and appear to be the excitable component of the sense organ. For a short distance from their origins, the processes have a typical ciliary structure but then narrow abruptly, and the microtubules decrease in number and change from doublets to singlets. Basally, the olfactory cells narrow to a thin process, an axon, that passes into the lamina propria, where, with similar fibers, it forms small nerve bundles (see Fig. 14-4). These collect as the **fila olfactoria** that pass directly to the olfactory bulb.

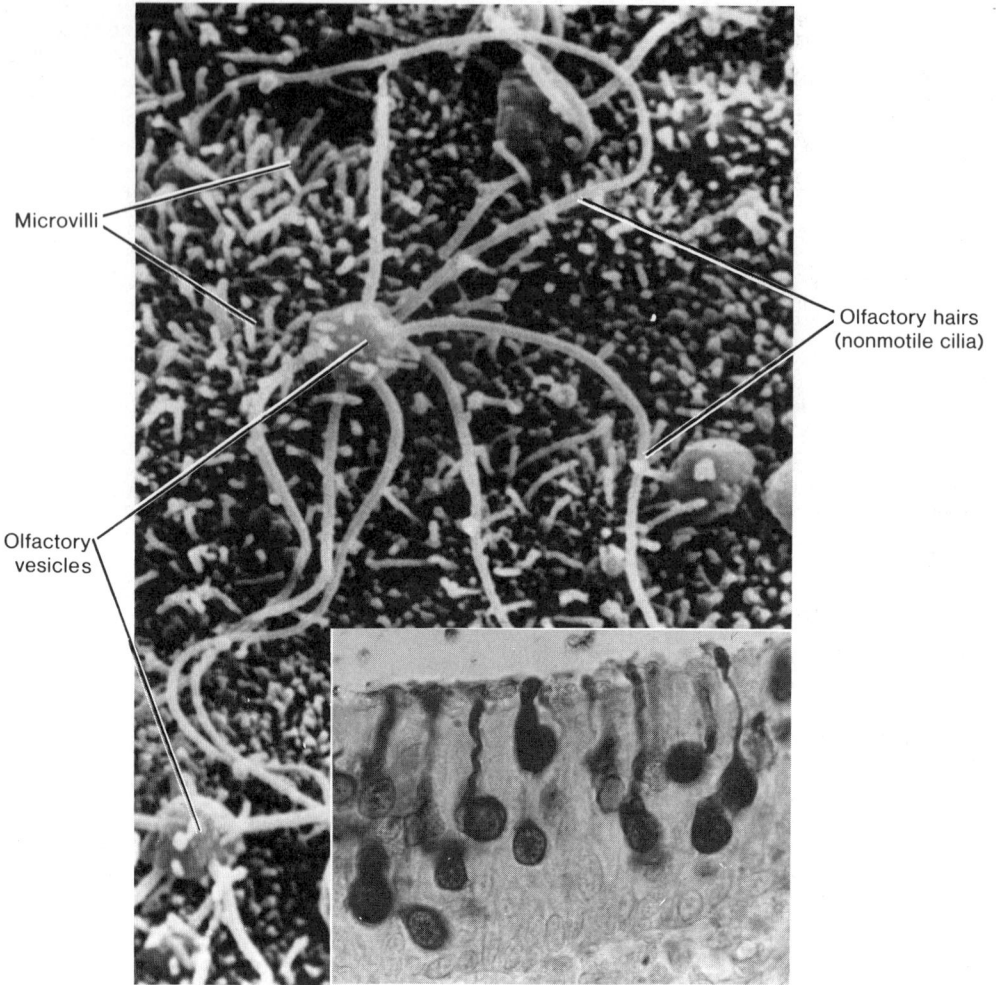

Fig. 14-6. Surface features of olfactory epithelium showing olfactory vesicles and cilia from underlying sensory bipolar neurons (SEM, × 5,700). Inset: Cell bodies of olfactory bipolar neurons stained by immunohistochemistry for olfactory marker protein (LM, × 500).

Basal cells are short, pyramidal cells crowded between the bases of sustentacular and olfactory cells. They are undifferentiated cells that give rise to either of the other two types of cells.

The olfactory cells are unique among neurons in that they are in direct contact with the external environment and, unlike other neurons, constantly are replaced every 30 to 60 days. New olfactory cells arise from differentiation of the basal cells. Between the surface and the basal cells lie vari-able numbers of progressively more differentiated receptor neurons, accounting in part for the thickness of the olfactory epithelium.

Rich plexuses of blood vessels and branched, serous, tubuloalveolar glands are present in the lamina propria. The glands are the **olfactory glands** (of Bowman), whose watery secretions flush the surfaces of the olfactory epithelium to prevent continuous stimulation by a single odor and also serve as a solvent for odorous materials (see Fig. 14-4).

In contrast to reports in the literature, vomeronasal organs (VNO) do exist in the nasal cavity of most adult humans. In humans, the VNO consist of two small sacs about 2 mm deep that open into shallow depressions on either side of the nasal septum. The epithelium lining these sacs appears similar to olfactory epithelium and contains supporting cells and cells that resemble olfactory receptor bipolar neurons. These bipolar neurons bind to antibodies against neuron-specific enolase and protein gene product 9.5, as do most neurons. The neuronal connection of these cells to the brain is unknown in humans. Likewise, the function of the vomeronasal organ in humans is unknown. In other mammals that have been studied in detail, the VNO has neural connections to an accessory olfactory bulb, which, in turn, is linked to the limbic system of the brain. In these mammals, the VNO functions in the detection of pheromones that mediate sexual and territorial behaviors.

Pharynx and Larynx

KEY WORDS: nasopharynx, pseudostratified ciliated columnar epithelium, goblet cell

The nasal cavities continue posteriorly into the pharynx, which has nasal, oral, and laryngeal parts. The most superior part, the **nasopharynx,** is directly continuous with the nasal cavities and is lined by the same respiratory passage epithelium — that is, **pseudostratified ciliated columnar epithelium** with **goblet cells.** In areas subject to abrasion, a nonkeratinizing stratified squamous epithelium may occur, such as on the edge of the soft palate and posterior wall of the pharynx, where these surfaces make contact during swallowing. The underlying connective tissue contains mucous, serous, and mixed mucoserous glands and abundant lymphatic tissue. The lymphatic tissue is irregularly scattered throughout the connective tissue and also forms tonsillar structures: the pharyngeal tonsils in the posterior wall of the nasopharynx, the tubal tonsils around the openings of the eustachian tubes into the nasopharynx, and the palatine and lingual tonsils at the junctions of the oral cavity and oropharynx. This lymphatic tissue often is referred to as *mucosa-associated lymphoid tissue (MALT)* and constitutes Waldeyer's ring.

The larynx connects the pharynx and trachea. Its framework consists of several cartilages, of which the thyroid, cricoid, and arytenoid are hyaline, while the epiglottis, corniculate, and tips of the arytenoid are elastic. This scaffold of cartilages is held together and to the hyoid bone by sheets of dense connective tissue that form the cricothy-roid and thyroid membranes. The lining epithelium varies with location. The anterior surface and about half the posterior surface of the epiglottis are covered by stratified squamous nonkeratinizing epithelium that may contain a few taste buds. The vocal cords also are covered by stratified squamous epithelium, but elsewhere the larynx is lined by the respiratory passage type of epithelium. The lamina propria of the larynx is thick and contains mucous and some serous or mucoserous glands.

Trachea and Extrapulmonary Bronchi

KEY WORDS: rings of hyaline cartilage, smooth muscle, pseudostratified ciliated columnar epithelium, goblet cell, brush cell, small granule cell, short cell, adventitia

The walls of the trachea and its terminal branches (the primary bronchi) are characterized by the presence of C-shaped **rings of hyaline cartilage** that maintain patency of the airway. The spaces between successive rings are filled with fibroelastic connective tissue, and the gaps between the arms of the cartilages are filled with bundles of **smooth muscle.** The gaps are directed posteriorly, and in section, the trachea and extrapulmonary bronchi appear flattened. The lining consists of **pseudostratified ciliated columnar epithelium** with numerous **goblet cells** that rests on a thick basal lamina (Figs. 14-7 and 14-8). Cilia beat toward the pharynx.

In addition to the ciliated and goblet cells, several other cell types have been described in the tracheobronchial epithelium. **Brush cells** contain glycogen granules, show long, straight microvilli on their apical surfaces, and basally, make contact with nerve processes. They are rare in humans and their function is unknown, but they may be sensory cells, exhausted goblet cells, or intermediates between short cells and mature ciliated cells. **Small granule cells** are few and scattered. They appear to be part of a population of diffuse neuroendocrine cells able to take up and store amines and amine precursors. **Short cells** occur in the depths of the epithelium between the bases of the other cell types and are undifferentiated cells that give rise to goblet and ciliated cells. A network of dendritic cells lies along the base of the tracheal epithelium. They function as antigen-presenting cells capable of binding and presenting antigens to T-lymphocytes. The lamina propria contains many small seromucous glands and occasional accumulations of lymphatic tissue. Myoepithelial cells are associated with the secretory units of the tracheal glands. External to the cartilage is a fibroconnective tissue coat, the **adven-**

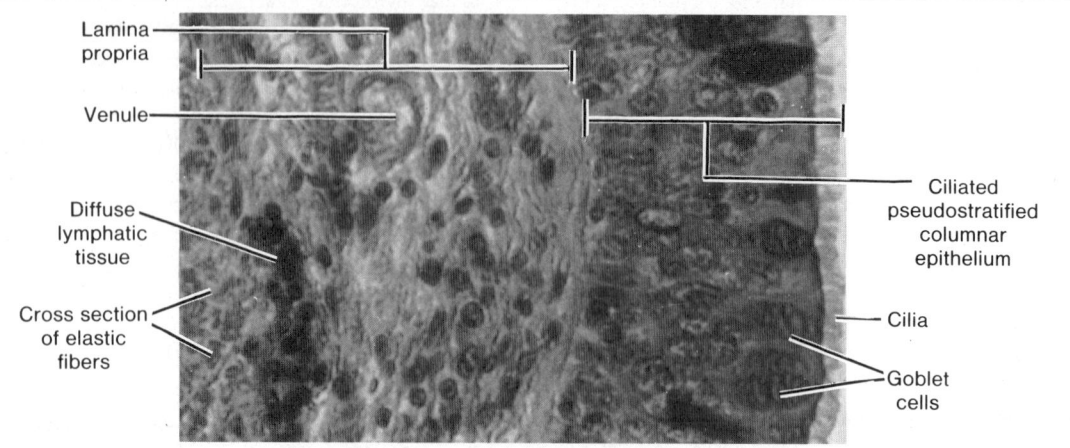

Lamina propria
Venule
Diffuse lymphatic tissue
Cross section of elastic fibers
Ciliated pseudostratified columnar epithelium
Cilia
Goblet cells

Fig. 14-7. The mucosal lining of a human trachea (LM, × 400).

Apex of nonciliated cell
Cilia
Erythrocyte
Lymphocyte

Fig. 14-8. Apices of ciliated and nonciliated cells from the epithelial lining of the trachea. Note the erythrocyte and lymphocyte on the epithelial surface (SEM, × 2,000).

titia, which contains elastic, collagenous and reticular fibers, blood vessels, and nerves.

Intrapulmonary Air Passages

KEY WORDS: intrapulmonary bronchus, plates of hyaline cartilage, pseudostratified ciliated columnar, goblet cell, submucosa, simple ciliated columnar, bronchiole, ciliated cuboidal, terminal bronchiole, Clara cell, surfactant, neuroepithelial body

The primary bronchi divide to give rise to several orders of **intrapulmonary bronchi** in which the C-shaped cartilaginous rings of the primary bronchi are replaced by irregular **plates of hyaline cartilage** that completely surround the structure. Thus the intrapulmonary bronchi are cylindrical and are not flattened on one side as are the main bronchi and trachea. Internally, the large intrapulmonary bronchi are lined by a mucous membrane that is continuous with and identical to that of the trachea and primary bronchi: **pseudostratified ciliated columnar** epithelium with **gob-**

let cells. The lamina propria contains some diffuse lymphatic tissue and is separated from the epithelium by a prominent basal lamina. Beneath the lamina propria, a sheet of irregularly arranged smooth muscle fibers runs around the bronchus in open left- or right-handed spirals. The muscle layer separates the lamina propria from the connective tissue of the **submucosa,** which lies immediately internal to the plates of cartilage. Mucous and mucoserous glands are present in the submucosa, their ducts penetrating the muscle layer to open onto the epithelial surface (Fig. 14-9). Secretions from these glands, as well as those throughout the respiratory tree, contain IgA, acquired from adjacent lymphoid tissue, which provides the larger respiratory passages with a degree of immunologic protection.

With successive divisions, the intrapulmonary bronchi progressively decrease in size, and although they retain the basic structure outlined above, the layers of their walls become thinner. The smallest of the intrapulmonary bronchi show only isolated cartilage plates and no longer are surrounded by cartilages. The epithelium is reduced to **simple ciliated columnar** with goblet cells. Mucous and mucoserous bronchial glands are present as far down as cartilages extend.

When the diameter of the tube reaches about 1 mm, cartilage disappears from the wall and the structure becomes a **bronchiole.** Glands and lymphatic tissue also disappear, but smooth muscle is fairly prominent and becomes the major supporting element (Figs. 14-10 and 14-11). The

Fig. 14-9. A section through a human intrapulmonary bronchus (LM, × 100).

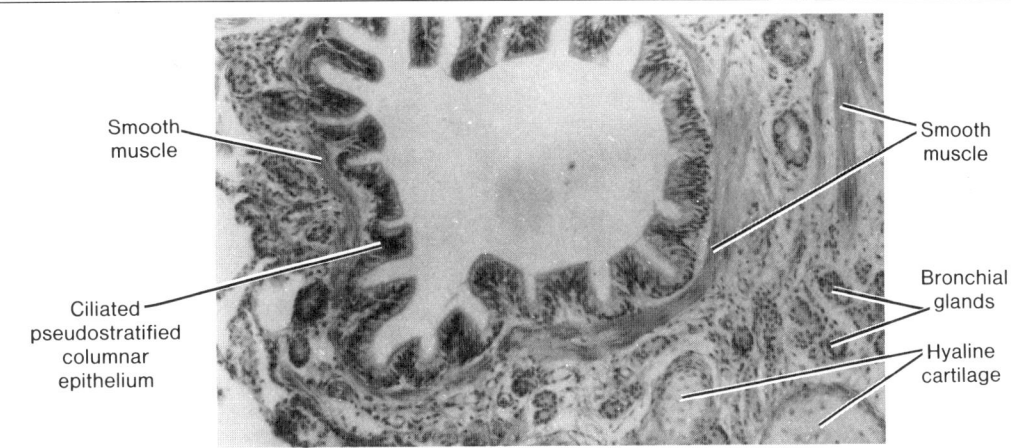

Fig. 14-10. A cross section of a bronchiole in a human lung (LM, × 100).

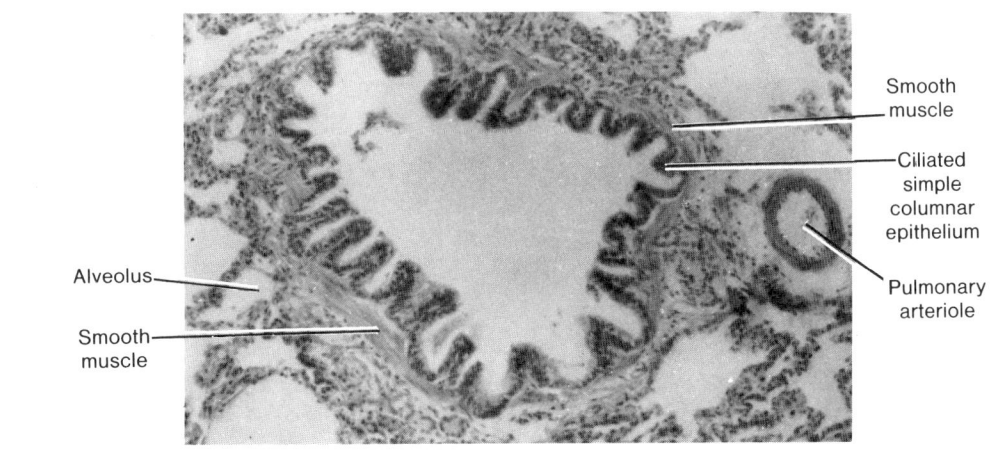

Fig. 14-11. Surface features of a bronchiole. Note the folded appearance as well as ciliated and nonciliated cells. Compare with 14-10 (SEM, × 1,000).

lining epithelium varies from ciliated columnar with goblet cells in the large bronchioles to **ciliated cuboidal** with no goblet cells in the **terminal bronchioles.** Terminal bronchioles are the smallest branches of the purely conducting system (Fig. 14-12). Scattered among the ciliated cells are a few nonciliated cells whose apical surfaces bulge into the lumen and bear a few microvilli. These are the **Clara cells,** also called *bronchiolar secretory cells,* which secrete **surfactant,** a phospholipid that alters the surface tension of the fluid layer covering the cell surface.

Neuroepithelial bodies are innervated epithelial cor-

puscles scattered throughout the intrapulmonary airways and even extending into alveoli. They are not present in the trachea. The corpuscles appear as ovoid or triangular bodies, 20 to 40 μm wide, set into the respiratory epithelium, where they extend from lumen to basement membrane. Basally, the neuroepithelial body is closely related to a capillary. The whole structure is richly innervated, and many of the nerve endings make contact with cells of the corpuscle.

The neuroepithelial bodies contain 4 to 10 tall, nonciliated cells with a slightly acidophilic cytoplasm that gives a

Fig. 14-12. Human lung showing a terminal bronchiole and several alveoli (LM, × 40).

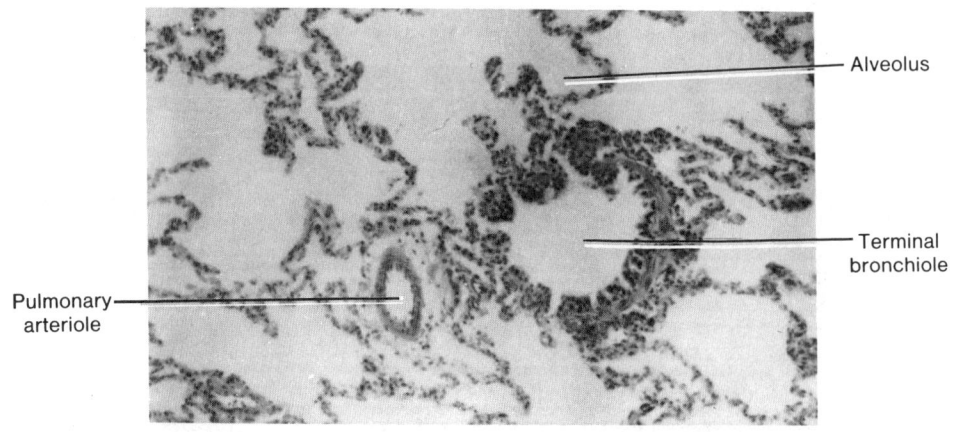

positive argyrophil (silver) reaction. The oval nuclei are basally located and oriented in the long axis of the cells, which contain numerous mitochondria, moderately well-developed granular endoplasmic reticulum, small Golgi complexes, and some glycogen and multivesicular bodies. Characteristic of these cells are numerous dense-cored vesicles. The neuroepithelial bodies are believed to be chemoreceptors that react to the composition of inhaled air. When the oxygen tension of the inspired air decreases, the neuroepithelial body releases dense-cored vesicles that contain serotonin, a potent vasoconstrictor. Thus, during local hypoxia, blood could be shunted from poor to better oxygenated and ventilated areas of the lungs.

Respiratory Tissue

KEY WORDS: respiratory bronchiole, alveolar duct, alveolar sac, alveolus, interalveolar septum, alveolar pore, blood-air barrier, pulmonary epithelial cell (type I pneumonocyte), septal cell (type II pneumonocyte), alveolar macrophage

Exchange of gases between air and blood occurs only when air and blood are in close relation. Such a condition occurs first in the **respiratory bronchioles,** which form a transition between the purely conducting and purely respiratory regions of the tract. The walls of respiratory bronchioles consist of collagenous connective tissue with interlacing bundles of smooth muscle and elastic fibers. The larger respiratory bronchioles are lined by simple cuboidal epithelium with only a few ciliated cells; goblet cells are lacking. Many of the cuboidal cells are Clara cells. In the smaller respiratory bronchioles, the epithelium becomes low cuboidal without cilia. Alveoli bud from the walls of the respiratory bronchioles and represent the respiratory portions of these airways. The alveoli become more numerous distally. Respiratory bronchioles end by branching into alveolar ducts. **Alveolar ducts** are thin-walled tubes from which numerous alveoli or clusters of alveoli open around the circumference so that the wall becomes little more than a succession of alveolar openings. Appearances of a tube persist only in a few places, where small groups of cuboidal cells intervene between successive alveoli and cover underlying bundles of fibroelastic tissue and smooth muscle. The alveolar ducts end in irregular spaces surrounded by clusters of alveoli called **alveolar sacs** (Fig. 14-13).

Alveoli are thin-walled, polyhedral structures that are open at one side to allow air into their cavities. Adjacent alveoli are separated by a common **interalveolar septum,** whose most conspicuous feature is a rich network of capillaries that bulge the septal wall to expose most of the capillary surface to alveolar air. Reticular and elastic fibers form a tenuous framework for the septa. Small openings in the septal wall, **alveolar pores,** permit communication and equalization of air pressure between alveoli. On each side, the alveolar wall is covered by an attenuated epithelium beneath which is a basal lamina. In many areas the epithelial basal lamina is separated from the basal lamina of the capillary by a space of only 15 to 20 nm; in other regions

Fig. 14-13. A segment of human lung illustrating the progression of terminal bronchiole to respiratory bronchiole to alveolar duct (LM, × 40).

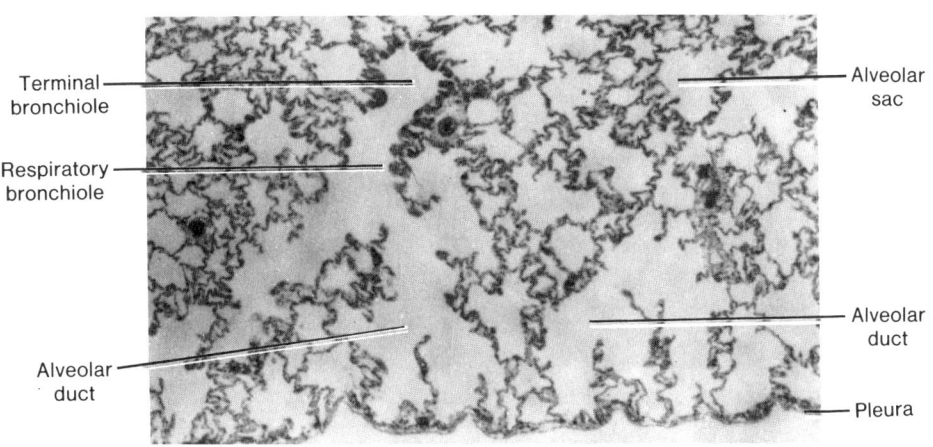

the two laminae are fused. Thus, at its thinnest, the **blood-air barrier** consists of a thin film of fluid, the attenuated epithelium of the alveolar lining cell, the fused basal laminae, and the endothelium of the capillaries within the septal wall.

Several types of cells are present in the septa. The attenuated squamous cells that form a continuous lining for the alveolar wall are called **pulmonary epithelial cells** or **type I pneumonocytes** (Figs. 14-14 and 14-15). In addition to these are septal cells, alveolar macrophages and endothelial cells that line blood capillaries. **Septal cells (type II pneumonocytes)** are rounded or cuboidal cells that may lie deep in the surface epithelium or bulge into the alveolar lumen between pulmonary epithelial cells. Their free surfaces have short microvilli, and laterally, the cells are united to pulmonary epithelial cells by junctional complexes (see Fig. 14-15). The most distinctive feature of the septal cells is the presence of multilamellar bodies in their cytoplasm (Fig. 14-16). These bodies consist of thin, concentric lamellae and are the storage sites

Fig. 14-14. A section of alveolar wall demonstrates the close association between the type I pneumonocyte and the endothelial cell. Note the fused basal laminae (TEM, × 4,000).

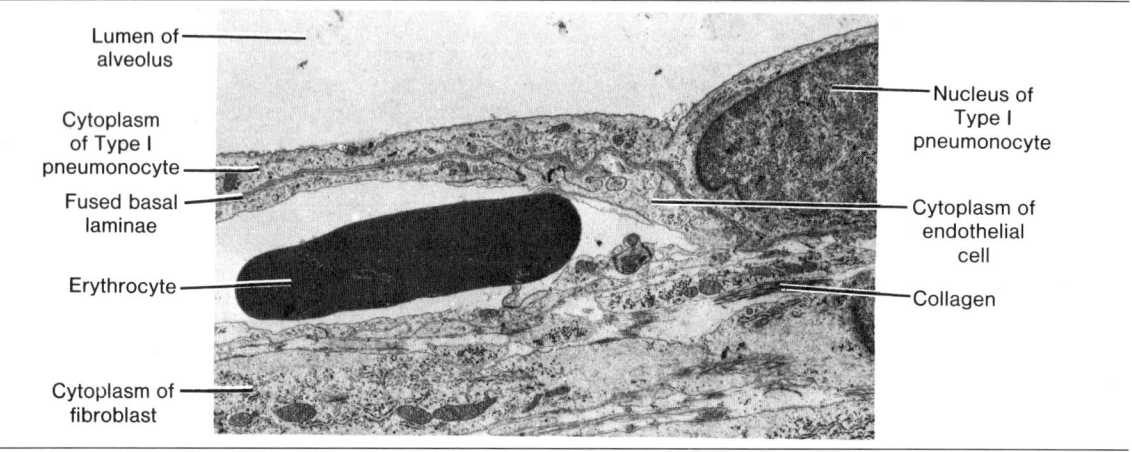

Fig. 14-15. The interior of an alveolus demonstrating types I and II pneumonocytes (SEM, × 2,000).

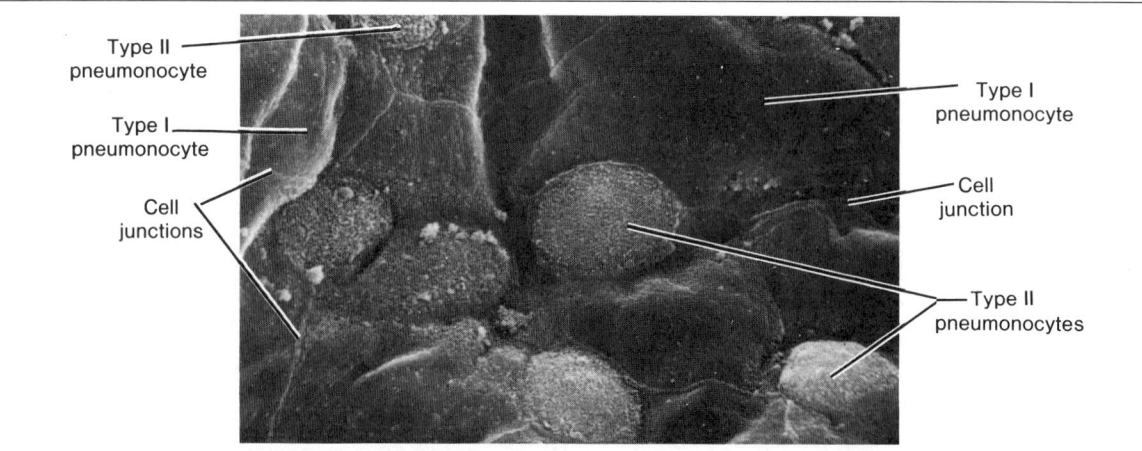

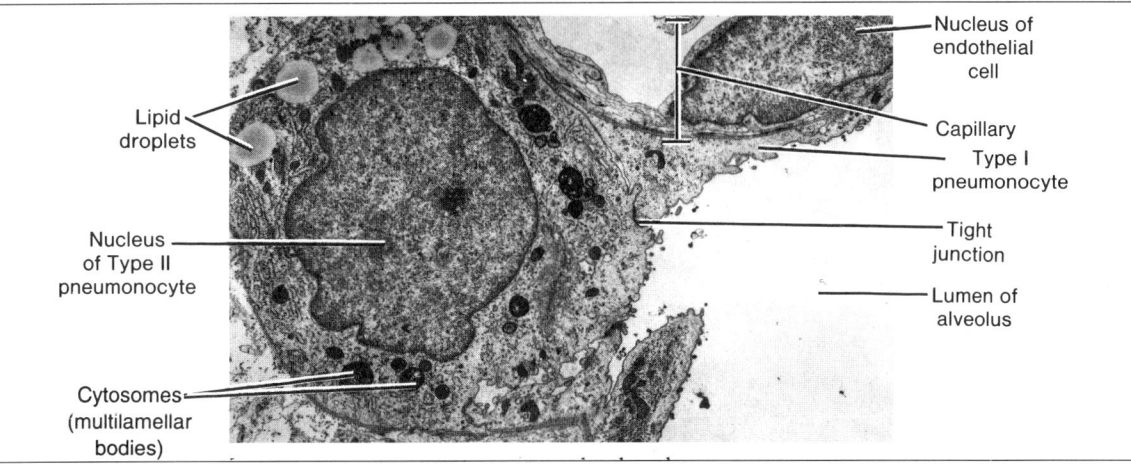

Fig. 14-16. A section through an alveolus shows type I and II pneumonocytes and an adjacent capillary (TEM, × 4,000).

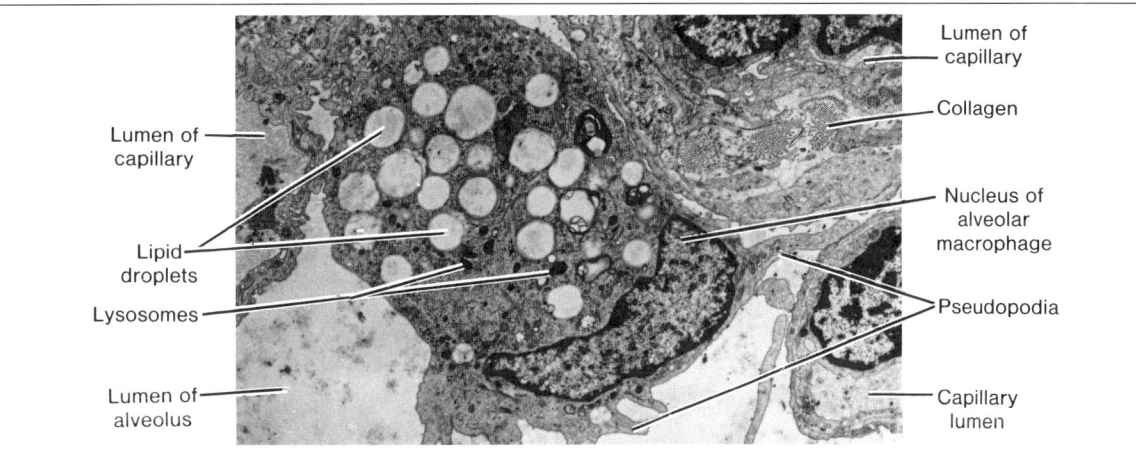

Fig. 14-17. An alveolar macrophage within the lumen of an alveolus (TEM, × 2,500).

of surfactant. Surfactant is about 90 percent phospholipid and consists mainly of phosphatidylcholine. It is synthesized by septal cells and released by exocytosis. The released surfactant forms a monolayer over the thin film of fluid that coats alveoli and thereby lowers surface tension at the liquid-air interphase. The lowering of surface tension aids in preventing the collapse of alveoli at the end of expiration.

Alveolar macrophages are present within the interalveolar septa and alveolar lumina (Fig. 14-17). Many contain particles of inhaled material and have been called "dust" cells. As with all macrophages, the alveolar macrophages originate from blood monocytes. Most of these cells ultimately are eliminated via the air passages and appear in the sputum; a few may migrate into lymphatics and escape the lungs by this route.

Pleura

The pleura are thin membranes made up of collagenous and elastic fibers, covered by a single layer of mesothelial cells (Fig. 14-18). The layer lining the wall of the thoracic cavity is the parietal pleura, which reflects from the thoracic wall onto the surface of the lungs, where it becomes the visceral pleura. The pleura secrete a small amount of fluid between the two layers to permit friction-free movement.

Fig. 14-18. A section of visceral pleura stained with orcein to selectively demonstrate the elastic membrane (LM, × 3,000).

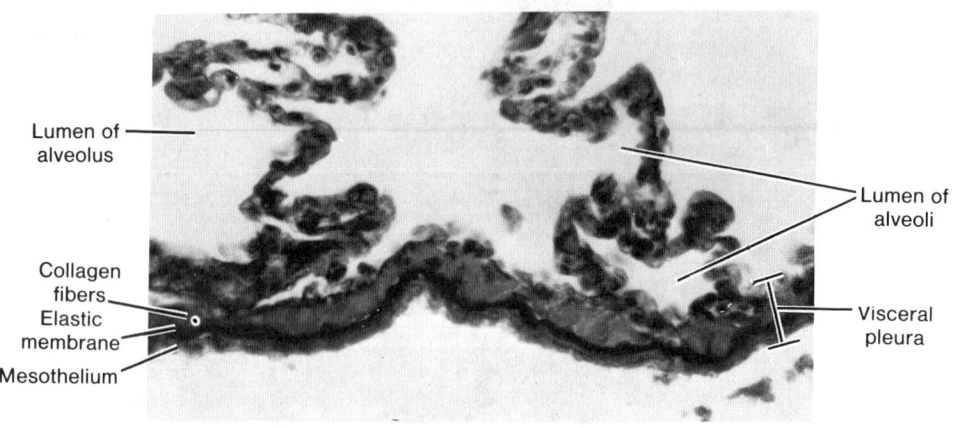

Review Table 14-1. Key Features of the Respiratory System

Division	Part	Epithelium	Support	Other Features
Extrapulmonary conducting	Nose Main portion	Pseudostratified ciliated columnar with goblet cells (respiratory lining epithelium)	Hyaline carilage and bone	Lamina propria forms a mucoperiosteum; contains seromucous glands and large venous sinuses
	Olfactory region	Pseudostratified columnar with 3 cells types: basal, supporting, olfactory	Bone of nasal concha	Lamina propria contains serous glands, venous sinuses, and nerves; forms a mucoperiosteum
	Pharynx Nasal	Respiratory lining epithelium	Skeletal muscle	Lamina propria contains mucous, serous, and mucoserous glands; abundant lymphatic tissue; pharyngeal tonsils
	Oral	Stratified squamous, nonkeratinized	Skeletal muscle	As above; palatine tonsils

Review Table 14-1 (continued)

Division	Part	Epithelium	Support	Other Features
	Laryngeal	Stratified squamous to middle of posterior surface of epiglottis, and vocal cords; respiratory lining epithelium elsewhere	Cartilage; dense regular connective tissue	Epiglottis, corniculate cartilage, tips of arytenoid are elastic cartilage, rest are hyaline; epithelium over epiglottis may contain taste buds
	Trachea and main bronchi	Respiratory lining epithelium	C-shaped hyaline cartilages; fibro-elastic connective tissue	Supporting connective tissue divided into lamina propria and submucosa by elastic tissue; seromucous glands in submucosa; scattered lymphatic tissue
Intrapulmonary conducting	Intrapulmonary bronchi	Respiratory lining epithelium in largest; ciliated columnar with goblet cells in smallest	Irregular, discontinuous plates of hyaline cartilage decreasing in smaller bronchi	Smooth muscle increasingly prominent as cartilage disappears; mucoserous glands and lymphatic tissue in submucosa
	Bronchioles	Ciliated columnar with goblet cells in largest to ciliated columnar with no goblet cells in smallest; Clara cells appear	Smooth muscle	Connective tissue decreased in amount; no glands or lymphatic tissue present
Transitional from conducting to respiratory	Respiratory bronchioles	Simple cuboidal, some ciliated, no goblet cells; Clara cells	Thin layer of fibroelastic connective tissue; few smooth muscle cells	Alveoli bud out from wall
	Alveolar ducts	Cuboidal, nonciliated between successive alveoli	Thin, delicate connective tissue	
Respiratory	Alveolar sacs	Thin squamous epithelium; type II pneumonocytes, phagocytes	Elastic and reticular fibers	Formed by clusters of alveoli
	Alveoli	Attenuated squamous cells, type II pneumonocytes, macrophages	Reticular and elastic fibers	Closely approximated to capillaries; blood-air barrier consists of cytoplasm of type I pneumonocyte, common basement membrane, cytoplasm of endothelial cells

Functional Review: Respiratory System

The stiff hairs in the vestibule of the nose help prevent particulate matter in the inspired air from reaching the rest of the respiratory passages. Mucus secreted by goblet cells in the respiratory lining and by glands in the lamina propria of the conducting portion provides a sticky blanket that traps particles and, along with the serous secretions from these glands, prevents drying of the mucosal surface. The secretions also humidify the air. The beat of the cilia moves the fluid blanket toward the mouth, in both the nose and lower respiratory tract. The large venous plexuses in the nasal cavities warm the incoming air. In several species, including humans, periodic engorgement of the venous plexus occurs, alternating on either side of the nasal cavity. This results in temporary obstruction to flow on the one side, allowing the mucosa to recover from the drying effects of air. In humans the periods of occlusion occur approximately hourly.

The olfactory epithelium serves for the sense of smell, the sensory component being the olfactory cells. The olfactory hairs borne by these cells form the excitable element of the sense organ. Glands in this area secrete a watery fluid that serves as a solvent for materials and flushes the olfactory epithelium to prevent continuous stimulation by a single odor.

Patency of the conducting portion, from the trachea to the smallest intrapulmonary bronchus, is maintained by cartilage in the walls. In the trachea and main bronchi, the fibroelastic tissue between cartilages gives pliability and longitudinal extensibility to these structures. This is essential for the accommodations these structures make for the incursions of the lung during breathing.

Beginning with the small bronchi, the epithelium progressively simplifies and becomes thinner. Ciliated cells persist farther along the respiratory tree than do mucus-secreting cells, and this may ensure that the sites of gaseous exchange do not become coated with a viscid, mucoid material that would impede or prevent the passage of gases. Significantly, cells that produce surfactant begin to appear at about the level at which goblet cells disappear. Surfactant reduces the surface tension of the fluid layer. During late phases of expiration, bronchioles close, so provision of a nonsticky fluid of decreased surface tension may be important for the reopening of bronchioles.

The small granule cells are part of a widespread diffuse endocrine system, with the cells apparently secreting various agents (perhaps hormones) that may have a role in regulating the secretion of mucus or surfactant.

The neuroepithelial bodies appear to be specifically adapted to respond to oxygen tension and adjust capillary flow to areas of high oxygen content.

The mucous secretions and cilia of the respiratory tree serve the same protective functions as in the nasal passages. Additionally, the mucosubstances have antiviral and antibacterial properties. The lymphoid tissue associated with the larger respiratory passages gives immunologic protection by producing immunoglobulin A (IgA) antibodies. Secretory IgA is assembled by mucosal cells of bronchial glands and mucosa; the secretory protein is produced by the serous cells of the submucosal glands.

Exchange of gases occurs where air and blood are closely approximated — in the alveoli. The barrier between air in the alveoli and blood in the capillaries of the alveolar septa consists only of a thin film of fluid, an extremely attenuated cytoplasm of the alveolar lining cell, the conjoined basal laminae plus the endothelium of the capillary. Surfactant-secreting cells also are present in the alveolar lining. Without surfactant, the alveolus tends to collapse because of the surface tension of the fluid that bathes the alveolar epithelium. Particulate matter that reaches the alveoli is removed by alveolar macrophages that, through their role as phagocytes, provide the main defense against microorganisms. During phagocytosis the cells release hydrogen peroxide and other peroxide radicals to destroy foreign organisms. Cigarette smoking triggers an inappropriate release of peroxide by alveolar macrophages, resulting in damage to normal respiratory and connective tissues. Phagocytosis by macrophages is greatly depressed in smokers, and the cells show twice the normal volume with a marked reduction in surface-to-volume ratios. The density of lysosomes also decreases.

Each lung is enclosed in a sac formed by the pleural membranes. The outer (parietal) layer lines the inner surface of the thoracic wall; the inner (visceral) layer covers the surface of the lungs. The space between the two layers is under negative pressure, and the two layers are separated only by a thin film of fluid. As the thoracic wall moves out to increase the size of the thoracic cage, the parietal pleura is taken along, passively. The combined effects of surface tension and negative pressure result in the visceral pleura being drawn outward also. Since the visceral pleura is firmly attached to the lungs, they are expanded, the air pressure within the lungs decreases, and air flows into the lungs. Expiration of air occurs as the result of elastic recoil of the lungs when the thoracic wall relaxes.

Review Questions and Answers

Questions

74. What structures constitute the extrapulmonary conducting portion of the respiratory tract? What types of epithelium are present and where do they occur?
75. Describe the changes in the air that occur in the nose and relate the changes to structures in the nasal cavity.
76. Where does gas exchange occur? What features are necessary for gas exchange and how are these met?

Answers

74. The extrapulmonary air passages consist of the nasal cavities; nasal, oral, and laryngeal pharynx; larynx; trachea; and main bronchi. The main type of epithelium present is pseudostratified ciliated columnar epithelium with goblet cells. Other types include keratinizing and nonkeratinizing stratified squamous in the vestibule of the nose, olfactory epithelium in the roof of the nasal cavities, and stratified squamous nonkeratinizing in the oro- and laryngopharynx, anterior surface of the epiglottis, margins of soft palate, and posterior wall of nasopharynx where this wall contacts the soft palate. The remainder of the passage is lined by the pseudostratified columnar type.

75. As the air enters the nostrils it is strained by stiff hairs that, with their associated sebaceous glands, help remove particulate matter. As it passes along the nasal passages, the air is warmed, moistened, and further depleted of particles. Warming results from the numerous venous sinuses in the lamina propria. Moisture is added from the fluid bathing the mucosal surface, contributed to by goblet cells and glands in the lamina propria. The fluid film also traps particulate materials, which are carried toward the mouth by the cilia. In the roof of the nasal cavities, any odorous materials stimulate the olfactory epithelium, resulting in the sense of smell.

76. Gas exchange occurs in the alveoli. Requirements for gas exchange are a close approximation of blood and air and a large surface area. These conditions are met in the alveolus by the presence, in the alveolar septum, of a rich capillary network that is closely apposed to the septal wall. The barrier between air in the alveolus and blood in the capillary is exceedingly thin and is composed of a molecular layer of fluid, an extremely attenuated alveolar epithelium, the endothelium of the blood capillary, and their fused basal laminae.

15 Digestive System

The digestive or alimentary tract consists of the oral cavity, pharynx, esophagus, stomach, small and large intestines, rectum, and anal canal. In addition to the digestive tract proper, there are associated glands that lie outside the tract but that are connected to it by ducts.

General Structure

KEY WORDS: mucous membrane, mucosa, lining epithelium, lamina propria, muscularis mucosae, submucosa, muscularis externa, adventitia, serosa

Throughout its extent, the alimentary tract shows several features that represent basic components. The innermost tissue layer of the digestive tract is called a **mucous membrane** or **mucosa** and consists of a **lining epithelium** and an underlying layer of fine, interlacing connective tissue fibers that form the **lamina propria.** A muscular component of the mucosa, the **muscularis mucosae,** is found only in the tubular portion of the tract. Immediately beneath the mucosa is a layer of connective tissue consisting of coarse, loosely woven collagen fibers and scattered elastic fibers. This layer is the **submucosa,** which houses larger vessels and nerves and provides the mucosa with considerable mobility. Where it is absent, the mucosa is immobile and firmly attached to underlying structures. In the tubular portion of the tract, a thick layer of smooth muscle, the **muscularis externa,**

forms an outer supporting wall that in turn is surrounded either by connective tissue, an **adventitia,** or a layer of connective tissue covered by mesothelium called a **serosa.**

Oral Cavity

The irregularly shaped oral cavity consists of the lips, cheeks, tongue, gingiva, teeth, and palate. The mucosa consists of a stratified squamous epithelium and a rather dense lamina propria. A submucosa may be present in some regions of the oral cavity.

Lip

KEY WORDS: vermillion border, papilla, labial gland

The external surface of the lip is covered by thin skin and contains sweat glands, hair follicles, and sebaceous glands. At the **vermillion border,** the lip is covered by a nonkeratinized, translucent, stratified squamous epithelium that lacks hair follicles and glands. Its lamina propria projects into the overlying epithelium to form numerous tall connective tissue **papillae** that contain abundant capillaries. The combination of tall vascular papillae and translucent epithelium accounts for the red hue of this part of the lip. The red portion is continuous with the oral mucosa internally and with thin skin externally (Fig. 15-1). Because

Fig. 15-1. A section of lip from an infant that illustrates the vermillion border (LM, × 100).

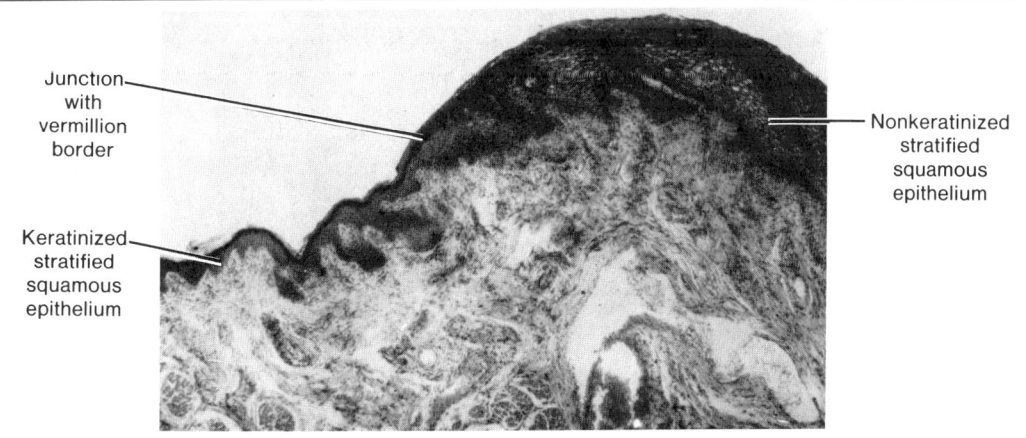

Junction with vermillion border

Nonkeratinized stratified squamous epithelium

Keratinized stratified squamous epithelium

glands are lacking, the red portion of the lips is kept moist by licking.

The inner surface is lined by nonkeratinized (wet) stratified squamous epithelium that overlies a compact lamina propria with numerous connective tissue papillae. Between the oral mucosa and the skin are the skeletal muscle fibers of the orbicularis oris muscle. Coarse fibers of connective tissue in the submucosa connect it to the underlying muscle. The ducts of numerous mixed mucoserous **labial glands** in the submucosa empty onto the internal surface of the lip and provide moisture and lubrication for this region. They can be identified as small bumps when the tongue is pressed firmly against the interior of the lip.

Cheek, Palate, and Gingiva

A similar mucous membrane lines the interior of the cheek, soft palate, floor of the mouth, and underside of the tongue. The mucosa of the cheek and soft palate, like that of the lips, is attached to underlying skeletal muscle fibers by coarse connective tissue elements of the submucosa, allowing considerable mobility to the oral mucosa in these regions. Mucoserous glands, adipose tissue, larger blood vessels, and nerves are present in the submucosa. The cheek and soft palate contain a core of skeletal muscle, the buccinator and palatine muscles, respectively. Externally, the cheek is covered by thin skin. Where the noncornified, stratified squamous epithelium of the oral cavity reflects

over the posterior aspect of the soft palate to enter the nasopharynx, it changes to a typical respiratory passage epithelium.

The hard palate, gingiva, and dorsum of the tongue, which are subjected to mechanical trauma during chewing, lack a submucosa and are covered by cornified, stratified squamous epithelium (Fig. 15-2). In these areas, the lamina propria is immobilized by its attachment to underlying structures. Where the mucosa attaches directly to bone, it forms a mucoperiosteum. The mucosa of the gingiva (gums) is attached directly to the periosteum of alveolar bone, and a similar attachment occurs in the midline raphe of the hard palate and where the hard palate joins the gingiva. A submucosa is present under the rest of the hard palate and contains abundant adipose tissue anteriorly and small mucous glands posteriorly.

Tongue

KEY WORDS: papilla, lingual tonsil, filiform papilla, fungiform papilla, taste buds, circumvallate papilla, serous glands (of von Ebner), foliate papilla

The mucous membrane of the tongue is firmly bound to a core of striated muscle, and there is no submucosa. The skeletal muscle forms interwoven fascicles that run in three planes, each perpendicular to the other. The undersurface of the tongue is smooth, whereas the anterior two-thirds of the upper surface shows numerous small protuberances

Fig. 15-2. The epithelial lining of a human hard palate (LM, × 250).

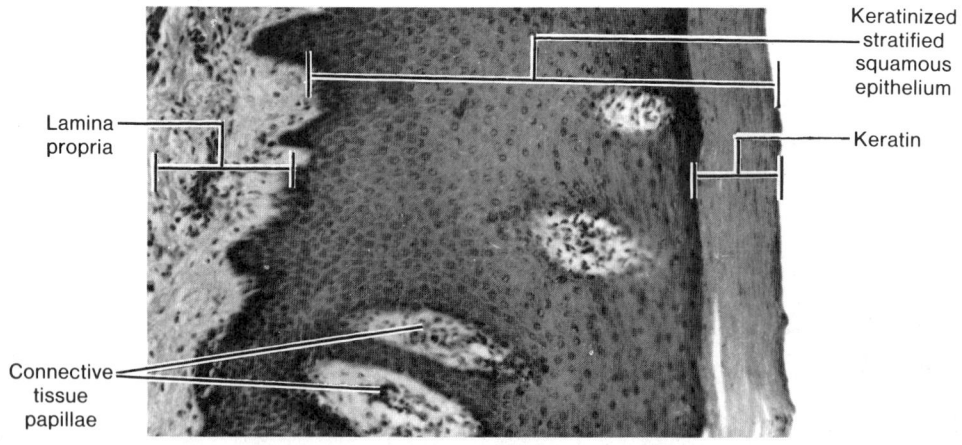

called **papillae.** The posterior one-third lacks papillae but shows mucosal ridges and moundlike elevations. The latter result from the underlying lymphatic tissue of the **lingual tonsils.**

Four types of papillae are associated with the anterior portion of the tongue. **Filiform papillae** are 2 to 3 mm long and contain a conical core of connective tissue that is continuous with the underlying lamina propria. The covering stratified squamous epithelium shows variable degrees of cornification. The mushroom-shaped **fungiform papillae** are scattered singly between the filiform type (Fig. 15-3). They are most numerous near the tip of the tongue, where they appear as small red dots. The color is due to the richness of capillaries in their connective tissue cores. **Taste buds** may be associated with this type of papilla. **Circumvallate papillae,** the largest of the papillae, are located along the V-shaped sulcus terminalis that divides the tongue into anterior and posterior parts. They number 10 to 14 in humans. The circumvallate papillae appear to have sunk into the mucosa of the tongue and are surrounded by a wall of lingual tissue, from which they are separated by a deep furrow (Fig. 15-4). Each contains a large core of connective tissue with numerous vessels and small nerves. Occasionally, small serous glands may be present. The epithelium over the lateral surfaces of these papillae contains numerous taste buds that may number 250 or more per papilla. **Serous (von Ebner's) glands** lie in the lamina propria and open into the bottom of the furrow. Their thin serous secretions continuously flush the furrow and provide an environment suitable for sensory reception by the taste buds. **Foliate papillae** are rudimentary in humans but in species such as the rabbit are well developed and contain many taste buds. They form oval bulges on the posterior, dorsolateral aspect of the tongue and consist of parallel ridges and furrows. Taste buds lie in the epithelium on the lateral surfaces of the ridges, and serous glands drain into the bottoms of the adjacent furrows.

Taste Buds

KEY WORDS: taste pore, supporting (sustentacular) cell, neuroepithelial cell, taste hair

Taste buds are present in fungiform, circumvallate, and foliate papillae and may be scattered in the epithelium of the soft palate, glossopharyngeal arches, pharynx, and epiglottis. They appear as lightly stained, oval structures that extend from the basement membrane almost to the surface of the lining epithelium (Fig. 15-5). A small **taste pore** allows for communication with the external environment. Taste buds consist of **supporting (sustentacular) cells,** between which are **neuroepithelial cells;** the cells surround the taste pore somewhat like the segments of an orange. Both types of cells have large microvilli called **taste hairs** that project into the taste pore and are embedded in an amorphous, polysaccharide material. Neuroepithelial cells are stimulated by substances that enter the taste pore.

Fig. 15-3. Dorsal surface of tongue demonstrating the surface features of filiform papillae (LM, × 100).

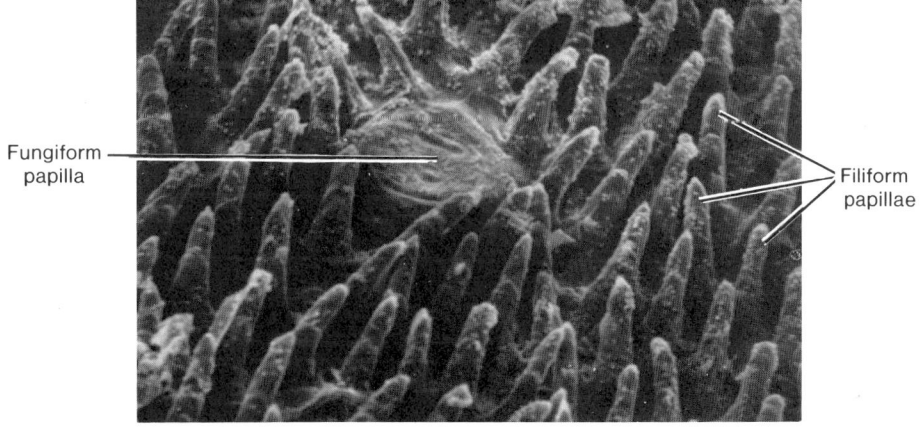

Fungiform papilla

Filiform papillae

Fig. 15-4. A circumvallate papilla. Note its position relative to the dorsal surface of the tongue and the position of von Ebner's glands (LM, × 10).

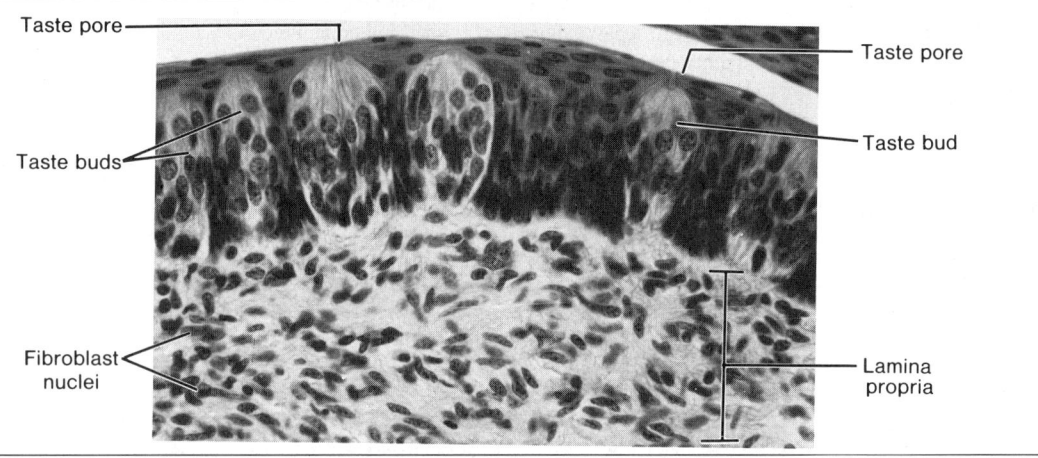

Fig. 15-5. Taste buds from the lateral surface of a circumvallate papilla of a human tongue (LM, × 250).

The sensation is transmitted to club-shaped nerve endings that pass between both cell types of the taste bud but apparently make synaptic contact only with neuroepithelial cells. Peripheral and basal cells, also associated with taste buds, are thought to represent undifferentiated progenitors of the supporting and neuroepithelial cells.

Four taste sensations are perceived: bitter, sweet, salty, and sour (or acid). The sensations may be detected regionally in the tongue — sweet and salty at the tip of the tongue, sour at the sides of the tongue, and bitter in the area of the circumvallate papillae — but structural differences in taste buds in these areas have not been seen.

Salivary Glands

KEY WORDS: major salivary glands, minor salivary glands, saliva, lysozyme, IgA, secretory piece

Salivary glands of the oral cavity can be divided into major and minor glands; all are compound tubuloalveolar glands. The **major salivary glands** are the parotid, submandibular, and sublingual glands and lie outside the oral cavity. Numerous smaller, intrinsic salivary glands are present in the oral cavity and make up the **minor salivary glands** (Table 15-1), which secrete continuously to moisten the mucosa of the oral cavity, vestibule, and lips.

Table 15-1. Minor Salivary Glands

Name	Location	Type of Gland
Labial glands	Upper and lower lips	Mixed
Buccal glands	Cheeks	Mixed
Anterior lingual glands	Tip of tongue	Mixed
von Ebner's glands	Near circumvallate papillae	Serous
Posterior lingual glands	Root of tongue	Mucous
Palatine glands	Palate	Mucous

Saliva is a mixture of secretions from both groups of salivary glands and contains water, ions, mucins, immunoglobulins, desquamated epithelial cells, and degenerating granulocytes and lymphocytes (salivary corpuscles). Amylase and maltase also are present to begin digestion of some carbohydrates. Saliva moistens the oral cavity, softens ingested materials, cleanses the oral cavity, and acts as a solvent to permit materials to be tasted. Some heavy metals are eliminated in the salivary secretions, and decreased secretion during dehydration helps initiate the sensation of thirst.

Saliva also is important in controlling the bacterial population in the mouth. In addition to producing **lysozyme,** the serous cells of the salivary glands participate in the production of immunoglobulin to suppress bacterial growth. Immunoglobulin A (**IgA**) synthesized and released by B-lymphocytes is taken up by the serous cells and complexed to a protein called the **secretory piece** before being released into the oral cavity. The secretory piece is synthesized by the epithelial cells and prevents the IgA from being broken down. The submandibular gland also secretes epidermal growth factor (EGF) and nerve growth factor (NGF) into the oral cavity.

Major Salivary Glands

The major salivary glands — parotid, submandibular, and sublingual — secrete only in response to nervous stimulation. The response is reflexive and can be stimulated by the smell, sight, or even thought of food.

Parotid Gland

KEY WORDS: compound tubuloalveolar, serous, myoepithelial cell, intercellular secretory canaliculi, intercalated duct, striated duct, basal striation, intralobular duct system, interlobular duct

The parotid is the largest of the major salivary glands and is located below and in front of the external ear. The main excretory duct passes through the cheek to open into the vestibule of the mouth opposite the upper second molar tooth. It is a **compound tubuloalveolar** gland and in humans is entirely **serous.** The adult human parotid contains abundant, scattered fat cells.

The parotid is enclosed in a fibrous capsule and subdivided into lobes and lobules by connective tissue septa (Fig. 15-6). A delicate stroma surrounds the secretory units

Fig. 15-6. Lobules comprising a parotid gland (LM, × 75).

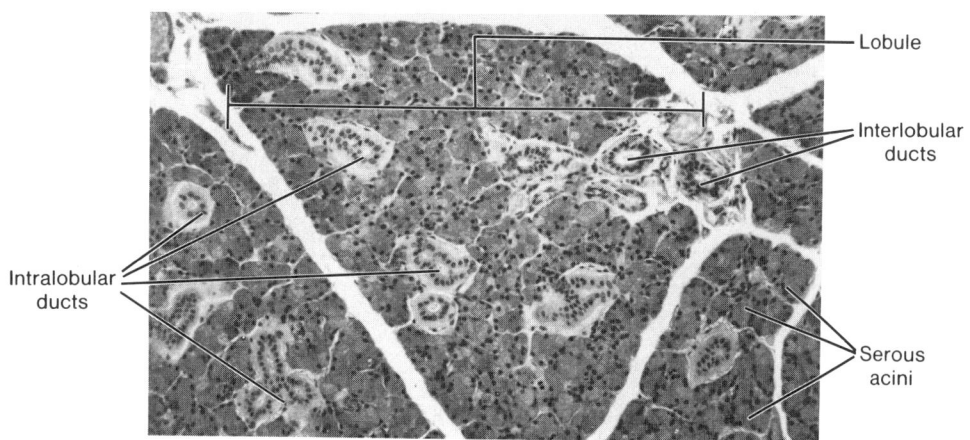

Labels: Lobule, Interlobular ducts, Intralobular ducts, Serous acini

and ducts and contains numerous blood capillaries and scattered nerve fibers. **Myoepithelial cells** lie between the limiting basement membrane and the bases of the secretory cells and may aid in expressing secretions out of the secretory units and into the duct system. Alveoli are composed of pyramidal-shaped, serous cells with basally placed, oval nuclei, basophilic cytoplasm, and discrete, apical secretory granules. Small channels, the **intercellular secretory canaliculi,** are found between serous cells and provide an additional route secretory products can take to reach the lumen.

The initial segment of the duct system is the **intercalated duct,** which is especially prominent in the parotid gland (Fig. 15-7). It is lined by a simple squamous or low cuboidal epithelium and may be associated with myoepithelial cells. Intercalated ducts are continuous with **striated ducts,** which are lined by columnar cells that show numerous **basal striations.** These elaborate infoldings of the basolateral plasmalemma increase the surface area, while the mitochondria associated with the infoldings provide energy at the base of the cell and play a significant role in fluid and ion transport. The intercalated and striated ducts constitute the duct system within the lobule and collectively form the **intralobular duct system.** The remaining ducts are found in the connective tissue between lobules and are referred to as **interlobular ducts.** They are continuous with the intralobular ducts and at first are lined by a simple columnar epithelium that becomes pseudostratified and then stratified as the diameter of the duct in-

creases. The surrounding connective tissue becomes more abundant as these ducts join to form the major excretory duct. The distal part of the duct is lined by stratified squamous epithelium that becomes continuous with the lining epithelium of the cheek.

Submandibular Gland

KEY WORDS: compound tubuloalveolar gland, mixed gland, serous demilune (crescent), intercellular canaliculi, striated duct

The submandibular gland, like the parotid, is a **compound tubuloalveolar gland** and has a fibrous capsule with septa, lobes, lobules, myoepithelial cells, and a prominent duct system. It is a **mixed gland,** with the majority of the secretory units being serous in humans. The mucous tubules present usually show **serous demilunes (crescents)** at their blind ends. Small channels, the **intercellular secretory canaliculi,** pass between the mucous cells and extend between the serous cells of the demilune. Thus the secretory product of the demilunes has direct access to the lumen of the mucous tubule. Myoepithelial cells lie between the secretory cells and the basement membrane and invest the secretory units as well as the initial portions of the ductal system. Generally, the duct system is similar to that of the parotid, but the **striated ducts** are longer and hence more conspicuous in sections of the submandibular gland (Fig. 15-8). The major excretory duct of the submandibular gland empties onto the floor of the oral cavity.

Fig. 15-7. A region of a parotid gland showing several serous acini, an intercalated duct, and an intralobular (striated) duct. Note the abundance of capillaries adjacent to the intralobular duct (LM, × 250).

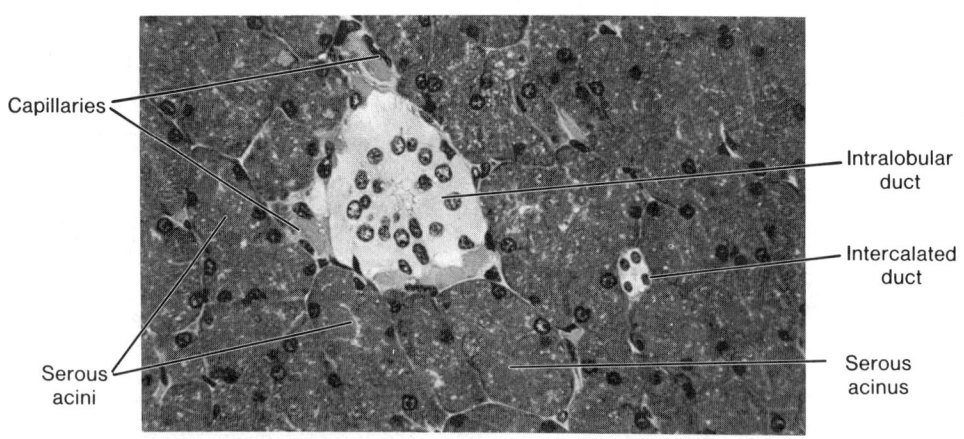

Capillaries

Intralobular duct

Intercalated duct

Serous acini

Serous acinus

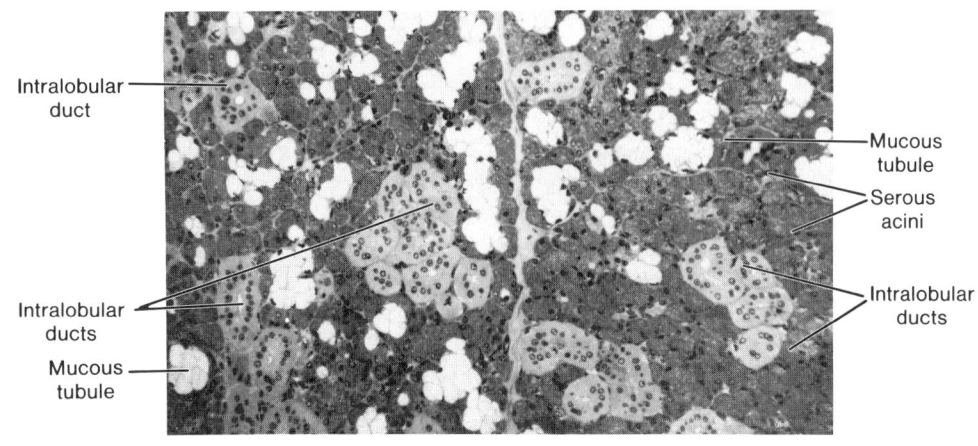

Fig. 15-8. A section of a submandibular gland showing abundant profiles of the intralobular ducts (LM, × 160).

Sublingual Glands

KEY WORDS: composite gland, mixed compound tubuloalveolar gland

The sublingual is a **composite gland,** formed by a number of individual glands of variable size. Each gland opens independently onto the floor of the mouth or into the excretory ducts of the submandibular gland. The capsule is not well defined, although septa are present. It is a **mixed compound tubuloalveolar gland** with most of the secretory units being mucous. Serous cells are, for the most part, arranged as demilunes. Myoepithelial cells are found in relation to the secretory units. Segments of the intralobular duct system are short and not as readily seen as they are in the other two major salivary glands.

Teeth

In humans, the teeth appear as two distinct sets: primary (deciduous) and permanent. Primary teeth erupt 6 to 8 months after birth and form a complete set of 20 by 2 years of age. They are shed between the sixth and thirteenth year and gradually are replaced by the permanent set of 32.

Structure

KEY WORDS: crown, root, neck, socket (alveolus), pulp cavity, apical foramen, enamel, dentin, cementum, pulp, periodontal membrane, gingiva

All teeth show the same histologic organization (Fig. 15-9). Each has a **crown** and a **root,** and the point where the two meet is called the **neck.** The root fits into a **socket** (or **alveolus**) of the mandible or maxilla. Each tooth contains a small **pulp cavity** that corresponds in shape to the external form of the tooth. The pulp cavity communicates with the alveolar cavity and periodontal membrane through the **apical foramen,** a small opening at the tip of

Fig. 15-9. Structure of a tooth.

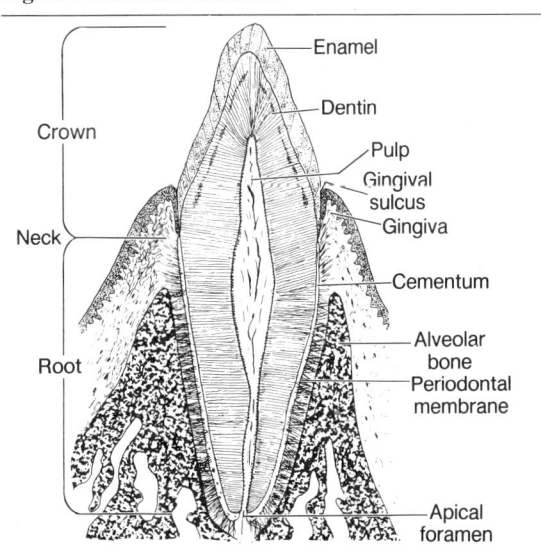

the root. The hard tissues of the tooth consist of **enamel, dentin,** and **cementum.** Soft tissues associated with the tooth are the **pulp, periodontal membrane,** and **gingiva.**

Dentin

KEY WORDS: dentinal tubule, dentinal (Tomes') fibers, odontoblast, Neumann's sheath

Dentin forms the bulk of the tooth and surrounds the pulp chamber (Fig. 15-10). It is harder than compact bone and consists of 80 percent inorganic material and 20 percent organic substance. Most of the organic material is collagen, while the inorganic component is in the form of hydroxyapatite crystals. Dentin has a radially striated appearance due to numerous minute canals called **dentinal tubules** that pursue an S-shaped course as they extend from the pulp chamber to the dentinoenamel junction. The dentinal tubules contain fine cytoplasmic processes, called **dentinal (or Tomes') fibers,** from a cell type known as the **odontoblast.** The thin layer of dentin that immediately surrounds each dentinal tubule shows greater refringence than the remaining dentin and is called **Neumann's sheath.** The cell body of the odontoblast lies in the pulp cavity adjacent to the dentin. The odontoblast lays down the organic matrix of dentin (predentin) and is active throughout life so that there is a progressive narrowing of the pulp cavity with

age. Predentin is rich in collagen, contains glycosaminoglycans, and is unmineralized. After its extracellular formation, predentin becomes mineralized. Some areas of dentin remain incompletely calcified and form the interglobular spaces. Dentin is sensitive to cold, pain, touch, and hydrogen ion concentration. Sensation is thought to be perceived by the process of odontoblast, which in turn transmits the sensory stimulation to adjacent nerves in the pulp chamber.

Enamel

KEY WORDS: enamel prism, interprismatic substance, ameloblast

Enamel covers the dentin of the crown and is the hardest substance of the body. It is acellular and consists primarily of calcium salts in the form of apatite crystals. Only 1 percent of the enamel substance is organic material. Enamel consists of thin rods called **enamel prisms** that are perpendicular to the surface of the dentin and extend from the dentinoenamel junction to the surface of the tooth. Each prism, 6 to 8 μm in diameter, follows a spiraling, irregular course to the surface of the tooth. A small amount of organic matrix surrounds each enamel prism and is called the prismatic rod sheath. The organic matrix of enamel consists primarily of proteins called *enamelins* that bind to

Fig. 15-10. The root of a molar tooth within the alveolar bone of a human jaw. Note the periodontal membrane and gingiva (LM, $\times$ 30).

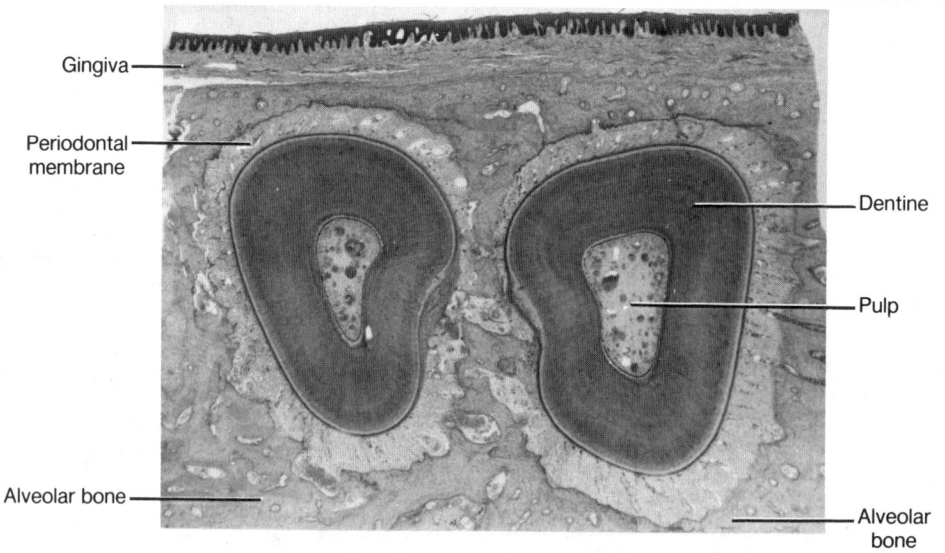

crystallites of the enamel prisms. Between the enamel prisms is the **interprismatic substance,** which also consists of apatite crystals in a small amount of organic matrix. Each enamel prism is the product of a single **ameloblast,** the enamel-producing cells that are lost during eruption of the tooth. Thus new enamel cannot be formed after the tooth has erupted.

Cementum

KEY WORDS: acellular cementum, cementocyte, cellular cementum, Sharpey's fibers

Cementum covers the dentin of the tooth root. Nearest the neck of the tooth, the cementum is thin and lacks cells, forming the **acellular cementum** (Fig. 15-11). The remainder, which covers the apex of the tooth, contains the **cementocytes** that lie in lacunae and are surrounded by a calcified matrix similar to that of bone. This type of cementum is referred to as **cellular cementum** (Fig. 15-12). The organic matrix of cementum increases with age. Coarse bundles of collagen fibers penetrate the cementum as **Sharpey's fibers,** which anchor the root of the tooth to the surrounding alveolar bone of the socket.

Fig. 15-11. Periodontal membrane uniting a human tooth root to an alveolar bone (LM, × 250).

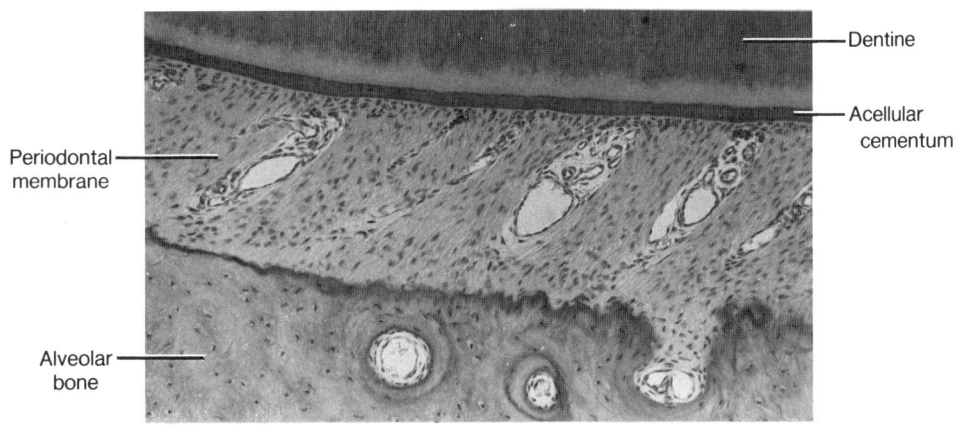

Fig. 15-12. A tooth root deep within a human jaw shows cellular cementum as well as Sharpey's fibers (compare with 15-11) (LM, × 250).

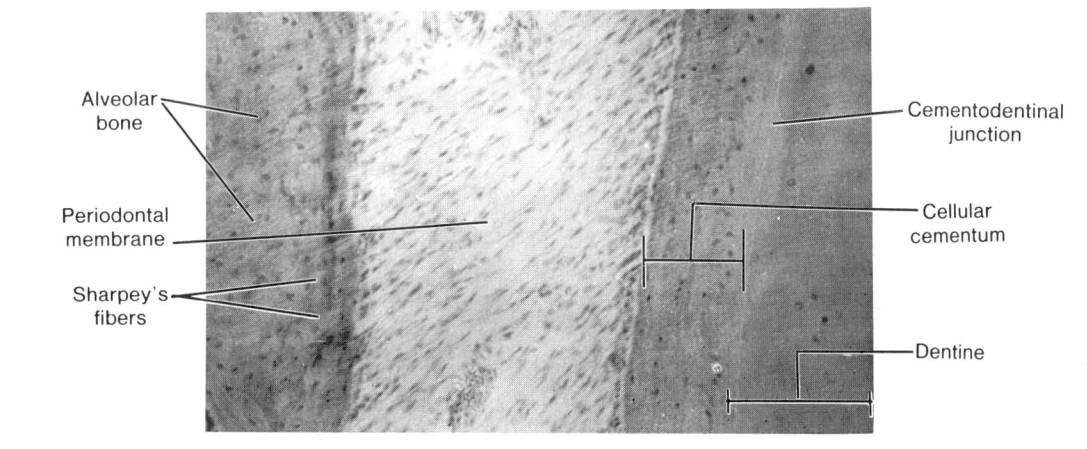

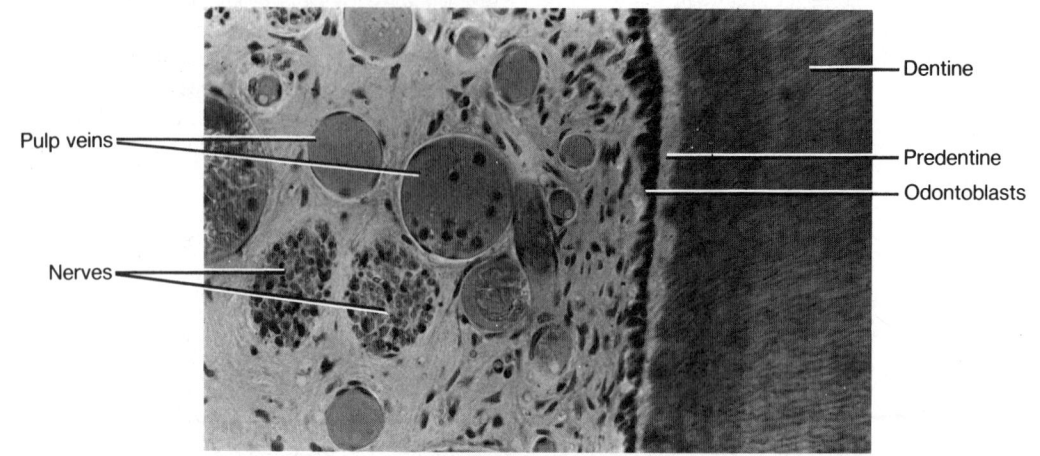

Fig. 15-13. The interior of a human tooth root demonstrates the elements of the pulp as well as the position of odontoblasts, predentin, and dentine (LM, × 250).

Pulp

KEY WORDS: connective tissue, apical foramen

Pulp is the **connective tissue** that fills the pulp cavity. It contains numerous thin, collagenous fibrils embedded in an abundant gelatinous ground substance. Stellate fibroblasts are the most prominent cells of the pulp, although mesenchymal cells, macrophages, and lymphocytes are found in limited numbers. The cell bodies of the odontoblasts also are found in the pulp, lining the perimeter of the pulp cavity immediately adjacent to the dentin (Fig. 15-13). Blood vessels, lymphatics, and nerves enter and exit the pulp cavity through the **apical foramen.**

Periodontal Membrane

KEY WORDS: bundle of collagen fibers, Sharpey's fibers

The periodontal membrane consists of thick **bundles of collagen fibers** that run between the cementum covering the root of the tooth and the surrounding alveolar bone (see Figs. 15-11 and 15-12). The fibers extend into the bone and cementum as **Sharpey's fibers.** The orientation of the fibers in the periodontal membrane varies at different levels in the alveolar socket. Although firmly attached to the surrounding alveolar bone, the fibers are not taut, and the tooth is able to move slightly in each direction. The periodontal membrane forms a suspensory ligament for the tooth. In addition to typical connective tissue cells, osteoblasts and osteoclasts may be found where the periodontal membrane enters the alveolar bone. The periodontal membrane has a rich vascular supply and is sensitive to pressure changes.

Gingiva (Gum)

KEY WORDS: gingival ligament, epithelial attachment cuff

The gingiva surrounds each tooth like a collar and is attached to the periosteum of the underlying alveolar bone (see Fig. 15-10). Near the tooth, collagenous fibers of the gingival lamina propria blend with the uppermost fibers of the periodontal membrane. Some collagenous fibers extend from the lamina propria into the cervical (upper) cementum and constitute the **gingival ligament,** which provides a firm attachment to the tooth. The keratinized stratified squamous epithelium of the gingiva also is attached to the surface of the tooth and at this point forms the **epithelial attachment cuff.** Attachment of the cuff to the tooth is maintained by a thickened basal lamina and hemidesmosomes that seal off the dentogingival junction.

Pharynx

The oral cavity continues posteriorly into the pharynx, which extends from the base of the skull to the level of the cricoid cartilage, where it is continuous with the esophagus. The digestive and respiratory systems merge briefly in the pharynx, which is subdivided into nasal, oral, and laryngeal parts. The pharyngeal walls basically consist of three strata: a mucosa, a muscularis, and an adventitia.

Oropharynx

KEY WORDS: nonkeratinized stratified squamous epithelium, lamina propria, muscularis, skeletal muscle, adventitia

The oropharynx is lined by a **nonkeratinized stratified squamous epithelium** and a dense, fibroelastic **lamina propria.** Immediately beneath the lamina propria is a well-developed layer of elastic fibers that is continuous with the muscularis mucosae of the esophagus. Proximally, the elastic layer blends with the connective tissue between the muscle bundles of the pharyngeal wall. A submucosa is present only where the pharynx is continuous with the esophagus and in the lateral walls of the nasopharynx. The **muscularis** of the pharyngeal wall consists of the **skeletal muscle** of the pharyngeal constrictor muscles, which, in turn, are covered by connective tissue of the **adventitia.**

Tubular Digestive Tract

The tubular digestive tract consists of the esophagus, stomach, small intestine, colon, and rectum. Each region of the digestive tube consists of the four basic layers, but the components of these layers vary according to the functions of the region.

Esophagus

The esophagus is a relatively straight muscular tube about 25 cm long. It is continuous above with the pharynx at the inferior border of the cricoid cartilage and below becomes continuous with the cardia of the stomach.

Mucosa

KEY WORDS: nonkeratinized stratified squamous epithelium, lamina propria, muscularis mucosae, smooth muscle

The mucosa of the esophagus consists of a lining epithelium, a lamina propria, and a muscularis mucosae. The epithelial lining is thick and in humans consists of **nonkeratinized stratified squamous epithelium** that is continuous with a similar epithelium lining the oropharynx (Figs. 15-14 and 15-15). Antigen-presenting (Langerhan's) cells have been identified in the esophageal epithelium. At the junction with the stomach, the epithelium shows an abrupt transition

Fig. 15-14. A region of mucosa from a human esophagus. Note the lymphatic nodule (LM, × 100).

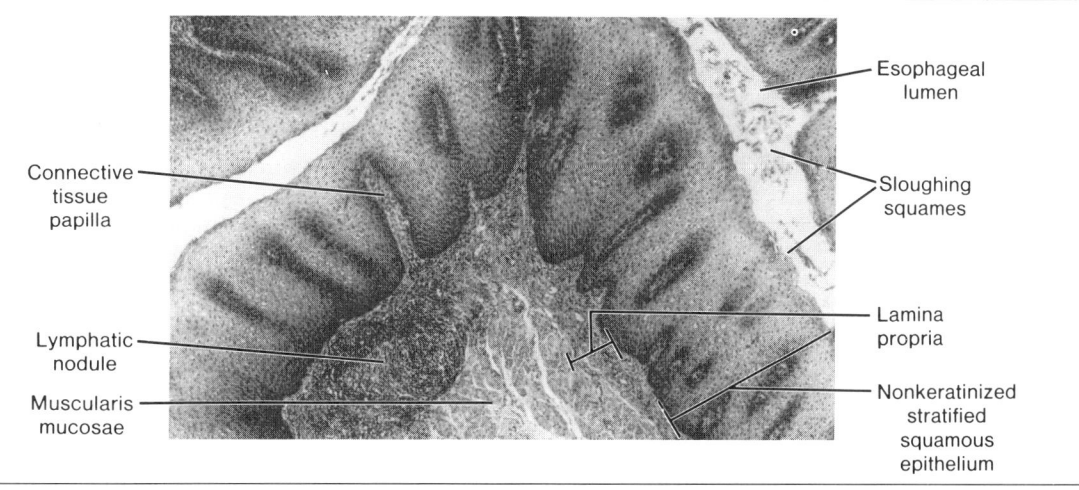

Connective tissue papilla

Lymphatic nodule

Muscularis mucosae

Esophageal lumen

Sloughing squames

Lamina propria

Nonkeratinized stratified squamous epithelium

Sloughing squames

Surface cells of stratified squamous epithelium

Fig. 15-15. Surface features of the nonkeratinized stratified squamous epithelium of a human esophagus (compare with 15-14) (SEM, × 200).

to a simple columnar epithelium. The **lamina propria** is a loose areolar connective tissue with diffuse and nodular lymphatic tissue scattered throughout its length. The **muscularis mucosae** consists of longitudinal **smooth muscle** cells in a fine, elastic network. Circularly arranged smooth muscle cells also may be present. The muscularis mucosae is continuous with the elastic layer of the pharynx at the level of the cricoid cartilage.

Submucosa

KEY WORDS: collagenous fiber, elastic fiber, longitudinal fold

The submucosa consists primarily of coarse **collagenous fibers** and **elastic fibers.** Within it are larger blood vessels, lymphatics, nerve fibers, occasional autonomic ganglia, and glands. Extensive **longitudinal folds** of the submucosa give the mucosa of the nondistended esophagus a characteristic pleated appearance. During swallowing the bolus of food smoothes out these folds and allows the lumen to increase in size temporarily to accommodate the material swallowed.

Muscularis Externa

KEY WORDS: skeletal muscle, mixed smooth and skeletal muscle, smooth muscle

The upper one-fourth of the muscularis externa consists of **skeletal muscle** only. From the lower border of the inferior

constrictor muscle of the pharynx, the muscle progressively becomes more regularly arranged into inner circular and outer longitudinal layers. In the second one-fourth of the human esophagus, **mixed skeletal and smooth muscle** is found, while in the distal one-half, only **smooth muscle** is present. Histochemical studies suggest that the skeletal muscle fibers of the esophageal wall are type IIa, fast-contracting, fatigue-resistant fibers. Cells of autonomic ganglia are present between the inner and outer layers of smooth muscle.

Zones of intraluminal high pressure occur at the ends of the esophagus and are referred to as the *upper esophageal sphincter* (UES) and the *lower esophageal sphincter* (LES). Both are physiologic rather than anatomic sphincters, and normally, the intraluminal pressure of the LES is sufficient to prevent reflux of gastric contents into the esophageal lumen. In some individuals, the LES fails, and the mucosa of the lower esophagus is chronically subjected to gastric acid and pepsin. A chemically induced chronic esophagitis results, and the esophageal epithelium undergoes metaplasia, changing from wet stratified squamous to a simple columnar epithelium similar to gastric lining epithelium. Such a condition (Barrett's esophagus) may result in severe ulceration of this area.

Adventitia

A layer of loosely arranged connective tissue, the adventitia, covers the outer surface of the esophagus and binds it to surrounding structures. Below the diaphragm a short segment of the esophagus is covered by a serosa.

Glands

KEY WORDS: esophageal cardiac gland, lamina propria, esophageal glands proper, submucosa

Two types of glands are present in the esophagus: esophageal cardiac glands and esophageal glands proper. Two groups of **esophageal cardiac glands** are present, one group located in the proximal esophagus near the pharynx, the other in the distal esophagus near the cardia of the stomach. Both are compound tubuloalveolar glands of mucous type and occur only in the **lamina propria.** The **esophageal glands proper** also are mucous compound tubuloalveolar glands but are confined to the **submucosa** throughout the remaining length of the esophagus (Fig. 15-16). Lysozyme and prosinogen have been identified in some but not all of the secretory cells. Myoepithelial cells are associated with the secretory units of human esophageal glands.

Stomach

The stomach forms an expanded portion of the tubular digestive tract between the esophagus and small intestine. In humans, the stomach is divided into the cardia, a short region where the esophagus joins the stomach; the fundus, a dome-shaped elevation of the stomach wall above the esophagogastric junction; the corpus, the large central part; and the pylorus, a narrow region just above the gastrointestinal junction. The digestive functions of the stomach involve mechanical and chemical breakdown of ingested materials.

Mucosa

KEY WORDS: ruga, gastric lining epithelium, simple columnar, gastric pit (foveola)

The mucosa of the empty stomach has numerous folds or ridges called **rugae** that mostly disappear when the stomach is filled or stretched. The **gastric lining epithelium** is **simple columnar** and begins abruptly at the junction with the stratified squamous epithelium of the esophagus (Fig. 15-17) and ends just as abruptly at its junction with intestinal epithelium. The lining epithelium has a similar structure throughout the stomach and consists of secretory columnar cells (Fig. 15-18) that collectively form a secretory sheet. The neutral mucin produced by the surface epithelium is secreted continuously and forms a mucous film that helps protect the mucosa from the acid pepsin in the gastric lumen and lubricates the surface.

The apices of the columnar cells are held in close apposition by tight junctions, and the lateral cell membranes show numerous desmosomes. Scattered, stubby microvilli are present on the apical surfaces, and numerous discrete granules fill the apical cytoplasm. Granular endoplasmic reticulum is present in the basal cytoplasm, and well-developed Golgi complexes occupy the supranuclear region.

The gastric mucosa contains about 3 million minute tubular infoldings of the surface epithelium that form the **gastric pits** or **foveolae** (Fig. 15-19), which are lined by the same simple columnar epithelium that covers the surface. The epithelial cells on the surface are replaced every 4 to 5 days. New cells are derived from a small population

Fig. 15-16. Distal esophagus containing esophageal glands (LM, × 40).

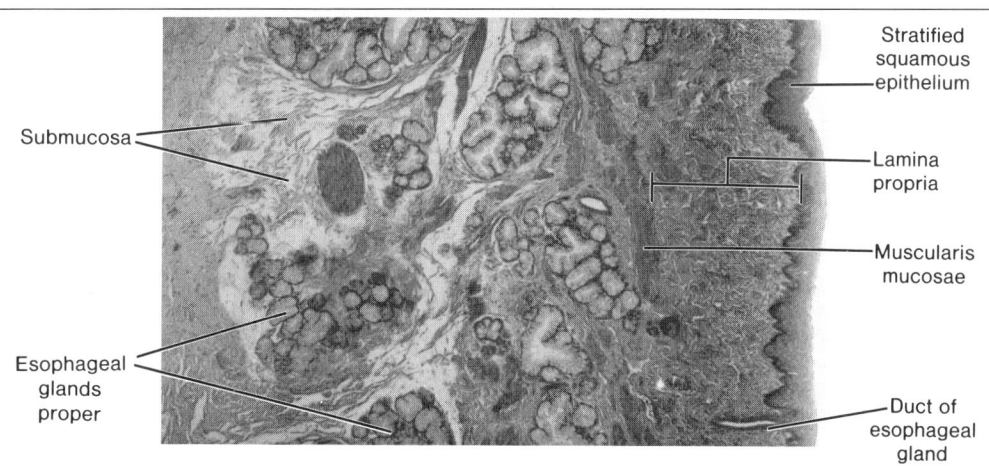

Submucosa

Esophageal glands proper

Stratified squamous epithelium

Lamina propria

Muscularis mucosae

Duct of esophageal gland

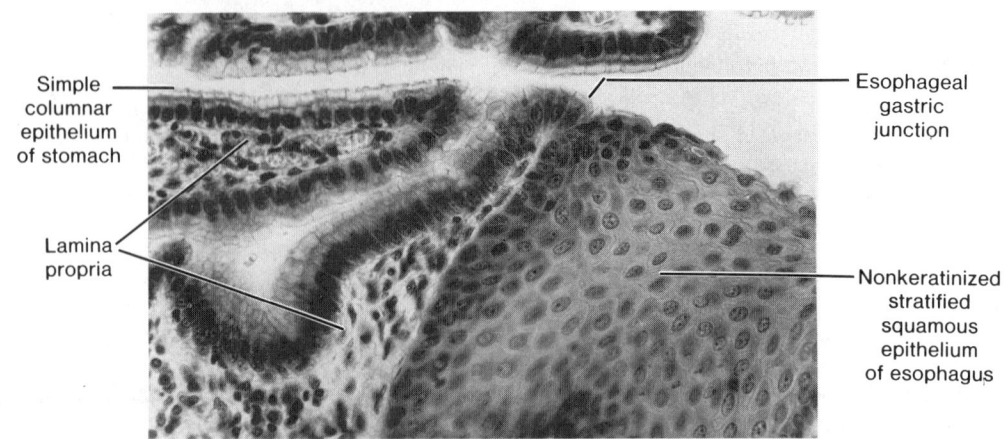

Simple columnar epithelium of stomach

Lamina propria

Esophageal gastric junction

Nonkeratinized stratified squamous epithelium of esophagus

Fig. 15-17. The abrupt transition of simple columnar to nonkeratinized stratified squamous epithelium at the esophageal gastric junction (LM, × 250).

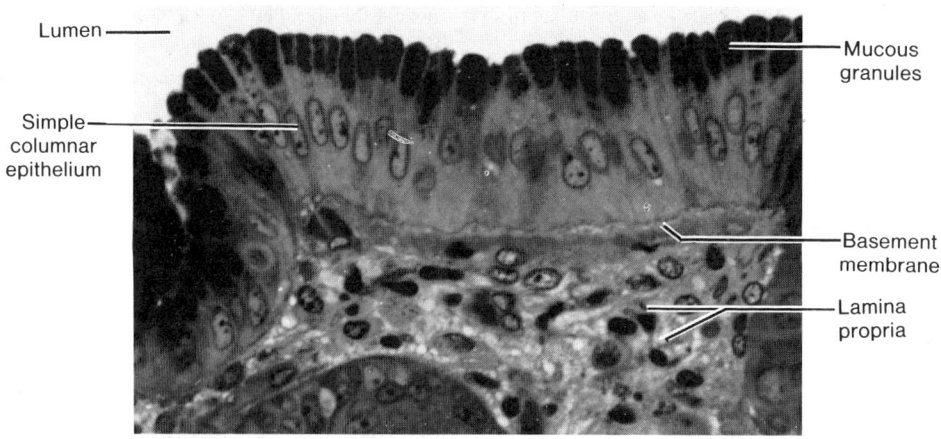

Lumen

Simple columnar epithelium

Mucous granules

Basement membrane

Lamina propria

Fig. 15-18. Surface lining epithelium of a human stomach (LM, × 500).

of relatively undifferentiated cells located in the bottoms of the gastric pits. Cells from this region gradually migrate upward along the gastric pit to replace the cells of the surface epithelium.

Three types of glands usually occur in the gastric mucosa: cardiac, gastric, and pyloric.

Cardiac Glands

KEY WORDS: Simple branched tubular gland, mucous cell, endocrine cell

The cardiac glands begin immediately around the esophageal orifice and extend about 3.4 cm along the proximal stomach. They attain their greatest depth near the esophageal-gastric junction. Here, the tubular glands branch freely and appear to be aggregated into lobule-like complexes by the surrounding connective tissue of the lamina propria. Near the junction with the fundus, the cardiac glands show less branching, the distinct grouping disappears, and the thickness of the glandular area decreases. For the most part, cardiac glands are **simple branched tubular glands** that open into the overlying gastric pits (Fig. 15-20). The depth

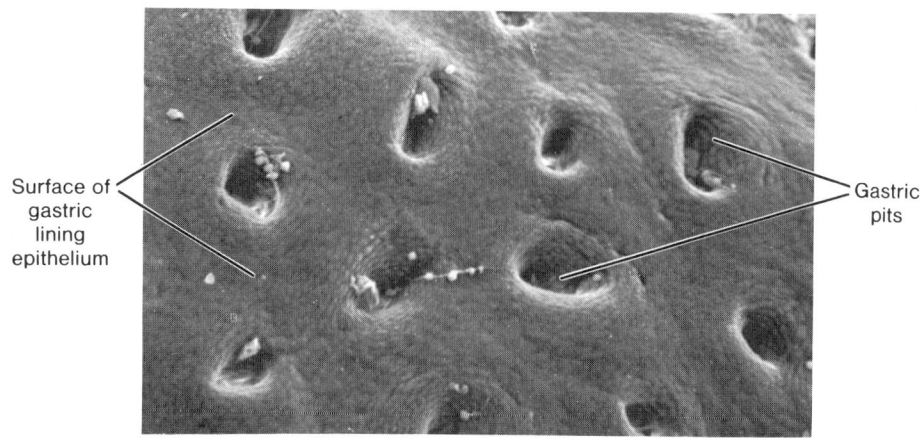

Fig. 15-19. The surface of a human fundic stomach exhibiting numerous gastric pits (SEM, × 200).

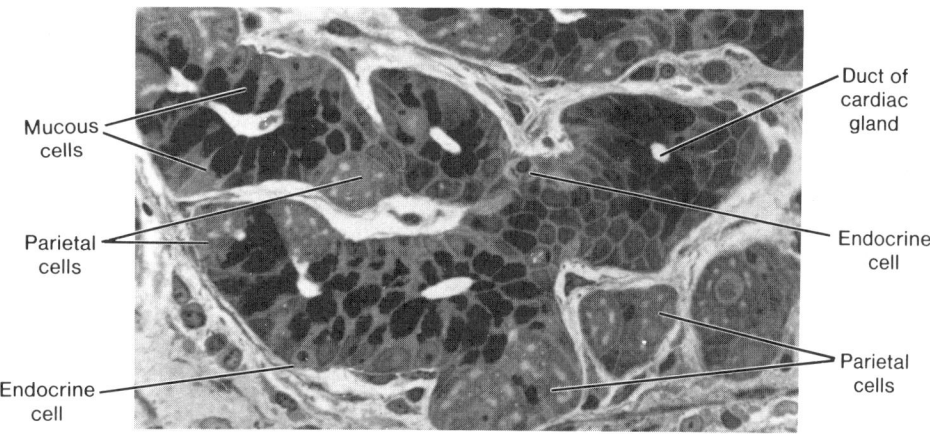

Fig. 15-20. Cardiac gland from a human stomach (LM, × 400).

of the gastric pit and the length of the cardiac gland are approximately equal. The secretory units consist mainly of **mucous cells,** but occasional parietal cells are present and appear to be identical to those of gastric glands. A number of **endocrine cells** of undetermined nature also are present.

Gastric (Oxyntic) Glands

KEY WORDS: simple branched tubular gland, mucous neck cell, parietal cell, intracellular canaliculus, tubulovesicular system, hydrochloric acid, gastric intrinsic factor, chief (zymogen) cell, pepsinogen, pepsin, endocrine cell

The gastric (oxyntic) glands are the most abundant glands in the stomach (about 15 million) and are found along the entire corpus of the stomach. They are mainly **simple branched tubular glands** in humans (Fig. 15-21). The glands run perpendicular to the surface of the mucosa, and one or several glands may open into the bottom of each gastric pit. The gastric glands are about four times as long as the pits into which they open. Each gland contains mucous neck cells, parietal cells, chief (zymogen) cells, and endocrine cells (Fig. 15-22).

Mucous neck cells occur primarily in the upper regions of the gastric glands, where the glands open into the

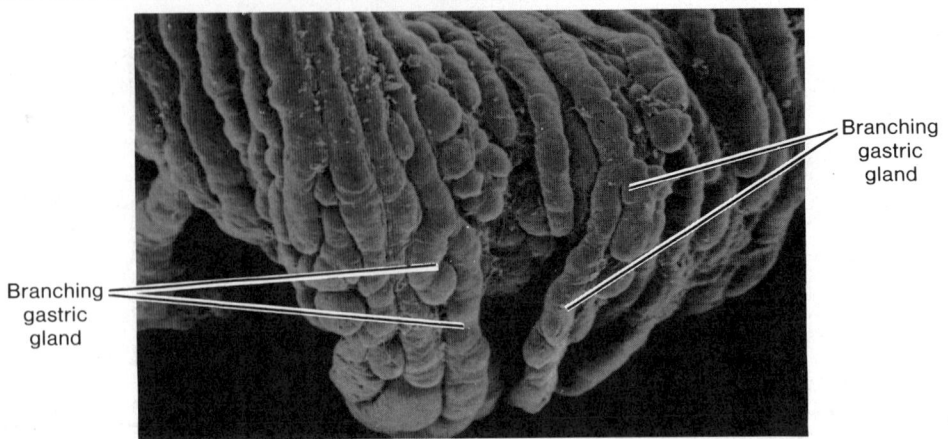

Fig. 15-21. A dissected preparation illustrating the tubular nature of the gastric (oxyntic) glands from a human stomach. Note that several of the glands branch (SEM, × 250).

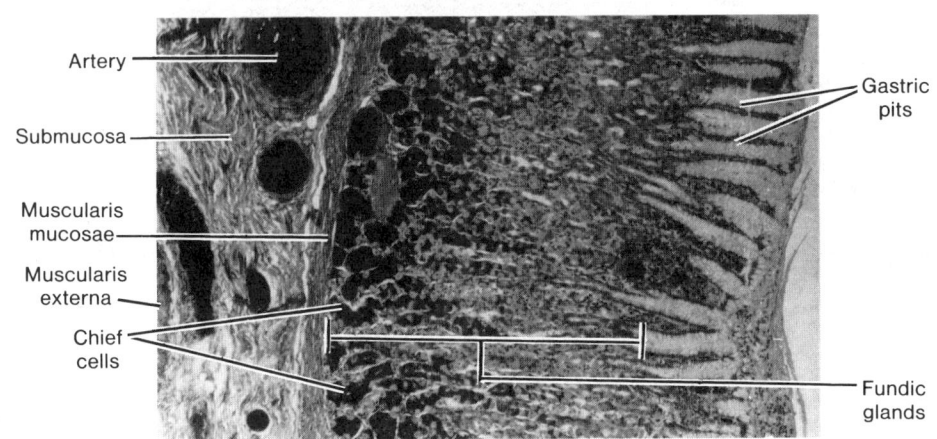

Fig. 15-22. The appearance of fundic (oxyntic) glands from a human stomach as they appear in section (compare with 15-21). Note the position of the chief cells at the bases of the oxyntic glands (LM, × 50).

gastric pits. The cells often appear to be sandwiched between other cell types, and their irregular shape is characteristic. Some have broad apices and narrow bases; others show narrow apices and wide bases. The nucleus is confined to the base of the cell, surrounded by basophilic cytoplasm. Secretory granules fill the apical cytoplasm. Unlike the mucous cells that line the gastric surface, mucous neck cells produce an acidic mucin.

Although most numerous in the central area of the gastric glands, **parietal cells** also are scattered among mucous neck cells in the upper parts of the gland (Fig. 15-23). They are large, spherical cells whose bases often appear to bulge from the outer margin of the glands into the lamina propria. In routine sections, parietal cells are distinguished by their acidophilic cytoplasm. Each cell contains a large, centrally placed nucleus and numerous mitochondria and

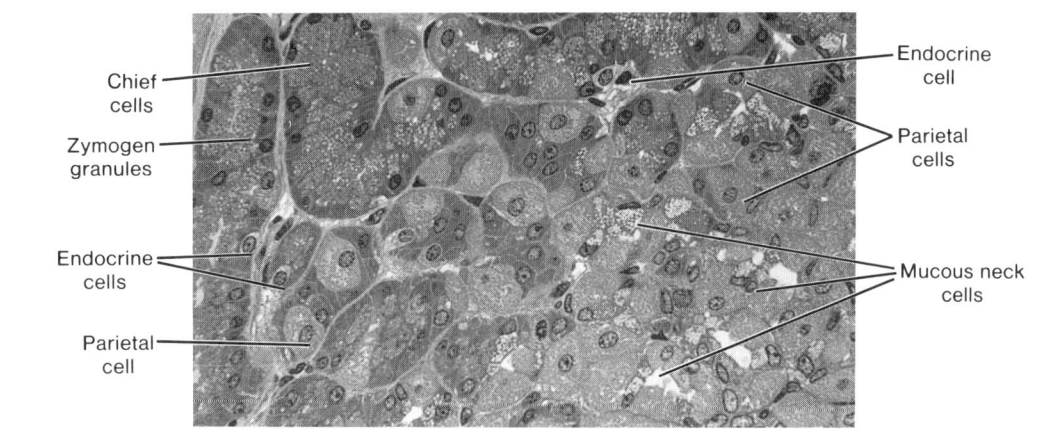

Fig. 15-23. A region of gastric mucosa showing chief, parietal, endocrine, and mucous neck cells, which together make up the oxyntic glands (LM, × 250).

is characterized by **intracellular canaliculi.** Mitochondria make up nearly 40 percent of the parietal cell volume and account for the eosinophilia in hematoxylin and eosin–stained preparations. The canaliculi are invaginations of the plasmalemma that form channels near the nucleus and open at the apex of the cell into the lumen of the gland. Numerous microvilli project into the lumina of the canaliculi, their number and length varying according to the secretory activity of the cell. Immediately adjacent to each canaliculus, a series of smooth cytoplasmic membranes

forms small tubules and vesicles that make up the **tubulovesicular system** (Fig. 15-24). The tubulovesicular membranes are rich in a unique hydrogen, potassium ATPase, which acts as a proton pump in the stimulated parietal cell. The membranes also contain potassium and chloride ion channel proteins. Secretion of **hydrochloric acid** occurs along the cell membrane that lines the canaliculi. During stimulation of parietal cells, the tubulovesicular membranes diminish, canalicular microvilli become more abundant, and canaliculi elongate; during acid

Fig. 15-24. The ultrastructural features of a human parietal cell (TEM, × 10,000).

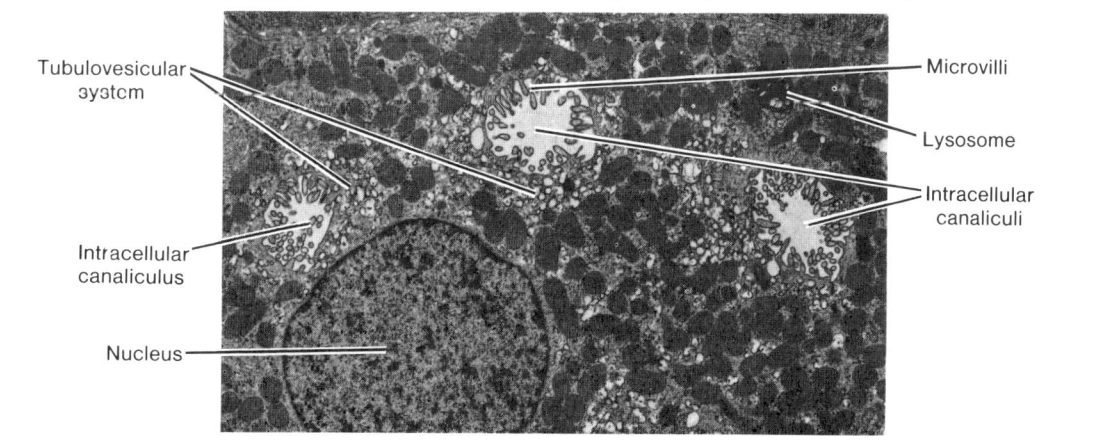

inhibition, the reverse occurs. Following parietal cell stimulation, actin microfilaments polymerize, and their interaction with myosin promotes the movement and fusion of tubulovesicular with canalicular membranes, resulting in the increased canalicular membrane area containing hydrogen, potassium-ATPase and ion channel proteins. Parietal cells contain abundant carbonic anhydrase, which catalyzes the joining of carbon dioxide to water, forming carbonic acid (H_2CO_3). Once formed, H_2CO_3 immediately dissociates into hydrogen ions (H^+) and bicarbonate ions (HCO_3^-). Hydrogen ions are actively pumped into the lumen of the canaliculus, and bicarbonate ions pass back into the bloodstream. Ion channels promote the movement of potassium and chloride ions into the canalicular lumen. The hydrogen, potassium-ATPase moves the potassium ion back into the parietal cell in exchange for hydrogen ion. The net result of this activity is the formation of hydrochloric acid in the canalicular lumen. Water enters the canaliculi osmotically driven by the movement of ions. The increase in canalicular membrane surface area during parietal cell stimulation makes these events possible.

In addition to hydrochloric acid, parietal cells of the human stomach secrete a glycoprotein, **gastric intrinsic factor,** which complexes with dietary vitamin B_{12} in the gastric lumen; the complex is absorbed in the distal small intestine. A deficiency of intrinsic factor results in decreased absorption of vitamin B_{12}, an essential vitamin for the maturation and production of red blood cells. Lack of the vitamin results in pernicious anemia and is a condition often associated with atrophic gastritis.

Chief (zymogen) cells are present mainly in the basal half of the gastric glands and in routine sections are distinguished by their basophilia. The cells contain abundant granular endoplasmic reticulum in the basal cytoplasm, well-developed Golgi complexes in the supranuclear cytoplasm, and apical zymogen granules, features that characterize cells involved in protein (enzyme) secretion (Fig. 15-25).

Chief cells secrete **pepsinogen,** a precursor of the enzyme **pepsin** that reaches its optimal activity at pH 2.0. The acid environment in the gastric lumen is created by the parietal cells. Pepsin is important in the gastric digestion of protein, hydrolyzing proteins into peptides.

Also scattered within the bases of the gastric glands are **endocrine cells,** peptide-producing cells that contain specific granules enclosed by smooth membranes. The polarization of these cells suggests that they secrete into the bloodstream or intercellular space rather than into the lumina of gastric glands. The gastrointestinal hormones somatostatin, glucagon, and pancreatic peptide and the amine 5-hydroxytryptamine (5-HT, serotonin) have been identified in the endocrine cells of the gastric glands.

Kinetics of the Gastric (Oxyntic) Glands.   Pluripotent stem cells located in the region where a gastric gland joins the bottom of a gastric pit (the isthmus) divide to maintain themselves as well as give rise to several committed cell types: prepit, pre-mucous neck, preparietal, and preen-

Fig. 15-25. The ultrastructural features of a human chief cell (TEM, × 2,000).

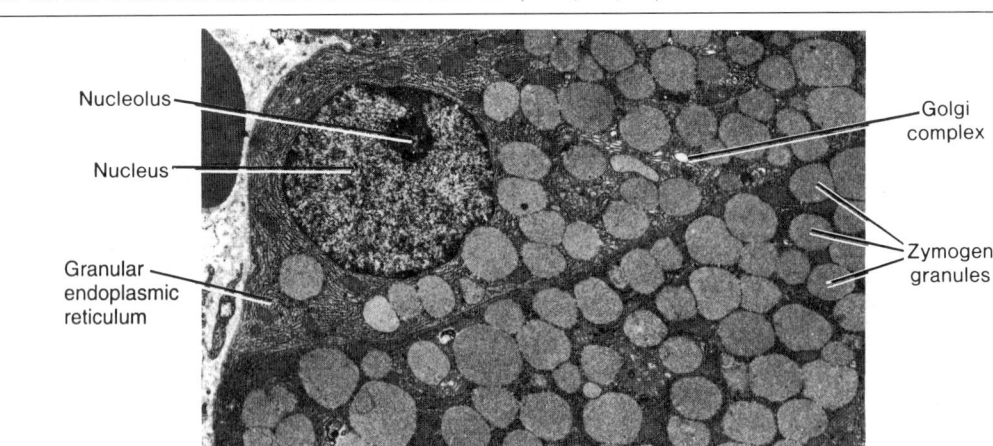

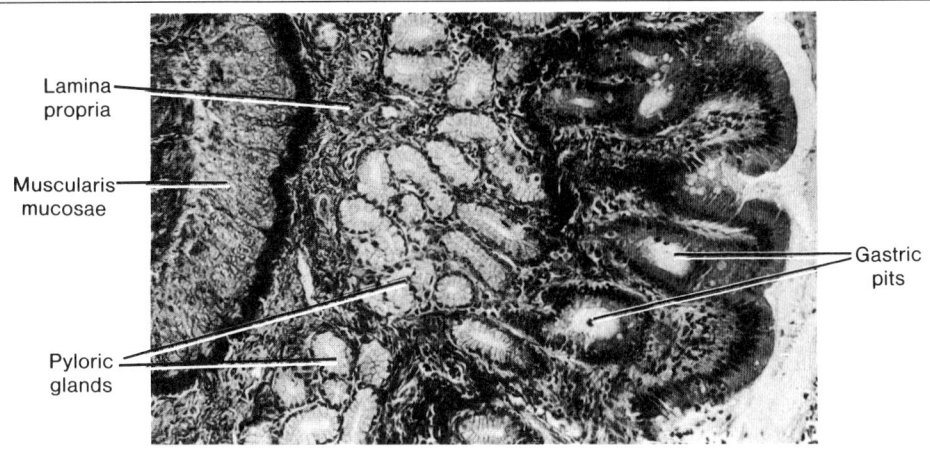

Fig. 15-26. Pyloric glands from a segment of human pyloric stomach. Note that these glands are confined to the mucosa (LM, × 250).

docrine. Chief cells are derived from mucous neck cells through a prezymogenic stage.

All pit cells migrate in the direction of the gastric lumen to become gastric lining epithelial cells and are eventually exfoliated into the lumen in about 5 days. Mucous neck cells, however, migrate inward toward the base of the glands and in about 2 weeks become prechief (zymogenic) cells. As fully differentiated chief cells form, they migrate to occupy the bottom of the gastric glands and remain active for up to 6 months. Thereafter, they are either lost by necrosis and shed into the lumen or, if apoptosis occurs, phagocytosed by adjacent chief cells or macrophages entering the area. Endocrine cells also arise at the isthmus and migrate toward the base of the glands where they are abundant. Their turnover time is about 3 months. In contrast, differentiating parietal cells migrate in both directions with about equal numbers migrating toward either the gastric lumen or the bottoms of the gastric glands. They have a turnover rate of about 60 days and, if they become necrotic, are lost by extrusion in the lumen of the gland. If the parietal cells become apoptotic, they are phagocytosed by neighboring cells or macrophages.

Pyloric Glands

KEY WORDS: simple or branched tubular gland, mucous cell, endocrine cell, G cell, gastrin

The pyloric glands are confined to the distal 4 or 5 cm of the stomach. They are **simple or branched tubular glands** composed of **mucous cells** similar in appearance to mucous neck cells of the fundic region (Fig. 15-26). An occasional parietal cell may be present. The pyloric glands empty into pits that are about the same length as the glands. Numerous **endocrine cells,** mainly **G cells,** are present. These unicellular endocrine glands produce **gastrin,** a peptide hormone that stimulates acid secretion by the parietal cells in the remainder of the stomach. Endocrine cells that produce somatostatin and 5-HT are present in pyloric glands also. Thus, in addition to the exocrine function of mucus production, pyloric glands have important endocrine functions as well.

Lamina Propria

The lamina propria of the stomach is obscured by the close approximation of the glands in the gastric mucosa. It consists of a delicate network of collagenous and reticular fibers that surrounds and extends between the glands and gastric pits. The lamina propria contains numerous capillaries. Numerous lymphocytes, some plasma cells, eosinophils, and mast cells are found within the lamina propria. Lymphatic nodules also may occur.

Muscularis Mucosae

The muscularis mucosae consists of an inner circular, an outer longitudinal and, in some places, an additional outer circular layer of smooth muscle. Slips of smooth muscle extending into the lamina propria form a loose network around individual glands and extend to the gastric lining epithelium. The muscularis mucosae gives additional mobility to the gastric mucosa.

Submucosa

The submucosa consists of a coarse connective tissue rich in mast cells, eosinophils, and lymphatic cells. This layer contains lymph vessels and the larger blood vessels. Parent vessels pierce the stomach wall at the lesser and greater curvatures and supply smaller branches that run about the circumference of the stomach in the submucosa. Smaller branches of the submucosal vessels enter the mucosa to provide its vascular supply.

Muscularis Externa

The muscularis externa may consist of three layers of smooth muscle: outer longitudinal, middle circular, and sometimes an inner oblique in the fundic region. In the pylorus, the middle circular layer thickens to form the pyloric sphincter, which is thought to aid in controlling emptying of the stomach. Strong contractions of the muscle wall of the stomach create a churning action that contributes mechanically to the breakdown of ingested material and gastric emptying.

Serosa

A thin layer of loose connective tissue covers the muscularis externa and is covered on its outer aspect by the mesothelial lining of the peritoneal cavity, forming the serosa.

Small Intestine

The small intestine extends between the stomach and colon and is divided into the duodenum, jejunum, and ileum. Although there are minor microscopic differences among these subdivisions, all have the same basic organization as the rest of the digestive tube — mucosa, submucosa, muscularis externa, and serosa. The transition from one segment to another is gradual. The small intestine moves chyme from the stomach to the colon and completes the digestive processes by adding enzymes secreted by the intestinal mucosa and accessory glands (liver and pancreas). Its primary function, however, is absorption.

Specializations for Absorption

KEY WORDS: **plicae circulares, villus, microvillus, striated border, terminal web**

Three specializations — plicae circulares, intestinal villi, and microvilli — markedly increase the surface area of the intestinal mucosa to enhance the absorptive process.

Plicae circulares are large, permanent folds that consist of the intestinal mucosa and a core of submucosa. These shelflike folds spiral about the lumen of the intestine for one-half to two-thirds of its circumference and may branch. Plicae begin in the upper duodenum, reach their maximum development in the proximal jejunum, and thereafter diminish in size and disappear in the distal half of the ileum.

The intestinal mucosa also presents numerous finger-like evaginations, the **villi,** that cover the entire surface of the small intestine (Fig. 15-27). They consist of a lamina propria

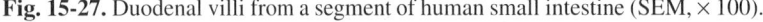

Fig. 15-27. Duodenal villi from a segment of human small intestine (SEM, × 100).

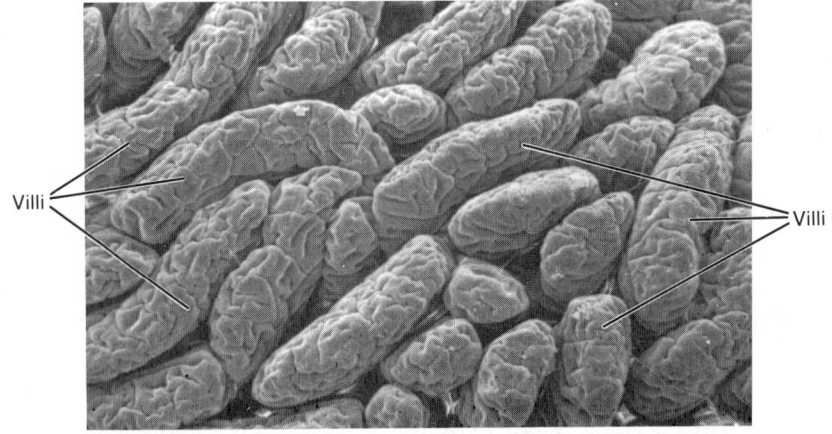

Villi

Villi

core and a covering of intestinal epithelium. Villi are 0.5 to 1.5 mm long and are best developed in the duodenum and jejunum, where they form broad, spatulate structures. In the ileum they become shorter and more finger-like.

The surface of the intestine is increased almost 30-fold by the large numbers of closely packed **microvilli** on the luminal surfaces of the intestinal absorptive cells (Fig. 15-28). Microvilli make up the **striated border** seen by light microscopy. Each microvillus represents a cylindrical extension of the plasma membrane enclosing a small core of cytoplasm. Each microvillus contains numerous thin actin filaments that extend into the apical cytoplasm of the cell and contribute to the **terminal web,** a network of fine filaments that lies just beneath and parallel to the microvillus border. The actin filaments within the microvillus core are linked together to form a bundle by the proteins fimbrin and villin. A brush border myosin links the actin bundle to the surrounding plasmalemma. Near the lateral surface of the cell, the filaments merge with those associated with junctional complexes. The filaments and terminal web contribute to the cytoskeleton of the cell and give stability to the microvillus border. Contraction of the terminal web region may act to spread the tips of microvilli and increase absorptive efficiency.

Fig. 15-28. Apices of three intestinal epithelial cells as shown by a freeze-fracture replica. Note the border of microvilli ($\times$ 25,000).

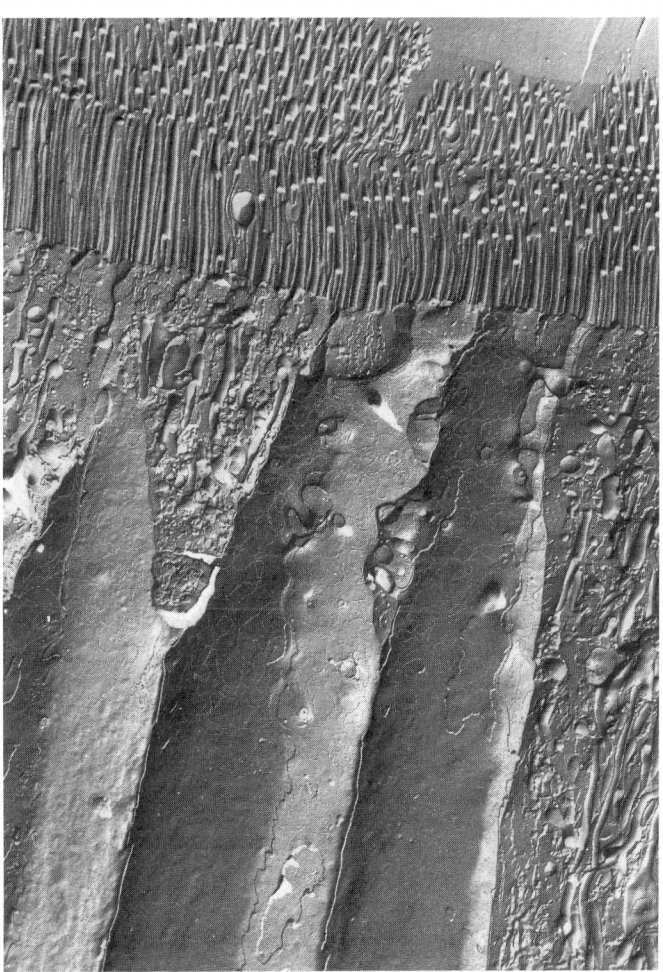

Mucosa

KEY WORDS: **simple columnar epithelium, villus, intestinal gland**

The intestinal mucosa consists of a **simple columnar epithelium,** a lamina propria, and a muscularis mucosae. The numerous **villi** of the mucosal lining give a velvety appearance to the interior of the small intestine. Simple tubular **intestinal glands** empty between the bases of the villi (Figs. 15-29 and 15-30). The glands measure 0.3 to 0.5 mm in depth and extend through the mucosa to the level of the muscularis mucosae, separated from one another by the connective tissue of the lamina propria. The intestinal glands contribute secretions that aid in the digestive process, but more important, they are the sites of renewal of intestinal epithelial cells as they are shed at the tips of the villi.

Epithelial Lining

KEY WORDS: **enterocyte, striated (microvillus) border, terminal bar, surface coat, goblet cell, endocrine cell, caveolated cell**

The epithelium that lines the intestinal lumen contains a heterogeneous population of simple columnar cells. The principal type is the intestinal absorptive cell, or **enterocyte,** a tall, cylindrical cell that rests on a delicate basement membrane and shows a prominent **striated (microvillus) border** at its free surface (see Fig. 15-28). Prominent **terminal bars** unite the intestinal absorptive cells at their apices, and desmosomes scattered along the lateral surfaces help hold the cells in close apposition. The microvilli that make up the striated border are covered by a layer of glycoprotein (glycocalyx) that is elaborated by the intestinal epithelial cells. This glycoprotein has been called the **surface coat** and is resistant to mucolytic and proteolytic agents. It is believed to have a protective function and to be involved in the digestive process.

Although considerable digestion occurs in the intestinal lumen due to the presence of pancreatic enzymes and bile salts, a significant amount of digestion also occurs on the microvillus surface prior to absorption. Enzymes that take part in the breakdown of disaccharides and polypeptides are present in or along the plasma membrane of the microvilli. In addition to contributing to the digestive process, the intestinal absorptive cells are actively involved in the absorption of protein, amino acids, carbohydrates, and lipids from the intestinal lumen. Absorbed materials passing through the epithelium must enter the cell through the apical plasmalemma and are prevented from passing between the cells by the zonulae occludens, which form a tight seal around the apices of the cells. Sugars and proteins pass through the basal cell membrane to enter blood capillaries in the connective tissue cores of the villi.

Lipid absorption is peculiar in its pathway through the intestinal absorptive cells and in the fact that lipids enter lymphatic channels rather than the blood vascular system. Triglycerides are broken down to fatty acids and *S*-monoglycerides by pancreative lipase in the intestinal

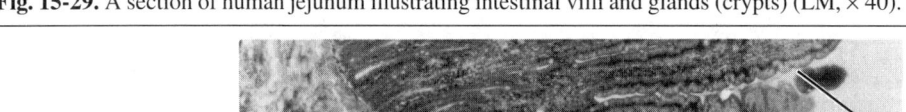

Fig. 15-29. A section of human jejunum illustrating intestinal villi and glands (crypts) (LM, × 40).

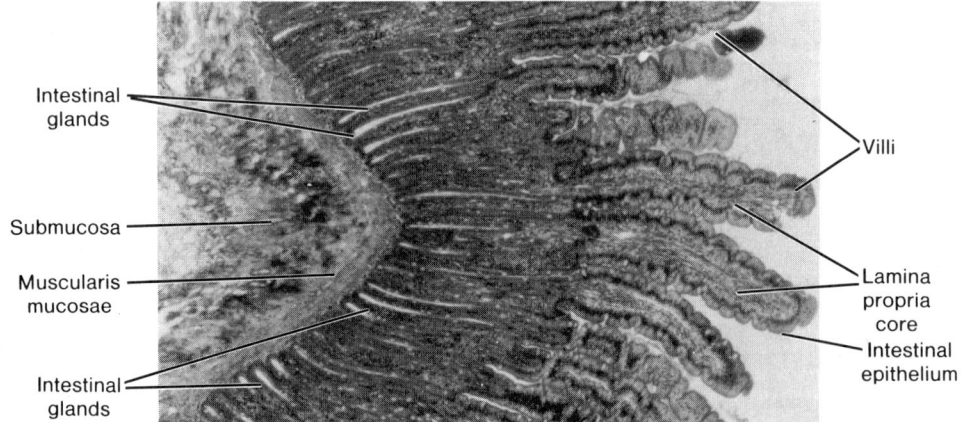

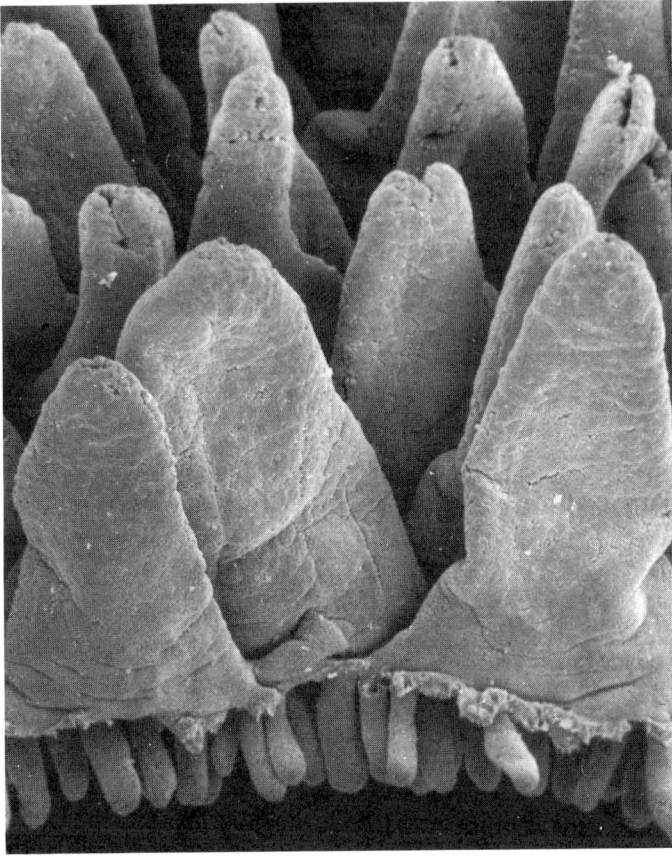

Fig. 15-30. Isolated villi and intestinal glands from a duodenum (SEM, × 200).

lumen. Resynthesis of triglycerides occurs in the smooth endoplasmic reticulum of the apical cytoplasm after they pass through the apical cell membrane. The re-formed lipid makes its way to Golgi complexes, where it is complexed with protein to form lipoprotein droplets called *chylomicrons*. Chylomicrons have no limiting membranes and are discharged through the lateral cell membranes of the intestinal absorptive cells into the intercellular space at the level of the nucleus, thus bypassing the basal regions. The chylomicrons pass down the intercellular spaces, through the basal lamina, and enter lymphatic channels (lacteals) after passing between endothelial cells.

As in other transporting epithelia, the zonula occludens junction not only tightly unites the apices of enterocytes into a continuous epithelial layer but also functions to clearly separate the enterocyte plasmalemma into apical and basolateral domains. Each contains specific transmembrane proteins unique to its domain. The apical cell membrane domain is rich in transmembrane proteins concerned with electrolyte transport and ion-coupled transport processes that bring materials into the cell from the intestinal lumen. Much of enterocyte absorption is linked to the transport of sodium. Sodium can enter the cell via a transmembrane protein ion channel or by transmembrane proteins that must bind to both sodium and chloride before either ion enters the cell, a coupled cotransport process. In addition, sodium can be absorbed by a sodium-solute–coupled cotransport mechanism. Solutes such as glucose, amino acids, water-soluble vitamins, and some bile acids are cotransported with sodium through the apical plasmalemma and into the cell by this mechanism. Once within the cytosol the solute follows a concentration gradient and is discharged from the cell through the basolateral plasmalemma. Sodium-hydrogen ion and chloride-bicarbonate ion exchange mechanisms also exist in the apical plasmalemma and are usually coupled to the absorption of sodium chloride. The absorptive sodium gradient established at the apical plasmalemma is maintained by a sodium, potassium-ATPase transmembrane protein located in the basolateral plasmalemma that functions as a sodium pump moving sodium into the extracellular space in exchange for potassium ions.

Goblet cells are scattered between the intestinal absorptive cells and increase in number distally in the intestinal tract (Fig. 15-31). The mucus secreted by these unicellular exocrine glands lubricates and protects the mucosal surface. The apical regions of the goblet cells often are expanded by the accumulation of secretory granules; the base forms a slender region that contains the nucleus, scattered profiles of granular endoplasmic reticulum, and occasional Golgi complexes. Goblet cells lie within the lining epithelium and are united to intestinal absorptive cells by apical tight junctions and scattered desmosomes along the lateral surfaces. Goblet cells also migrate from the intestinal glands, eventually to exfoliate at the tip of the villus. In humans, the life span of goblet cells is about 4 to 5 days. They may secrete continuously but are believed to pass through only one secretory cycle. Mucin granules are released by exocytosis and within a few seconds expand

several hundred-fold in volume due to the hydrophilic nature of the mucin granules. The resulting mucus is a viscid fluid consisting of glycoproteins. It is less viscid in the intestinal gland region.

Endocrine cells also are present in the epithelial lining, scattered within the epithelium of the villi and glands, but are far fewer in number than goblet cells. Endocrine cells secrete peptide hormones that influence gastric and intestinal secretion and motility, gallbladder contraction, and pancreatic and liver function (Table 15-2 and Fig. 15-32). They are characterized by dense secretory granules, scattered profiles of granular endoplasmic reticulum, and an electron-lucent cytoplasm. In addition to the hormones of the endocrine cell, numerous peptide hormones (neurotensin, substance P, vasoactive intestinal polypeptide, somatostatin, enkephalins) occur in the neurons of the enteric plexus. Whether these peptides serve as neurotransmitters or neuromodulators is not known.

Another cell type, the **caveolated cell,** may be seen occasionally in the epithelial lining. It is a pear-shaped cell with a wide base and narrow apex that protrudes slightly into the lumen. These cells are held in close apposition to neighboring epithelial cells by tight junctions and desmosomes. Caveolated cells have large microvilli that typically contain bundles of actin filaments that extend deep into the supranuclear region. Between the bundles of filaments, caveolae form small elongated channels that extend from the apical cell membrane between microvilli into the deep, apical cytoplasm. This cell type occurs elsewhere in the

Fig. 15-31. A segment of ileum from a child. Note the concentration of goblet cells (LM, × 40).

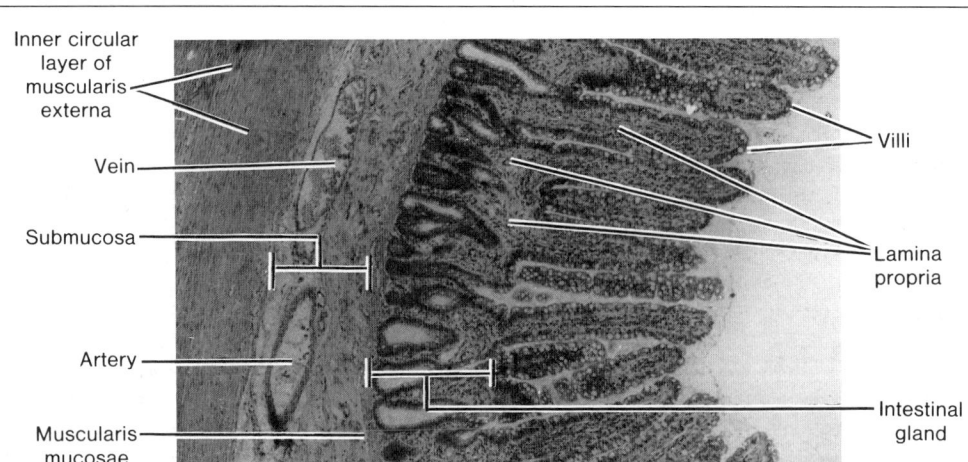

Table 15-2. Hormones and Related Substances of the Alimentary Canal

Hormone	Primary Location of Endocrine Cells	Primary Action
Gastrin	Antrum of stomach (G cell)	Stimulates acid secretion
Secretin	Duodenal and proximal jejunal mucosae (S cell)	Controls pancreatic H_2CO_3 secretion and biliary H_2O/ion secretion
Cholecystokinin	Mucosae of duodenum and jejunum (I cell)	Stimulates pancreatic enzyme secretion, gallbladder contraction
Enteroglucagon	Mucosa of ileum (L cell) and fundus (A cell)	Stimulates hepatic glyconeogenesis
Gastric inhibitory peptide	Mucosae of duodenum and jejunum (K cell)	Inhibits acid secretion
Motilin	Mucosa of jejunum (one type of enterochromaffin cell, EC_2)	Strong stimulator of motor activity in intestines
Vasoactive intestinal polypeptide	Gastric, small intestinal, and colonic mucosae (D_1 cell)	Stimulates intestinal motility, H_2O/ion secretion
Somatostatin	Mucosae of stomach, duodenum, and jejunum (D cell)	Inhibits the action or release of other hormones (paracrine functions)
Neurotensin	Mucosa of ileum (N cell)	Stimulates contractions of muscularis externa
Serotonin	Mucosae of small intestine and colon (enterochromaffin (EC) cells)	Stimulates intestinal motility
Histamine	Fundic mucosa of stomach (enterochromaffin-like (ECL) cell)	Stimulates secretion and intestinal motility and is a vasodilator
Bombesin-like peptide	Mucosae of stomach and jejunum (P cell)	Stimulates gastric acid secretion and exocrine pancreas secretion
Substance P	Gastric mucosa (EC_1 cell)	Stimulates intestinal motility

gastrointestinal tract and other organs of endodermal origin. They may be chemoreceptors in some regions, as in the respiratory system, but their function in the intestine is unknown.

Intestinal Glands

KEY WORDS: undifferentiated cell, oligomucous cell, Paneth cell, intraepithelial lymphocyte

All epithelial cells that line the intestinal surface arise from cells in the intestinal glands. The upper regions of the glands are lined by simple columnar epithelium that is similar to and continuous with that lining the intestinal lumen. It contains absorptive cells, goblet cells, scattered endocrine cells, and a few caveolated cells. **Undifferentiated cells** line the basal half of the intestinal glands and undergo frequent mitoses. Intestinal absorptive cells (enterocytes) and **oligomucous cells,** which represent intermediate stages of goblet cell formation, lie in the midgland region. New intestinal epithelial cells are formed continually in the basal region of the glands, migrate by upward displacement of cells, and ultimately come to cover the villi and the entire intestinal surface. The oldest intestinal

epithelial cells are shed at the tips of the villi. Complete turnover of the epithelium of the glands and lining of the intestine occurs every 4 to 5 days. As absorptive intestinal cells migrate up the villus, they increase in height and the striated border becomes more prominent. The function of the enterocyte differs depending on its location along the villus-crypt axis. Those enterocytes deep within the intestinal glands are relatively undifferentiated, whereas enterocytes that line the upper one-third of the intestinal glands and cover the lower one-third of villi are involved primarily in secretion. Enterocytes covering the upper two-thirds of villi are involved primarily in absorption, although not exclusively, as these enterocytes also are involved in secretion but to a lesser extent. Thus enterocyte function changes during its lifespan as it migrates along the villus-crypt axis to eventually be shed at the villus tip.

Chloride ion is moved into enterocytes active in secretion via a sodium, potassium-2 chloride cotransport mechanism. These transmembrane enzyme proteins are located in the basolateral plasmalemma. Potassium protein channels and sodium, potassium pumps also localized within the basolateral plasmalemma participate in a cooperative way so that intracellular chloride levels are maintained

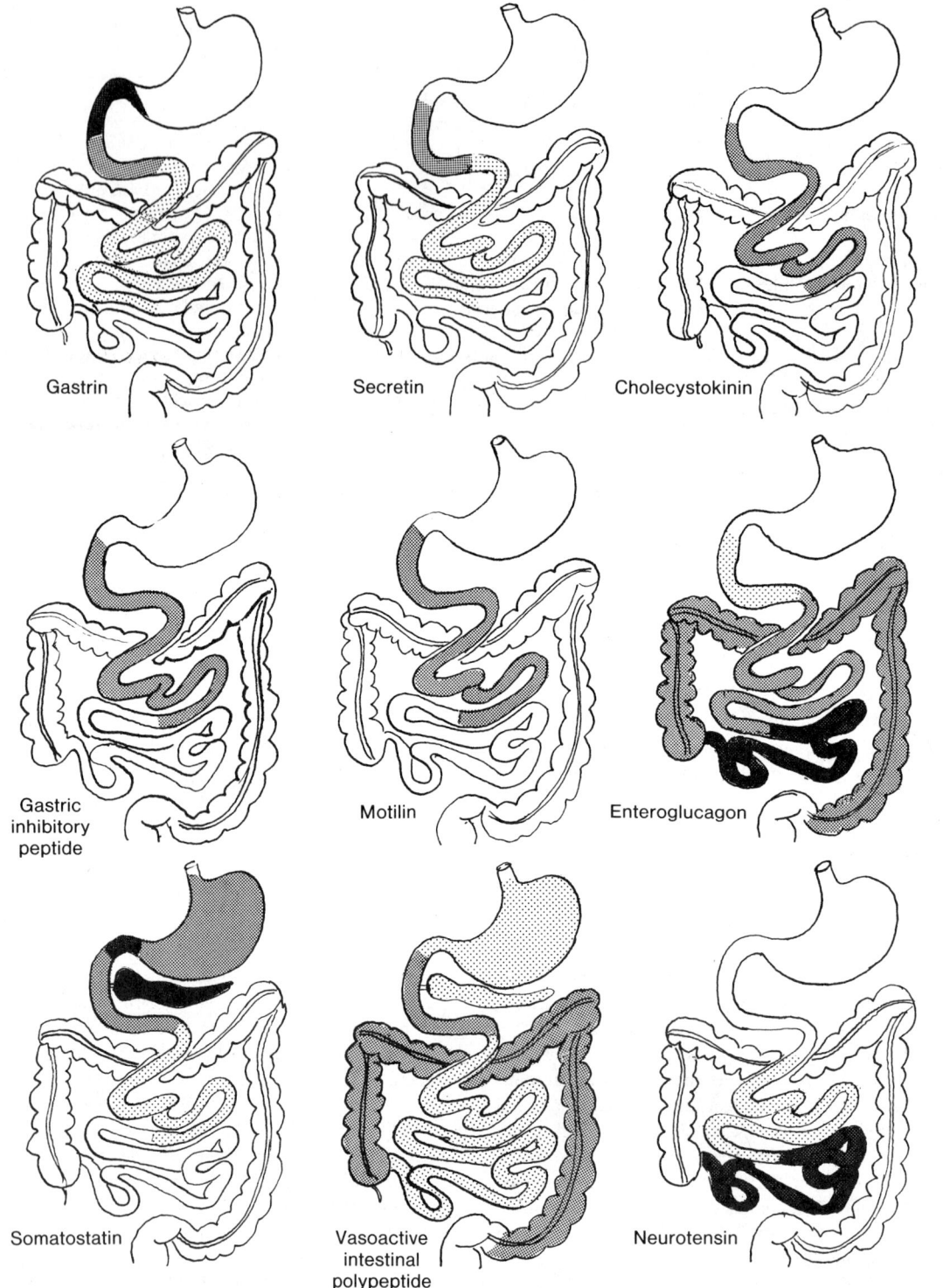

Fig. 15-32. Distribution of endocrine cells in humans.

sufficiently high to allow for passive chloride outflow through chloride channel proteins in the apical plasmalemma. Transcellular chloride secretion is thought to be the driving force for the movement of water and sodium through the paracellular route, the net result of which is secretion rich in sodium chloride and water entering the intestinal lumen. The transmembrane conductance regulator protein that regulates the passage of chloride through the apical plasmalemma is thought to be controlled by two intestinal peptide hormones called *guanylin* and *uroguanylin*. Certain pathogenic strains of enteric bacteria secrete heat-stable enterotoxins that mimic the actions of these peptides, which results in "traveler's secretory diarrhea" and is illustrative of the magnitude of fluid production through this system if unchecked.

Small groups of pyramidal-shaped cells with conspicuous granules are found at the bases of the intestinal glands (Fig. 15-33). These are **Paneth cells,** which, unlike other cells of the intestinal epithelium, form a relatively stable population with a low rate of turnover. Paneth cells have been shown to contain lysozyome and IgA; however, their role in the function of the human small intestine is unknown.

Intraepithelial lymphocytes often are seen between and within intestinal epithelial cells that cover villi or line glands.

Lamina Propria

KEY WORDS: lymphatic nodule, Peyer's patch, M-cell, central lacteal

The connective tissue of the lamina propria forms the cores of the villi and fills the areas between intestinal glands. The intestinal lamina propria is rich in reticular fibers that form a delicate meshwork containing numerous reticular cells, lymphocytes, plasma cells, macrophages, and eosinophils, giving the lamina propria a very cellular appearance. The lamina propria of the intestinal tract represents a special type of lymphatic tissue and is part of the gut-associated lymphatic tissue (GALT). The free cells of the lamina propria provide a protective sleeve that encircles the intestinal lumen immediately below the epithelium. Lymphocytes are the most numerous cells present and provide a vast reserve of immunocompetent cells. Plasma cells elaborate immunoglobulin A (IgA), which is taken up by the intestinal absorptive cells and complexed to a protein (secretory piece) that is synthesized by the enterocyte. The complex is then released from the cell and provides a protective surface for the epithelial cells — an immunologic border — against viruses and bacteria. The secretory piece acts as a carrier for the IgA and possibly protects it from lysosomal digestion while it is within the cell and from enzymatic digestion on the luminal surface. Lymphocytes often leave the lamina propria to migrate through the intestinal epithelium into the lumen of the intestine.

Lymphatic nodules of variable size lie in the lamina propria, scattered along the entire intestinal tract, but become larger and more numerous distally. They are particularly numerous and well developed in the ileum and often extend the full thickness of the mucosa. The lymphatic nodules may be solitary or grouped together in aggregates

Fig. 15-33. The bases of human intestinal glands exhibit scattered Paneth cells. Note the underlying muscularis mucosae (LM, × 250).

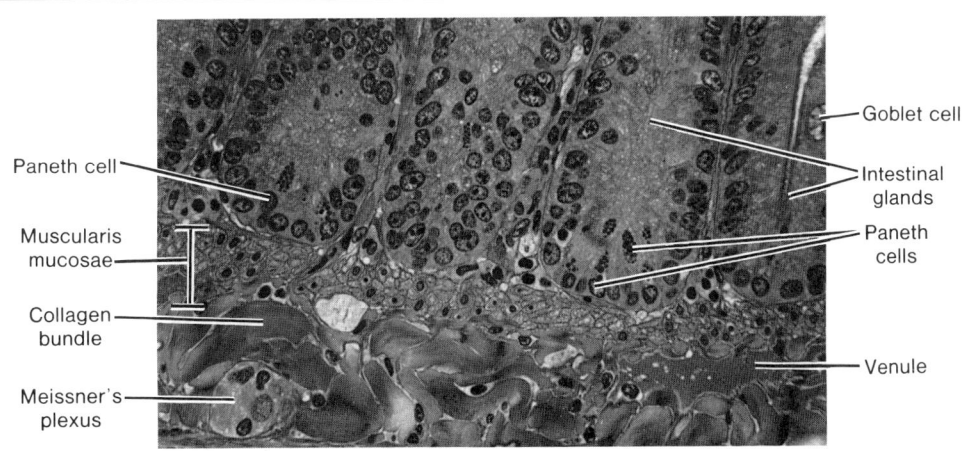

Paneth cell

Muscularis mucosae

Collagen bundle

Meissner's plexus

Goblet cell

Intestinal glands

Paneth cells

Venule

called **Peyer's patches** (Fig. 15-34). These oval structures may be quite large — 20 mm in length — and are visible to the naked eye. Peyer's patches may occupy the full depths of the mucosa and submucosa.

A peculiar type of cell, the **M-cell,** lies in the intestinal epithelium that covers the lymphatic nodules. The cells show characteristic microplicae on the luminal surface, rather than microvilli. M-cells sequester intact macromolecules from the intestinal lumen and transport them in membrane-bound vesicles to intraepithelial lymphocytes. The sensitized lymphocytes then migrate to various aggregates of lymphatic tissue in the lamina propria or mesenteric lymph nodes, transporting information to sites of antibody production. M-cells provide a way for the immune system to maintain immunologic surveillance of the environment in the gut lumen and to respond appropriately to any changes. M-cell monitoring of the intestinal lumen is not unique. Langerhans' cells of the epidermis perform a similar function in monitoring the external environment and relaying information to the immune system. Both are considered to be antigen-presenting cells.

Lymphatic vessels occur in the lamina propria and are important in fat absorption. Dilated lymphatic channels begin blindly in the cores of villi near their tips and form the **central lacteals.** Their walls consist of thin endothelial cells surrounded by basal lamina and reticular fibers. At the bases of the villi, the lymphatic capillaries anastomose with those coursing between intestinal glands and form a plexus before passing through the muscularis mucosae. In the submucosa they join larger lymphatic vessels. Vessels that pierce the muscularis mucosae may contain valves.

Muscularis Mucosae

KEY WORDS: **inner circular layer, outer longitudinal layer**

The muscularis mucosae is well defined and consists of an **inner circular layer** and an **outer longitudinal layer** of smooth muscle. It is intimately associated with thin elastic fibers. Slips of smooth muscle leave the muscularis mucosae and enter the cores of villi to provide the means by which villi contract. Contraction of villi continuously provides the intestinal epithelium covering the villi with a fresh environment for maximum absorptive efficiency. The pump-like action during contraction also aids in moving absorbed materials into the blood and lymphatic capillaries, out of vessels in the lamina propria, and into the larger vessels of the submucosa and thus further aids the absorptive process.

Small arteries run on the inner surface of the muscularis mucosae and break up into capillary networks that form a dense network of fenestrated capillaries just beneath the intestinal epithelium. Near the tips of villi, the capillaries drain into small veins that run downward to anastomose with a venous plexus surrounding the intestinal glands, before joining veins in the submucosa.

Submucosa

KEY WORDS: **duodenal (Brunner's) gland, branched tubular gland, urogastrone, epidermal growth factor, submucosal plexus**

The submucosa consists of coarse collagenous fibers and numerous elastic fibers. It often contains lobules of adi-

Fig. 15-34. A portion of human ileum illustrating a Peyer's patch (LM, × 100).

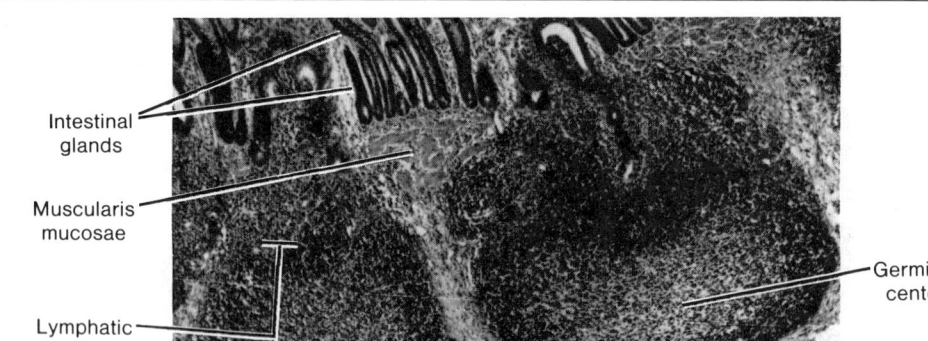

Intestinal glands

Muscularis mucosae

Lymphatic nodule

Germinal center

Submucosa

pose tissue as well as the cells normally associated with loose connective tissue. In the ileum it may contain large aggregates of lymphoid cells derived from Peyer's patches. The submucosa of the duodenum houses the **duodenal (Brunner's) glands,** the ducts of which pierce the muscularis mucosae and drain into overlying intestinal glands (Fig. 15-35). These **branched tubular glands** are first encountered immediately distal to the pyloric sphincter and also may be found in the proximal jejunum. They often occur in the submucosal cores of the plicae circulares. The duodenal glands are the only submucosal glands present in the gastrointestinal tract.

The duodenal glands consist of secretory cells with ultrastructural features of serous and mucous glands. The duodenal glands secrete a mucin-containing alkaline fluid that protects the duodenal mucosa proximal to the entrance of the pancreatic duct from the erosive effects of acid pepsin from the stomach. Cells of the human duodenal glands show immunoreactivity for **urogastrone,** a peptide that is a potent inhibitor of gastric acid secretion, and **epidermal growth factor,** a peptide that may control proliferation and migration of intestinal epithelial cells in the proximal small intestine.

Nerve fibers and parasympathetic ganglia form a **submucosal plexus** found in the submucosa of the esophagus, stomach, and small and large intestines. This plexus provides motor innervation for the muscularis mucosae and the slips of smooth muscle that enter villi or other regions of the mucosa. Large blood and lymphatic vessels also are found in the submucosa. The large mesenteric vessels pierce the muscularis externa at its attachment to the mesentery and then enter the submucosa, branch, and run around the circumference of the intestinal tube. The submucosal vessels supply and receive small tributaries from the overlying mucosa.

Muscularis Externa

KEY WORDS: inner circular layer, outer longitudinal layer, myenteric plexus

The muscularis externa consists of **inner circular** and **outer longitudinal layers** of smooth muscle. Both layers take a helical course around the intestinal tube. The fibers in the outer layer form a more gradual helix, giving the impression of a longitudinal arrangement. Between the two layers of smooth muscle are the nerve fibers and parasympathetic ganglia of the **myenteric plexus.** This supplies the motor innervation to the muscularis externa of the entire gastrointestinal tract.

Serosa

The muscularis externa is invested by a thin layer of loose connective tissue covered by a layer of mesothelial cells of the visceral peritoneum. Where the mesentery attaches to the intestinal wall, the serosa becomes continuous with both sides of the mesentery and encloses connective tissue elements, blood vessels, and nerves.

Ileocecal Junction

The lumen of the ileum becomes continuous with that of the large intestine at the ileocecal junction. Here the lining

Fig. 15-35. A region of human duodenum illustrating the duodenal glands. Note that these glands empty into intestinal glands shown at the right (LM, × 250).

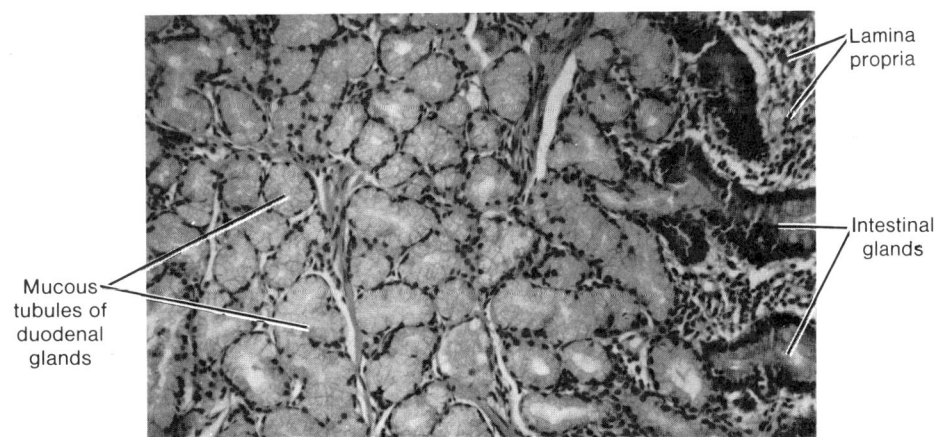

Lamina propria

Intestinal glands

Mucous tubules of duodenal glands

is thrown into anterior and posterior folds called ileocecal valves, which consist of both the mucosa and submucosa surrounded by a thickening of the inner circular layer of the muscularis externa.

Large Intestine

The large intestine is approximately 180 cm long and is divided into several regions. The cecum is continuous with the ileum at the ileocecal junction and forms a blind pouch at the proximal end of the colon. It bears the appendix, a small, slender diverticulum. The remainder of the colon is divided into ascending, transverse, and descending parts that, structurally, show the same features. The large intestine is continuous with the rectum and anal canal, the latter terminating as the anus.

Mucosa

KEY WORDS: lacks villi, absorptive cell (enterocytes), goblet cell, fluid absorption, mucus secretion

The mucosa of the large intestine is similar to that of the small intestine but **lacks villi** and has a smooth interior surface (Figs. 15-36 and 15-37). The lumen is lined by a simple columnar epithelium that consists of intestinal **absorptive cells (enterocytes)** and **goblet cells.** The latter increase in number toward the rectum so that in the distal region of the colon the epithelium consists mainly of goblet cells.

Intestinal glands of the colon are larger and more closely packed than those in the small intestine, and they increase in length distally to reach their maximum depth (0.7 mm) in the rectum. The glands contain numerous goblet cells, and in the basal half of the glands, proliferating and undifferentiated epithelial cells and endocrine cells are present. Paneth's cells usually are absent. The turnover rate for intestinal epithelial cells in the colon is about 6 days.

The composition of the lamina propria and muscularis mucosae is the same as that of the small intestine. Scattered lymphatic nodules are present and may protrude through the muscularis mucosae to lie in the submucosa.

Ingested material enters at the cecum as a semifluid and becomes semisolid in the colon. Although no digestive enzymes are secreted by the colon, some digestion does occur as a result of enzymes present in the developing fecal mass and through the action of bacterial flora normally associated with the large intestine. The primary functions of the large intestine are the **fluid absorption** and the **mucus secretion** to protect the mucosa from abrasion by the developing feces.

The cellular mechanisms involved with enterocyte absorption and secretion in the colon are thought to be similar to those occurring in enterocytes of the small intestine. The secretory activities of enterocytes in both regions of the gut appear to be influenced by the peptides guanylin and uroguanylin. The latter, which also occurs in relatively high concentrations in the blood, has been suggested to act as an intestinal natriuretic factor. It is well established that

Fig. 15-36. A region of mucosa from the human colon. Note the absence of villi (LM, × 100).

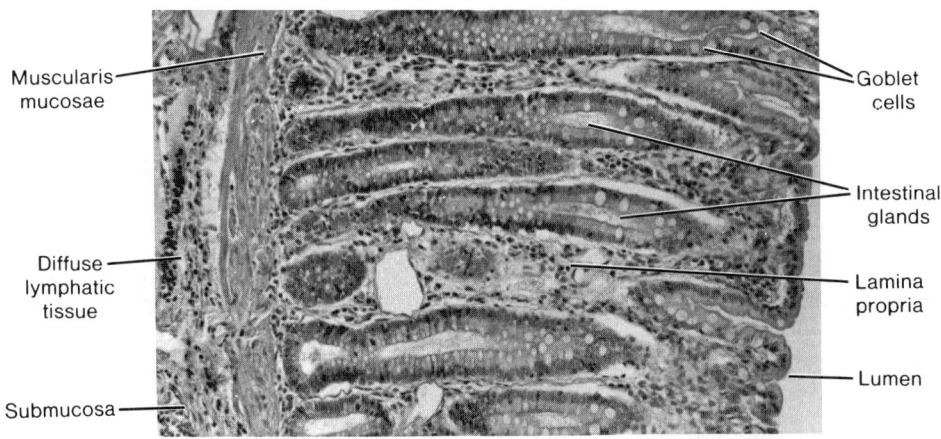

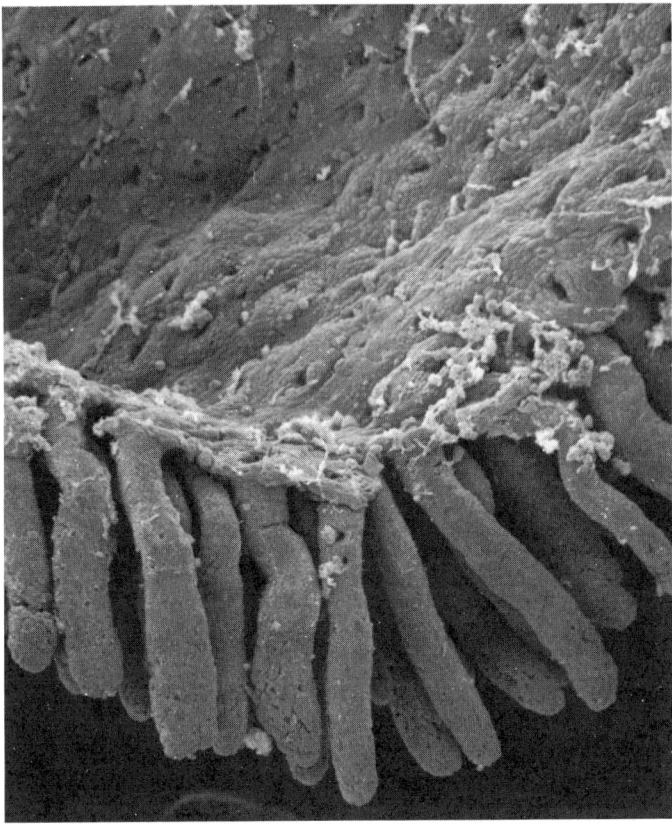

Fig. 15-37. A region of isolated colonic mucosa showing intestinal glands and their openings on the mucosal surface. Note that villi are absent (SEM, × 120).

when large amounts of salt are taken orally, a large and rapid increase in urinary salt excretion occurs. The same amount of salt given intravenously causes only small changes in urinary salt excretion. Aldosterone and the atrial natriuretic peptides do not appear to be involved in this phenomenon, suggesting the presence of intestinal natriuretic factors such as uroguanylin.

Submucosa

The submucosa of the colon is similar to that of the small intestine and contains the larger vessels and the submucosal nerve plexus. Glands are not present. As in the remainder of the digestive tube, blood vessels pierce the muscularis externa and course around the circumference of the intestinal wall. The large vessels supply tributaries that penetrate the mus-

cularis mucosae and form capillary networks in the lamina propria, around intestinal glands, and beneath the lining epithelium. The veins follow the same course as the arteries.

Muscularis Externa

KEY WORDS: taeniae coli, haustrum, plicae semilunares

The outer, longitudinal coat of the muscularis externa of the cecum and colon differs from that of the small intestine in that it forms three longitudinal bands called the **taeniae coli.** Between the taeniae the muscle coat is thinner and may be incomplete. The inner circular layer is complete and appears similar to that of the small intestine. Components of the myenteric plexus lie just external to the circular layer, as in the small intestine. Due to the tonus of the taeniae coli, the wall of the colon is gathered into outwardly bulging pockets, the **haustra.** Crescentic folds called **plicae semilunares** project into the lumen of the colon between the haustra.

Serosa

KEY WORDS: adventitia, appendices epiploicae

The serosa is incomplete in the colon, and the ascending and descending limbs are retroperitoneal. Here the muscular wall of the colon is attached to adjacent structures by an **adventitia.** Where a serosa is present, it may contain large, pendulous lobules of fat called **appendices epiploicae.**

Appendix

The structure of the appendix resembles that of the colon, but in miniature. The lumen is small and irregular in outline, and taeniae coli are absent. The lamina propria is extensively infiltrated with lymphocytes, and details of the mucosa often are obscured by the many lymphatic nodules that may fill the mucosa and submucosa.

Rectoanal Junction

KEY WORDS: pectinate line, circumanal gland, rectal column, anal valve, hemorrhoidal plexus, internal anal sphincter, external anal sphincter

The mucosa of the first part of the rectum is similar to that of the colon except that the intestinal glands are slightly longer and the lining epithelium is composed primarily of goblet cells. The distal 2 to 3 cm of the rectum forms the anal canal, which ends at the anus. Immediately proximal to the **pectinate line,** the intestinal glands become shorter and then disappear. At the pectinate line, the simple columnar epithelium makes an abrupt transition to noncornified stratified squamous epithelium (Fig. 15-38). After a short transition, the noncornified stratified squamous epithelium becomes continuous with the keratinized stratified squamous epithelium of the skin at the level of the external anal sphincter. Beneath the epithelium of this region are simple tubular apocrine sweat glands, the **circumanal glands,** which secrete an oily material.

Proximal to the pectinate line, the mucosa of the anal canal forms large longitudinal folds called **rectal columns** (of Morgagni). The distal ends of the rectal columns are united by transverse mucosal folds, the **anal valves.** The recess above each valve forms an anal sinus. It is at the level of the anal valves that the muscularis mucosae becomes discontinuous and then disappears.

The submucosa of the anal canal contains numerous veins that form a large **hemorrhoidal plexus.** When distended (varicosed), these vessels protrude into the overlying mucosa and form internal hemorrhoids (piles). The inner circular layer of the muscularis externa increases in thickness and ends as the **internal anal sphincter.** The thin outer muscular layer breaks up and ends by blending with the surrounding connective tissue. Skeletal muscle fibers circumscribe the distal anal canal and form the **external anal sphincter,** which is under voluntary control.

Fig. 15-38. The epithelium lining the human anal canal (LM, × 250).

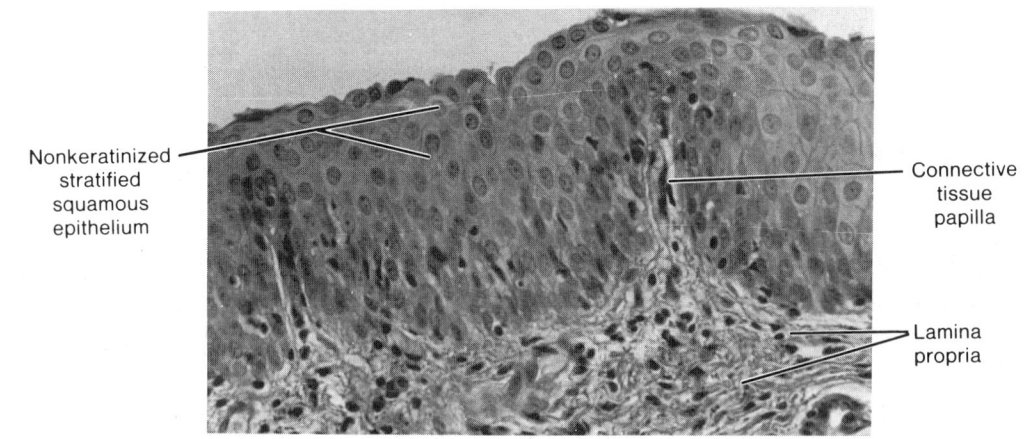

Nonkeratinized stratified squamous epithelium

Connective tissue papilla

Lamina propria

Liver

The liver is the largest organ in the body and is both an endocrine and exocrine gland, releasing several substances directly into the bloodstream and secreting bile into a duct system.

The liver is uniquely situated with respect to the venous blood flow from the gastrointestinal tract. Blood from the portal vein, which drains the gastrointestinal tract, passes through the substance of the liver before entering the systemic circulation. The liver therefore receives all of the materials absorbed by the gastrointestinal tract except lipids, which are carried by lymphatics to enter the general circulation. Tributaries of the portal vein empty into hepatic sinusoids that drain into hepatic veins. The arrangement in the liver of a set of sinusoids between veins is called the *hepatic portal system*. The liver also has an arterial supply from the hepatic artery, and both the venous (portal) and arterial (hepatic) bloods percolate through the liver sinusoids and exit by way of the hepatic vein.

Structural Organization

KEY WORDS: **capsule, hepatocyte, sinusoid, central vein, hepatic lobule, portal area, portal vein, hepatic artery, bile duct, lymphatic channel, liver acinus, portal lobule, subcapsular limiting plate**

The liver is invested by a delicate connective tissue **capsule** that is continuous with the peritoneum. The capsule contains numerous elastic fibers and is covered by a mesothelium except for a small bare area where the liver abuts the diaphragm.

The liver is composed of epithelial cells, the **hepatocytes,** arranged in branching and anastomosing plates separated by blood **sinusoids.** Both form a radial pattern about a **central vein** that is the smallest tributary of the hepatic vein. The spokelike arrangement of hepatic plates about a central vein constitutes the basis of the classic **hepatic lobule,** which appears somewhat hexagonal in cross section, with a central vein at the center and portal areas at the corners (Fig. 15-39). The liver consists of about 1 million such units, each about 0.7 mm wide and 2.0 mm long.

A **portal area** contains a branch of the **portal vein,** a branch of the **hepatic artery,** a **bile duct,** and occasionally a **lymphatic channel** (Fig. 15-40). All are enclosed in a common investment of connective tissue. Blood passes from small branches of the hepatic artery and portal vein into the sinusoids that lie between plates of hepatocytes. Blood flows slowly through the sinusoids toward the center of the lobule and exits through the central vein. Hepato-

cytes nearest the branches of the portal vein and hepatic artery — that is, at the periphery of the lobule — receive blood with the highest nutrient and oxygen content. Both diminish as blood flows toward the central vein. Due to this arrangement, three zones can be recognized in a hepatic lobule according to the metabolic activity: a zone of permanent function at the periphery, a zone of intermittent activity near the center of the lobule, and a zone of permanent repose near the central vein. The boundaries of the hepatic lobule can be estimated by observing the central vein and noting the portal areas at the corners of the lobule.

Also related to the blood supply is a smaller unit of liver structure called the *liver acinus* or *functional unit* (see Fig. 15-39). The **liver acinus** is defined as the hepatic tissue supplied by a terminal branch of the hepatic artery and portal vein and drained by a terminal branch of the bile duct. It is represented by a diamond-shaped area with central veins at two of the opposite corners. Branches of the vessels and a branch of the bile duct lie between the two portions of adjacent hepatic lobules that they supply.

An additional lobule is related to the exocrine secretion (of bile) in the liver. This unit is the **portal lobule** and has a portal canal at its center. It consists of the hepatic tissue that is drained by a bile duct of a portal area. A portal lobule is triangular in shape and contains parts of three adjacent hepatic lobules. A central vein is located at each corner of this unit (see Fig. 15-39).

A single layer of small, dark hepatocytes limits the liver parenchyma beneath the capsule and is called the **subcapsular limiting plate.** A similar wall of hepatic cells surrounds the portal areas and forms the periportal limiting plate, which is pierced by tributaries of the hepatic artery, portal vein, and bile ductules.

Hepatic Sinusoids

KEY WORDS: **discontinuous endothelium, fenestration, perisinusoidal space, hepatic macrophage, Kupffer cell, lipocyte**

Hepatic sinusoids are larger and more irregular in shape than ordinary capillaries (Fig. 15-41). The sinusoidal lining consists of a simple layer of squamous epithelium supported by very little connective tissue. Three types of cells are associated with the sinusoidal lining: endothelial cells, stellate cells (Kupffer cells or hepatic macrophages), and fat-storing cells (lipocytes).

Endothelial cells constitute the major cellular element of the sinusoidal lining and form a **discontinuous endothelium.** There is no basal lamina, and the cells are separated by gaps 0.2 to 0.5 μm wide. The endothelial cells

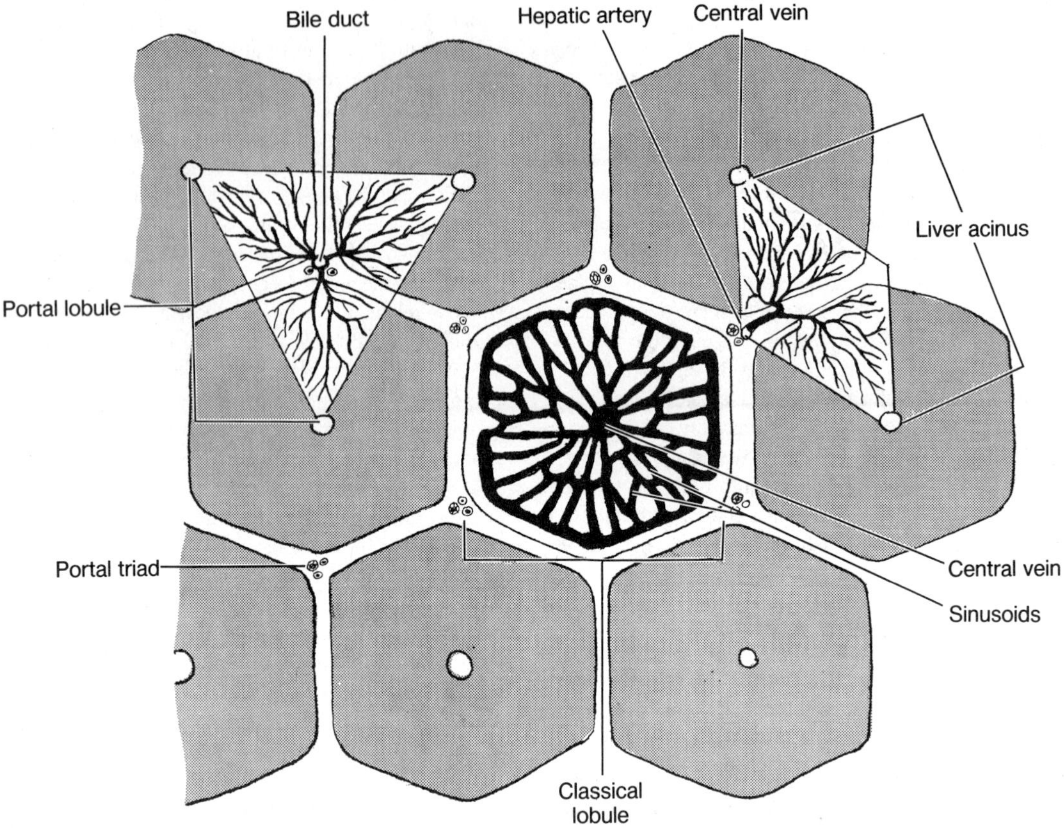

Bile duct   Hepatic artery   Central vein

Portal lobule

Liver acinus

Portal triad

Central vein

Sinusoids

Classical
lobule

Fig. 15-39. Diagram of liver lobules.

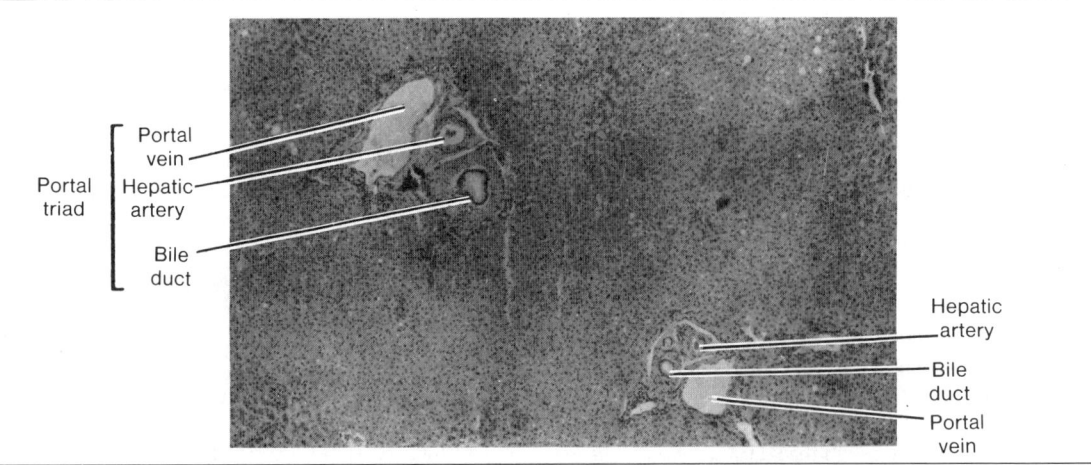

Portal
vein

Portal
triad

Hepatic
artery

Bile
duct

Hepatic
artery

Bile
duct

Portal
vein

Fig. 15-40. A region of human liver showing two portal triads (LM, × 40).

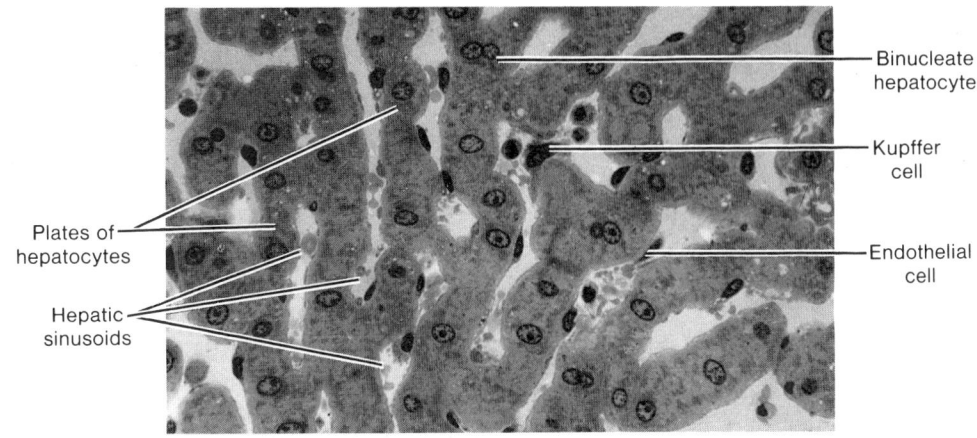

Binucleate hepatocyte

Kupffer cell

Endothelial cell

Plates of hepatocytes

Hepatic sinusoids

Fig. 15-41. Hepatic cords and sinusoids of the liver. Note the Kupffer cell near the center of the photomicrograph (LM, × 300).

also show numerous intracellular **fenestrations** or pores. The sinusoidal lining is separated from the liver cells by a narrow **perisinusoidal space** (Fig. 15-42). Blood plasma flows freely through the endothelium and into the space, but the sinusoidal lining does hold back erythrocytes. Although occasional bundles of reticular fibers and fine collagenous fibers are present in the perisinusoidal space, there is no ground substance, and the flow of blood plasma is unhindered. Because the plasma has direct access to the perisinusoidal space, the liver cells are constantly bathed on one surface by fluid that is rich in the nutrients absorbed by the intestinal tract. Thus the perisinusoidal space has considerable significance in the exchange of materials between the liver and plasma. The plasmalemma of the hepatocytes that face the perisinusoidal space bear numerous well-developed microvilli that project into the space and greatly increase the surface area and facilitate absorption.

Also present in the sinusoidal lining are actively phagocytic cells, variously called *stellate cells,* **hepatic macrophages,** or **Kupffer cells.** These form part of the sinusoidal lining and are irregularly shaped cells that expose the greater part of their cytoplasm to the blood in the sinusoid and extend processes between the endothelial cells. Unlike the neighboring endothelial cells, the cytoplasm of the hepatic phagocytes contains vacuoles, lysosomes, Golgi bodies, and short profiles of granular endoplasmic reticulum. Phagocytized material also may be present. The cells are part of the mononuclear system of macrophages and arise from monocytes of the bone marrow.

The third type of cell is located on the side of the sinu-

soidal lining that faces the perisinusoidal space. This cell accumulates lipid and is most numerous in the peripheral and intermediate zones of the hepatic lobule. They have been called **lipocytes,** *fat-storing, stellate,* or *interstitial cells,* but their function is unknown.

Hepatocytes

KEY WORDS: polyploid, bile canaliculus

The parenchyma of the liver consists of large polyhedral hepatocytes arranged in plates that radiate from the region of the central vein. The surfaces of an individual hepatocyte either contact an adjacent liver cell or border on a perisinusoidal space. This latter surface bears numerous well-developed microvilli. The plates of hepatocytes are supported by a delicate stroma that consists mostly of reticular fibers.

The nuclei of hepatocytes are large and round and usually occupy the center of the cell. A single nucleus usually is present, but as many as 25 percent of hepatocytes are binucleate. There also is considerable variation in the size of nuclei from cell to cell, reflecting the **polyploid** nature of hepatocytes.

The cytoplasm of hepatocytes is rich in organelles — large arrays of granular endoplasmic reticulum, a moderate amount of smooth endoplasmic reticulum, mitochondria, Golgi complexes, peroxisomes, and lysosomes. Inclusions such as glycogen and lipid also are common. Glycogen often appears as dense rosettes (alpha particles) made up of smaller beta particles. The alpha particles measure 20 to

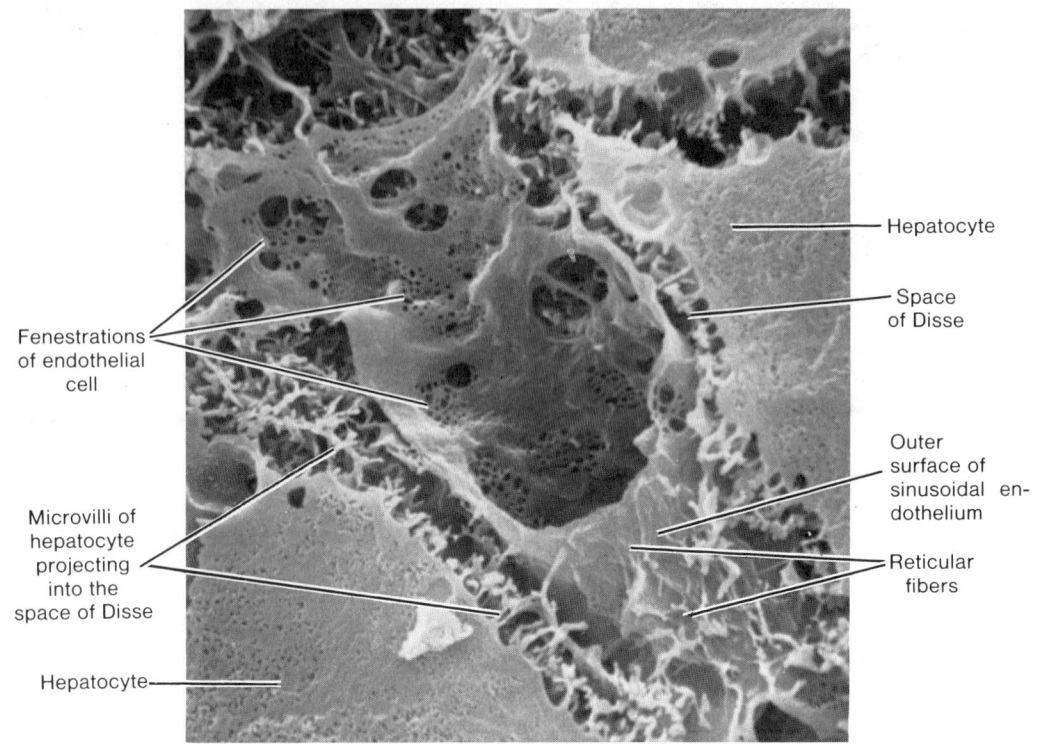

Fig. 15-42. A portion of a hepatic sinusoid. Note the surrounding hepatocytes and the perisinusoidal space of Disse (SEM, × 6,000).

30 nm in diameter. The cytoplasm is variable in appearance and changes with the nutritive state of the organ.

Tiny channels, the **bile canaliculi,** course through the parenchyma between hepatocytes to end in the bile ducts of the portal areas. Bile canaliculi represent an expansion of the intercellular space, and their walls are formed by the adjacent plasmalemma of two neighboring hepatocytes. Short microvilli extend from the cell membrane into the lumen. At the margins of the canaliculus, the plasma membranes are joined by occluding junctions similar to the zonula occludens in other epithelia. They form a seal to prevent bile from escaping into the intercellular spaces between hepatocytes. Golgi complexes of hepatic cells often lie adjacent to the canaliculi. The bile canaliculi show intermittent contractions.

Bile Ducts

KEY WORDS: bile ductule, interlobular bile duct, extrahepatic duct, common hepatic duct, common bile duct

Bile canaliculi unite with bile ducts in the portal canals via small, interconnecting channels called **bile ductules.** They are small and have thin walls, and their small lumina are surrounded by a low cuboidal epithelium that rests on a distinct basal lamina. The terminal ductules empty into **interlobular bile ducts** of the portal areas. The lumina of the bile ducts increase in diameter as they course toward the exterior, and the lining epithelium increases in height. Interlobular ducts unite to form the **extrahepatic ducts,** in which the surrounding layers of connective tissue become thicker and the lining epithelium becomes tall columnar. Two large extrahepatic ducts, the left and right hepatic ducts, unite to form the major excretory duct of the liver, the **common hepatic duct.** It is joined by the cystic duct from the gallbladder to form the **common bile duct** (ductus choledochus), which empties into the duodenum.

The major extrahepatic ducts are lined by a tall columnar, mucus-secreting epithelium. The remainder of the wall consists of a thick layer of connective tissue that is rich in elastic fibers and often contains numerous lymphocytes and occasional migrating granulocytes. Bundles of smooth muscle, running in longitudinal and oblique directions, are present in the common bile duct and form an incomplete layer that spirals around the lumen. Near the wall of the duodenum, the smooth muscle forms a complete investment and thickens to form a small sphincter, the sphincter choledochus. Distal to this region, the common bile duct and the major pancreatic duct merge as they pass through the intestinal wall and empty through a common structure, the hepatopancreatic ampulla. As the ducts pierce the duodenal wall, they are surrounded by a common sphincter of smooth muscle.

Bile is produced continuously by the liver and ultimately leaves the organ through the extrahepatic duct system. Resistance at the sphincters forces bile to enter the cystic duct and pass into the gallbladder, where it is stored. Bile formation occurs primarily at two anatomic sites: bile canaliculi and bile ductules. Secretin increases secretion of ductular bile, which is rich in bicarbonate. Glucagon increases canalicular bile secretion.

One function of the liver is to store carbohydrate in the form of glycogen. When needed, glycogen is released to the bloodstream as glucose. The liver has functions in protein metabolism; fat metabolism and storage; synthesis of fibrinogen, prothrombin, and albumin; storage of several vitamins (primarily A, D, B_2, B_3, B_4, and B_{12}); and detoxification of lipid-soluble drugs and alcohol. Detoxified materials are excreted by hepatocytes into the bile and conducted by the biliary duct system to the intestinal tract for elimination.

Bile is a complex exocrine secretion of the liver and contains bile acids, bile pigments, bile salts, cholesterol, lecithin, soaps, and neutral fats. Bile salts act as emulsifying agents and are important in the breakdown of fat in the intestinal lumen during digestion. Some bile salts may be reabsorbed in the intestinal tract to be secreted once again into the bile. This reuse of bile salts is referred to as the *enterohepatic circulatory system.*

Gallbladder

KEY WORDS: cystic duct, mucous membrane, simple columnar epithelium, lamina propria, muscularis, adventitia, serosa

The gallbladder is a saclike structure on the inferior surface of the liver, measuring about 4 cm wide and 8 cm long. It is joined to the common hepatic duct by the **cystic duct,** whose mucous membrane forms prominent spiraling folds that contain bundles of smooth muscle. These folds make up the spiral valve that prevents the collapse or distention of the cystic duct during sudden changes in pressure. The wall of the gallbladder consists of a mucous membrane, a muscularis, and a serosa or adventitia (Fig. 15-43).

The **mucous membrane** of the gallbladder wall consists of a **simple columnar epithelium** and an underlying lamina propria. The oval nuclei are located basally in the cells, and the luminal surfaces show numerous short microvilli. The apices of adjacent cells are joined, near the lumen, by typical zonula occludens junctions. The epithelium rests on a thin basal lamina that separates it from the

Fig. 15-43. A section through the wall of a human gallbladder shows the elaborately folded mucosa (LM, × 40).

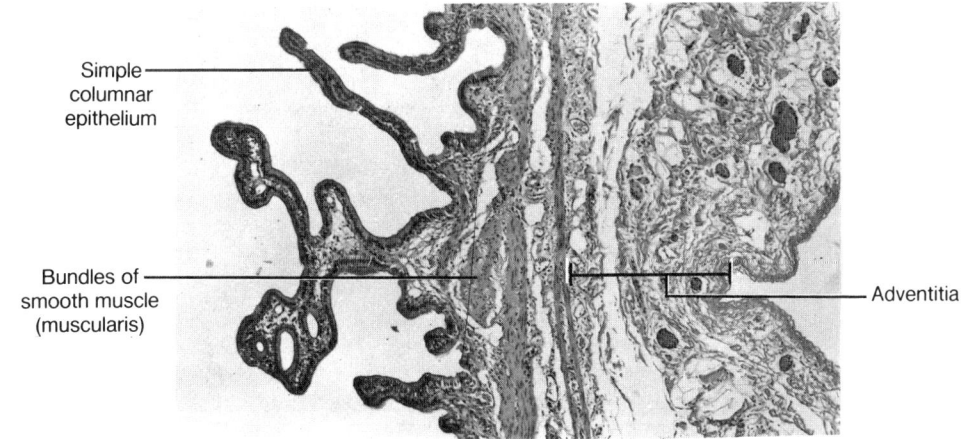

Simple columnar epithelium

Bundles of smooth muscle (muscularis)

Adventitia

delicate connective tissue of the **lamina propria,** which contains numerous small blood vessels. Occasional glands are found in the lamina propria, especially where the gallbladder joins the cystic duct. These small, simple tubuloalveolar glands are thought to secrete mucus. The mucosa of the nondistended gallbladder forms large irregular folds called rugae, which flatten out as the gallbladder fills with bile.

There is no submucosa in the gallbladder. The **muscularis** consists of interlacing bundles of smooth muscle that spiral around the lumen of the gallbladder. Gaps between the smooth muscle bundles are filled with collagenous, reticular, and elastic fibers. Because of the musculoelastic wall and the rugae, the gallbladder has considerable capacity for distention.

The surrounding fibroconnective tissue of the **adventitia** is fairly dense and is continuous with the connective tissue of the liver capsule. The free surface of the gallbladder (that exposed to the abdominal cavity) is covered by a **serosa.**

The gallbladder stores and concentrates bile, which is elaborated continuously by the liver. In humans the gallbladder may contain 30 to 50 ml of bile. On stimulation by cholecystokinin, the gallbladder wall contracts and the sphincters of the common bile duct and ampulla relax, allowing bile to be released into the duodenum.

Pancreas

The pancreas is the second largest gland associated with the gastrointestinal tract. In humans it is a retroperitoneal organ, 20 to 25 cm long, extending transversely from the duodenum across the posterior abdominal wall to the spleen. Macroscopically it consists of a head that lies in the C-shaped curve of the duodenum, a slightly constricted neck, and a body that forms the bulk of the pancreas. It lacks a definite capsule but is covered by a thin layer of areolar connective tissue that extends delicate septa into the substance of the pancreas and subdivides it into numerous small lobules. Blood vessels, nerves, lymphatics, and excretory ducts course through the septa. The major excretory duct runs the length of the pancreas, collecting side branches along its course. The duct and its branches give some structural support to the gland.

The pancreas consists of an exocrine portion, which elaborates numerous digestive enzymes, and an endocrine portion, whose secretions are important in carbohydrate metabolism. Unlike the liver, the exocrine and endocrine functions of the pancreas are performed by different groups of cells.

Exocrine Pancreas

KEY WORDS: compound tubuloacinar (alveolar) gland, pyramidal cell, zymogen granule, digestive enzyme, secretory canaliculus

The exocrine pancreas is a large, lobulated, **compound tubuloacinar (alveolar) gland** in which the secretory units are tubular or flask-shaped (Fig. 15-44). A delicate network of reticular fibers surrounds each secretory unit and forms the supporting stroma. Each acinus consists of a

Fig. 15-44. A section of pancreas showing both exocrine and endocrine components (LM, × 100).

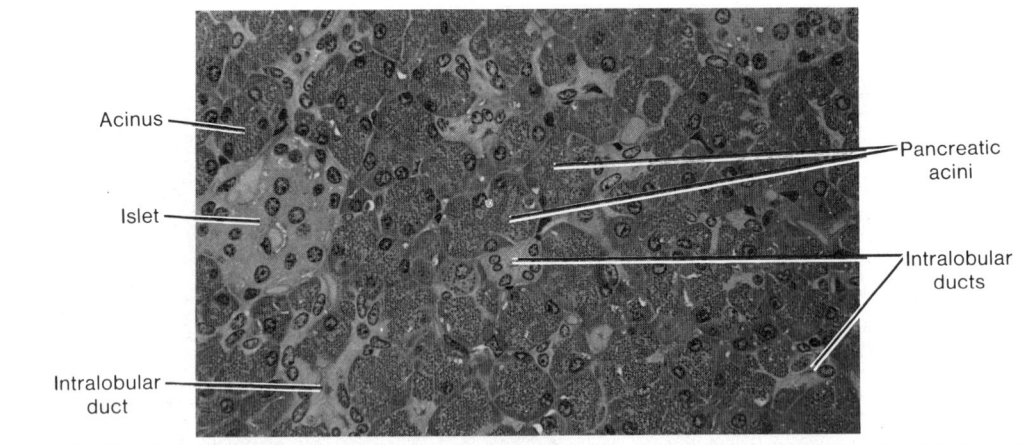

Acinus

Islet

Intralobular duct

Pancreatic acini

Intralobular ducts

single layer of large **pyramidal cells** whose narrow apices border on a lumen, while their broad bases lie on a thin basement membrane (Fig. 15-45). A single, spherical nucleus is located near the base of the acinar cell, and one or more nucleoli may be present. The basal and perinuclear cytoplasm is filled with granular endoplasmic reticulum and mitochondria. Extensive Golgi complexes occupy the supranuclear region and are associated with forming zymogen granules. Mature **zymogen granules** are large, spherical, homogeneous structures that are limited by a membrane and often fill the apical cytoplasm. The pancreatic acinar cell, as indicated by its morphology, is actively involved in the synthesis and release of proteins (enzymes).

Digestive enzymes of the pancreas are synthesized in the granular endoplasmic reticulum of the acinar cell and pass through the cisternae of this organelle to reach the Golgi complex in small transport vesicles. Precursors of the enzymes are packaged into zymogen granules by membranes of the Golgi complex. The granules then are liberated into the apical cytoplasm, where they accumulate. When the enzymes are to be released from the cell, the granules migrate to the apical cell membrane and discharge their contents into the acinar lumen by exocytosis. The contents of the granules also may be discharged into fine **secretory canaliculi** that lie between adjacent acinar cells and are continuous with the acinar lumen. Pancreatic acinar cells secrete amylases, endopeptidases (lipases, trypsin, chymotrypsin) that cleave central peptide bonds, and exopeptidases (carboxypeptidases A and B)

that cleave terminal peptide bonds. Together these proteolytic enzymes cleave proteins in the intestinal lumen into peptide fragments that then are reduced to amino acids prior to absorption.

Deoxyribonuclease and ribonuclease also are produced by pancreatic acinar cells and digest deoxyribonucleoprotein and ribonucleoprotein, respectively. Pancreatic amylase breaks down starch and glycogen to disaccharides, and pancreatic lipase cleaves neutral fat into fatty acid and glycerol.

Duct System

KEY WORDS: **centroacinar cell, intercalated duct, intralobular duct, interlobular duct, secretin, cholecystokinin, enterokinase**

An extensive duct system permeates the pancreas. At their beginnings, the ducts extend into the acini and are interposed between the acinar cells and the lumen. Ductal cells within the pancreatic acini are called **centroacinar cells** and appear as flattened, light-staining cells with few organelles. The wall of the short initial segment of the duct system is lined by centroacinar cells and is continuous, outside the secretory unit, with **intercalated** and **intralobular ducts** (see Fig. 15-44). These ducts are tributaries of the **interlobular ducts** found in the loose connective tissue between lobules; the transition between ducts is gradual. The epithelial lining begins as simple squamous in the intercalated ducts, increases in depth to cuboidal in the

Fig. 15-45. A section through a pancreatic acinus. Note the zymogen granules within acinar cells and the centroacinar cell (LM, × 1,000).

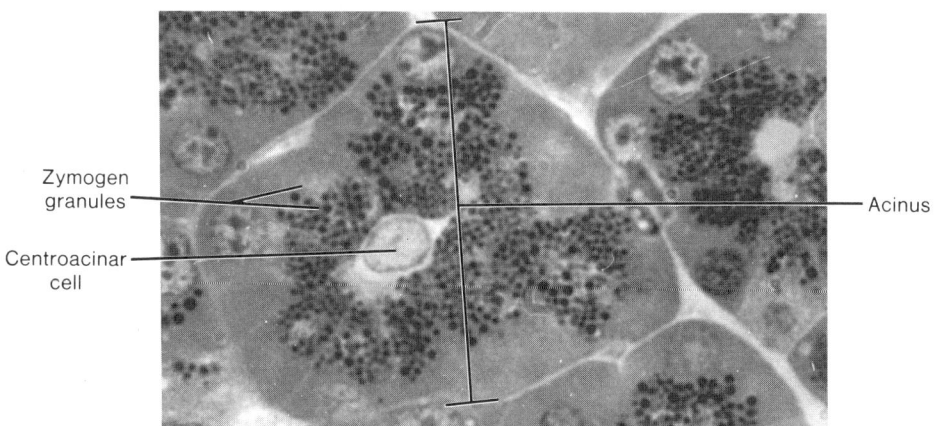

Zymogen granules —
Centroacinar cell —
Acinus —

intralobular ducts, and is columnar in the interlobular ducts. Scattered goblet cells and endocrine cells also are found in the ductal system. The cells lining the ducts stain lightly, and organelles are not prominent in their cytoplasm. Desmosomes are scattered along the lateral cell surfaces, and the apices of adjoining cells are united by tight junctions. The apical surfaces bear microvilli.

A scanty connective tissue consisting mainly of reticular fibers supports the intercalated and intralobular ducts. The interlobular and major secretory ducts are contained within the interlobular septae and thus are surrounded by a considerable amount of fibroconnective tissue. The interlobular ducts drain into the primary and accessory pancreatic ducts.

The primary duct runs the length of the pancreas, increasing in size near the duodenum, where it runs parallel to the common bile duct, with which it often shares a common opening at the greater duodenal papilla. The openings of both ducts are controlled by the sphincter of the papilla. An accessory duct may lie proximal to the main duct and opens independently into the duodenal lumen.

Secretion by the exocrine pancreas is under nervous and hormonal control. Two polypeptide hormones, secretin and cholecystokinin, are released from cells of the intestinal mucosa and influence pancreatic secretion. **Secretin,** elaborated by *S* cells in the mucosa of the duodenum and proximal jejunum, stimulates the ductal system of the pancreas to secrete a large volume of fluid that is rich in bicarbonate. In the intestinal lumen, this alkaline secretion neutralizes the acid chyme from the stomach and deactivates the gastric enzyme pepsin, whose optimal activity occurs at a low pH (2.0). It also establishes the neutral to alkaline conditions needed for the optimal activity of pancreatic enzymes.

Cholecystokinin, elaborated by I cells in the duodenal and jejunal mucosa, stimulates secretory units of the pancreas to synthesize and release digestive enzymes. The proteolytic enzymes are secreted as inactive precursors (zymogens) that, in part, are converted to their active forms by **enterokinase,** an enzyme from the intestinal mucosa. Pancreatic amylase and lipase are said to be secreted in their active forms. Cholecystokinin also stimulates the gallbladder to contract, thereby adding bile to aid in neutralizing the intestinal contents and providing bile salts that act as emulsifying agents in the breakdown of neutral fats.

Endocrine Pancreas

KEY WORDS: pancreatic islet, alpha cell, glucagon, beta cell, insulin, delta cell, somatostatin, pancreatic polypeptide, PP cell, exocytosis, fenestration

Scattered among the pancreatic acini are irregular, elongate masses of pale-staining cells called the **pancreatic islets,** which make up the endocrine portion of the pancreas (see Fig. 15-44). In humans they number about 1 million. The islets are separated from the surrounding exocrine pancreas by a delicate investment of reticular fibers, and as with all endocrine glands, the islets have a rich vascular supply. In ordinary tissue sections, the islets appear to be composed of a homogeneous population of pale polygonal cells, but with special stains and in electron micrographs, several distinct cell types can be identified (Table 15-3). The majority of cells are alpha and beta cells.

Alpha cells make up about 20 percent of the islets in humans and generally are located at the periphery of the islet. They contain large, dense, spherical granules that are lim-

Table 15-3. Endocrine Cells of the Pancreas

Lausanne Classification of Cell Types (1977)	Hormonal Substance	Primary Function
A (alpha) cell	Glucagon	Increases blood sugar
B (beta) cell	Insulin	Decreases blood sugar
D cell	Somatostatin	Inhibitor of hormone secretion (paracrine function)*
PP cell	Pancreatic polypeptide	Opposes action of cholecystokinin
G cell†	Gastrin	Increases HCl secretion in stomach

* Paracrine function. Secretion of somatostatin is directly into the intercellular space from which it diffuses to inhibit the secretion of adjacent endocrine cells.
† G cells are present only in some species or restricted to fetal life in others.

ited by a membrane. Mitochondria are long and slender with a typical internal structure. A few profiles of granular endoplasmic reticulum are scattered about the cytoplasm, and Golgi bodies usually are found near the nucleus. They contain secretory granules with diameters of about 250 nm. Alpha cells produce the peptide hormone **glucagon,** which elevates blood sugar.

Beta cells form about 78 percent of the islet cells and tend to locate near the center of the islet (Fig. 15-46). Their secretory granules are smaller than those in alpha cells and show species differences. In humans, beta cell granules contain small, dense crystals that give them a distinctive appearance. The granular endoplasmic reticulum is less extensive and the Golgi complexes are more distinctive than in alpha cells. Beta cells produce **insulin,** which acts on the plasmalemma of various cell types, especially liver, muscle, and fat cells, to facilitate the entry of glucose into the cell and thus lower blood glucose. Insulin also stimulates pancreatic acinar cells.

Several other types of endocrine cells are present in small numbers in the islets. **Delta cells** secrete the hormone **somatostatin,** which inhibits hormone secretion by surrounding endocrine cells. **Pancreatic polypeptide,** secreted by **PP cells,** opposes the action of cholecystokinin and inhibits pancreatic acinar secretion. Endocrine cells of the kind found in the islets, including alpha and beta cells,

also are scattered within the ducts and acini of the exocrine pancreas. The number of different endocrine cells may vary according to their locations in the head or body of the pancreas and may relate to the different origins of these two parts. There is considerable species variation in the distribution of cells within the islets. In humans beta cells tend to be located centrally, while alpha, delta, and PP-cells are concentrated at the periphery of the islets.

Islet cells are intimately associated with surrounding capillaries, and the secretory granules often appear to be located near the cell membrane that is adjacent to the vasculature. Secretory granules are released from islet cells by **exocytosis** into the surrounding blood vessels. The endothelium of the islet capillaries contains numerous **fenestrations** to facilitate entry of secretory products into the vasculature. In contrast, the endothelium of capillaries of the exocrine pancreas is not fenestrated.

A significant proportion of the arterial blood enters the pancreas via small interlobular and intralobular arteries that first supply the islets. Arterioles derived from these arteries give rise to capillaries at the periphery of the islets, in the region of the alpha, delta, and PP-cells. The capillaries then run centrally within the islet, after which they supply adjacent acinar cells of the exocrine pancreas. Thus the islet cells can interact with each other and influence the function of the exocrine pancreas.

Fig. 15-46. A pancreatic islet stained by immunohistochemical means to demonstrate insulin. Note that the insulin immunoreactivity is concentrated near the cell surfaces facing capillaries (LM, × 250).

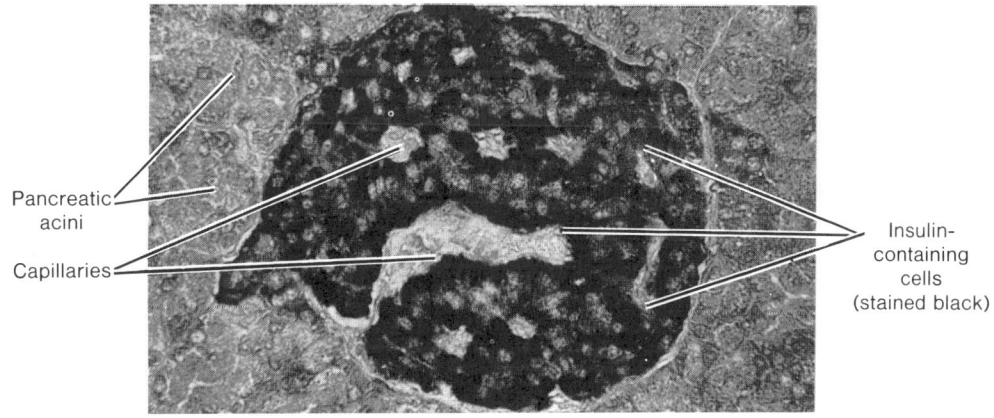

Pancreatic acini

Capillaries

Insulin-containing cells (stained black)

Review Table 15-1. Key Features of the Alimentary Tract

	Epithelium	Muscularis Mucosae	Muscularis Externa	Lymphatic Tissue	Glands
Esophagus	Stratified squamous nonkeratinizing	Prominent; longitudinal smooth muscle	Outer longitudinal, inner circular layers. Skeletal in upper one-fourth, mixed in middle one-fourth, smooth in lower one-half	Few nodules, diffuse in lamina propria	1. Esophageal proper, scattered along length of esophagus in submucosa, compound tubuloalveolar, mucus-secreting 2. Esophageal cardiac, proximal and distal ends in lamina propria; compound tubuloalveolar, mucus-secreting
Stomach Cardia	Simple columnar	Smooth muscle; inner circular, outer longitudinal; a third outer circular layer in some areas	Outer longitudinal, middle circular, inner oblique smooth muscle	Small patches of diffuse lymphatic tissue; rare nodules	Long pits (one-half the depth of mucosa); simple branched tubular; mucous, endocrine, and occasional parietal cells
Fundus	Simple columnar	Same	Same	Same	Short pits (one-fourth the depth of the mucosa); simple branched tubular; mucous neck, parietal, chief, and endocrine cells
Pylorus	Simple columnar	Inner circular, outer longitudinal smooth muscle	Same	Occasional small lymphatic nodules in lamina propria	Long pits (one-half the depth of mucosa); simple branched tubular; mucous, occasional parietal and numerous endocrine cells (G cells)
Duodenum	Villi covered by simple columnar absorptive cells with microvillus border; goblet and endocrine cells	Inner circular, outer longitudinal smooth muscle	Inner circular, outer longitudinal smooth muscle; myenteric plexus between layers	Solitary nodules and diffuse lymphatic tissue	1. Intestinal glands in lamina propria; simple tubular; absorptive, goblet, endocrine and Paneth 2. Brunner's glands in submucosa; branched tubular

Small intestine	Villi covered by simple columnar absorptive cells; goblet cells increase in number from upper to lower; endocrine cells present	Same	Same	Nodular lymphatic tissue forms aggregates (Peyer's patches) in ileum	Intestinal glands; same as for duodenum
Colon	Villi absent; simple columnar absorptive cells; goblet cells increase in number	Same	Outer longitudinal layer condensed into 3 bands — taeniae coli	Solitary nodules and diffuse lymphatic tissue	Intestinal glands; goblet, endocrine, and absorptive cells; Paneth cells absent
Rectum	Simple columnar; mainly goblet cells	Same	Inner circular and outer longitudinal	Solitary nodules	Intestinal glands; goblet cells more abundant
Anal canal	At pectinate line becomes stratified squamous, noncornified, then stratified squamous cornified at orifice	Disappears at level of anal columns	Inner circular layer thickens to form internal sphincter, outer longitudinal layer disappears	Solitary lymphatic nodules	1. Intestinal glands disappear at pectinate line 2. Apocrine sweat glands (circumanal glands)

Functional Review: Digestive System

The digestive system is designed for breaking down ingested materials to their basic constituents, which then are absorbed and used by the organism. The process of mechanical and chemical breakdown of food substances is called digestion.

The oral cavity is specialized to take in and mechanically break down food through the cutting and grinding action of the teeth. The pulverized food is moistened and softened by secretions of the major and minor salivary glands. Digestion of carbohydrates is initiated in the oral cavity by the enzymes amylase and maltase, which are secreted by the major salivary glands. This chemical digestion of carbohydrates is short-lived, however, since both enzymes are destroyed by the acid environment of the stomach. Epidermal growth factor (EGF) and nerve growth factors from the submandibular glands influence the kinetics of epithelial tissues and other tissues lining the alimentary canal. Salivary gland secretions also moisten and clean the mouth, provide a proper environment for taste, and act as vehicles to transport heavy metals out of the body. When salivary gland secretion decreases during dehydration, it initiates the sensation of thirst and thus is a factor in maintaining fluid balance. The stratified squamous epithelium that lines the oral cavity protects the underlying structures from abrasion during eating. In regions subjected to excessive abrasion (gingiva and palate), the epithelium is cornified.

The tongue is important for chewing and swallowing. The interior of the tongue, cheek, and lips consists of skeletal muscle and is under voluntary control. The three structures serve to direct food to the teeth during mastication, direct it to the back of the throat for swallowing, and provide the voluntary control for suckling and speech.

The mixture of saliva and ground food is formed into a semisolid bolus that, when of the right consistency, is swallowed. Saliva is important for this activity also; swallowing is impossible if the mouth is parched. During the initial stage of swallowing, the bolus is directed to the oropharynx by several simultaneous, coordinated events. The anterior part of the tongue is pressed firmly against the hard palate, and at the same time, the base of the tongue is retracted. Bone in the hard palate provides a rigid platform against which the tongue can press and prevents collapse of the palate due to the pressure exerted against it during swallowing. Formation of a mucoperiosteum prevents sliding and tearing of the mucosa that covers the palate. As these events occur, the larynx and pharynx are elevated to receive the bolus. Skeletal muscles associated with the soft palate contract, moving the soft palate upward to seal off the oropharynx and prevent food from entering the nasopharynx and nose. As the pharynx is elevated, its lumen dilates to receive the bolus from the oral cavity. Simultaneously, the musculature of the larynx contracts, closing the entrance to the respiratory tree. Skeletal muscle fibers of the three pharyngeal constrictor muscles contract about the entering bolus and quickly force it into the esophagus.

Skeletal muscle fibers in the muscularis externa of the upper esophagus are fast-acting, and their constriction carries the bolus of food into the central region of the esophagus, where there is a gradual transition to slow-acting smooth muscle. Peristaltic waves are formed by the smooth muscle in the lower half of the esophagus and move the bolus into the stomach. Because of peristalsis, materials from the esophagus enter the stomach in spurts. The lumen of the esophagus in humans is lined by a wet stratified squamous epithelium that, as in the oral cavity, protects the surrounding structures from the abrasive action of materials as they pass through the lumen. The esophageal glands provide a lubricant for passage of material. The main function of the esophagus is to transport food from the oral cavity to the stomach. It secondarily serves to warm or cool ingested materials.

In the stomach the arrangement of the muscle layers provides a churning action that promotes mixing with gastric secretions and mechanical breakdown of foodstuff. Protein is degraded by pepsin, an endopeptidase that splits peptide linkages near the center of the molecule. In addition to secreting hydrochloric acid and pepsin, the gastric mucosa secretes a considerable amount of fluid (about 1000 ml of gastric juice after each meal in humans), most of which is reabsorbed in the intestinal tract. In humans the parietal cells also secrete gastric intrinsic factor, which binds to vitamin B_{12}, important in erythropoiesis. The complex is absorbed in the distal small intestine. If gastric intrinsic factor is not produced or is present in insufficient amount, much of the vitamin passes through the intestinal tract and is lost in the feces, resulting in development of pernicious anemia.

The gastric mucosa of many suckling mammals contains rennin, an enzyme that curdles milk, and a lipase that begins the digestion of fat.

The contents of the proximal stomach are semisolid, whereas those in the distal region form a pulplike, fluid mass called chyme. After reaching the proper consistency, chyme enters the duodenum in small portions. The smooth muscle of the pyloric sphincter helps to control the evacuation of the stomach.

The simple columnar epithelium that lines the stomach forms a glandular sheet that secretes a neutral mucin. The epithelium and its secretions protect the stomach from the erosive effect of acid pepsin in the gastric lumen. Bicarbonate produced during hydrochloric acid formation by parietal cells enters the mucosal microcirculation beneath the surface lining and contributes to its protection also. The mucus secreted by the cardiac and pyloric glands helps to protect the mucosa where the stomach joins the esophagus and duodenum, respectively. Gastrin-producing endocrine cells are present in large numbers in the pyloric glands; gastrin stimulates acid secretion and gastric motility. Because of the large population of G cells, the pyloric region is not merely a mucus-secreting area but also assumes the status of an endocrine organ.

As the acid chyme enters the duodenum it is subjected to a change of pH. Duodenal glands secrete an alkaline fluid that protects the proximal duodenal mucosa from the acid pepsin from the stomach. The duodenal glands, like the submandibular glands, are involved in the secretion of EGF. The alkaline secretions of the pancreas and liver elevate the pH to neutral or alkaline and provide an environment optimal for the activities of digestive enzymes secreted by the pancreas; the proteolytic enzyme of the stomach, pepsin, is deactivated. Bile salts also are added to the intestinal contents and act as detergents or emulsifying agents to aid in the digestion of fats by pancreatic lipase.

Most of the digestive process takes place in the intestinal lumen. The final degradation of materials to their basic components (amino acids, monosugars, monoglycerides, fatty acids) occurs on the luminal surface of the intestinal epithelial cells just prior to absorption.

The intestinal wall is specialized for absorption of digested materials, and the surface area available is greatly increased by the plicae circulares, villi, and microvilli. The efficiency of absorption is further increased by movement of the villi and by segmental movement of the luminal contents by contractions of the muscularis externa, which move the intestinal contents back and forth. Peristaltic waves of the muscularis externa move the contents through the intestine and along the remainder of the tract. As materials are absorbed, the luminal contents become more concentrated. Correspondingly, there is an increase in the number of goblet cells distally in the small intestine to provide mucus for lubrication.

Goblet cells are among the most numerous cells in the colon, and they increase in number toward the rectum. Villi are absent in the colon, but intestinal glands remain a prominent feature of the mucosa. The primary function of the colon is to absorb fluid that remains in the luminal contents. As the fecal mass develops, secretions from the large population of goblet cells provide the lubrication necessary to prevent damage to the colonic mucosa. In the upper rectum, the lining epithelium is made up primarily of goblet cells.

The anal canal is lined by a wet stratified squamous epithelium that protects this area from abrasion by the feces during defecation. Smooth muscle of the inner circular layer of the muscularis externa forms the internal anal sphincter. The surrounding skeletal muscle fibers adjacent to the anal orifice form the external sphincter, which is under voluntary control.

The organism is protected from a vast, ever-present population of microorganisms in the digestive system by immunoglobulins (mainly IgA) that form a barrier against fungal, bacterial, and viral invasions. The immunologic barrier extends throughout the digestive tract and is the joint product of epithelial and lymphatic cells. The epithelial component, the secretory piece, is thought to protect the immunoglobulin from digestion both within and outside the cell. In the small intestine and colon, this barrier together with phagocytes forms a protective sleeve around the intestinal lumen. Large amounts of IgA also enter the intestinal lumen from the liver via the hepatobiliary pathway and play an important role in keeping the population of microorganisms in check.

From its exocrine portion, the pancreas provides enzymes that digest carbohydrates, proteins, and lipids. The endocrine portion secretes insulin and glycogen, hormones that are instrumental in regulating blood glucose levels, the flow of glucose across cell membranes, and glyconeogenesis in the liver. Carbohydrate is stored by the liver as glycogen, which can be released into the bloodstream as glucose. The liver contributes bile salts for the emulsification of fats during digestion, plays a vital role in detoxification, and supplies proteins that are important in blood coagulation. The liver functions in lipid transport and maintains lipid levels in blood. Lipid is complexed to protein in the hepatocytes, secreted into the perisinusoidal space, and transported in the blood as plasma lipoprotein.

The endocrine cells of the gastrointestinal tract and pancreatic islets produce hormones that influence or control gastric secretions, gastrointestinal motility, gastric emptying, gallbladder contraction, and pancreatic and liver secretion. How the specific hormones interact with one another or are influenced by other major endocrine organs is unknown.

Review Questions and Answers

Questions

77. List in order the different types of surface lining epithelium encountered in the digestive tract beginning with the oral cavity and ending with the anal canal. Briefly discuss the functional significance of each type listed.
78. List and briefly discuss three morphologic adaptations for absorption found in the small intestine.
79. Discuss the classic and portal hepatic lobules in relation to blood and bile flow of the liver lobule.
80. Discuss the morphology and functional significance of the perisinusoidal space.
81. Briefly discuss the relationship of two endocrine cells in the gastrointestinal mucosa to the exocrine pancreas and its function.
82. Briefly discuss the ultrastructure of a pancreatic acinar cell as related to its function.

Answers

77. The oral cavity is lined throughout by a wet stratified squamous epithelium that provides a protective surface resistant to abrasion. The dorsum of the tongue, the hard palate, and the gingiva are subjected to a greater degree of abrasion by foodstuff, and in these regions the lining epithelium may become cornified. The esophagus also is lined by stratified squamous epithelium, which in humans is noncornified. As in the oral cavity, it provides protection against the abrasive action of materials that pass through the esophagus to the stomach. At the esophageal-gastric junction there is an abrupt transition from nonkeratinized stratified squamous epithelium to the simple columnar epithelium of the stomach.

The epithelial lining of the stomach is organized into a secretory sheet that secretes continuously. The neutral mucin produced by the surface epithelium of the stomach forms a mucous film that protects the gastric mucosa chemically from the acid pepsin in the lumen. The mucus secretion also lubricates the mucosal surface. This type of lining epithelium ends abruptly at the gastrointestinal junction.

The small intestine is lined by a heterogeneous population of simple columnar cells. The principal cell type is the intestinal absorptive cell, which shows a prominent microvillus border on its apical (luminal) surface. This cell serves primarily in the absorption of digested materials from the intestinal lumen. Goblet cells within the lining epithelium represent unicellular exocrine glands that secrete mucus to lubricate and protect the mucosal surface. As materials pass through the intestinal tract, fluids are absorbed and the luminal contents become firmer and more abrasive. Thus the number of goblet cells increases distally in the intestinal canal, and in the distal colon they become the most prominent cell type. A variety of endocrine cells (unicellular endocrine glands) are found in limited numbers, scattered throughout the lining epithelium of the small intestine and colon. They secrete a number of peptide hormones that influence gastric secretion, motility, gallbladder contraction, and hepatic and pancreatic activity.

An abrupt transition from simple columnar to noncornified stratified squamous epithelium occurs at the junction between the rectum and anal canal. The wet stratified squamous epithelium of the anal canal protects this region from the abrasive effects of the fecal mass during defecation.

78. Three morphologic adaptations for absorption are found in the small intestine: plicae circulares, villi, and microvilli. The plicae circulares are large, permanent, shelflike folds that may branch and that spiral about the lumen of the intestine for one-half to two-thirds of

its circumference. They consist of intestinal mucosa with a central core of submucosa. Plicae begin in the duodenum, reach their maximum development in the proximal jejunum, and diminish in size and disappear in the distal half of the ileum.

Villi are best developed in the duodenum and jejunum, where they form broad, spatulate structures; in the ilium they are shorter and more finger-like. Villi consist of a central core of lamina propria covered by intestinal epithelium. They are mobile structures and contract four to five times a minute, extending and retracting from the intestinal lumen.

Microvilli are finger-like evaginations of the apical plasmalemma that extend from the intestinal absorptive cells. They are closely packed and form the striated border of light microscopy.

All three structures greatly increase the surface area of the intestinal mucosa and increase its absorptive capacity.

79. A classic liver lobule is related to the blood flow through the liver. It consists of a central vein about which radiate branching and anastomosing plates of hepatocytes. The plates are separated by blood sinusoids and thus assume a spokelike arrangement. The hepatic lobule itself is somewhat hexagonal in cross section, with a central vein at the center and portal areas at the corners. Blood passes from small branches of the hepatic artery and portal vein into the sinusoids that lie between the plates of hepatocytes. Blood flows slowly through the hepatic sinusoids toward the center of the lobule and leaves through the central vein.

The lobule associated with the exocrine secretion of the liver (bile) is termed the portal lobule and has at its center a portal canal. A portal lobule consists of the surrounding hepatic tissue that is drained by a bile duct of a portal area. It is triangular in shape and contains portions of three adjacent hepatic lobules. A central vein is located at each corner of a portal lobule.

Thus blood flow in the liver is directed toward the central vein of the classic lobule, whereas bile flows in the opposite direction — that is, toward the periphery of the lobule to bile ducts in surrounding portal areas.

80. Hepatic sinusoids are wider and have a more irregular shape than most capillaries. In those species where a discontinuous, fenestrated type of endothelium is pre-

sent, the sinusoidal lining is separate from hepatocytes by a narrow space called the perisinusoidal (Disse's) space. It is devoid of ground substance and most connective tissue elements. Blood plasma flows freely through the endothelium and into and through the perisinusoidal space. Because the blood plasma has direct access to it, liver cells are constantly bathed on one surface by plasma that is rich in nutrients absorbed by the gastrointestinal tract. The plasmalemma of hepatocytes adjacent to the perisinusoidal space exhibits numerous well-developed microvilli that increase its surface area and facilitate absorption. The perisinusoidal space has considerable functional significance in the active exchange of nutrients between the blood plasma and hepatocytes.

81. Two distinct polypeptide hormones, secretin and cholecystokinin, are released from endocrine cells in the intestinal mucosa. These hormones influence secretion by the exocrine pancreas. Secretin, produced by the S cells of the duodenal and proximal jejunal mucosae, stimulates the ductal system of the exocrine pancreas to secrete a large volume of fluid that is rich in bicarbonate. Cholecystokinin, produced by the I cells of the duodenal and jejunal mucosa, stimulates the synthesis and release of digestive enzymes by the pancreatic acinar cells.

82. The pancreatic acinar cell is actively involved in the synthesis and release of digestive enzymes. Ultrastructurally, the acinar cell is characterized by an abundance of granular endoplasmic reticulum, which fills the basal and perinuclear cytoplasm. Extensive Golgi complexes occupy the supranuclear region and are associated with forming zymogen granules that accumulate in and often fill the apical cytoplasm. An acinar cell contains a single, spherical nucleus located at its base, and one or more nucleoli may be present. A number of digestive enzymes (proteins) are synthesized in the granular endoplasmic reticulum and pass through the cisternae of this organelle to reach the Golgi complex via small transport vesicles. These precursors of the active digestive enzymes are packaged into membrane-bound zymogen granules by the Golgi complex and then are released into the apical cytoplasm. The zymogen granules migrate to the apical cell membrane, fuse with it, and discharge their contents into the acinar lumen by the process of exocytosis.

16 Urinary System

The urinary system consists of the kidneys, ureters, urinary bladder, and urethra. The kidneys constitute the glandular component; the remainder of the urinary system forms the excretory passages. The ureters conduct urine from the kidneys to the bladder, where it is stored temporarily. In turn, the bladder is drained by the urethra, through which the urine ultimately is voided from the body.

Kidneys

The kidneys are a pair of compound tubular glands that clear the blood plasma of metabolic wastes, regulate fluid, osmolality, and volume, regulate electrolyte balance, eliminate foreign chemicals, and help maintain the acid-base balance of the body. Excretion of metabolic products includes creatinine (from muscle creatine phosphate), uric acid (from nucleic acids), urea (from amino acids), metabolites of hormones and hemoglobin, as well as other substances. The kidneys also are instrumental in eliminating drugs, pesticides, and other environmental factors from the blood plasma. In addition to excretory functions, the kidneys have properties of endocrine organs and release two substances, renin and erythropoietin, directly into the bloodstream. Renin is important in regulating blood pressure and sodium ion concentration; erythropoietin influences hemopoietic activity. The kidneys also are important in activating circulating vitamin D to an active form, 1,25-dihydroxyvitamin D_3, necessary for normal absorption of calcium ion in the gastrointestinal tract.

Macroscopic Features

KEY WORDS: hilum, renal sinus, renal pelvis, major calyx, minor calyx, renal papilla, cortex, medulla, medullary pyramid, renal column, medullary ray, renal lobe, renal lobule, multilobular kidney

Human kidneys are bean-shaped organs that lie in a retroperitoneal position against the posterior abdominal wall, one on either side of the upper lumbar vertebrae. Each is contained within a thin but strong connective tissue capsule that contains much fat. The renal artery and nerves enter the kidney on the medial border at the **hilum,** a concavity that also serves as the point of exit for the renal vein, lymphatics, and ureter. The hilum is continuous with

the **renal sinus,** a large central cavity surrounded by the parenchyma of the kidney and filled with loose areolar connective tissue that normally contains much fat. Nerves, lymphatics, and branches of the renal artery and vein run through the sinus. The **renal pelvis** is a funnel-shaped expansion of the ureter where it joins the kidney; it also passes through the sinus, dividing into two or three short tubular structures called the **major calyces.** These in turn divide into eight to twelve smaller units called **minor calyces.** Each minor calyx forms a cylindrical attachment around a conical projection of renal tissue called a **renal papilla.**

When a hemisection of the kidney is examined macroscopically, the organization of the parenchyma into two distinct regions can be seen readily. The darker, granular-appearing outer region is the **cortex,** which forms a continuous layer beneath the capsule. The inner region, or **medulla,** is paler and smoother and consists of 8 to 20 cone-shaped structures called **medullary pyramids,** which are separated from each other by inward extensions of cortical tissue (Fig. 16-1). The cortex that separates adjacent pyramids makes up a **renal column.** The bases of the pyramids are directed toward the overlying cortex, while their apices are oriented toward the renal sinus and form the renal papillae. From the bases of the pyramids, groups of tubules extend into the cortex, giving it a striated appearance. These striations represent a continuation of medullary tissue into the cortex and constitute the **medullary rays.**

The arrangement of cortex and medulla allows subdivision of the kidney into smaller units, the lobes and lobules. A medullary pyramid, together with its closely associated cortical tissue, forms a **renal lobe,** while a medullary ray, together with its associated cortical tissue, forms a **renal lobule.** When the kidney consists of several lobes, it is referred to as a **multilobular kidney;** the kidneys of humans are this type.

Microscopic Features

KEY WORDS: uriniferous tubule, nephron, collecting duct

Each renal lobule is made up of numerous epithelial tubules called **uriniferous tubules** (Fig. 16-2) that collectively

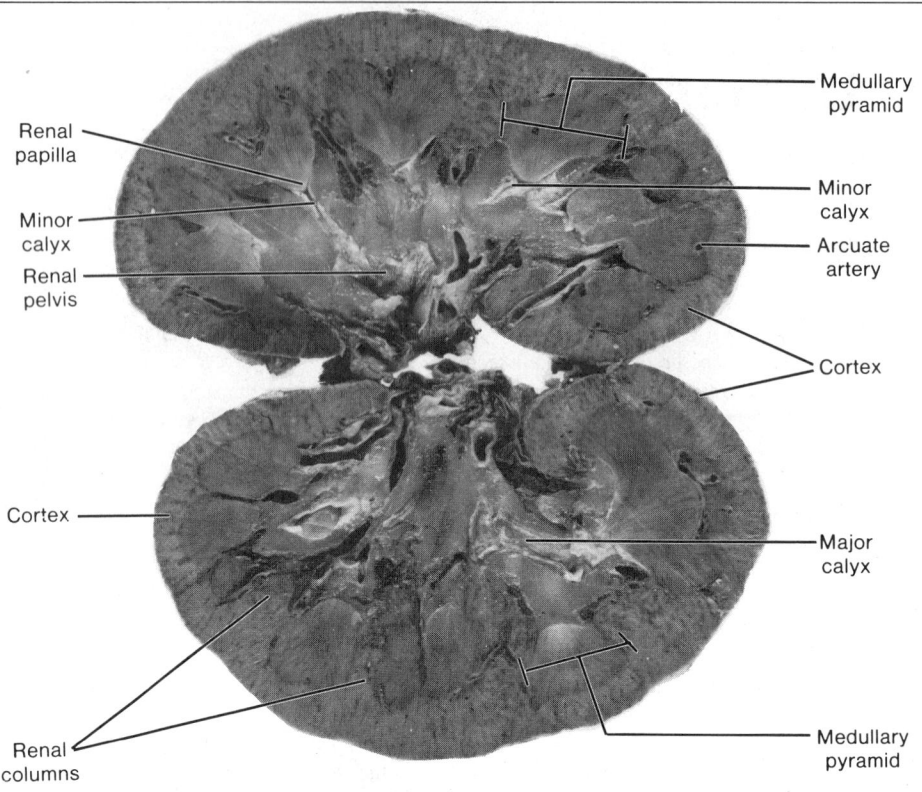

Fig. 16-1. A dissected human kidney illustrating its anatomical features (actual size).

form the parenchyma of the kidney. Each uriniferous tubule can be divided into a **nephron,** which is the secretory portion, and a **collecting duct,** which is the excretory portion that carries urine to the renal pelvis.

Nephron

KEY WORDS: renal corpuscle, proximal tubule, thin segment, distal tubule, loop of Henle

There are 1 to 2 million nephrons in the human kidney, each representing a functional unit of the kidney. A nephron is a blindly ending epithelial tubule that can be subdivided into several components, each differing in structure, function, and position in the kidney. During urine formation, the various segments of a nephron take part in filtration, secretion, and resorption. A typical nephron consists of a **renal corpuscle,** a **proximal tubule** with convoluted and straight

portions, a **thin segment,** and a **distal tubule** that also has straight and convoluted parts. Those segments of the nephron between the proximal and distal convoluted tubules (i.e., the straight descending part of the proximal tubule, the thin segment, and the straight ascending portion of the distal tubule) are collectively referred to as the **loop of Henle.**

The renal corpuscle and the proximal and distal convoluted tubules occur only in the cortex, whereas Henle's loop generally is confined to the medulla or to a medullary ray. The size of the nephrons and the length of their various segments vary according to the position of the parent renal corpuscle in the cortex. Renal corpuscles near the medulla are larger and the tubules are longer than are those of nephrons whose renal corpuscles are located in the periphery of the cortex. Nephrons from the subcapsular region have small renal corpuscles, and their thin segments of the loop of Henle extend for only a short distance into the

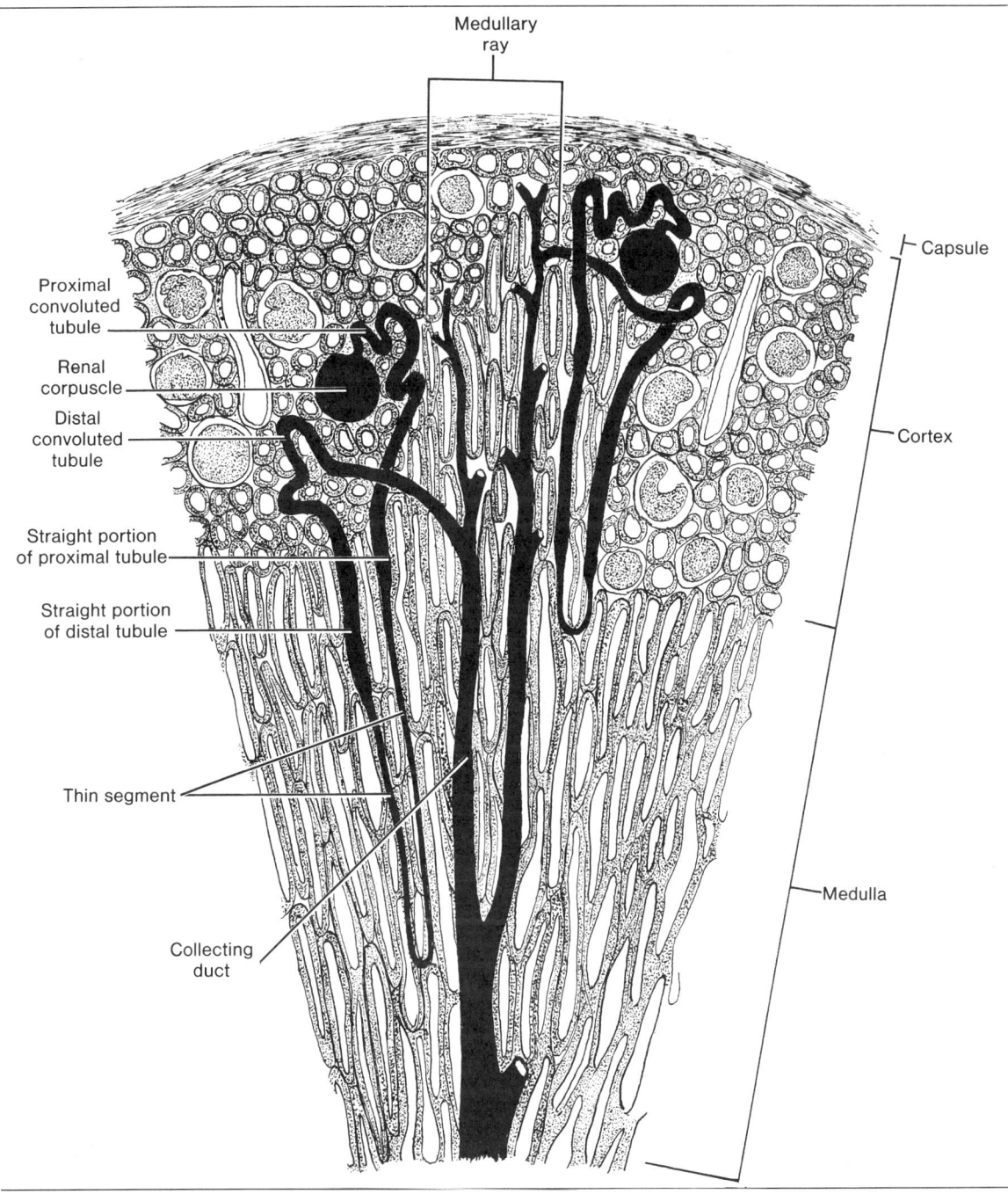

Fig. 16-2. Diagram of uriniferous tubules.

medulla. Nephrons whose renal corpuscles occupy an intermediate position in the cortex show intermediate features.

Renal Corpuscle

KEY WORDS: glomerulus, glomerular capsule, parietal layer, visceral layer, capsular epithelium, glomerular epithelium, capsular space, vascular pole, urinary pole, podocyte, primary process, secondary (foot) process (pedicle), slit pore (filtration slit), slit membrane, common basal lamina, fenestrated glomerular endothelium, mesangial cell, extraglomerular mesangium, intraglomerular mesangial cells, filtration barrier

A renal corpuscle is roughly spherical and measures 150 to 250 μm in diameter. It consists of a capillary tuft, the **glomerulus,** which projects into a blind expansion of the uriniferous tubule called the **glomerular capsule** (Figs. 16-3 and 16-4). The outer layer of the capsule surrounds the glomerulus as the **parietal layer,** which then reflects

onto the glomerulus, where it is intimately applied to the glomerular capillaries to form a complete covering called the **visceral layer** of the glomerular capsule. The parietal layer also is known as the **capsular epithelium,** while the visceral layer frequently is referred to as the **glomerular epithelium.** The visceral and parietal layers of the capsule are separated only by a narrow **capsular space.**

The **vascular pole** is that area of the renal corpuscle at which the afferent and efferent arterioles enter and leave. As the afferent arteriole enters the renal corpuscle, it immediately divides into several primary branches, each of which forms a complex of capillaries (a capillary lobule) that then reunite to form the efferent arteriole. Each glomerulus is made up of several such capillary lobules. On the side opposite the vascular pole, the capsular space becomes continuous with the lumen of the proximal convoluted tubule; this region is known as the **urinary pole** (Figs. 16-4 and 16-5).

The parietal layer of the glomerular capsule consists of

Fig. 16-3. The cast of a renal glomerulus (SEM, × 800).

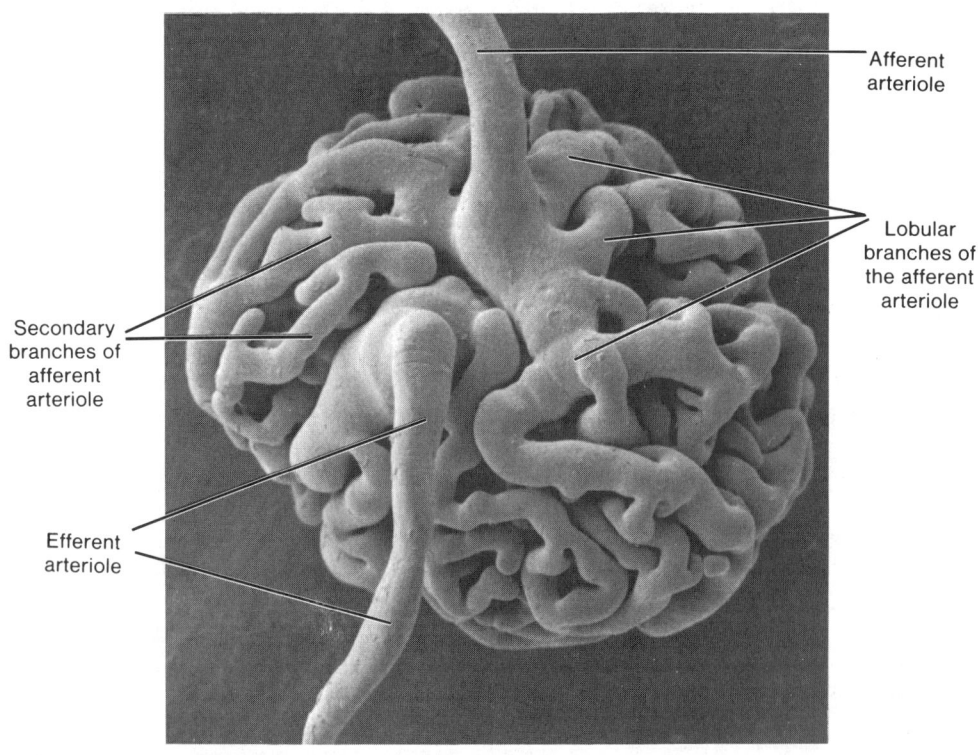

Afferent arteriole

Lobular branches of the afferent arteriole

Secondary branches of afferent arteriole

Efferent arteriole

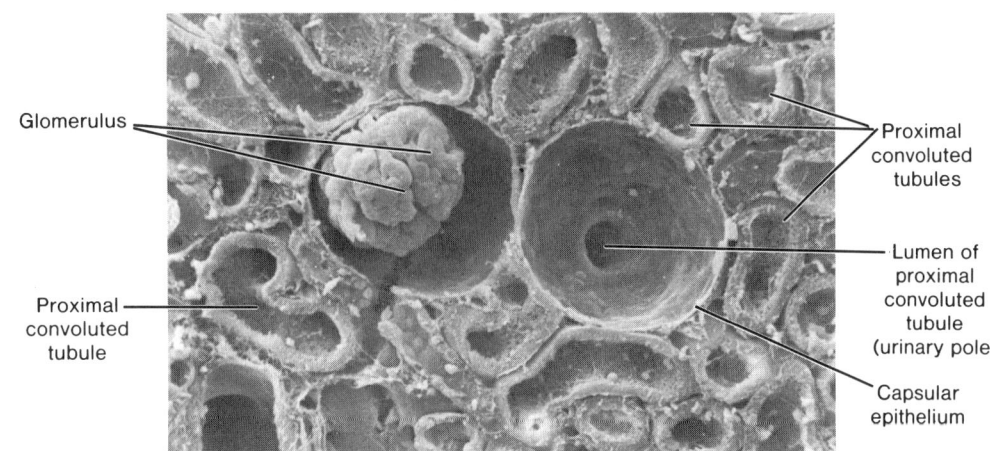

Fig. 16-4. A region of renal cortex showing a renal corpuscle and several proximal convoluted tubules (SEM, × 250).

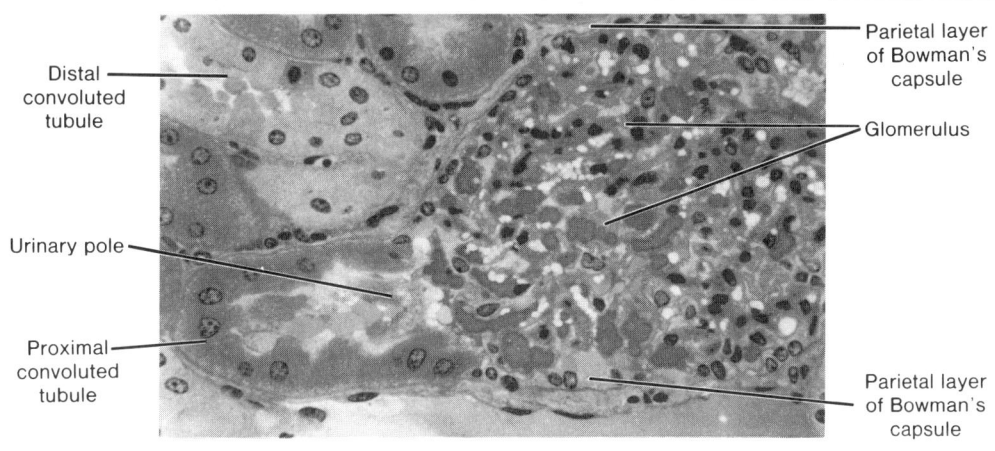

Fig. 16-5. A renal corpuscle showing a urinary pole (compare with 16-4) (LM, × 250).

a single layer of tightly adherent squamous cells that contain few organelles. Near the urinary pole there is an abrupt transition from this simple squamous form to the large pyramidal cells that line the proximal convoluted tubule. At the vascular pole, the simple squamous capsular epithelium changes to a specialized cell type called the podocyte.

Podocytes are large, stellate cells whose bodies lie some distance from the underlying capillaries, separated from them by several cytoplasmic extensions called the **pri-mary processes** (Fig. 16-6). These processes wrap around the glomerular capillaries and give rise to numerous **secondary (foot) processes (pedicles)** that interdigitate with similar processes from adjacent podocytes to completely invest the capillary loops of the glomerulus. The plasmalemma of the foot processes exhibits a prominent glycocalyx rich in a sialoglycoprotein called *podocalyxin.* The narrow clefts between the interdigitating processes form the **slit pores (filtration slits),** which measure about

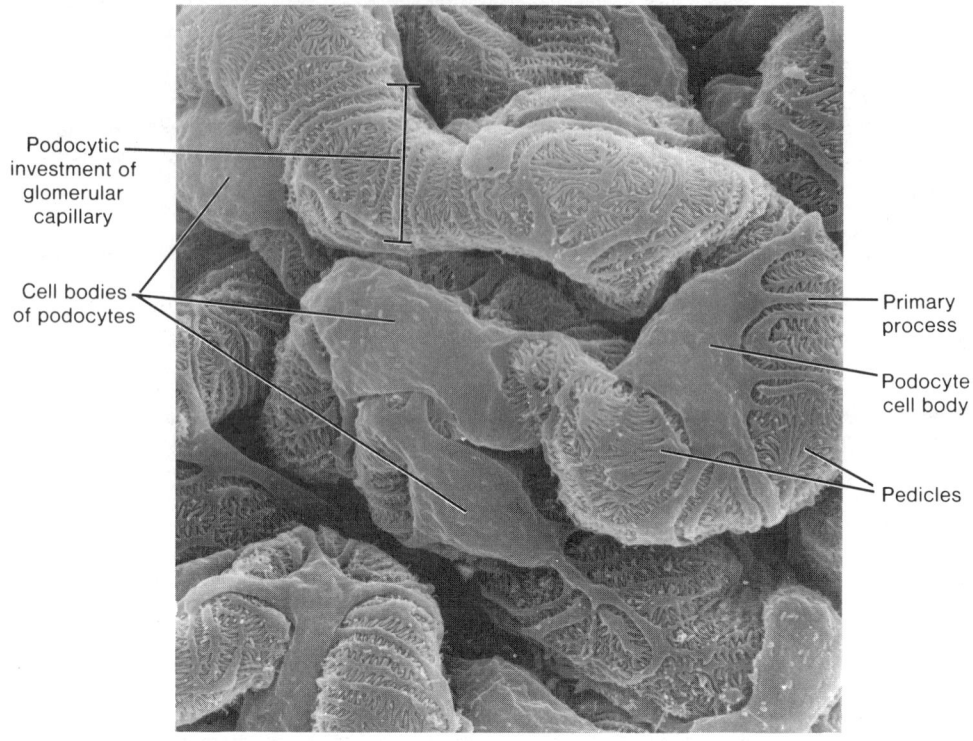

Podocytic investment of glomerular capillary

Cell bodies of podocytes

Primary process

Podocyte cell body

Pedicles

Fig. 16-6. Podocytes comprise the glomerular epithelium and form a complete investment around glomerular capillaries (SEM, × 2,300).

25 nm in width (Fig. 16-7). The gaps may be bridged by a thin **slit membrane** (Fig. 16-8).

The foot processes of the podocytes and the endothelial cells of the glomerular capillaries share a continuous **common basal lamina** that measures 0.1 to 0.5 μm thick. Ultrastructurally, the basal lamina consists of a central electron-dense lamina densa, a lamina rara externa adjacent to the podocyte foot processes, and a lamina rara interna adjacent to the glomerular endothelium. The lamina densa consists of type IV collagen and laminin and heparan sulfate proteoglycan. The lamina rara externa and interna consist primarily of fibronectin, which is thought to firmly attach both epithelial and endothelial cells to the lamina densa. Both cell populations are essential to establish the common basal lamina and thereafter continue to contribute to its production and maintenance.

The glomerular capillaries are lined by a **fenestrated glomerular endothelium** (Figs. 16-8 and 16-9). The at-

tenuated squamous cells show numerous large pores (fenestrae) that measure 50 to 100 nm in diameter.

The renal corpuscle also shows stellate cells with long cytoplasmic processes that contain numerous filaments. These are **mesangial cells,** which occupy the area between the afferent and efferent arterioles at the vascular pole, where they lie in a matrix of amorphous material. The cells constitute the **extraglomerular mesangium** and are continuous with cells of similar appearance, the **intraglomerular mesangial cells,** that lie between the glomerular endothelium and the basal lamina. The latter cells are thought to clear away large protein molecules that become lodged on the common basal lamina during filtration of blood plasma. Mesangial cells may participate in the removal of older portions of the basal lamina from the endothelial side as it is added to by the podocytes. Mesangial cells are contractile and respond to angiotensin II and other vasoconstrictors. They also have receptors for the

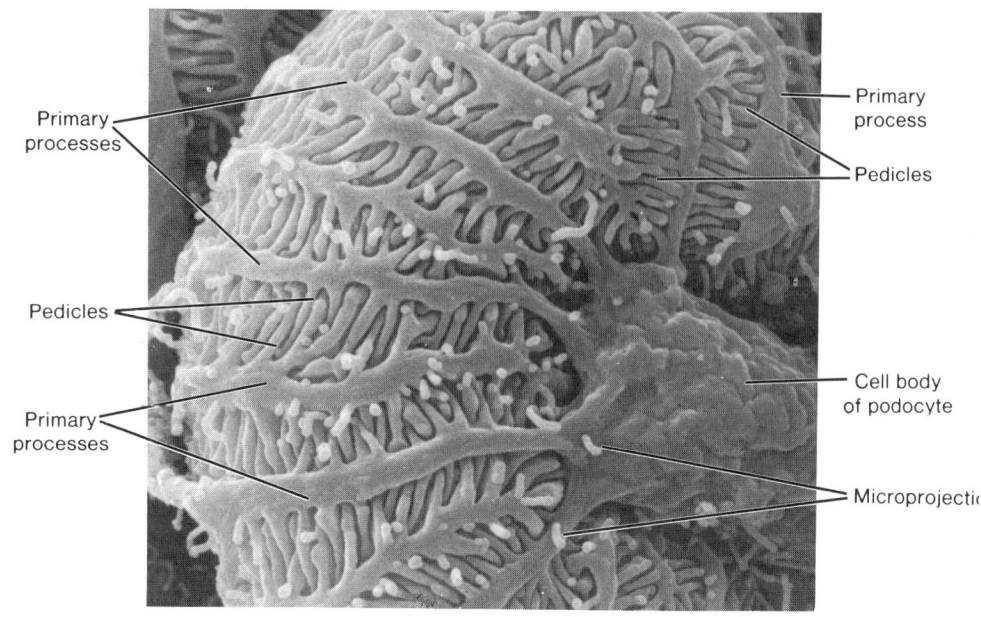

Fig. 16-7. Details of podocytes forming the glomerular epithelium. Note the uniformity of the interdigitating pedicles (SEM, × 7,600).

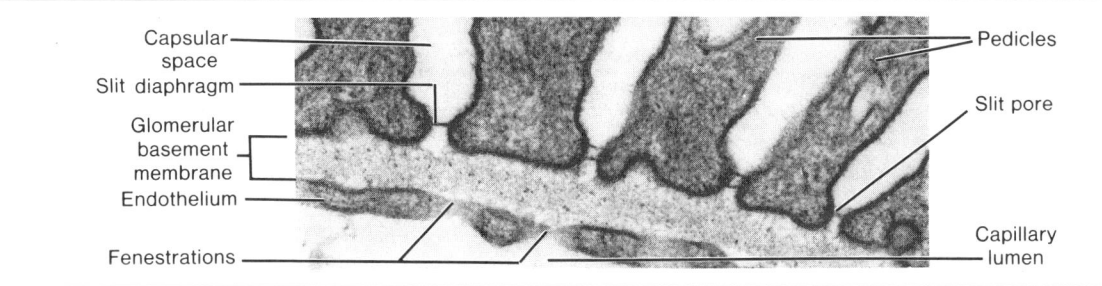

Fig. 16-8. A section through the renal filtration barrier (TEM, × 80,000).

atriopeptides. The contractile activity of the mesangial cells is thought to mediate blood flow through the glomerular capillaries. Mesangial cells are of clinical importance in some kidney disease because of their tendency to proliferate.

The renal corpuscle is the filter of the nephron. The fenestrated glomerular endothelium, the common basal lamina, and the foot processes of the glomerular epithelium form the **filtration barrier** of the renal corpuscle (see Fig. 16-8). This barrier permits passage of water, ions, and small molecules from the capillaries into the capsular space, but larger structures such as the formed elements of the blood and large, irregular molecules are retained. The capillary endothelial cells prevent passage of formed elements; the common basal lamina restricts passage of molecules with a molecular weight greater than 400,000. Substances of small molecular weight (40,000 or less) cross the basal

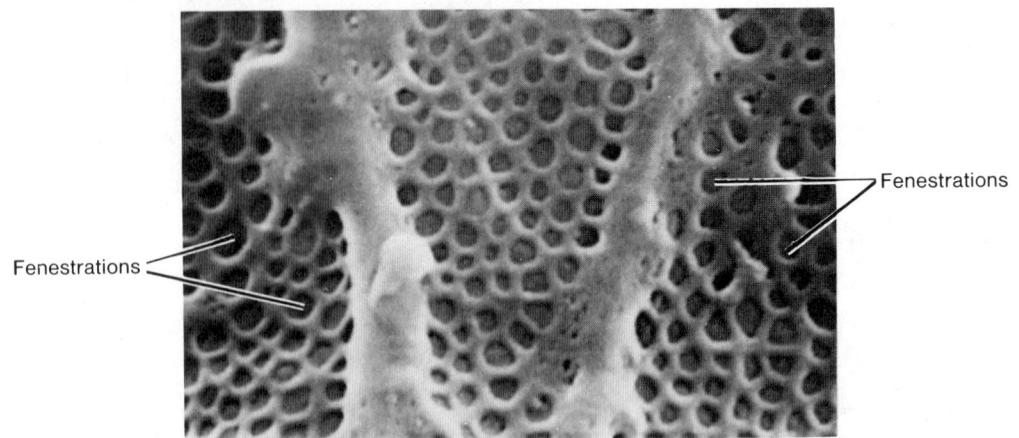

Fig. 16-9. A portion of a fenestrated glomerular endothelial cell viewed from the capillary lumen (SEM, × 35,000).

lamina and pass through the filtration slits of the surrounding glomerular epithelium to enter the capsular space. Material that collects in the capsular space is not urine but a filtrate of blood plasma.

Although materials with molecular weights larger than 45,000 or that have highly irregular shapes may pass through the endothelium and common basal lamina, they are unable to traverse the barrier provided by the foot processes of the podocytes. The filtration barrier limits passage of materials not only on the basis of size and shape but also with respect to their charge. Anionic molecules are more restricted in their passage through the filtration barrier than are neutral molecules of similar size. Heparan sulfate is a negatively charged (polyanionic) molecule of the glomerular basal lamina. The sialoprotein (podocalyxin) coats the podocyte foot processes and together with heparan sulfate gives the filtration barrier a net negative charge. Thus the glomerular epithelium and the common basal lamina are important in limiting the kinds of materials that pass from the blood into the capsular space.

The energy for the filtration process is supplied by the hydrostatic pressure of the blood in the glomerular capillaries. The pressure (about 70 mm Hg) provides sufficient force to overcome the colloidal osmotic pressure of substances in the blood (approximately 33 mm Hg) and the capsular pressure of the filtration membrane (about 20 mm Hg). The resulting filtration pressure (approximately 18 mm Hg) is great enough to force filtrable materials through all three layers of the filtration barrier and into the capsular space.

The hydrostatic pressure exerted within the glomerular capillaries results from the unusual vascular arrangement of the glomerulus. Most vascular areas of the body are supplied by arterioles that form capillaries that then reunite into venules, but the glomerular capillaries are interposed between an afferent arteriole that conducts blood to the glomerulus and an efferent arteriole that conducts blood away from the glomerulus. This arrangement results in considerable pressure being exerted on the capillary walls and can be regulated by contraction of either arteriole.

As more filtrate from the blood enters the capsular space, the rise in pressure forces the filtrate into the lumen of the proximal convoluted tubule. The capsular epithelium forms a tight seal around each renal corpuscle, preventing leakage of filtrate into the cortical tissue.

Proximal Tubule

KEY WORDS: proximal convoluted tubule, microvillus (brush) border, straight portion, apical canaliculus, endocytic complex

The proximal tubule begins at the urinary pole of the renal corpuscle and is the largest segment of the human nephron. It is about 17 mm long and makes up most of the cortex. The proximal tubule is divided into a convoluted portion (pars convoluta) and a straight portion (pars recta). The convoluted portion is the longer of the two parts and is the one most frequently seen in sections of the cortex. After a tortuous course through the cortex in the region of its parent renal corpuscle, the proximal tubule takes a more direct route through the cortex to become the straight portion of the proximal tubule. It then enters the medulla or a medullary ray, where it turns toward a renal papilla as the first part of the loop of Henle.

The **proximal convoluted tubule** consists of a single layer of large, pyramidal cells that have a well-developed **microvillus (brush) border** (Fig. 16-10). The lateral surfaces of the cells form an intricate system of interdigitating processes and ridges that often extend beneath neighboring cells to interdigitate with similar processes of adjacent cells. Tight junctions fuse the apices of the adjacent cells together around the tubular lumen and clearly delineate apical and basolateral cell membrane domains. The resulting separation of transmembrane (transporter) proteins is vital to the function of proximal tubular cells. Thus a complex labyrinth of cell membranes extends from the basal region of the epithelium to near the luminal surface between individual epithelial cells. Compartmentalization of the lateral and basal regions results in a greater surface area of cell membrane, which facilitates the transport of ions (Fig. 16-11).

In typical tissue sections, profiles of proximal convoluted tubules are the most common structures seen in the cortex and usually show a stellate lumen bounded by a distinct brush border. Each cell of the proximal tubule has a large spherical nucleus, but not all cells show a nucleus because of the large size of the cells. The cells contain a supranuclear Golgi complex and numerous rod-shaped mitochondria parallel to the long axis of the cell, closely associated with the lateral ridges and processes. The cytoplasm usually is darkly staining and granular.

Structurally, the convoluted and straight parts of the proximal tubule are similar, but the cells of the **straight portion** are shorter, and the brush border and lateral interdigitations are less well defined than in the convoluted part. Mitochondria are abundant but are smaller and more randomly scattered.

One of the main functions of the proximal tubule is absorption of the glomerular filtrate. The brush border, which consists of closely packed, elongated microvilli, increases the surface area available for absorption by about 30-fold. The microvilli are embedded in a coat of extracellular glycoprotein (glycocalyx). The apical surfaces of cells absorb sugars and amino acids from the luminal contents in a manner similar to that of intestinal epithelial cells. Normally, all the glucose in the glomerular filtrate is absorbed in the proximal convoluted tubule. If blood glucose levels exceed the absorptive capacity of the enzymes that control the absorption of glucose, the excess spills into the urine (glycosuria).

Protein is absorbed in the proximal tubule by a system of invaginations called **apical canaliculi** that give rise to a series of small vesicles containing protein sequestered from the lumen. The tubular invaginations, the vesicles, and the vacuoles together make up the **endocytic complex,** which is actively involved in protein absorption (see Fig. 16-10). The vesicles coalesce to form large vacuoles that condense

Fig. 16-10. The brush (microvillus) border and apical endocytic complex of a cell lining the proximal convoluted tubule (TEM, × 10,000).

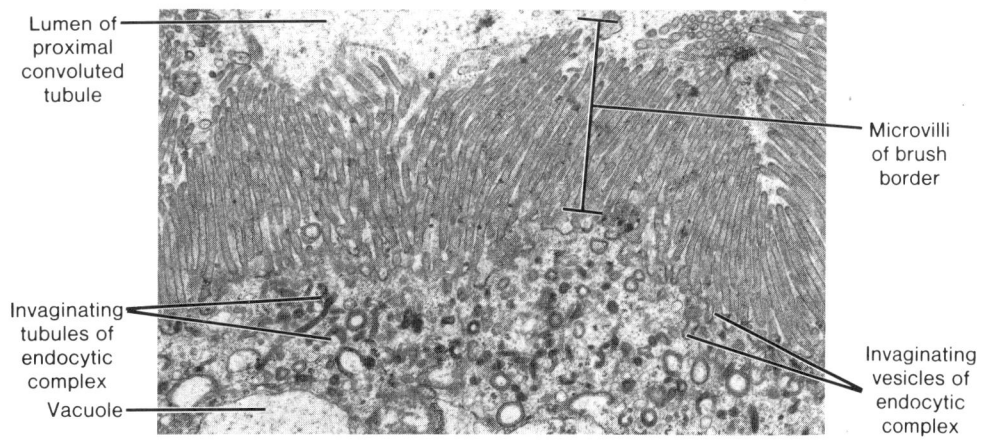

Lumen of proximal convoluted tubule

Microvilli of brush border

Invaginating tubules of endocytic complex

Vacuole

Invaginating vesicles of endocytic complex

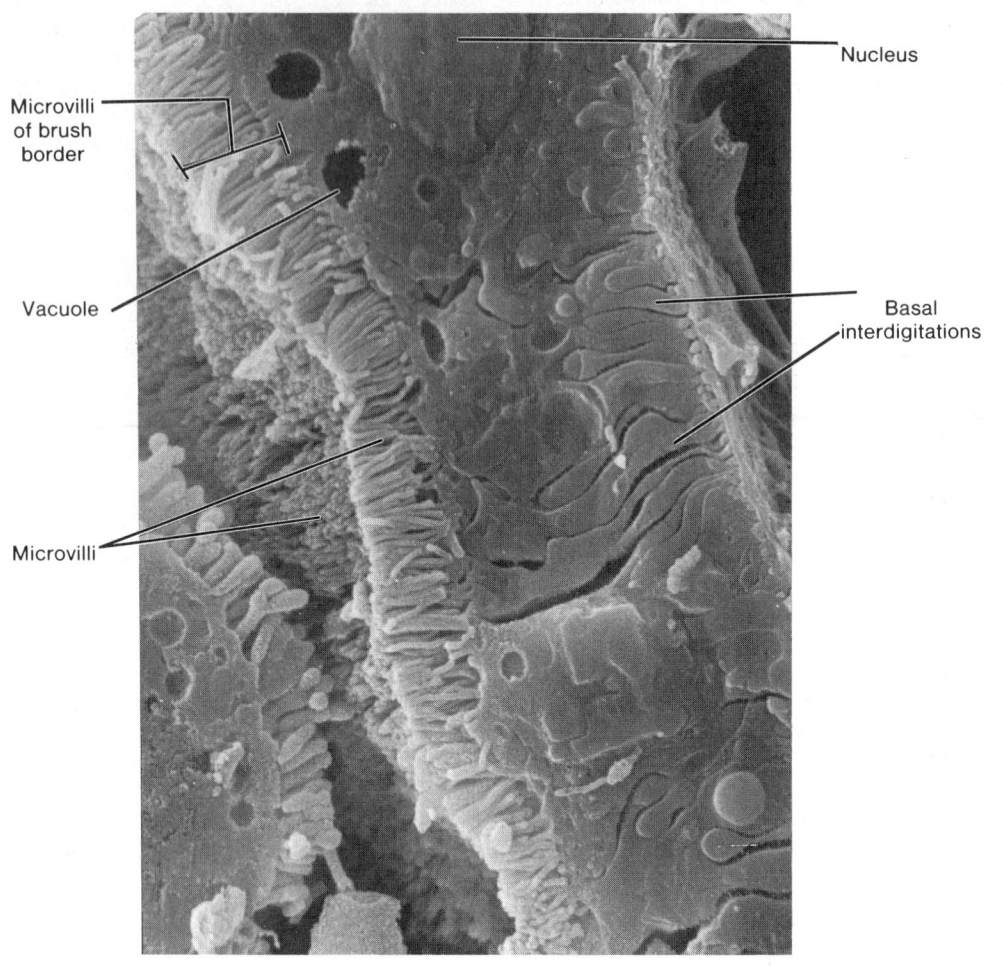

Microvilli
of brush
border

Vacuole

Microvilli

Nucleus

Basal
interdigitations

Fig. 16-11. A fractured preparation of proximal convoluted tubule illustrating microvilli forming the brush border and infoldings of the basolateral membrane (SEM, × 7,000).

and ultimately fuse with lysosomes whose acid hydrolases reduce the protein to amino acids that are released back into the bloodstream.

Over 80 percent of the sodium chloride and water of the glomerular filtrate is absorbed in the proximal tubule. Sodium ion is actively transported from the lumen, and chloride ion and water follow passively with no expenditure of energy by the cell to maintain the osmotic balance. The key factor in proximal tubular cell reabsorption is a Na^+,K^+-ATPase enzyme located within the basolateral cell membranes that limit the complex intercellular space. The

reabsorption of most substances including water is linked to the function of this Na^+,K^+-ATPase system. Glucose, amino acids, lactate, phosphate, and bicarbonate are coupled to sodium entry across the apical plasmalemma of the cell by specific symporter (cotransport) and antiporter (countertransport) transmembrane proteins. Sodium is then moved out of the cell and into the intercellular space by the Na^+,K^+-ATPase mechanism, and glucose, amino acids, phosphate, and lactate follow by passive mechanisms. Glucose, amino acids, and lactate are usually completely removed from the tubular fluid in the first half of the proximal tubule by this mechanism. The remainder of the proximal tubule is involved primarily in the reabsorption of sodium chloride. Chloride ion concentration is high in this region because most bicarbonate, glucose, and organic anions have been preferentially reabsorbed proximally, resulting in an enriched chloride ion solution in the distal half of the proximal tubule. Here sodium is reabsorbed with chloride by parallel Na^+-H^+ and Cl^--base$^-$ antiporter transmembrane proteins within the apical plasmalemma that also is dependent on the Na^+,K^+-ATPase enzyme system of the basolateral membranes. Other inorganic ions and vitamin C also are absorbed by the proximal convoluted tubule.

In addition to absorption, the proximal convoluted tubule secretes organic acids and bases that are destined to be eliminated in the urine. Hence this region of the nephron serves as an exocrine gland. Hydrogen ion also is secreted into the lumen in exchange for bicarbonate ions. Parathyroid hormone acts to decrease phosphate resorption in the proximal convoluted tubule.

Thin Segment

KEY WORDS: simple squamous epithelium

The simple cuboidal epithelium of the straight descending part of the proximal tubule (the initial segment of the loop of Henle) changes abruptly in the thin segment to an attenuated **simple squamous epithelium** (Fig. 16-12). The luminal diameter of the nephron also decreases markedly, from 65 μm in the descending portion to about 20 μm in the thin segment. The thin segment is confined largely to the

medulla of the kidney and forms the thin limb of the loop of Henle.

The nuclei of the lining epithelial cells are centrally placed and cause this area of the cell to bulge into the lumen of the tubule. At the luminal surface, the brush border of the proximal tubule is replaced by short, scattered microvilli. Ultrastructurally, the cytoplasm shows fewer organelles; the lateral cell membranes have elaborate folds that interdigitate with those of adjacent cells. The cells are united at their apices by typical junctional complexes. The lateral interdigitations are less well developed near the hairpin loops. The cells rest on a relatively smooth basal lamina of moderate thickness. The length of the thin segment varies with each nephron and depends on the position of the parent renal corpuscle in the cortex. Thin segments are short or even absent from nephrons whose renal corpuscle is located near the capsule and are longest when the renal corpuscle is located near the medulla. In these nephrons, the thin segments may extend almost to the tip of a renal papilla.

Distal Tubule

KEY WORDS: straight portion, macula densa, pars maculata, juxtaglomerular cell, convoluted portion, basal striation

The distal tubule is divided into a straight part (pars recta), a portion that contains the macula densa (pars maculata), and a convoluted portion (pars convoluta).

The **straight portion** of the distal tubule begins as a

Fig. 16-12. Epithelial tubules forming the renal medulla (LM, × 250).

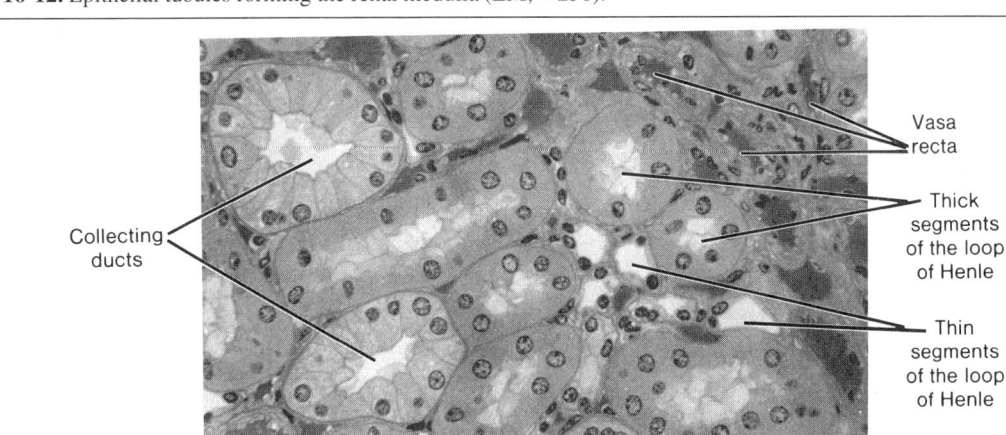

gradual transition from the thin segment of the nephron. The epithelium becomes cuboidal, and while a few microvilli are present, the cells lack the brush border and apical canaliculi of the proximal tubule. Interdigitating processes are present along the basolateral surfaces of the cells, and these contain numerous mitochondria. The straight part of the distal tubule constitutes the thick ascending limb of the loop of Henle and completes the looping course of this structure as it ascends through the medulla. It then reenters the cortex to return to the parent renal corpuscle. Sodium reabsorption by cells of the thick ascending limb, as in

cells forming the proximal tubule, is dependent on the Na+,K+-ATPase enzyme system confined to the basolateral cell membrane. Movement of sodium across the apical cell membrane and into the cell is controlled by a transmembrane protein (symporter) that couples the movement of sodium, chloride, and potassium ions.

As it reaches the renal corpuscle, the distal tubule assumes a close relationship with the afferent arteriole, and where it contacts the arteriole, the tubule contains a group of specialized cells that form the **macula densa** (dense spot) (Figs. 16-13 and 16-14). This segment of the tube is

Fig. 16-13. A renal corpuscle from the outer kidney cortex (LM, × 250).

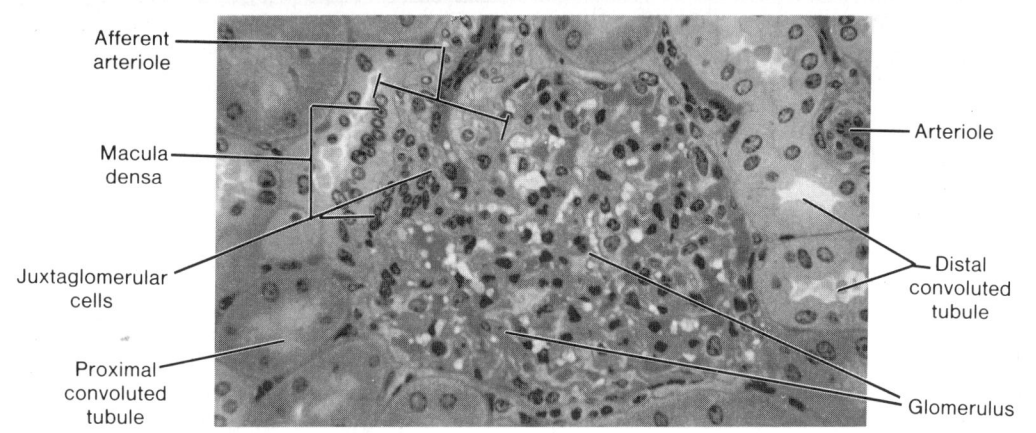

Fig. 16-14. A macula densa lying immediately adjacent to an afferent arteriole, the wall of which contains juxtaglomerular cells (LM, × 1,000).

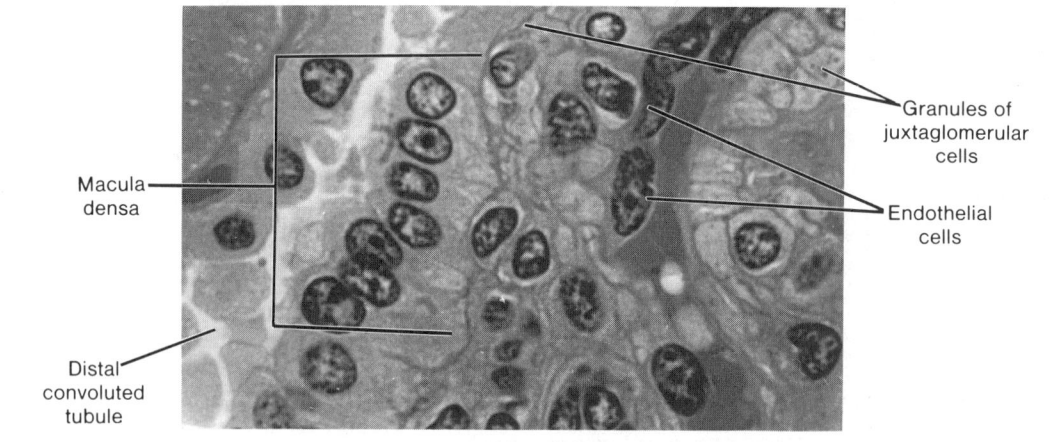

called the **pars maculata.** The cells are taller and narrower than in adjacent tubules so that their nuclei are closer together and are more prominent. Cells of the macula densa are polarized toward the basal surface — that is, the Golgi membranes and occasional granules are found basal to the nucleus, facing the basal cell membrane. The macula densa is related to a modified region of the afferent arteriole that contains the **juxtaglomerular cells** and is considered to represent an endocrine portion of the kidney. This feature and the polarization of the cell suggest that the macula densa monitors the luminal fluid in the distal tubule. Cells of the macula densa produce the enzyme nitric oxide synthase. The nitric oxide released by the macula densa is thought to control vascular tone of the afferent and/or efferent arterioles, thereby influencing kidney function.

The macula densa marks the division between the ascending and **convoluted portion** of the distal tubule. The convoluted portion runs a tortuous course in the cortex near the renal corpuscle of its origin. The distal convoluted tubule is shorter than the proximal convoluted tubule, so fewer sections of the distal tubule are seen in histologic preparations. The lumen of the distal tubule generally is wider than that of the proximal tubules, the cells are shorter and lighter staining, and nuclear profiles usually are seen in each cell, partly because many are binucleate. A brush border is lacking.

Electron micrographs show deep, elaborate infoldings of the basolateral cell membrane associated with numerous elongated mitochondria that lie parallel to the long axis of the cell, producing the prominent **basal striations**

of light microscopy (Fig. 16-15). The cells of the distal tubules show a few luminal microvilli.

The distal tubule is the main site for acidification of urine, and it is here that further absorption of bicarbonate occurs in exchange for hydrogen ions. About one-half the bicarbonate in the glomerular filtrate has already been absorbed in the proximal tubule. The distal tubule also is the site where ammonia is converted to ammonium ion. Thus this segment of the nephron plays an important role in maintaining the acid-base balance of the body. Aldosterone, a steroid hormone, promotes transport of sodium ion from the distal tubule in exchange for potassium ion. Inadequate concentrations of aldosterone result in a serious loss of sodium in the urine. The last portion of the distal tubule can become permeable to water under the influence of antidiuretic hormone (ADH).

Loop of Henle

KEY WORDS: straight portion of proximal tubule, thin segment, straight portion of distal tubule, countercurrent multiplier system

The loop of Henle is interposed between the proximal and distal convoluted tubules and consists of the **straight portion of the proximal tubule,** the **thin segment,** and the **straight portion of the distal tubule.** The loops descend in the medulla for variable distances, form hairpin loops, and then ascend through the medulla, parallel to the descending limb, to the cortex.

The loop of Henle is essential for the production of

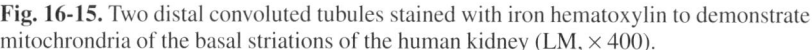

Fig. 16-15. Two distal convoluted tubules stained with iron hematoxylin to demonstrate mitochrondria of the basal striations of the human kidney (LM, × 400).

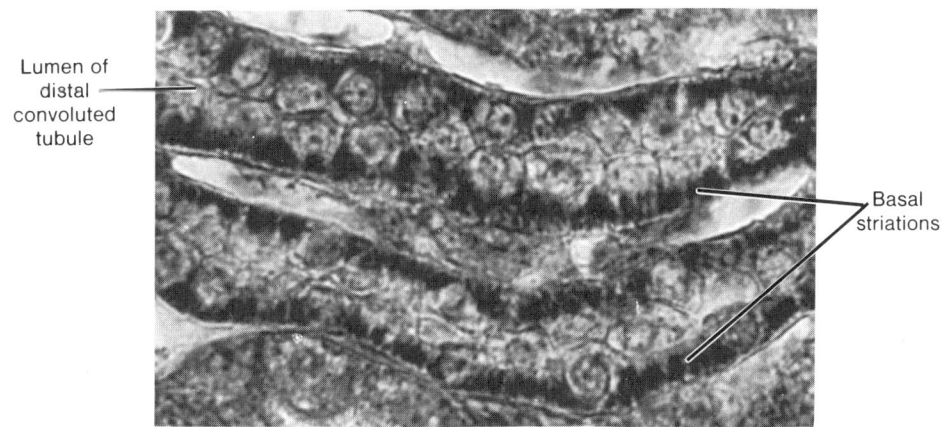

Lumen of distal convoluted tubule

Basal striations

hypertonic urine and is important in the conservation of body water. The loop consists of parallel limbs with tubular fluid flowing in opposite directions that act as a **countercurrent multiplier system** to aid in the concentration of urine. The descending thin segment is highly permeable to water but much less so to sodium and chloride ions and urea. In contrast, the ascending thin segment is impermeable to water but permeable to sodium chloride, which diffuses out of the tubule and contributes to the high osmolarity of the surrounding interstitium of the inner medulla in a passive fashion. Some ions diffuse back into the lumen of the adjacent descending limb. Cells comprising the thick ascending limb (straight portion of the distal tubule) of the loop of Henle also are impermeable to water but actively transport sodium and chloride ions out of the tubule. Such activity results in a dilute tubular fluid and contributes to an increasing hyperosmotic interstitium. Urea follows a route similar to that of these ions and makes a significant contribution to the hyperosmolarity of the interstitium. Thus sodium and chloride ions, as well as urea, are trapped in the interstitial substance of the medulla, resulting in an increase in the osmotic concentration around the tubules in the medulla. All three passively diffuse into the descending limb but either diffuse or are actively pumped out again by the ascending limb as the cycle repeats itself over and over. The sodium chloride–urea trap thus formed establishes an osmotic gradient that increases in strength toward the renal papillae. This gradient is important for the conservation of water and formation of hypertonic urine by the distal tubules and collecting ducts.

Collecting Ducts

KEY WORDS: **initial segment, arched portion, medullary collecting duct, papillary duct, area cribrosa, principal (light) cell, intercalated (dark) cell, renal interstitium**

The collecting ducts can be subdivided according to their location in the kidney. The **initial segment** is found in the cortex and includes short connecting portions that unite the distal tubules of cortical nephrons to collecting ducts and **arched portions** that are formed by the confluence of several connecting portions from juxtamedullary nephrons. The arched portions originate deep in the cortex, which they ascend through before arching to descend within a medullary ray.

Medullary collecting ducts occur primarily in the medulla. As the ducts pass through the medulla they converge to form larger, straight collecting ducts called **papillary ducts,** which end at the tip of a renal papilla. The numerous openings of the papillary ducts give a sievelike appearance to the external surface of the papillae; the area

has been called the **area cribrosa.** The external surfaces of the papillae are covered by transitional epithelium.

The lining epithelium of the collecting ducts consists of principal and intercalated cells. Similar-appearing cells also are found in the last segment of the distal tubule. The **principal (light) cells** generally are cuboidal with centrally placed, round nuclei and lightly staining cytoplasm with characteristic, distinct cell boundaries. Ultrastructurally, the light cells show scattered, short microvilli on their apical surfaces, scattered mitochondria, and some infolding of the basal plasmalemma. They also possess a single, centrally located cilium (Fig. 16-16). Principal cells reabsorb sodium ions and secrete potassium ions; their function is dependent on Na^+,K^+-ATPase activity in the basolateral cell membrane. The activity of this cell type is influenced by the hormones aldosterone and antidiuretic hormone (ADH) in fluid/volume homeostasis.

Scattered between the light cells are the **intercalated (dark) cells,** which have more mitochondria, stain more deeply, and show a large number of vesicles in the apical cytoplasm. The cells lack central cilia, but the apical plasmalemma shows microplicae (see Fig. 16-16). The intercalated cells secrete hydrogen ions and reabsorb bicarbonate and potassium ions. For each hydrogen ion secreted into the tubular lumen, a bicarbonate ion is released from the basolateral cell membrane destined for the vasculature. Thus intercalated cells play an important role in regulating acid-base balance. As the collecting tubules pass through the medulla, the cells increase in height to become tall columnar in the papillary ducts.

The collecting ducts function to conserve water and produce hypertonic urine. As the ducts traverse the medulla to the tips of the pyramids, they pass through the increasingly hypertonic environment established by the loop of Henle. The permeability of the collecting ducts to water is controlled by ADH. In the presence of ADH, water is drawn from the collecting ducts by osmosis because of the hypertonic environment maintained in the medullary interstitium. The loss of water from the tubular contents results in a concentrated, hypertonic urine. If ADH is lacking due to injury or disease, the kidney is unable to concentrate urine, and copious amounts of dilute urine are produced, resulting in severe dehydration. This condition is known as *diabetes insipidus.*

The **renal interstitium** fills the spaces between the tubular elements of the kidney. The cortical interstitium is relatively scant except around blood vessels and consists of fine bundles of collagen, fibroblasts, and scattered phagocytes. The medullary interstitium is more plentiful, and its cells lie parallel to the long axis of the tubules. These interstitial cells possess long, branching processes that encircle

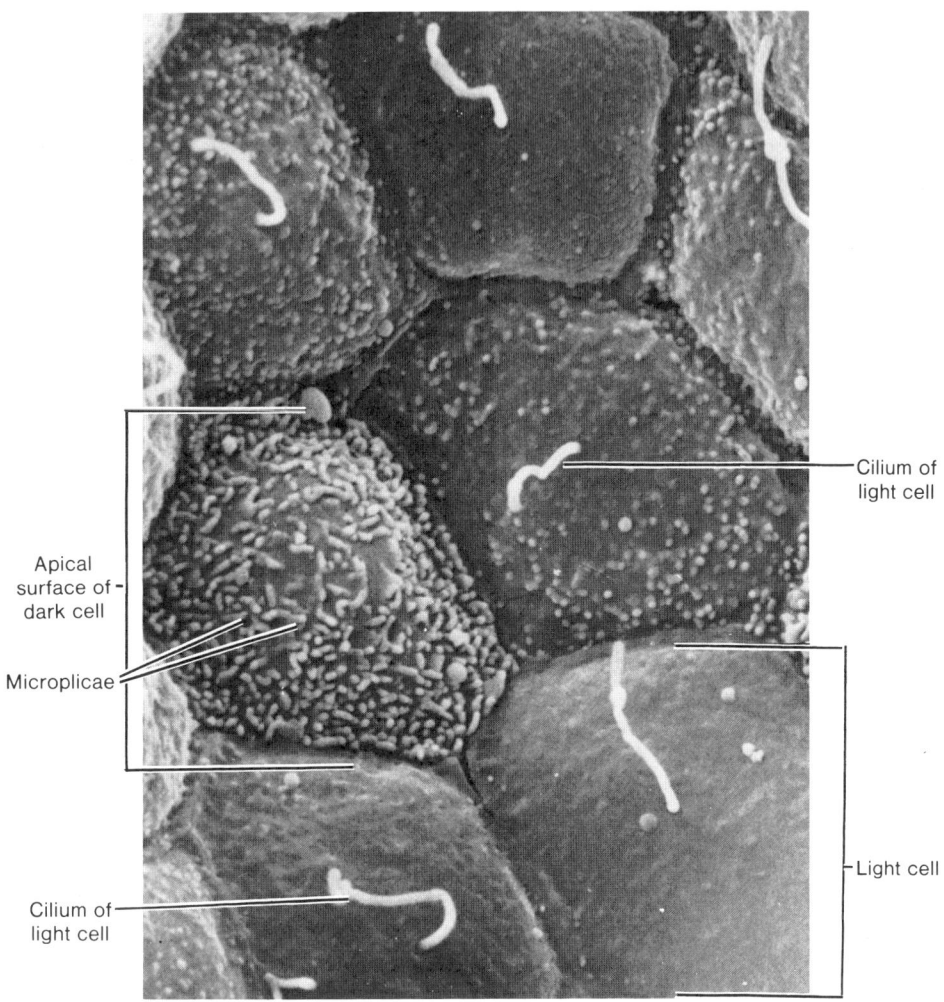

Apical
surface of
dark cell

Microplicae

Cilium of
light cell

Cilium of
light cell

Light cell

Fig. 16-16. Surface features of apices of light and dark cells in a human collecting tubule (SEM, × 8,000).

adjacent tubules and blood vessels. Pleomorphic interstitial cells filled with small lipid droplets also occur in the medullary interstitium at regular intervals between the epithelial tubules and vessels. The cells are suspected to be endocrine cells and may secrete an antihypertension factor. The medullary interstitium contains abundant intercellular matrix and small bundles of collagen.

Juxtaglomerular Apparatus

KEY WORDS: juxtaglomerular (JG) cell, renin, angiotensinogen, angiotensin I, angiotensin II, aldosterone, erythropoietin, extraglomerular mesangium

The juxtaglomerular apparatus consists of the macula densa of the distal tubule, the extraglomerular mesangium, and the juxtaglomerular cells in the wall of the afferent arteriole. The **juxtaglomerular (JG) cells** appear to be highly modified smooth muscle cells in the wall of the afferent arteriole as it enters the renal corpuscle. The cells contain a number of secretory granules, well-developed Golgi complexes, and abundant granular endoplasmic reticulum. The JG cells produce **renin,** which, when released into blood, converts the plasma protein **angiotensinogen** to **angiotensin I.** A converting enzyme in the blood converts angiotensin I to the polypeptide **angiotensin II,** which acts on the zona glomerulosa of the adrenal cortex, stimulating it to release aldosterone. **Aldosterone** influences the distal tubule to transport sodium ion from the lumen in exchange for potassium ion. Angiotensin II also is a potent vasoconstrictor and elevates blood pressure. The juxtaglomerular apparatus also has been implicated in the production of **erythropoietin,** an agent that stimulates erythropoiesis in the bone marrow.

The location of the macula densa, its basal polarization, and its orientation toward adjacent JG cells suggest that the two groups of cells may interact. The macular densa may "sense" the sodium ion concentration in the distal tubule and thereby influence the activity of JG cells.

The **extraglomerular mesangium** forms a loose mass of cells between the afferent and efferent arterioles. Component cells may contain granules, but their exact role in the function of the juxtaglomerular apparatus is unknown.

Vascular Supply

KEY WORDS: renal artery, segmental artery, interlobar artery, arcuate artery, interlobular artery, intralobular artery, afferent arteriole, efferent arteriole, peritubular capillary network, vasa recta, vascular countercurrent exchange

The vascular pattern of the kidneys is complex and shows regional specializations related to the organization and function of the various parts of the nephron. The **renal arteries** arise from the abdominal aorta and before reaching the renal sinus usually divide into branches that pass ante-

rior and posterior to the renal pelvis to enter the renal sinus. Here they give rise to **segmental arteries,** which also divide to form **interlobar arteries** that run between renal lobes. At the corticomedullary junction, these arteries divide into several **arcuate arteries** that arch across the base of each medullary pyramid and give off **interlobular arteries** that pass into the cortex between lobules. The interlobular arteries run peripherally in the cortex, giving rise to a system of **intralobular arteries** that enter renal lobules and provide **afferent arterioles,** which supply the glomeruli of renal corpuscles (Fig. 16-17). As the **efferent arteriole** leaves the renal corpuscle of a cortical nephron, it immediately breaks up into a **peritubular capillary network** that supplies the convoluted tubules.

The main circulation of the renal cortex is unique in that the arterioles give rise to two distinct, sequential capillary beds: the glomerular and peritubular capillaries. Efferent arterioles from juxtamedullary nephrons, on the other hand, form several long, straight vessels, the **vasa recta,** that descend into the medullary pyramid and form hairpin loops. Like the loop of Henle, the loops of the vasa recta are staggered throughout the medulla. The walls of the vasa recta are thin, and the endothelium of the ascending (venous) limb is fenestrated. The vasa recta pass in close proximity to the loop of Henle, permitting passive interchange between the two elements. The vasa recta form a **vascular countercurrent exchange** system that removes excess water and ions. The osmotic gradient is not disrupted due to the slower flow rate and the smaller volume in the vasa recta.

Fig. 16-17. A region of renal cortex from a kidney injected with colored gelatin to demonstrate the vasculature (LM, × 100).

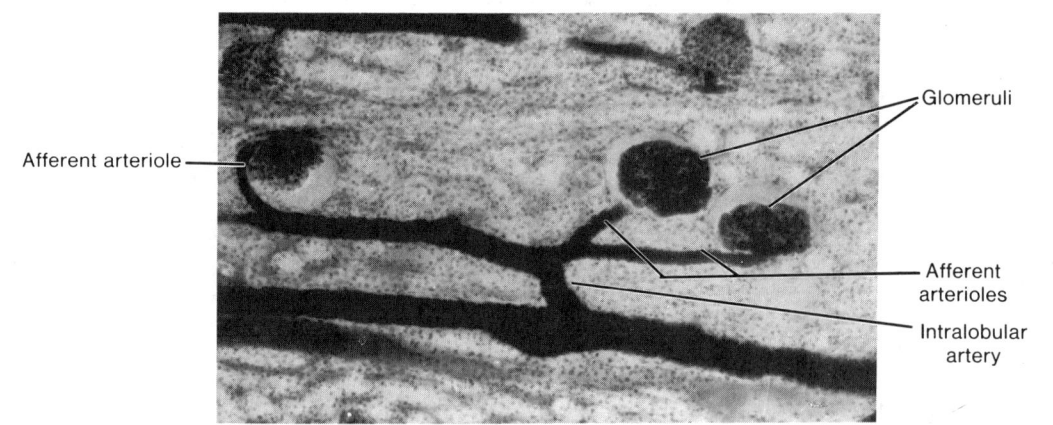

Afferent arteriole

Glomeruli

Afferent arterioles

Intralobular artery

The venous drainage of the kidney is similar to and follows the same course as the arterial supply. However, there is no venous equivalent of the glomerulus or the afferent and efferent arterioles. The venous system of the medulla begins in the ascending limb of the vasa recta, which drains into interlobular or arcuate veins. In the peripheral cortex, capillaries unite to form small veins that assume a starlike pattern as they drain into interlobular veins. The renal veins drain into the inferior vena cava. The left renal vein differs in that it is much longer and receives venous drainage from the left gonad.

Extrarenal Passages

The extrarenal passages consist of the minor and major calyces, renal pelvis, ureter, urinary bladder, and urethra. They convey urine to the outside of the body or, in the case of the bladder, store it temporarily. Except for the urethra, all have a similar basic structure with a mucosa, muscularis, and adventitia. The layers are thinnest in the minor calyces and increase in depth distally to reach their maximum development in the bladder.

Calyces, Renal Pelvis, Ureters, and Bladder

KEY WORDS: mucosa, transitional epithelium, lamina propria, muscularis, adventitia

The thickness of the wall of the excretory passage gradually increases from the upper to lower parts; except for this, all parts of the extrarenal passages show a similar structure. The lumen is lined by a **mucosa** consisting of **transitional epithelium** that rests on a lamina propria. There is no submucosa, and the lamina propria blends with the connective tissue of the well-developed muscular coat.

Transitional epithelium covers the external surfaces of the renal papillae and reflects onto the internal surfaces of the surrounding minor calyces (Fig. 16-18). It also is continuous with the epithelium of the papillary ducts, thus providing a complete epithelial lining that prevents escape of urine into the neighboring tissues. Transitional epithelium forms a barrier to the diffusion of salts and water into and out of the urine. In the major and minor calyces, the epithelium is two to three cells thick, increasing in the ureter to four or five layers and to six, eight, or more layers in the bladder (Fig. 16-19). The surface cells are large and rounded and in the relaxed bladder have convex or dome-shaped borders that bulge into the lumen. The superficial cells sometimes contain large polyploid nuclei. The epithelium in the distended organ shows considerable changes in morphology. In the filled bladder, or as urine is propelled down the ureter, the epithelium is stretched and flattened and temporarily assumes the appearance of a thin stratified squamous epithelium. When the intraluminal pressure is relieved, the epithelium again assumes its nondistended appearance. In the relaxed bladder, the apical cytoplasm of the superficial cells contains fine filaments and fusiform

Fig. 16-18. A renal papilla together with its surrounding minor calyx from a human kidney (LM, × 40).

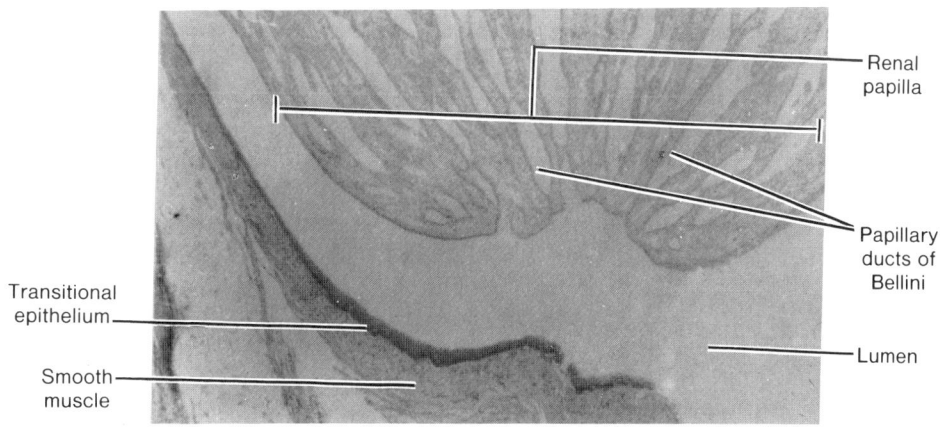

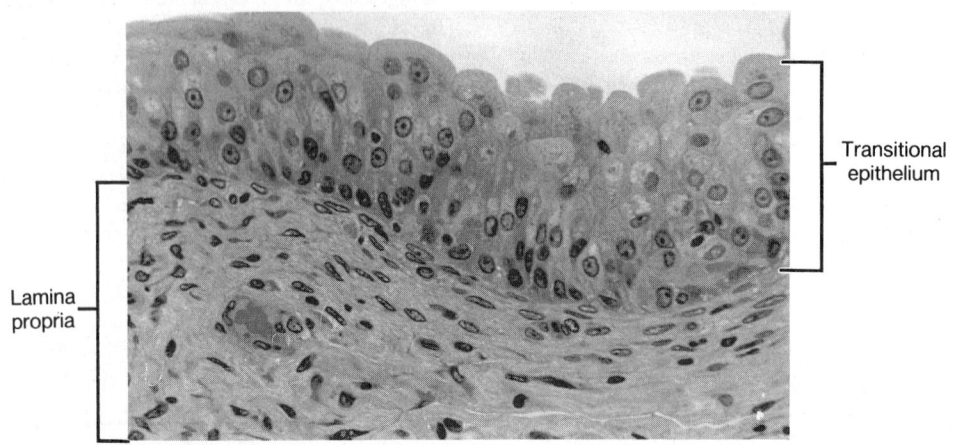

Fig. 16-19. Transitional epithelium lining a urinary bladder (LM, × 250).

vesicles that are limited by a membrane of the same thickness as the cell membrane. These vesicles are thought to represent reserve surface membrane for use during distension. Transitional epithelium lies on a very thin basement membrane that usually is not seen with the light microscope.

The epithelium lies on a **lamina propria** that consists of a compact layer of fibroelastic connective tissue. Some diffuse lymphatic tissue also may be present.

The **muscularis** of the urinary passageways consists of bundles of smooth muscle separated by abundant fibroconnective tissue and begins in the minor calyces as two thin layers of smooth muscle. The inner, longitudinal layer begins at the attachment of the minor calyces to the renal papillae; the outer layer of muscle spirals around the renal papilla forming a thin coat. The walls of the minor calyces contract periodically around the renal papillae and aid in moving urine from the papillary ducts to the calyces and into the renal pelvis. The muscularis of the renal pelvis and upper two-thirds of the ureter consist of the same two layers and differ only in thickness (Fig. 16-20). An additional outer, longitudinal layer of smooth muscle is added in the lower one-third of the ureter.

Fig. 16-20. A portion of the wall of a human ureter (LM, × 100).

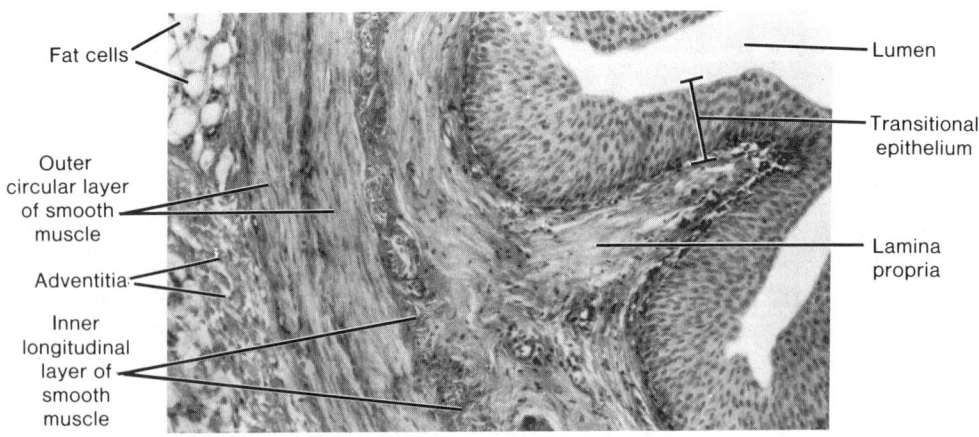

The distal ends of the ureters, the intramural portions, pass obliquely through the wall of the bladder to empty into its lumen. The circular layer of smooth muscle disappears, and the contractions of the longitudinal layers help dilate the lumen of the distal ureter so that urine can enter the bladder. Peristaltic waves periodically pass along the ureter to convey urine to the bladder, into which it empties in small spurts. As the bladder fills, the pressure of its contents keeps the intramural portions of the ureters closed due to their oblique course in the bladder wall. The ureters open only when urine is forced through them. Reflux of urine into the ureters is prevented by a valvelike flap of bladder mucosa that lies over the ureteral openings.

The muscularis of the urinary bladder is moderately thick and consists of inner longitudinal, middle circular, and outer longitudinal layers of smooth muscle (Fig. 16-21). The middle layer is the most prominent and spirals around each ureteral opening. Around the internal urethral orifice, the muscle increases in thickness to form the internal sphincter of the bladder.

A coat of fibroelastic tissue, the **adventitia,** surrounds the muscularis and attaches the extrarenal passages to surrounding structures. In the renal pelvis it blends with the capsule of the kidney. On the superior surface of the bladder the fibroelastic coat is covered by peritoneum. The adventitia contains numerous blood vessels, lymphatics, and nerves. Blood vessels pierce the muscularis, provide it with capillaries, and then form a plexus of small vessels in the lamina propria. A rich capillary layer lies immediately beneath the epithelium. Nerve fibers and small ganglia

also occur in the adventitia and represent the sympathetic and parasympathetic divisions of the autonomic nervous system. Parasympathetic fibers are important for control of micturition.

Male Urethra

KEY WORDS: prostatic urethra, membranous urethra, sphincter urethrae, penile urethra (pars cavernosa), fossa navicularis, glands of Littre, intraepithelial nests

In the male, the urethra conveys urine from the bladder to the exterior and serves for the passage of seminal fluid during ejaculation. The male urethra is 18 to 20 cm long and has three segments. The first segment is 3 to 4 cm long and lies within the prostate, an accessory sex gland. This part forms the **prostatic urethra** (pars prostatica) and is lined by transitional epithelium similar to that of the bladder. The **membranous urethra** (pars membranacea) is very short (1.5 cm) and extends from the apex of the prostate to the root of the penis. The membranous urethra pierces the skeletal muscle of the urogenital diaphragm immediately before it enters the penis. Skeletal muscle surrounding this part of the urethra forms the external sphincter (**sphincter urethrae**) and is under voluntary control during micturition.

The third and longest segment, about 15 cm, is the **penile urethra (pars cavernosa),** which runs longitudinally through the corpus cavernosa urethrae to end at the tip of the glans penis. The membranous and penile parts of the urethra are

Fig. 16-21. A section through the urinary bladder wall (LM, × 100).

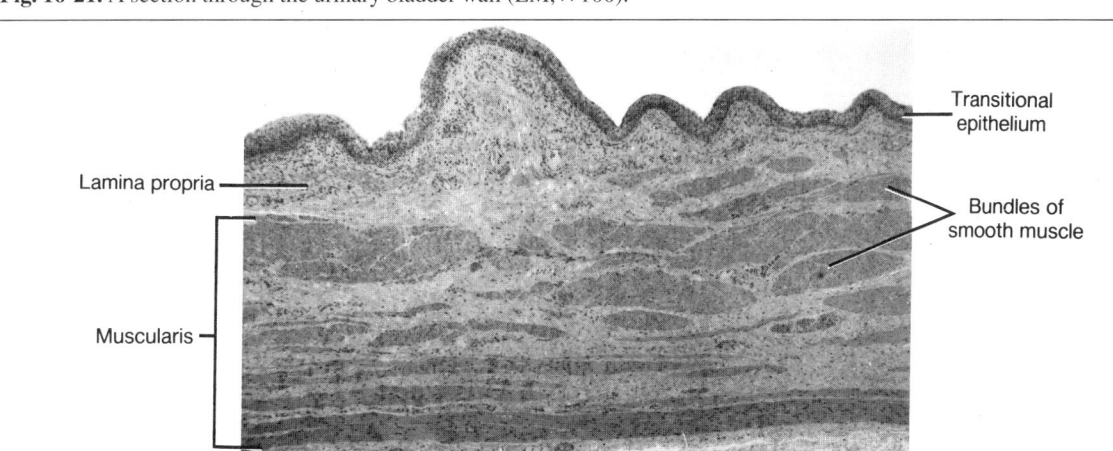

Transitional epithelium

Lamina propria

Bundles of smooth muscle

Muscularis

lined by stratified or pseudostratified columnar epithelium. Stratified squamous epithelium often occurs in patches in the penile portion and also lines a distal enlargement, the **fossa navicularis.** The mucous membrane of the urethra shows small depressions or invaginations called the *lacunae of Morgagni,* which are continuous with the branched tubular **glands of Littre** (Fig. 16-22). The epithelium lining these glands is the same as that of the luminal surface, which contains **intraepithelial nests** of clear, mucus-secreting cells. The lamina propria beneath the urethral epithelium is a highly vascular, loose fibroconnective tissue rich in elastic fibers. The mucosa is bounded by inner longitudinal and outer circular layers of smooth muscle.

Female Urethra

The female urethra is shorter than the male (3 to 5 cm long) and is lined by stratified squamous epithelium, although patches of stratified or pseudostratified columnar may be found. Glands of Littre are present throughout its length. As in the male, the lamina propria is a vascular, fibroelastic connective tissue that contains numerous venous sinuses. The surrounding muscularis consists of an inner longitudinal layer of smooth muscle bundles and an outer circular layer. The female urethra is surrounded by skeletal muscle of the urogenital diaphragm that forms the sphincter urethrae at its orifice.

Fig. 16-22. A region of human penile urethra illustrating the glands of Littre (LM, × 100).

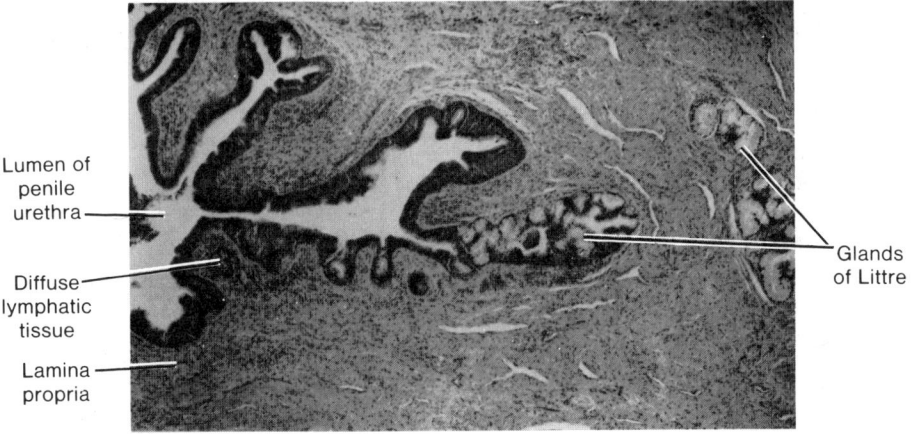

Review Table 16-1. Key Features of the Urinary System

	Primary Location	Features of Epithelium	Other Features
Kidney			
Proximal convoluted tubule	Cortex	Large granular, dark-staining cells; brush border; not all cells show a nuclear profile	Stellate or irregular lumen
Distal convoluted tubule	Cortex	Light-staining columnar cells; each cell shows nuclear profile	Smooth luminal surface, large, circular in outline
Renal corpuscle poles, (Bowman's capsule, glomerulus)	Cortex	Glomerular epithelium — light-staining nuclei; glomerular endothelium — dark-staining nuclei	Vascular and urinary thick basal lamina
Thin segment (loop of Henle)	Medulla	Light-staining tubule lined by simple squamous epithelium	Narrow lumen
Thick segment (loop of Henle)	Medulla	Light-staining cuboidal cells	Wider lumen
Collecting tubule	Medulla	Light-staining columnar cells, distinct	

Review Table 16-1 (continued)

	Epithelium	Supporting Wall
Extrarenal Passages		
Ureter (proximal two-thirds)	Transitional (4–5 cell layers)	Inner longitudinal and outer circular smooth muscle
Ureter (distal one-third)	Transitional (4–5 cell layers)	Inner longitudinal, middle circular, and outer longitudinal smooth muscle
Urinary bladder	Transitional (8 or more layers)	Inner longitudinal, middle circular, and outer longitudinal smooth muscle
Male urethra		
Pars prostatica	Transitional	Prostate
Pars membranacea	Transitional	Skeletal muscle (sphincter urethrae)
Pars cavernosa	Stratified and/or pseudostratified columnar; patches of wet stratified squamous in fossa navicularis	Erectile tissue of corpus cavernosum urethrae
Female urethra	Stratified columnar; stratified squamous	Inner longitudinal, outer circular smooth muscle

Functional Review: Urinary System

The kidney serves as an organ of excretion, is both an endocrine and exocrine gland, is a target organ of other endocrine glands, and plays an important role in the acid-base balance of the organism.

The uriniferous tubules perform three separate functions in the formation of urine: filtration, secretion, and selective absorption. The tubules do not synthesize and release new material in significant amounts but eliminate excess water and waste products of metabolism that are being transported in the blood plasma. After filtration of the plasma, metabolic wastes such as urea, uric acid, and creatinine are not absorbed but remain in the tubular lumen and form constituents of the urine that will be eliminated.

The renal corpuscles filter blood plasma, and in humans, the kidneys produce about 125 ml of glomerular filtrate each minute. About 124 ml of this is absorbed by the rest of the uriniferous tubule, resulting in the formation of approximately 1 ml of urine. The total glomerular filtrate in humans during a 24-hour period is 170 to 200 liters, of which about 99 percent is absorbed. The filtration process is driven by the hydrostatic pressure of blood, which is sufficient to overcome the colloidal osmotic pressure of plasma and the capsular pressure at the filtration membrane. The resulting filtrate contains ions, glucose, amino acids, small proteins, and the nitrogenous wastes of metabolism. Blood cells and proteins of large molecular weight are prevented from entering the capsular space by the filtration barrier.

The glomerular filtrate is reduced to about 15 percent of its original volume in the proximal convoluted tubule. In addition to the obligatory absorption of sodium chloride and water, glucose, amino acids, proteins, and ascorbic acid are actively absorbed in the proximal convoluted tubules. Bicarbonate also is absorbed here in exchange for the secretion of hydrogen ions. Exogenous organic bases and acids are actively secreted into the lumen by the epithelium of the proximal convoluted tubule, thus fulfilling the requirements of an exocrine gland. Most of the materials that have passed through the filtration barrier are immediately resorbed by the epithelium of the uriniferous tubules and put back into the circulation. Materials in excess of the body's needs as well as any metabolic wastes (urea, uric acid, creatinine) pass along the tubular lumen and are excreted in the urine.

The loop of Henle is essential for the conservation of water and production of hypertonic urine. An active sodium pump mechanism resides in the cells of the thick ascending limb of the loop and creates and maintains a gradient of osmotic pressure that increases from the base of the medulla to the papillary tip. Henle's loops, with their hairpin loops, act as a countercurrent multiplier system to maintain the gradient.

The distal tubule is the principal site for acidification of urine and is the site for further absorption of bicarbonate in exchange for secretion of hydrogen ions. It therefore plays an important role in acid-base balance. The conversion of ammonia to ammonium ions also occurs in the distal tubule. Absorption of sodium ions in the distal tubule is referred to as facultative absorption and is controlled by the steroid hormone aldosterone, which increases the rate of absorption of sodium ion and excretion of potassium ion. If aldosterone is absent, large amounts of sodium are lost in the urine. Parathyroid hormone also acts on the distal convoluted tubule of the nephron to promote absorption of calcium ion and inhibit absorption of phosphate ion from the developing urine.

Collecting ducts pass through the hypertonic environment of the medulla. The permeability of the collecting ducts as well as the last portion of the distal tubule to water is controlled by ADH. In the presence of ADH, the collecting ducts and the last segment of the distal tubule become permeable to water, which, by osmosis, leaves the lumen to enter the surrounding interstitium. The flow of water out of the tubular fluid thus produces a hypertonic urine. If the concentration of ADH in the blood is low, a larger volume of dilute urine results because the permeability of the collecting ducts to water is decreased.

Atrial natriuretic factor (ANF) belongs to a family of atrial peptides stored in atrial-specific granules within atrial cardiac myocytes. When released, ANF acts on the kidney to increase sodium excretion and to increase the glomerular filtration rate. Its precise mode of action is unknown.

The endocrine part of the kidney, the juxtaglomerular apparatus, elaborates renin, an enzyme that acts on angiotensinogen in the blood plasma, converting it to an inactive protein called angiotensin I. A converting enzyme in the blood converts angiotensin I to the angiotensin II that stimulates the zona glomerulosa of the adrenal to release aldosterone. Aldosterone stimulates the distal tubule of the nephron to absorb sodium ions in exchange for potassium ions. Angiotensin II is a potent vasoconstrictor also and raises blood pressure. The renin-angiotensin system is influenced by blood flow through the kidney and is an important factor in hypertension. The juxtaglomerular apparatus also produces a blood-borne factor called erythropoietin, which stimulates erythropoiesis in the bone marrow.

A minor calyx is attached around each renal papilla and represents the beginning of the extrarenal passageways. Minor calyces undergo periodic rhythmic contractions that aid in moving urine from the papillary ducts into the extrarenal system. The walls gently contract around each renal papillae and transport the urine to the renal pelvis. Periodic contractions of the muscular walls propel small amounts of urine from the renal pelvis through the ureter to the urinary bladder, where it is temporarily stored until a sufficient volume is obtained for urine to be evacuated through the urethra. Contraction of the muscularis of the bladder wall together with voluntary relaxation of the skeletal muscle that forms the sphincter urethrae accomplishes this process during micturition.

Review Questions and Answers

Questions

83. Describe the morphology of a renal corpuscle.
84. List the components of the filtration barrier and briefly discuss the functional significance of each.
85. List three hormones that influence the function of the uriniferous tubule and briefly discuss the effect of each.
86. Briefly discuss the morphology and function of the juxtaglomerular apparatus.
87. Discuss the morphology and function of the loop of Henle.

Answers

83. A renal corpuscle consists of a capillary tuft, the glomerulus, which projects into a blindly ending expansion of the uriniferous tubule called Bowman's capsule. Each renal corpuscle has a vascular pole, where afferent and efferent arterioles enter and leave the glomerulus, and a urinary pole, where the capsular space becomes continuous with the lumen of the proximal convoluted tubule.

The epithelium of the parietal layer of Bowman's capsule consists of a single layer of tightly adherent squamous cells that prevent the leakage of glomerular filtrate into surrounding cortical tissue. At the vascular pole, component cells of the investing glomerular epithelium dramatically change their morphology from the simple squamous cells of the capsular epithelium to large, stellate-shaped podocytes. Cytoplasmic extensions from the podocytes, the primary processes, wrap around the underlying glomerular capillaries and give rise to numerous secondary processes that interdigitate with similar processes from adjacent podocytes to completely surround the capillary loops of the glomerulus. Narrow clefts called filtration slits lie between the interdigitating secondary process. The secondary processes of the podocytes and the heavily fenestrated endothelium share a continuous, common basal lamina. The renal corpuscle is that portion of the nephron that functions in filtration.

84. The filtration barrier of a renal corpuscle consists of three entities: a fenestrated glomerular endothelium, a common basal lamina, and foot processes of the glomerular epithelium. The glomerular endothelium prevents the passage of formed blood elements (blood cells and platelets), while the common basal lamina restricts the passage of substances with a molecular weight greater than 400,000 or highly charged molecules. Substances of smaller molecular weight (40,000 or less) traverse the endothelial barriers and the common basal lamina and pass through the filtration slits of the surrounding glomerular epithelium to enter the capsular space. Although materials with a molecular weight larger than 45,000 or that have a highly irregular shape may pass through the fenestrated endothelium and common basal lamina, they are unable to traverse the barrier provided by the foot processes of the podocytes. The barrier also limits passage of materials with respect to charge. Heparan sulfate and podocalyxin give the barrier a net negative charge; therefore, anionic molecules are more restricted in their passage through the basal lamina of the barrier than are neutral molecules of similar size. Hydrostatic pressure of blood in the glomerular capillaries is sufficient to overcome the colloidal osmotic pressure of substances in the blood and the capsular pressure of the filtration membrane.

85. Three hormones that influence the function of the uriniferous tubule are aldosterone, parathyroid hormone, and antidiuretic hormone (ADH). Aldosterone, a steroid hormone from the adrenal cortex, promotes sodium ion transport from the lumen of the distal tubule in exchange for potassium ion. Inadequate concentrations of aldosterone result in a serious loss of sodium in the urine. The presence of parathyroid hormone also acts on the tubular components of the nephron to promote absorption of calcium ion and to inhibit the absorption of phosphate ion from the forming urine. Antidiuretic hormone acts primarily on distal tubules and collecting ducts and increases their permeability to water. In the presence of ADH, water leaves the lumen of the tubules by osmosis, being drawn by the hypertonic environment of the surrounding interstitium. The flow of water out of the tubular fluid and into the interstitium produces a hypertonic urine. If the concentration of ADH in the blood is low, the permeability of the collecting ducts to water is decreased, resulting in a larger volume of dilute urine.

86. The juxtaglomerular apparatus is the endocrine portion of the kidney and consists of the juxtaglomerular (JG) cells, the macula densa, and the extraglomerular mesangium. The JG cells are found within the wall of the afferent arteriole and are thought to be modified smooth muscle cells. They are cuboidal in shape and contain secretory granules, well-developed Golgi complexes, and numerous profiles of granular endoplasmic reticulum. The JG cells synthesize, at least in part, an enzyme called renin that activates the angiotensin-aldosterone system. The location of the macula densa and its basal polarization and orientation toward the JG cells suggest a functional relationship between these two cell groups. The macula densa is thought to sense the sodium ion concentration in the distal tubule and may influence the activity of the JG cells. The extraglomerular mesangium forms a loose mass of cells between the afferent and efferent arterioles. Component cells may contain granules, but their exact role in the function of the juxtaglomerular apparatus is unknown. The juxtaglomerular apparatus in some species also produces a blood-borne factor called erythropoietin, which stimulates erythropoiesis in the bone marrow.

87. The loop of Henle is that part of the nephron that is interposed between the proximal and distal convoluted tubules and consists of the straight portion of the proximal tubule, the thin segment, and the straight portion of the distal tubule (the ascending thick limb). The loops of Henle descend in the medulla for variable distances, form a hairpin loop, and then ascend parallel to the descending limb, back through the

medulla to the cortex. The loop of Henle is essential for the conservation of water and the production of hypertonic urine. An active sodium pump mechanism resides in the cells of the ascending limb and provides and maintains a gradient of osmotic pressure that increases from the base of the medulla to the papillary tip. The loops of Henle with their hairpin turns play an important role by serving as a countercurrent multiplier system to maintain this osmotic gradient.

17 Meiosis and Male Reproductive Organs

In mammals the male and female reproductive organs produce the gametes, provide a mechanism by which the gametes can be brought together to yield a new individual, and in the female, house and nourish the zygote until birth of that new individual.

The generative organs for the production of male gametes (sperm) are the testes, while in the female the ovary is the site of the female gametes, the ova. Gametes cannot contain the diploid (somatic) number of chromosomes, since their fusion would result in a doubling of the chromosomes with each new generation. Thus the gametes can contain only half the somatic number of chromosomes, and reduction to the haploid number occurs through a special form of cell division called *meiosis.*

Meiosis

KEY WORDS: **two cell divisions, one DNA replication, first meiotic division, separation of homologous chromosomes, second meiotic division, separation of chromatids**

Meiosis is characterized by **two cell divisions** but only **one replication of deoxyribonucleic acid (DNA).** Chromosomal DNA is duplicated during interphase, prior to the first division, and results in a double or tetraploid amount of DNA within the diploid number of chromosomes. This is no different from what occurs during replication of DNA prior to ordinary mitosis. The **first meiotic division** brings about **separation of homologous chromosomes,** and while this results in halving the number of chromosomes, each chromosome contains a double amount of DNA so that the total content of DNA within the resulting cells remains diploid. A **second meiotic division** then occurs without a period of DNA replication. Sister **chromatids are separated** at this time to yield cells with the haploid number of chromosomes and the haploid amount of DNA.

The stages of meiotic division are the same as those of mitosis, namely, prophase, anaphase, metaphase, and telophase.

Prophase I

KEY WORDS: **leptotene, zygotene, synapsis (conjugation), bivalent, pachytene, tetrad, diplotene, crossing over, chiasma, recombination, diakinesis**

The first meiotic prophase is long and complex and varies markedly from that of mitosis. It customarily is divided into five stages: leptotene, zygotene, pachytene, diplotene, and diakinesis.

In **leptotene** the chromosomes begin to condense and become visible as individual, slender threads that resemble those of early prophase of mitosis. The ends of the chromosomes become oriented to the side of the nucleus nearest the centrosome, with the bodies of the chromosomes extending in loops into the interior of the nucleus.

During **zygotene,** the homologous chromosomes come together in pairs in which the chromosomes lie side by side, aligned point for point along their lengths. Because of the close apposition of the homologous chromosomes at this stage, they appear to be present in the haploid number. This pairing is called **synapsis** (or **conjugation**), and each pair of homologous chromosomes forms a **bivalent.**

The chromosomes continue to contract and at **pachytene** are short and thick. Each chromosome of the bivalent begins to split lengthwise and then can be seen to consist of two chromatids; the bivalent therefore consists of four chromatids and is frequently called a **tetrad.**

In the **diplotene** stage, each chromosome splits into its constituent chromatids, which remain attached only at their centromeres (kinetochores). The homologous chromatids move apart slightly, and the tetrad formation becomes more obvious. At points along their lengths, the homologous chromatids make contact with each other and exchange segments; this exchange constitutes **crossing over.** The physical regions where the contacts are made are called **chiasmata.** The exchange of segments between homologous chromatids results in reassortment or **recombination** of the genetic material. The chromosomes continue to shorten and thicken, and the nucleolus begins to fragment and disappear.

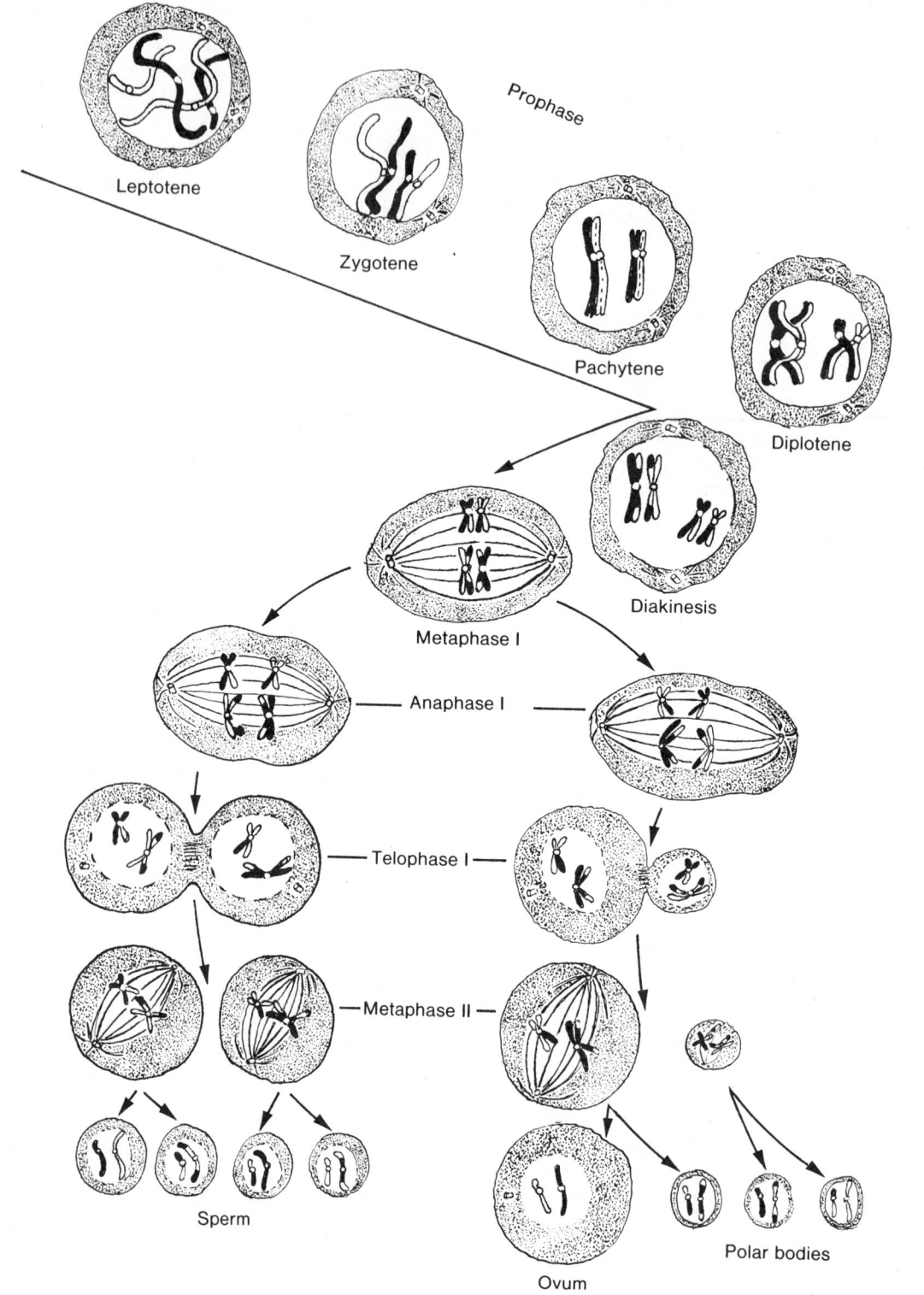

Fig. 17-1. Diagrammatic representation of meiotic division.

Diakinesis is marked by an even greater contraction of the chromosomes that are scattered throughout the cell. The homologous chromosomes still are paired but are joined only at the chiasmata. At the end of prophase I, the nuclear envelope and nucleoli have disappeared, and the tetrads have begun their migration to the equator of the cell.

Metaphase I

Metaphase I is similar to the metaphase of mitotic division except that pairs of chromosomes (bivalents), not single chromosomes, are arranged about the spindle. The homologous pairs are aligned so that the members of each pair lie at either side of the equatorial plate, with the centromeres of homologous chromosomes facing opposite poles.

Anaphase I and Telophase I

Anaphase I and telophase I also are similar to those of mitotic cell division. However, at anaphase I of meiosis, the centromeres have not split so that rather than chromatids separating and moving to opposite poles, whole chromosomes, each consisting of two chromatids, are separated. Reconstruction of the nuclei and cytoplasmic separation occur at telophase I to yield two new cells. However, since the orientation of the bivalents on the equatorial plate is random, there is a random assortment of maternal and paternal chromosomes in each of the telophase nuclei.

Meiotic Division II

KEY WORDS: no DNA synthesis, splitting of centromeres, separation of chromatids

The second meiotic division is more like that of mitosis and occurs after a brief interphase during which there is **no synthesis of DNA.** In some species, the cells produced at the end of the first division enter directly into metaphase II and bypass a second prophase. The **centromeres split** at metaphase, and the **chromatids separate** during anaphase. Unlike mitosis, the sister chromatids are not identical, due to the crossing over that occurred during the first meiotic prophase and to the random alignment of the homologous chromosomes at the first division. Following telophase and cytokinesis, four haploid cells result.

Differences in Meiosis of Male and Female Germ Cells

The mechanics of meiosis are the same regardless of whether sperm or ova are produced, but there are marked differences in the net yield of the two types of gametes and in the times at which meiosis is initiated. In the male, four viable, functioning sperm are produced from each germ cell that enters meiosis, whereas the same events in the female yield only a single functioning ovum. After the first and second meiotic divisions, the cytoplasm is distributed evenly among the developing sperm, but in the formation of ova, the bulk of the cytoplasm is passed to only one cell. The remaining ova, called *polar bodies,* receive so little cytoplasm that they are unable to survive.

The time at which meiotic division is initiated in the male and female germ cells is remarkably different. In the human female, the germ cells begin their meiotic divisions in the embryo, and by the fifth month of intrauterine life, the developing ova are in the diplotene stage of the first meiotic division, but the division is not complete until just before ovulation. Hence many of the ova remain suspended in the diplotene stage for several decades. The second meiotic division and the formation of the second polar body occur only after fertilization. In contrast, meiotic division in the male germ cells is not initiated until puberty, and once the cell has entered meiosis, the process is carried to completion with no prolonged interruption. A diagrammatic representation of meiosis is given in Figure 17-1.

Male Reproductive System

The male reproductive system consists of the gonads (testes), their excretory ducts, the accessory glands, and the penis. The gonads supply the male germ cells or gametes (sperm), which are conducted to the exterior by the excretory system, which includes the penis. The accessory glands contribute secretions that, together with the sperm, form the semen.

Testes

KEY WORDS: tunica vaginalis, tunica albuginea, mediastinum testis, septula testis, lobuli testis, seminiferous tubules, tunica vasculosa, interstitial tissue, interstitial cells (of Leydig)

The testes are compound tubular glands that lie within a scrotal sac, suspended from the body by a spermatic cord. The testes are dual organs that act as exocrine glands producing a holocrine secretion, the sperm, and as endocrine organs that secrete male sex hormone, testosterone.

Each testis is covered anteriorly and laterally by a simple squamous epithelium (mesothelium) called the visceral layer of **tunica vaginalis.** On the posterior aspect of the testis, this mesothelium reflects onto the scrotal sac and lines it as the parietal layer of the tunica vaginalis. The serous cavity between visceral and parietal layers allows the testes to move freely and reduces the chance of injury from increased pressure on the exterior of the scrotum.

A thick, fibrous capsule, the **tunica albuginea,** lies beneath the visceral layer of tunica vaginalis, separated from it only by a basal lamina. Tunica albuginea consists of a dense fibroelastic connective tissue that contains scattered smooth muscle cells. In humans, the muscle cells are concentrated in the posterior region, where the tunica albuginea thickens and projects into the testis to form the **mediastinum testis.** Connective tissue partitions, the **septula testis,** extend from the mediastinum into the interior of the testis and subdivide it into approximately 250 pyramid-shaped compartments called **lobuli testis.** The apices of the compartments are directed toward the mediastinum, and each lobule contains one to four convoluted **seminiferous tubules** that represent the exocrine portion of the testis. Their product is whole cells (spermatozoa).

The inner region of the tunica albuginea, the **tunica vasculosa,** consists of loose connective tissue and contains numerous small blood vessels that supply the testis. The connective tissue extends into each lobule and fills the spaces between the seminiferous tubules, forming the **interstitial tissue** of the testis. The interstitial tissue is rich in extracellular fluid and contains abundant small blood vessels and lympathics that form a plexus around the seminiferous tubules. In addition to fibroblasts, the interstitial tissue contains macrophages, mast cells, mesenchymal cells, and large polyhedral cells 15 to 20 μm in diameter. These are the **interstitial cells (of Leydig),** which commonly occur in groups and make up the endocrine portion of the testis. They secrete the male steroid hormone testosterone. Interstitial cells usually contain single, large, spherical nuclei, but binucleate cells are not uncommon. In electron micrographs, the cytoplasm shows abundant smooth endoplasmic reticulum, well-developed Golgi complexes, and numerous mitochondria, which contain tubular rather than lamellar cristae.

Testosterone and its metabolites are essential for the proliferation and the differentiation of excretory ducts and male accessory sex glands and for maintaining these structures in a functional state. Testosterone influences other tissues and is responsible for beard growth, low pitch of the voice, muscular build, and male distribution of hair. Production of testosterone by the interstitial cells is controlled by a gonadotropic hormone called interstitial cell–stimulating hormone (ICSH), which is secreted by cells in the anterior pituitary.

Most of the enzymes involved in the synthesis of testosterone are located in the smooth endoplasmic reticulum, although the enzymes for one step (conversion of cholesterol to pregnenolone) are found in the mitochondria of interstitial cells. In humans, interstitial cells are characterized by large cytoplasmic crystals called the *crystals of Reinke.* These proteinaceous bodies are highly variable in shape and size but are readily seen with the light microscope (Fig. 17-2). In electron micrographs they show a highly ordered structure. The crystals appear in the testes of most postpubertal males and vary considerably in number. Their significance is unknown.

Seminiferous Tubules

KEY WORDS: peritubular tissue, myoid (peritubular) cell, germinal epithelium, supporting (Sertoli) cell, spermatogenic cell

Fig. 17-2. Interstitial (Leydig) cells that exhibit crystals of Reinke from human testis (LM, × 400).

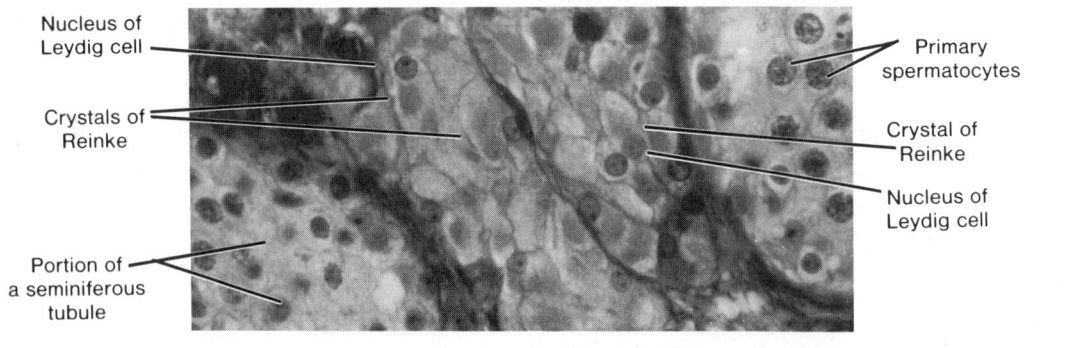

Nucleus of Leydig cell

Crystals of Reinke

Portion of a seminiferous tubule

Primary spermatocytes

Crystal of Reinke

Nucleus of Leydig cell

Each testicular lobule contains one to four highly convoluted seminiferous tubules that measure 150 to 250 μm in diameter and 30 to 70 mm long. A seminiferous tubule consists of a complex stratified germinal (seminiferous) epithelium surrounded by a layer of peritubular tissue (Fig. 17-3).

The **peritubular tissue** is separated from the germinal epithelium by a basal lamina and consists of collagenous fibers and flattened cells that, depending on the species, may contain numerous actin filaments. These are the **myoid (peritubular) cells.** In humans, the myoid cells are said to be noncontractile. The production of testicular fluid by the germinal epithelium and its flow through the seminiferous tubules move the spermatozoa into the excretory duct system.

The **germinal epithelium** of the adult is unique among epithelia in that it consists of a fixed, stable population of **supporting (Sertoli) cells** and a proliferating population of differentiating **spermatogenic cells.** Developing germ cells slowly migrate upward along the lateral surfaces of the supporting cells to be released at the free surface into the lumen of the seminiferous tubule.

Sertoli Cells

KEY WORDS: tall columnar, basal compartment, adluminal compartment, blood-testis barrier, androgen-binding protein, inhibin

The Sertoli cell is a **tall columnar** cell that spans the germinal epithelium from the basal lamina to the luminal surface. The cell has an elaborate shape with numerous lateral processes with recesses or concavities that surround differentiating spermatogenic cells. The apical portion of the cells also envelops developing germ cells and releases them into the lumen of the seminiferous tubule. The expanded portion of the cell contains an irregularly shaped nucleus distinguished by a large, prominent nucleolus that is read-

ily seen with the light microscope. The basal cytoplasm contains abundant smooth endoplasmic reticulum, and a large, well-developed Golgi complex occupies the supranuclear region. The cytoplasm contains many lipid droplets, lysosomes, thin elongated mitochondria, scattered profiles of granular endoplasmic reticulum, and a sheath of fine filaments that envelops the nucleus and separates it from the organelles. Microtubules are present also, their numbers depending on the state of activity of the cell. Human Sertoli cells feature membrane-bound inclusions called the *crystalloids of Charcot-Böttcher;* their function is unknown.

Tight junctions occur between adjacent Sertoli cells near their bases and subdivide the germinal epithelium into basal and adluminal compartments, each of which has a separate, distinct population of spermatogenic cells. The **basal compartment** extends from the basal lamina of the germinal epithelium to the tight junctions; the **adluminal compartment** lies between the tight junctions and the lumen of the tubule (Fig. 17-4). The tight junctions between the Sertoli cells appears to form, at least in part, a **blood-testis barrier.** Germ cells in the basal compartment are contained within an environment that has access to substances in the blood, while the germ cells in the adluminal compartment reside in a specialized milieu that is maintained and controlled by the Sertoli cells. A number of plasma proteins are present in the basal compartment that are not found in the luminal contents of the seminiferous tubule that, however, are rich in other amino acids and ions. Sertoli cells are thought to provide all the nutrients for the avascular germinal epithelium.

In addition to secreting testicular fluid, Sertoli cells release **androgen-binding protein,** whose synthesis is thought to be stimulated by a pituitary gonadotropin, follicle-stimulating hormone (FSH) (Fig. 17-5). The protein binds testosterone, thus providing the adluminal compartment with the level

Fig. 17-3. Seminiferous tubules from a human testis (LM, × 100).

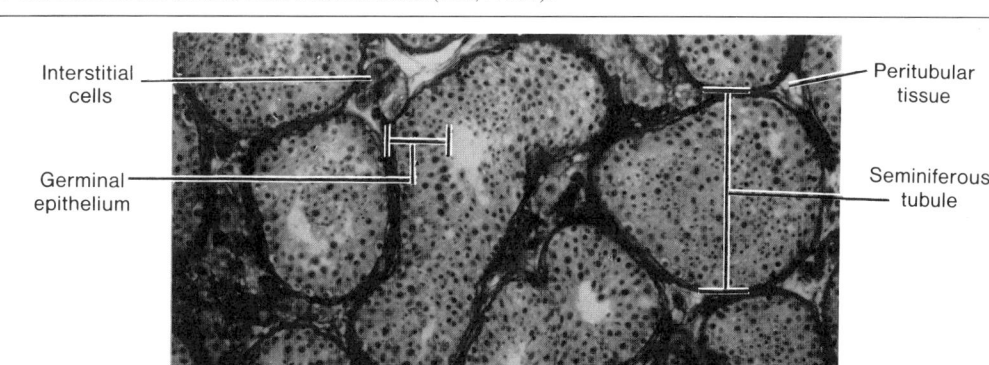

Interstitial cells

Germinal epithelium

Peritubular tissue

Seminiferous tubule

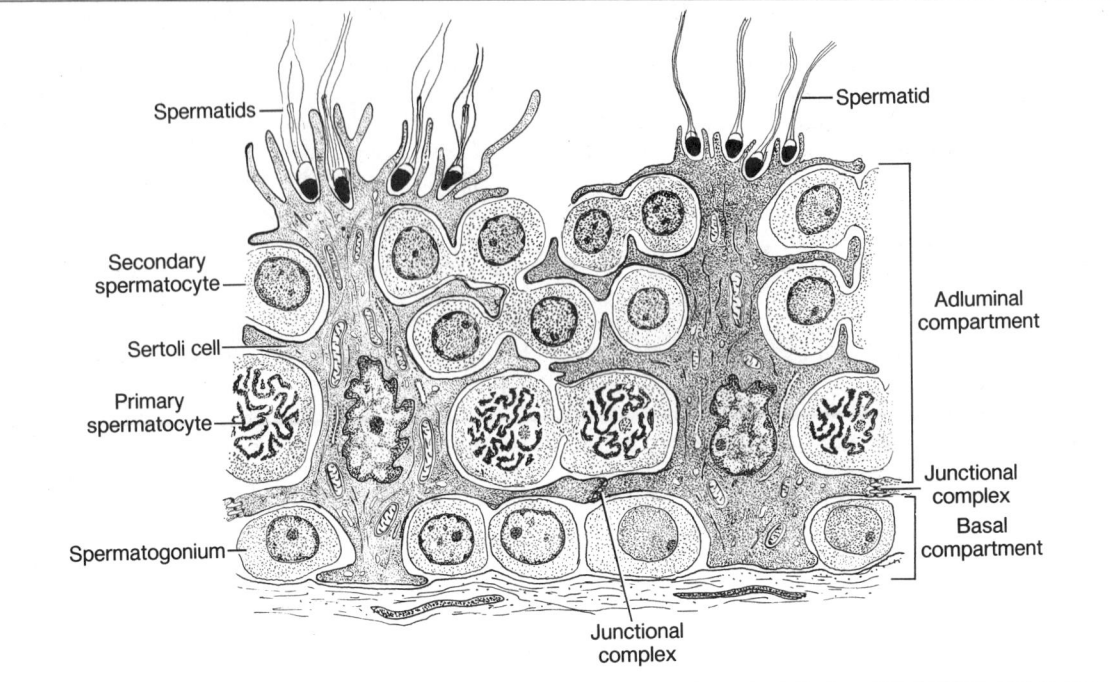

Fig. 17-4. Germinal epithelium.

of hormone needed for the normal differentiation and development of germ cells. The blood-testis barrier helps to confine the high concentration of testosterone to the adluminal compartment and thus allows a different environment to be established in the basal compartment. Many of the developing germ cells in the adluminal compartment are haploid and might be regarded as foreign material by the body if released into surrounding tissues. The tight junctions between Sertoli cells may prevent haploid germ cells from contacting general body tissues and thus prevent an antibody response to the organism's own germ cells. Although tight junctions probably contribute to the barrier, there may be other factors involved.

Sertoli cells phagocytize any degenerating germ cells and take up the residual cytoplasm that normally is shed during release of mature germ cells into the seminiferous tubule. In addition to providing mechanical and nutritional support for developing sperm, Sertoli cells also control the movement of germ cells from the basal lamina through the epithelium to the lumen and are important in the release of germ cells into the lumen. The microtubules and actin filaments in the cytoplasmic processes of the Sertoli cells pro-

vide these processes with the mobility they need to carry out their functions. The numerous gap junctions that occur between adjacent Sertoli cells facilitate communication between cells along specific segments of a seminiferous tubule during migration and release of germ cells. Sertoli cells also secrete a glycoprotein known as **inhibin.** Inhibin suppresses the secretion of FSH by the anterior pituitary and via this feedback loop acts to control the rate of spermatogenesis (see Fig. 17-5).

Spermatogenic Cells

KEY WORDS: spermatogenesis, type A spermatogonia, type B spermatogonia, primary spermatocyte, secondary spermatocyte, spermatid

The spermatogenic cells of the germinal epithelium consist of spermatogonia, primary and secondary spermatocytes, and spermatids. These are not separate cell types but represent stages in a continuous process of differentiation called **spermatogenesis.** The term includes the entire sequence of events in the transformation of diploid spermatogonia at the base of the germinal epithelium into haploid

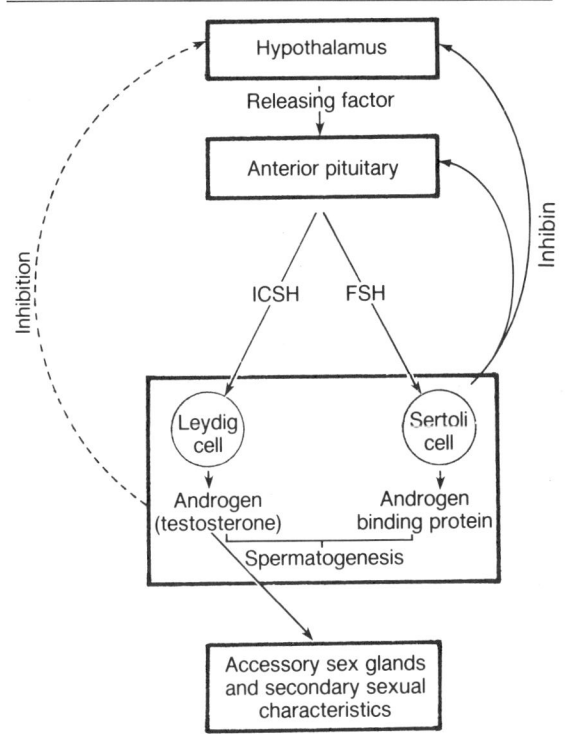

spermatozoa that are released into the lumen of the seminiferous tubule.

Spermatogonia lie in the basal compartment of the germinal epithelium, immediately adjacent to the basal lamina (Figs. 17-4 and 17-6). The cells are 10 to 20 μm in diameter and round or ellipsoidal in shape, and the nucleus of each cell contains the diploid number of chromosomes. Two types of spermatogonia can be differentiated. **Type A spermatogonia** replicate by mitosis and provide a reservoir of stem cells for the formation of future germ cells. During mitotic division, some type A spermatogonia give rise to intermediate forms that eventually produce **type B spermatogonia;** these are committed to the production of primary spermatocytes. The two types of spermatogonia can be differentiated by their nuclei. Type A spermatogonia have spherical or elliptical nuclei with fine chromatin granules and one or two nucleoli near the nuclear envelope. The lighter-stained, spherical nuclei of type B spermatogonia contain variably sized clumps of chromatin, most of which are arranged along the nuclear envelope. Only a single, centrally placed nucleolus is found.

Primary spermatocytes at first resemble type B spermatogonia, but as they migrate from the basal lamina of the germinal epithelium, they become larger and more spherical, and the nucleus enters the initial stages of division. Primary spermatocytes usually are found in the central zone of the germinal epithelium. How these large cells pass from the basal to the adluminal compartment is unknown.

Fig. 17-5. Endocrine control of testicular functions (ICSH = interstitial cell–stimulating hormone; FSH = follicle-stimulating hormone).

Fig. 17-6. The cell types comprising the germinal epithelium of seminiferous tubules (LM, × 250).

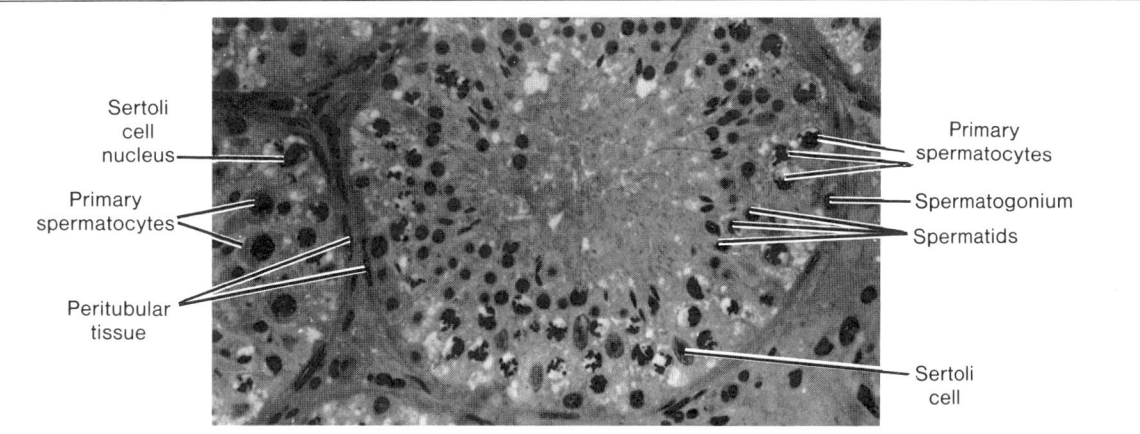

The primary spermatocyte undergoes the first meiotic division to produce **secondary spermatocytes.** The homologous chromosomes have separated and the number is reduced by half, but the cells contain the diploid amount of DNA. Secondary spermatocytes lie nearer the lumen than the primary forms and are about half their size. Unlike the extended division of primary spermatocytes, secondary spermatocytes divide quickly in the second meiotic division to produce **spermatids,** which contain the haploid number of chromosomes and the haploid amount of DNA. Because they divide so quickly after being formed, secondary spermatocytes are seen only rarely in the germinal epithelium. Spermatids are about half the size of secondary spermatocytes. Numerous spermatids in different stages of maturation border the lumen of the seminiferous tubule.

The cell divisions that take place during formation of male germ cells are unique in that not only is the genetic material reduced by half, the division of the cytoplasm is incomplete. Thus the cells resulting from a single spermatogonium remain in cytoplasmic continuity throughout the different stages of differentiation. The continuity is broken only when the sperm are finally released by the Sertoli cell.

Spermiogenesis

KEY WORDS: acrosome, acrosomal vesicle, head cap, axoneme, implantation fossa, connecting piece, annulus, middle piece, head

The elaborate events by which spermatids differentiate into slender, motile sperm are called *spermiogenesis.* Newly formed spermatids are round cells with central, spherical nuclei, prominent Golgi complexes, numerous mitochondria, and a pair of centrioles. Each of these components undergoes changes during spermiogenesis.

At the onset of spermiogenesis, many small granules appear in the Golgi membranes and eventually coalesce to form a single structure called the **acrosome.** The developing acrosome is bounded by a membrane, the **acrosomal vesicle,** which also is derived from the Golgi complex and is closely associated with the outer layer of the nuclear envelope. The acrosomal vesicle expands and then collapses over the anterior half of the nucleus to form the **head cap.** The acrosome, which contains hydrolytic enzymes, remains within the acrosomal membrane. As these events occur, the two centrioles migrate to a position near the nucleus on the side opposite of the forming acrosome. Nine peripheral doublets plus a central pair of microtubules develop from the distal centriole and begin to form the **axoneme** of the tail. The proximal centriole becomes closely

associated with a caudal region of the nucleus called the **implantation fossa.**

As the axoneme continues to develop, nine longitudinal coarse fibers extend around it and blend with nine short, segmented columns that form the **connecting piece,** which unites the nucleus (head) with the tail of the spermatozoon. The **annulus,** a ringlike structure, forms near the centrioles and migrates down the developing flagellum. Randomly distributed mitochondria migrate to the flagellum and become aligned in a tight helix between the centrioles and the annulus. This spirally arranged mitochondrial sheath characterizes the **middle piece** of the tail of a mature spermatozoon.

Simultaneously, marked changes occur in the nucleus: It becomes condensed, elongated, and slightly flattened. Together with the acrosome, it forms the sperm **head.** The bulk of the cytoplasm now is associated with the middle piece of the evolving spermatid, and as differentiation nears completion, the excess cytoplasm is shed as the residual body, leaving only a thin layer of cytoplasm to cover the spermatozoon. The residual cytoplasm is phagocytized by Sertoli cells as the spermatozoa are released into the lumen of the seminiferous tubule. Although morphologically the spermatozoa appear mature, they are nonmotile and incapable of fertilization at this time.

In most species, spermatids at specific stages of differentiation always are associated with spermatocytes and spermatogonia, which also are at specific stages of development. A series of such associations occurs along the length of the same seminiferous tubule, and the distance between two identical germ cell associations is called a "wave" of seminiferous epithelium. Human germinal epithelium exhibits a mosaic of such areas, and six different cell associations have been described. Such associations represent a fundamental pattern of cycling of the germinal epithelium during sperm production. The time taken for spermatogonia to become spermatozoa is relatively constant and species-specific: In humans it is about 64 days. If germ cells fail to develop at their normal rate, they degenerate and are phagocytosed by adjacent Sertoli cells.

Spermatozoa

KEY WORDS: head, tail, acrosomal cap, neck, capitulum, middle piece, principal piece, end piece, capacitation

Spermatozoa that lie free in the lumina of seminiferous tubules consist of a **head** containing the nucleus and a **tail,** which eventually gives motility to the free cell (Fig. 17-7). The chromatin of the nucleus is very condensed and re-

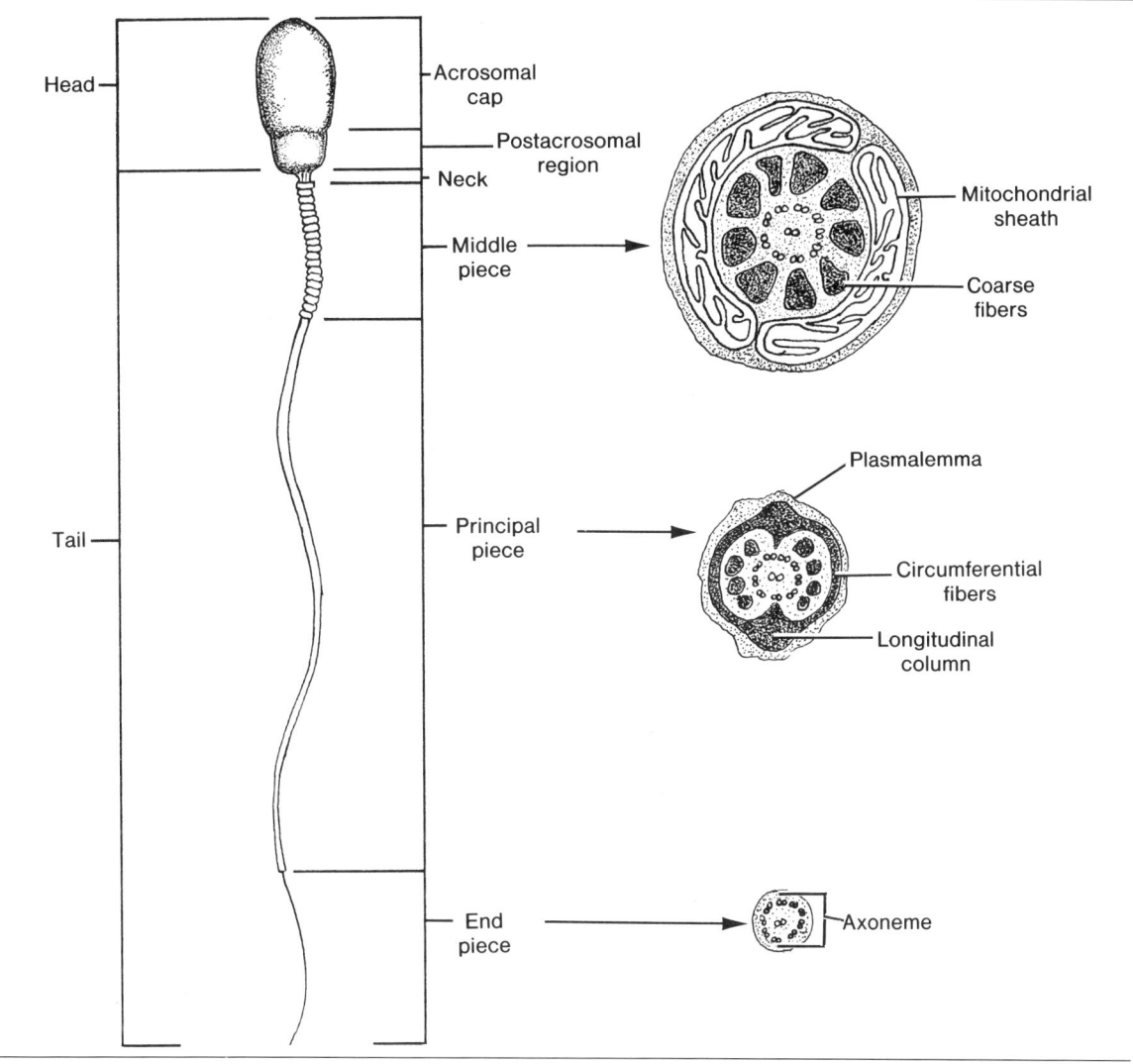

Fig. 17-7. Structure of a spermatozoon.

duced in volume, providing the functionally mature sperm with greater mobility. The condensed form of chromatin also protects the genome while the spermatozoon is en route to fertilize the ovum. The **acrosomal cap** covers the anterior two-thirds of the nucleus and contains lysosomes that are important for penetration of the ovum at fertilization. The size and shape of the nucleus vary tremendously in different species.

The sperm tail is about 55 μm long and consists of a neck, middle piece, principal piece, and end piece. The

structural details of the different segments are best seen with the electron microscope.

The **neck** is the region where the head and tail unite. It contains the connecting piece that joins the nine outer dense fibers of the tail to the implantation fossa of the nucleus. The region of the connecting piece that joins the implantation fossa is expanded slightly to form the **capitulum.** The **middle piece** extends from the neck to the annulus and consists of the axoneme, the nine coarse fibers, and the helical sheath of mitochondria. The **principal piece** is the longest part of the tail and consists of the axoneme and the nine coarse fibers (2 + 9 + 9) enclosed by a sheath of circumferential fibers. The circumferential fibers join two longitudinal thickenings of this sheath located on opposite sides. The **end piece** is the shortest segment of the tail and consists only of the axoneme surrounded by the cell membrane (see Fig. 17-7).

The last step in the physiologic maturation of spermatozoa takes place in the female reproductive tract and is called **capacitation.** This final activation requires 1 to 66 hours to complete, depending on the species, and is characterized by changes in the acrosomal cap and the respiratory metabolism of the spermatozoa. Capacitation substantially increases the number of spermatozoa capable of fertilization, but the mechanism of this activation is unknown.

Excretory Ducts

The excretory ducts consist of a complex system of tubules that link each testis to the urethra, through which the exocrine secretion, semen, is conducted to the exterior. The duct system consists of the tubuli recti (straight tubules), rete testis, ductus efferentes, ductus epididymis, ductus deferens, ejaculatory ducts, and prostatic, membranous, and penile urethra.

Tubuli Recti

KEY WORDS: **simple columnar epithelium**

Near the apex of each testicular lobule, the seminiferous tubules join to form short, straight tubules called the *tubuli recti.* The lining epithelium has no germ cells and consists

only of Sertoli cells. This **simple columnar epithelium** lies on a thin basal lamina and is surrounded by loose connective tissue. The lumina of the tubuli recti are continuous with a network of anastomosing channels in the mediastinum, the rete testis.

Rete Testis

KEY WORDS: **simple cuboidal epithelium**

The rete testis is lined by **simple cuboidal epithelium** in which each of the component cells bears short microvilli and a single cilium on the apical surface. The epithelium lies on a delicate basal lamina. The channels of the rete testis are surrounded by a dense bed of vascular tissue (Fig. 17-8).

Ductuli Efferentes

KEY WORDS: **initial segment, head of epididymis, tall and short columnar, ciliated and nonciliated cells**

In humans, 10 to 15 ductuli efferentes emerge from the mediastinum on the posterosuperior surface of the testis and unite the channels of the rete testis with the ductus epididymis. The ductules follow a convoluted course and, with their supporting tissue, make up the **initial segment** of the **head of the epididymis.** The luminal border of the efferent ductules shows a characteristic irregular contour due to the presence of alternating groups of **tall and short columnar** cells (Fig. 17-9). **Ciliated and nonciliated cells** are present. Each contains the organelles normally associated with epithelia as well as supranuclear granules that are lysosomal in nature. The nonciliated cells show numer-

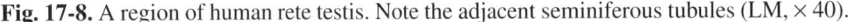

Fig. 17-8. A region of human rete testis. Note the adjacent seminiferous tubules (LM, × 40).

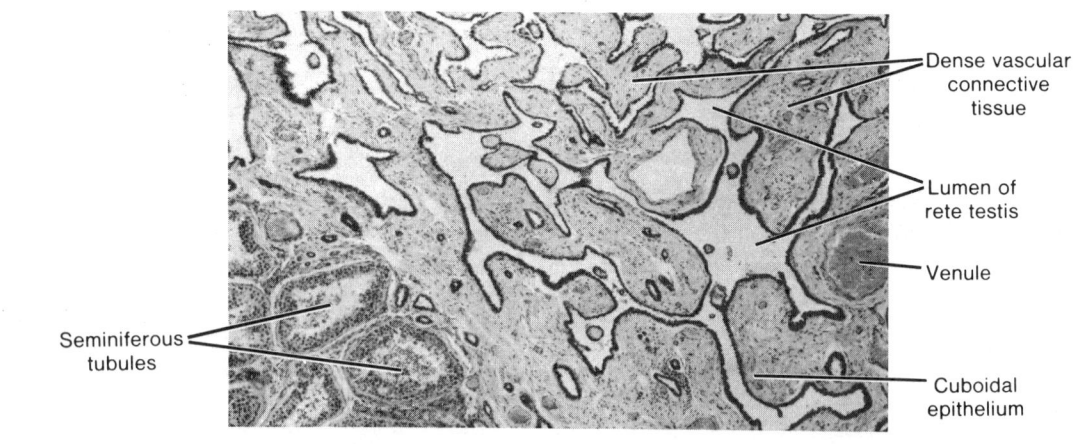

Dense vascular connective tissue

Lumen of rete testis

Venule

Cuboidal epithelium

Seminiferous tubules

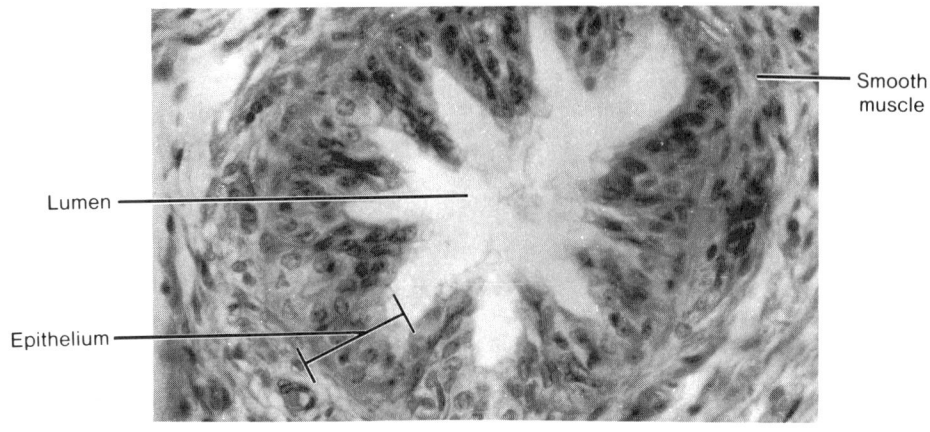

Fig. 17-9. A tubule from human ductuli efferentes (LM, × 250).

ous microvilli on their apical surfaces and are thought to absorb much of the testicular fluid that is present in this part of the ductal system. Cilia on the other cell type beat toward the epididymis and help move the nonmotile sperm and testicular fluid in that direction.

The epithelium lies on a basal lamina and is bounded by a layer of circularly arranged smooth muscle that thickens toward the duct of the epididymis. Cilia are not found beyond the ductuli efferentes, and the movement of sperm through the remainder of the duct system depends on the contractions of the muscular wall.

Ductus Epididymidis

KEY WORDS: pseudostratified columnar, principal cell, stereocilia, basal cell, smooth muscle, three-layered coat

The efferent ductules gradually unite to form a single ductus epididymidis that measures 5 to 7 m long. This highly coiled duct and its associated vascular connective tissue form the rest of the head, body, and tail of the epididymis. The epithelial lining is **pseudostratified columnar** and consists of principal and basal cells (Fig. 17-10).

Fig. 17-10. A portion of human ductus epididymidis showing both basal and principal cells of the lining epithelium. Stereocilia of the principal cells project into the lumen (LM, × 250).

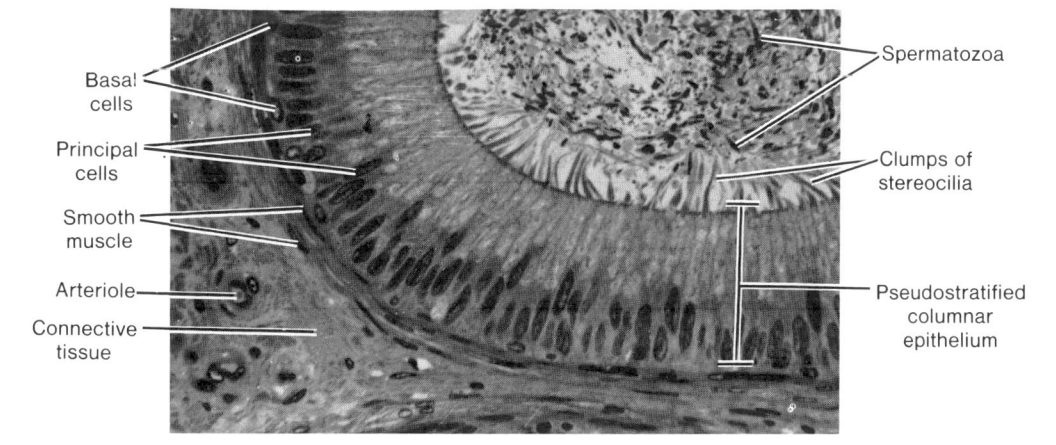

Principal cells are the most numerous and are very tall (80 μm) in the proximal segment but gradually decrease in height distally to measure only 40 μm near the junction with the ductus deferens. Long microvilli, inappropriately called **stereocilia,** extend from the apical surfaces of the principal cells. Like microvilli found elsewhere, they contain an axial bundle of actin microfilaments. The actin microfilaments extend from the stereocilia into the terminal web region of the principal cells. A very large Golgi complex takes up a supranuclear position in the cell, and the basal region is filled by an abundance of endoplasmic reticulum. Lysosomes, multivesicular bodies, and numerous coated vesicles also are present. One of the functions of the principal cell is absorption; together the proximal portion of the ductus epididymidis and the ductuli efferentes absorb over 90 percent of the fluid produced by the seminiferous epithelium.

Basal cells are small, round cells that lie on the basal lamina, interposed between the principal cells. They have a lightly staining cytoplasm with few organelles. Scattered intraepithelial lymphocytes often are present in the epithelium of the ductus epididymidis.

The epididymal epithelium lies on a basal lamina and is surrounded by a thin lamina propria and a thin layer of circularly arranged **smooth muscle** cells. Near the ductus deferens, the muscle layer thickens and becomes **three-layered** (inner longitudinal, middle circular, outer longitudinal); it is continuous with that of the ductus deferens. The regional differences reflect differences in the mobility of the ductus epididymidis. Proximally, the duct shows spontaneous peristaltic contractions that slowly move spermatozoa through the epididymis; distally, the peristaltic contractions are reduced, and this region serves to store sperm. Spermatozoa become physiologically mature as they pass through the epididymis. Those entering the proximal part of the duct mostly are incapable of fertilization and swim in a weak, random fashion. Spermatozoa from the distal portion are capable of fertilizing ova and show strong, unidirectional motility. How the ductus epididymidis contributes to the maturation of spermatozoa is unknown. Transit time through the epididymis is about 7 days for humans.

Ductus Deferens

KEY WORDS: **mucosa, pseudostratified columnar, stereocilia, lamina propria, muscularis, adventitia, ampulla**

The ductus deferens unites the ductus epididymidis and the prostatic urethra. It is characterized by a thick wall consisting of a mucosa, a muscularis, and an adventitia. The **mucosa** is thrown into longitudinal folds that project into the lumen and consists of **pseudostratified columnar** epithelium resting on a thin basal lamina. The epithelium bears **stereocilia** on the luminal surface. The surrounding **lamina propria** is dense and contains numerous elastic fibers. The **muscularis** is the dominant feature of the ductus deferens and consists of three layers of smooth muscle arranged longitudinally in the inner and outer layers and circularly in the middle layer (Fig. 17-11). The muscularis

Fig. 17-11. Three distinct layers forming the muscular wall of the human ductus deferens (LM, × 150).

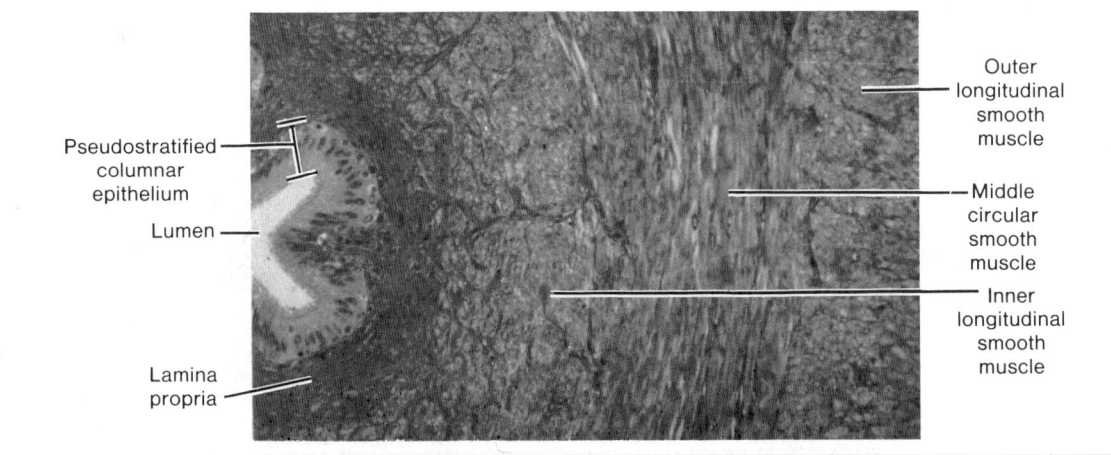

is surrounded by the loose connective tissue of the **adventitia,** which blends with neighboring structures and anchors the ductus deferens in place. Powerful contractions of the muscular wall propel sperm from the distal ductus epididymidis and rapidly transport them through the ductus deferens during ejaculation.

Near its termination, the ductus deferens dilates to form the **ampulla.** The lumen expands and the mucosa is folded, creating a labyrinth of pocket-like recesses (Fig. 17-12). The lining epithelium is pseudostratified columnar. The muscular coat of the ampulla is thinner and the layers are less distinct than in other parts of the ductus deferens. Each smooth muscle cell in the muscularis of the ductus deferens receives direct sympathetic innervation. This accounts for the rapid, forceful contractions of the ductus during ejaculation.

Ejaculatory Duct

KEY WORDS: pseudostratified or simple columnar epithelium

A short, narrow region of the ductus deferens extends beyond the ampulla and joins the duct of the seminal vesicle. This combined duct forms the ejaculatory duct, which empties into the prostatic urethra. It is about 1 cm long and is lined by a **pseudostratified or simple columnar epithelium** (Figs. 17-13 and 17-14). The mucosa forms outpockets that are similar to, but not as well developed as, those in the ampulla. The remainder of the wall of the ejaculatory duct consists only of fibrous connective tissue.

Prostatic Urethra

KEY WORDS: transitional epithelium, colliculus seminalis, prostatic utricle

The ejaculatory duct penetrates the prostate gland and opens into the prostatic part of the urethra. The prostatic urethra is lined by a thin **transitional epithelium** and bears a dome-shaped elevation, the **colliculus seminalis,** on its posterior wall. A small blind diverticulum called the **prostatic utricle** lies on the summit of the colliculus and represents a remnant of the müllerian duct in the male. The two ejaculatory ducts, one draining each testis, empty into the prostatic urethra on either side of the utricle. Numerous glands from the surrounding prostate also empty into this part of the urethra.

Accessory Sex Glands

Male accessory sex glands include the seminal vesicles, prostate, and bulbourethral glands. The secretion from each of these glands is added to the testicular fluid and forms a substantial part of the semen.

Seminal Vesicles

KEY WORDS: folds, honeycombed appearance, pseudostratified columnar epithelium

The seminal vesicles are elongated, saclike structures that lie posterior to the prostate gland. Each joins with the distal end of the ductus deferens to form the ejaculatory duct.

Fig. 17-12. The ampulla of human ductus deferens. Note the elaborate mucosal folds projecting into the lumen. Compare the mucosa of this region with that shown in 17-11 (LM, × 100).

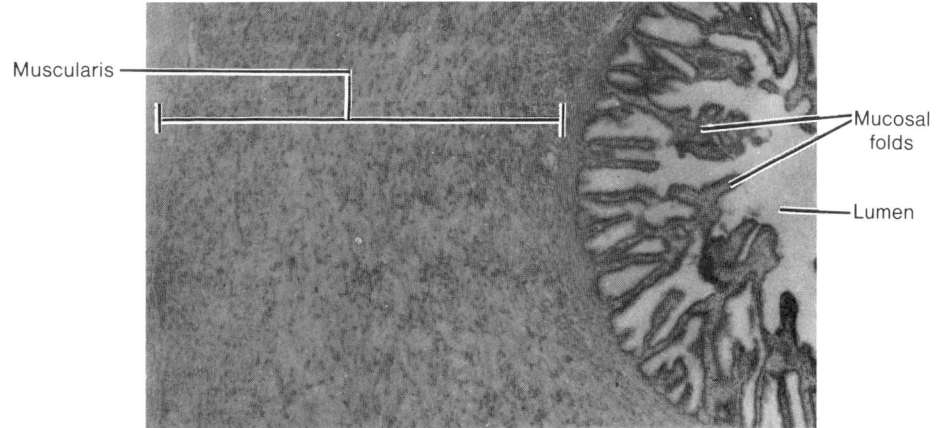

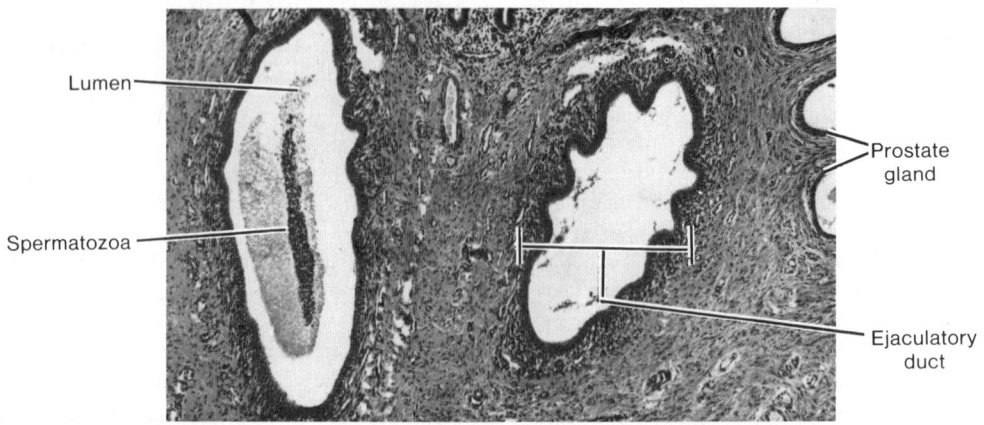

Fig. 17-13. A region of human prostate containing the ejaculatory ducts (LM, × 250).

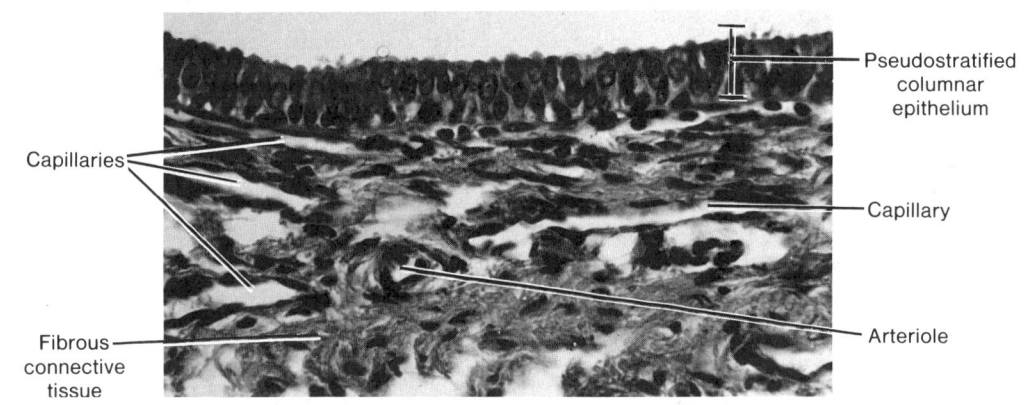

Fig. 17-14. Pseudostratified columnar epithelium lining a human ejaculatory duct (LM, × 250).

The mucosa of the seminal vesicles is thrown into numerous complex primary **folds** that give rise to secondary and tertiary folds (Fig. 17-15). These project into the lumen to subdivide it into many small, irregular compartments that give the lumen a **honeycombed appearance.** All the compartments communicate with the central lumen, although in sections the impression is one of individual chambers.

The mucosal folds are lined mainly by **pseudostratified columnar epithelium** consisting of rounded basal cells interposed between cuboidal or columnar cells. Regions of simple columnar epithelium may be present also. The mucosal cells contain numerous granules, some lipid droplets, and lipochrome pigment, which first appears at sexual ma-

turity and increases with age. The ultrastructural features indicate a cell that is active in protein synthesis. The cytoplasm contains abundant granular endoplasmic reticulum, a prominent supranuclear Golgi complex, and conspicuous, dense secretory granules in the apical cytoplasm.

The epithelium rests on a thin lamina propria of loose connective tissue with many elastic fibers. A muscular coat is present and consists of an inner layer of circularly arranged smooth muscle cells and an outer layer in which the muscle fibers have a longitudinal orientation. Both layers of muscle are thinner than those in the ductus deferens. External to the muscle coat is a layer of loose connective tissue rich in elastic fibers.

Secretions from the seminal vesicles form a substantial

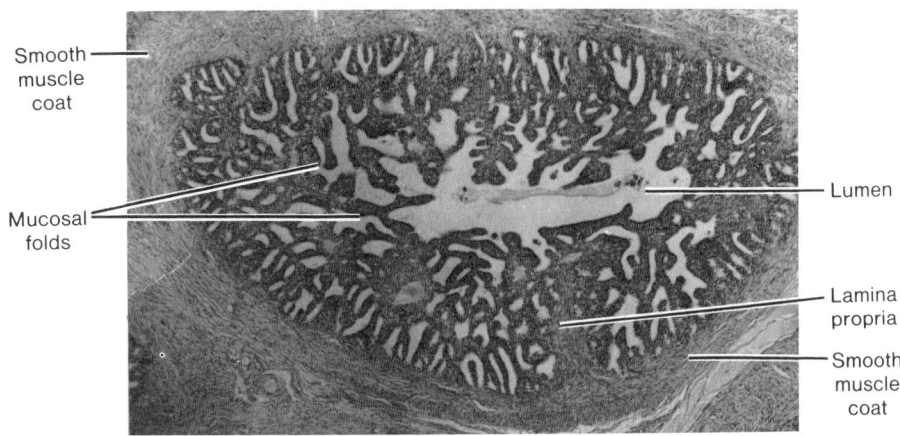

Fig. 17-15. Elaborate mucosal folds and a well-defined smooth muscle wall characterize the human seminal vesicle (LM, × 40).

part of the total ejaculate. It is a yellowish, viscid secretion that contains much fructose and prostaglandins. It provides a source of energy for sperm. In sections the secretions appear as deeply stained, coagulated masses, often with a netlike structure. The seminal vesicles depend on testosterone, and removal of the hormone, as by castration, results in their involution and loss of secretory function.

Prostate

KEY WORDS: composite gland, compound tubuloalveolar, capsule, septa, prostatic concretion, fibromuscular stroma

The prostate is the largest of the accessory sex glands in the male and surrounds the urethra at its origin from the bladder. It is a **composite gland,** made up of 30 to 50 small, **compound tubuloalvcolar** glands from which 20 or more ducts drain independently into the prostatic urethra. These small glands appear to form strata around the urethra and consist of the periurethral mucosal glands, submucosal glands, and the main or principal prostatic glands, which lie peripherally and make up the bulk of the prostate. The prostate is contained within a vascular, fibroelastic **capsule** that contains many smooth muscle fibers in its inner layers. Broad **septa** extend into the prostate from the capsule and become continuous with the dense fibroelastic tissue that separates the individual glandular elements.

The secretory units of the glands are irregular and vary greatly in size and shape. The glandular epithelium differs from gland to gland and even within a single alveolus. It usually is simple or pseudostratified columnar but may be low cuboidal or squamous in some of the larger saccular cavities. The epithelium is limited by an indistinct basal lamina and rests on a layer of connective tissue that contains dense networks of elastic fibers and numerous capillaries. The cells contain abundant granular endoplasmic reticulum and many apical secretory granules. The lumina of secretory units may contain spherical bodies, the **prostatic concretions,** that are thought to result from condensation of secretory material. The concretions appear to increase with age and may become calcified. The connective tissue surrounding the individual glandular units contains much smooth muscle (a **fibromuscular stroma**), which aids in the discharge of prostatic fluid at ejaculation (Fig. 17-16).

The prostatic secretion is a thin, milky fluid of pH 6.5 that is rich in zinc, citric acid, phosphatase, and proteolytic enzymes, one of which, fibrinolysin, is important in the liquefaction of semen. Like the seminal vesicles, the development and functional maintenance of the prostate depend on testosterone and its metabolites.

Bulbourethral Glands

KEY WORDS: compound tubuloalveolar, simple cuboidal to simple columnar

The bulbourethral glands are a pair of pea-sized structures located in the urogenital diaphragm, close to the bulb of the penis. They are **compound tubuloalveolar glands**

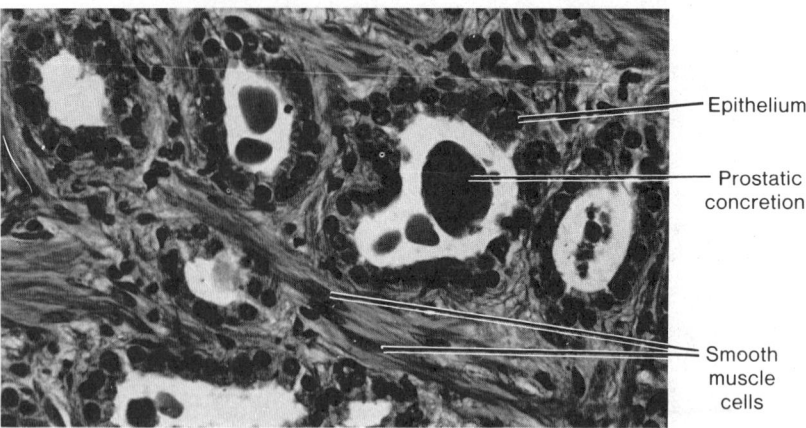

Fig. 17-16. Principal glands of the human prostate. Note the smooth muscle cells in the surrounding fibromuscular stroma (LM, × 250).

whose long ducts drain into the proximal part of the penile urethra. Each gland is limited by a connective tissue capsule from which septa, containing elastic fibers and smooth and skeletal muscle cells, extend into the glands, dividing them into lobules.

The ducts and secretory portions are irregular in shape and size, and at their terminations, the secretory parts may form cystlike enlargements. The glandular epithelium varies from **simple cuboidal to simple columnar** depending on the functional state, but in distended alveoli the epithelium may be flattened. Active cells show a lightly stained cytoplasm filled with mucinogen granules that confine the nucleus to the base of the cell. Excretory ducts are lined by simple columnar epithelium that becomes pseudostratified near the urethra. In smaller ducts, the epithelium appears to be secretory. The surrounding connective tissue contains an incomplete layer of circularly arranged smooth muscle.

Bulbourethral glands produce a clear, viscid fluid that is rich in amino sugars and contains sialoprotein. It is secreted in response to erotic stimulation and serves as a lubricant for the penile urethra.

Semen

The final product formed by the exocrine secretions of the testes and accessory glands is a whitish fluid called *seminal fluid* or *semen.* The average ejaculate in human males consists of about 3 ml of semen which, in addition to about 300 million sperm, contains degenerating cells exfoliated by the ductal system, occasional wandering cells from connective tissues, pigment granules, and prostatic concre-

tions. Hyaline bodies of unknown origin, lipid granules, fat, and protein also are present.

External Genitalia

In the male, the two structures that make up the external genitalia are the scrotum and penis.

Scrotum

KEY WORDS: **skin, tunica dartos, spermatic cord, pampiniform plexus**

The scrotum is a pendulous, cutaneous pouch situated at the base of the penis and below the symphysis pubis. It is divided into two compartments, each of which houses a testis, an epididymis, and the lower part of the spermatic cord. The scrotum consists only of skin and a closely associated tunica dartos.

The scrotal **skin** is thin, pigmented, and commonly thrown into folds. It contains many sweat glands, sebaceous glands that produce an odorous secretion, and some coarse hairs, the follicles of which are visible through the skin. The **tunica dartos** underlies the skin and forms the septum that divides the scrotum into its two compartments. It is firmly attached to the skin and consists largely of smooth muscle and collagenous connective tissue. The appearance of the scrotum varies with the state of contraction of the smooth muscle. Under influence of cold, exercise, or sexual stimulation, the muscle contracts and the scrotum becomes short and wrinkled.

The **spermatic cord** consists of several thin layers of

connective tissue that are acquired from the anterior abdominal wall as the testes descend from the abdominal cavity into the scrotal sac during development. It contains the ductus deferens, nerve fibers, lymphatic channels, testicular artery, and **pampiniform plexus** of testicular veins. As the testicular artery nears the testis, it becomes highly convoluted and is surrounded by the venous plexus. The proximity of the surrounding, cooler venous blood causes the arterial blood to lose heat and provides a thermoregulatory mechanism for precooling incoming arterial blood. In this way the temperature of the testes is maintained a few degrees below body temperature, a condition necessary for production of sperm. The temperature can be elevated by drawing the testes closer to the abdominal wall by contraction of the layer of striated muscle (cremaster muscle) that invests the spermatic cord. In humans, failure of the testes to descend (cryptorchidism) results in sterility.

Penis

KEY WORDS: erectile tissue, corpora cavernosa penis, pectiniform septum, tunica albuginea, cavernous spaces, corpus cavernosum urethrae (corpus spongiosum), glans penis, prepuce

The penile urethra serves as a common channel for conducting urine and seminal fluid to the exterior. It is contained within a cylinder of **erectile tissue,** the corpus cavernosum urethrae (corpus spongiosum) that lies ventral to a pair of similar erectile bodies called the *corpora cavernosa penis.* Together the three structures make up the bulk of the penis, the copulatory organ of the male.

The **corpora cavernosa penis** begin as separate bodies along the rami of the pubis on either side and join at the pubic angle to form the shaft of the penis. They are united by a common connective tissue septum called the **pectiniform septum,** and each corpus is surrounded by a thick fibrous sheath, the **tunica albuginea.** Trabeculae of collagenous and elastic fibers with numerous smooth muscle cells extend into the corpora from the tunica albuginea and divide the central regions of the corpora cavernosa into numerous **cavernous spaces,** those near the center being the larger. These spaces are endothelial-lined vascular spaces and are continuous with the arteries that supply them and with draining veins. The cavernous tissue of each corpus cavernosum penis communicates with the other through numerous slitlike openings in the pectiniform septum.

The ventrally placed **corpus cavernosum urethrae (corpus spongiosum)** ends in an enlargement, the **glans penis,** which forms a cap over the ends of the corpora cavernosa penis. Structurally, corpus spongiosum is similar to the corpora cavernosa penis, but the tunica albuginea is thinner and contains more elastic fibers and smooth muscle, and the trabeculae are thinner and contain more elastic tissue (Fig. 17-17).

The three corpora are bound together by subcutaneous connective tissue that contains numerous smooth muscle cells but is devoid of fat. The shaft of the penis is covered by a thin, mobile skin that shows a slight increase in pigmentation. The skin of the distal shaft, unlike that of the root, lacks hair but does contain scattered sweat glands. The hairless skin of the glans penis is fused to the underlying connective tissue and is nonmobile. Unusual sebaceous glands not associated with hair follicles occur in this region. The glans penis is covered by a fold of skin called the **prepuce** or foreskin, the inner surface of which is moist and resembles a mucous membrane. Numerous free nerve endings are present in the epithelium of the glans penis, prepuce, and subepithelial connective tissue of the urethra and skin. Meissner's corpuscles are associated with the epidermis of the genitals, and Vater-Pacini corpuscles are present in the deeper layers of the dermis.

A section through the penis is shown in Figure 17-18.

Arterial Supply

KEY WORDS: helicine arteries, intimal ridge

The principal arterial supply to the penis is from two arteries that lie dorsal to the corpora cavernosa penis and deep arteries that run within the erectile tissue of these structures. Branches from the dorsal arteries penetrate the tunica albuginea and enter the cavernous tissue, where the arteries branch and either form capillary plexuses or course distally in the cavernous tissue. These **helicine arteries** are highly convoluted in the flaccid penis and take a spiral course through the trabeculae of the cavernous tissue. The intima of most of these arteries, even before they enter the cavernous tissue, have long, ridgelike thickenings that project into and partially occlude the lumina. These **intimal ridges** are most frequent where vessels branch. They consist of loose connective tissue that contains many smooth muscle cells. Blood from the large central lacunae drains peripherally toward the smaller vascular spaces and finally into a plexus of veins at the periphery. The veins run along the interior of the tunica albuginea, pierce the limiting tunica, and drain into the deep dorsal vein of the penis.

The arterial blood supply to the corpus spongiosum is similar to that of the corpora cavernosa penis, except for the venous drainage. Beginning at the lacunae, the veins of the corpus spongiosum have large openings and immediately penetrate the tunica albuginea to drain to the exterior.

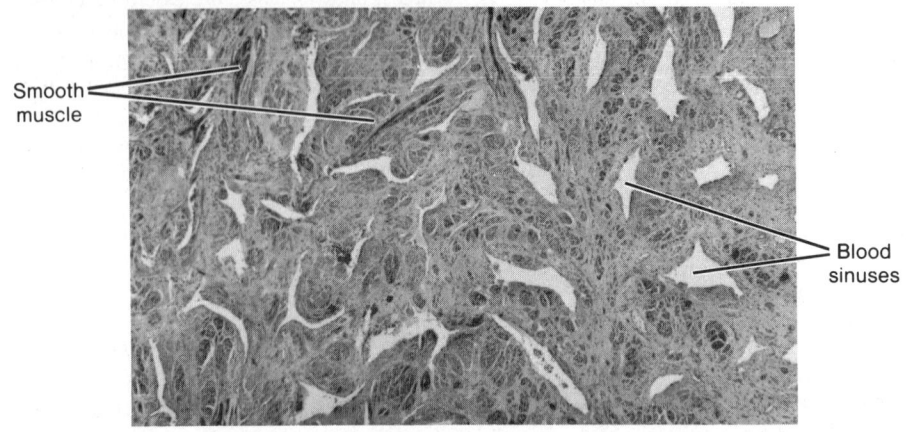

Smooth muscle

Blood sinuses

Fig. 17-17. Erectile tissue from the corpus spongiosum of a human penis (LM, × 30).

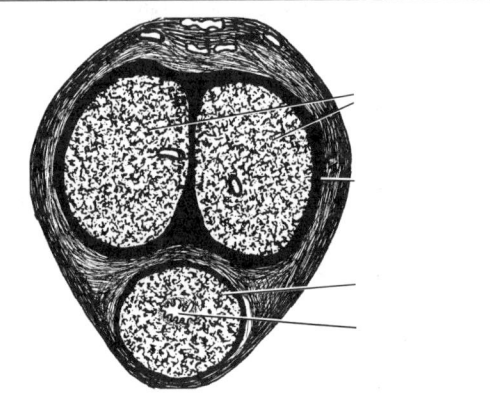

Fig. 17-18. Section through the penis.

The structure and arrangement of the blood vessels in the cavernous tissues provide the mechanism for erection. During erotic stimulation, the smooth muscle of the arterial and trabecular walls relaxes, and blood pressure overcomes the elastic resistance of the arteries. The helicine arteries dilate and straighten, and the vascular spaces of the cavernous tissues quickly fill with blood. The lacunae, especially near the center of the cavernous bodies, become engorged with blood, thus compressing the small peripheral spaces and veins against the tunica albuginea, retarding the egress of blood; the erectile tissues of corpora cavernosa penis become enlarged and rigid. Corpus spongiosum does not become as rigid as the corpora cavernosa penis because there is less compression of the venous drainage and the tunica albuginea is thinner and more yielding. The lesser rigidity of corpus spongiosum allows the urethra to remain patent, which is essential for the passage of semen during ejaculation.

After cessation of sexual activity, the smooth muscle in the arteries and cavernous tissue regains its tone. The intimal ridges once again partially occlude the lumina of the arteries, thus reducing the volume of incoming blood. Excess blood in the vascular spaces of the erectile tissue is forced out by contraction of smooth muscle cells in the trabeculae and by recoil of surrounding elastic tissue. Gradually, the normal route of blood flow through the penis is restored, and the penis returns to the flaccid condition.

Review Table 17-1. Key Features of the Male Reproductive Tract

	Epithelial Lining	Support	Muscle
Seminiferous tubule	Germinal epithelium containing spermatogenic cells (spermatogonia, primary and secondary spermatocytes, spermatids) and Sertoli cells	Peritubular connective tissue; collagen fibers and myoid cells	
Tubuli recti	Simple columnar	Loose connective tissue	
Rete testis	Simple cuboidal. Short microvilli, single cilium on each cell	Dense, vascular connective tissue	
Ductuli efferentes	Tall, ciliated columnar and short ciliated or nonciliated cuboidal. Lining has irregular contour	Loose connective tissue between tubules	Thin, circular smooth muscle
Ductus epididymidis	Pseudostratified columnar: tall principal cells with stereocilia; short basal cells	Thin lamina propria	Primarily circular smooth muscle, three layers appear distally
Ductus deferens	Pseudostratified columnar, thrown into longitudinal folds; stereocilia	Dense lamina propria; collagen and elastic fibers; adventitia of loose connective tissue	Three layers: inner longitudinal, middle circular, outer longitudinal smooth muscle
Ampulla of ductus deferens	Pseudostratified columnar with stereocilia; complex folds form extensive pocket-like recesses	Lamina propria and adventitia less well defined	Thinner, less distinct layers of smooth muscle
Ejaculatory duct	Pseudostratified or simple columnar; forms pockets but less prominent than in ampulla	Dense collagenous connective tissue with many elastic fibers	
Seminal vesicles	Honeycombed appearance; pseudostratified columnar with regions of simple columnar	Thin lamina propria with many elastic fibers; external layer of loose connective tissue rich in elastic fibers	Inner circular, outer longitudinal smooth muscle
Prostate	Variable; usually simple columnar or pseudostratified columnar. Low cuboidal or squamous may occur	Vascular connective tissue with dense network of elastic fibers	Smooth muscle fibers in connective tissue surround individual glandular units
Bulbourethral glands	Variable; simple cuboidal to simple columnar; flattened in distended alveoli; ducts lined by simple columnar, pseudostratified near urethra	Fibroelastic connective tissue around each glandular structure	Smooth and skeletal muscle fibers in interstitial tissue between glandular elements
Corpora cavernosa penis	Endothelium of vascular spaces	Fibroelastic connective tissue	Smooth muscle cells in trabeculae between cavernous spaces
Corpora cavernosa urethrae	Cavernous spaces lined by endothelium, penile urethrae lined by pseudostratified columnar, nonkeratinized stratified squamous	Fibroelastic connective tissue rich in elastic fibers	Smooth muscle cells more abundant

Functional Review: Meiosis and Male Reproductive Organs

The primary function of the male reproductive system is to provide haploid germ cells for procreation, and one of the consequences of meiotic division is to reduce the number of chromosomes to half that present in somatic cells. This ensures that on union of gametes, the normal diploid number of chromosomes will be maintained. A second consequence of meiosis that is of great importance is provision of genetic variation as the result of the exchange of genetic material between homologous chromosomes and from the random distribution of homologous chromosomes between daughter nuclei. The unequal division of cytoplasm among ova provides one cell with sufficient material to nourish the zygote until other forms of nourishment can be established.

Production of sperm occurs in the germinal epithelium of the seminiferous tubules and represents a holocrine (exocrine) secretion of the testes. Spermatogenesis and spermiogenesis depend on high concentrations of testosterone, the endocrine secretion of the testes produced by interstitial cells.

Germinal epithelium essentially contains two major classes of cells: a mobile population of germ or spermatogenic cells and a stable population of supporting (Sertoli) cells. Seminiferous epithelium is unique in that it is divided into basal and adluminal compartments, each differing in their cellular content and environments. The basal compartment contains stem cells (spermatogonia) and newly formed primary spermatocytes that are exposed to an environment similar to that of blood plasma. The primary spermatocytes soon migrate into the adluminal compartment, where, along with secondary spermatocytes, spermatids, and spermatozoa, they are exposed to an environment rich in ions that contains amino acids not found in the basal compartment and has a high concentration of testosterone. The two different environments are established and maintained by Sertoli cells, and the tight junctions between them constitute — at least in part — a blood-testis barrier. Any substances that enter or leave the adluminal compartment must pass through the cytoplasm of the Sertoli cells.

Under the influence of gonadotropic hormone (FSH), Sertoli cells release androgen-binding protein that complexes with testosterone to maintain a high concentration of hormone in the adluminal compartment. The blood-testis barrier also prevents escape of developing germ cells out of the adluminal compartment and into general body tissues. If this escape occurred, the haploid germ cells might be interpreted as foreign elements, resulting in immunologic responses directed against developing germ cells in the germinal epithelium and thus causing sterility.

Development of germ cells occurs in the adluminal compartment, in the embrace of the complex cytoplasmic processes of Sertoli cells. As spermatogenic cells develop and differentiate, the Sertoli cell is thought to move them actively along its lateral surfaces toward the lumen, where they are released from the apical surface of the Sertoli cell. Gap junctions permit communication between Sertoli cells and allow for coordinated release of spermatozoa from a given segment of a seminiferous tubule. Testosterone is present in the adluminal compartment of the seminiferous tubules and in the proximal duct system as far as the lower regions of the ductus epididymidis. Thus there exists the unusual circumstance of an endocrine secretion of the testis (testosterone) not only entering the bloodstream but also passing through the lumen of the proximal duct system to have effect on the ductus epididymidis. Systemic concentrations of testosterone are essential for development and functional maintenance of the rest of the excretory duct system and for the male secondary sex characteristics. In some instances, testosterone serves primarily as a prohormone and its metabolites are the active factors. Dihydrotestosterone, for example, is the active agent for the prostate and epididymis.

Much testicular fluid is produced by the germinal epithelium and serves as a vehicle for the transport of spermatozoa out of the seminiferous tubules and into the excretory duct system. Spermatozoa released into the seminiferous tubules are nonmotile and are moved into the tubuli recti by the flow of testicular fluid and the shallow peristaltic movements of the seminiferous tubules. The ciliated epithelium and the increased smooth muscle in the walls of the ductuli efferentes contribute to the movement of spermatozoa into the proximal epididymis. Most of the testicular fluid is absorbed in the ductuli efferentes and proximal ductus epididymidis.

Sperm entering the epididymis show weak, random movement, and for the most part they are incapable of fertilization; spermatozoa from the distal epididymis show strong, unidirectional motility and are capable of fertilization. Sperm are slowly propelled through the long ductus epididymidis by peristaltic contractions of its muscular walls. During their passage through the epididymis, spermatozoa show continued differentiation of the acrosome and sperm head, as well as progressive increase in the fertilizing capacity. Sperm require the environment provided by the epididymis to become physiologically mature and viable. Although the maturation process is androgen-dependent, how the various segments of the epididymis control and regulate the physiologic maturation of spermatozoa is unknown. The distal ductus epididymidis stores sperm. Because of its thick muscular wall, it can contract forcibly and, in coordination with the wall of the ductus deferens, empty stored sperm into the prostatic urethra.

During ejaculation, the primary accessory sex glands contribute successively to the seminal fluid. The bulbourethral and intraepithelial urethral glands secrete fluids that lubricate the urethra during sexual arousal and erection of the penis. At the onset of ejaculation, muscular tissue of the prostatic stroma contracts, discharging the secretory product into the prostatic urethra. The abundant, thin secretion is slightly acidic and contains acid phosphatase. Sperm and their suspending fluids are expelled from the distal ductus epididymidis and ductus deferens. The last component added to the ejaculate is the viscous secretion of the seminal vesicles. This secretion is rich in fructose, an important energy source for the active, motile spermatozoa. Seminal vesicles also secrete prostaglandins in high concentration, but the role of these agents in the female tract is unknown. The final phase of physiologic maturation of spermatozoa is capacitation and occurs in the female reproductive tract, substantially increasing the number of spermatozoa that are capable of fertilization.

Testosterone is vital for the male reproductive system to develop properly and to maintain normal structure and function. It is synthesized and released by interstitial (Leydig) cells of the testis. Secretion is controlled by circulating levels of testosterone that are monitored by neurons within the hypothalamic region of the brain that, through releasing factors, influence the secretion of ISCH (LH) by cells of the anterior pituitary. It is through this feedback system that normal testosterone levels are maintained. Sertoli cells, which are under the influence of FSH, produce androgen-binding protein that functions to concentrate testosterone within the seminiferous tubules and the genital duct system. Secretion of FSH by cells of the anterior pituitary also is controlled by releasing factors from neurons with the hypothalamus. Inhibin released by Sertoli cells suppresses FSH secretion and, through this negative feedback loop, functions to control the rate of spermatogenesis.

Review Questions and Answers

Questions

88. Briefly discuss the role of the Sertoli cells.
89. What is the effect of testosterone on the male reproductive system?
90. What is meant by the terms *spermatogenesis* and *spermiogenesis*?
91. Relate the ultrastructural features of the interstitial cells (of Leydig) to their function.
92. What role does the pampiniform plexus play in the physiology of the testis?

Answers

88. The Sertoli cell is a tall columnar cell that spans the entire depth of the germinal epithelium. Tight junctions occur between adjacent Sertoli cells near their bases and subdivide the germinal epithelium into basal and adluminal compartments. The basal compartment extends from the basal lamina of the germinal epithelium to the tight junctions, while the adluminal compartment lies between the tight junctions and the lumen of the seminiferous tubule. The basal compartment contains spermatogonia and newly formed primary spermatocytes, exposed to an environment similar to that of blood plasma. Cells of the adluminal compartment (primary and secondary spermatocytes, spermatids, and spermatozoa) differentiate in an environment that is rich in ions, contains amino acids not found in the basal compartment, and has a relatively high concentration of testosterone. Sertoli cells are responsible for establishing and maintaining these two different environments, and the tight junctions contribute to the blood-testis barrier between the two

compartments. Substances that enter or leave the adluminal compartment must first pass through the cytoplasm of the Sertoli cells.

Sertoli cells synthesize androgen-binding protein, which complexes with testosterone to maintain a high concentration of the hormone in the adluminal compartment. High levels of testosterone are necessary for the differentiation and development of germ cells. The Sertoli cells and blood-testis barrier also may prevent the escape of spermatogenic cells from the adluminal compartment and into the general body tissues, thus preventing an immunologic response against the forming germ cells.

In addition to providing nutritional and mechanical support for the developing germ cells, Sertoli cells also control the movement of germ cells through the epithelium and play an important role in their release into the lumen of the seminiferous tubules. Sertoli cells phagocytose degenerating germ cells and the residual cytoplasm that normally is shed during the release of germ cells. Inhibin released by Sertoli cells suppresses FSH secretion and through this negative feedback loop, controls the rate of spermatogenesis.

89. Testosterone, the male hormone produced by the interstitial cells of the testes, is essential for the proliferation and differentiation of germ cells, for the structural development of the excretory ducts and accessory sex glands, and for the maintenance of these structures in a functional state. In addition, testosterone influences other tissues of the body and is responsible for the development and maintenance of the male secondary sex characteristics: growth of the beard, low pitch of the voice, muscular build, and male pattern of pubic hair. Production of testosterone is under the control of a gonadotropic hormone called interstitial cell–stimulating hormone, secreted by cells in the anterior pituitary.

90. *Spermatogenesis* encompasses the entire sequence of events in the transformation of the diploid spermatogonia at the base of the germinal epithelium into the haploid spermatozoa that are released into the lumen of the seminiferous tubule. In humans this process takes about 64 days. Spermiogenesis, on the other hand, refers to the elaborate sequence of events by which the spermatids differentiate into slender, motile spermatozoa.

91. The interstitial cells (of Leydig) commonly are found in groups or nests and constitute the endocrine portion of the testis, producing the male hormone testosterone. They usually contain a single large, spherical nucleus, although binucleate cells are not uncommon. The cytoplasm is characterized by an abundance of smooth endoplasmic reticulum and well-developed Golgi complexes. Mitochondria are numerous, and their cristae are of the tubular rather than the lamellar form. These are features characteristic of a cell type that is actively engaged in steroid synthesis. Testosterone is a steroid hormone. All enzymes (except one) needed for the conversion of acetate to testosterone are associated with membranes of the smooth endoplasmic reticulum. Enzymes for one step — cleavage of the side chain of cholesterol — are found in the mitochondria. Thus, in this particular cell type, acetate is synthesized to cholesterol on the membranes of the smooth endoplasmic reticulum. Cholesterol then enters surrounding mitochondria, where its side chain is removed enzymatically and the resulting pregnenolone molecule then diffuses out of the mitochondria. The synthesis of pregnenolone to testosterone is completed on the membranes of the smooth endoplasmic reticulum.

92. Each testicular vein forms an extensive plexus, the pampiniform plexus, which surrounds the testicular artery as it nears the testis. The close proximity of the surrounding cooler venous blood causes the arterial blood to lose heat and provides a thermoregulatory mechanism for the precooling of the incoming arterial blood. In this way the temperature of the testis in the scrotal sac is maintained a few degrees below the body temperature, a condition necessary for the production of germ cells.

18 Female Reproductive System

The female reproductive system consists of internal organs — the ovaries, oviducts, uterus, and vagina — and external structures (external genitalia), which include the labia majora, labia minora, and clitoris. Although not genital organs, the mammary glands are important accessory organs of the female reproductive tract. For the first 10 or 11 years of life, the reproductive organs remain immature and growth parallels that of the body generally. During the 2 to 3 years prior to the first menstrual period, the generative organs increase in size, the breasts enlarge, and pubic and axillary hair appears. Following the first menses and thereafter throughout the reproductive period, the ovaries, oviducts, uterus, vagina, and mammary glands undergo cyclic changes in function and structure associated with the menstrual cycle and pregnancy. During menopause, the cycles become irregular and eventually cease; in the postmenopausal period the reproductive organs atrophy.

Ovaries

The ovaries are paired oval bodies that lie on each side of the uterus, suspended from the broad ligament by a mesentery, the mesovarium. The ovaries are homologous with the testes and are compound organs with exocrine and endocrine functions. The exocrine secretion consists of whole cells — the ova — and thus the ovary can be classed as a holocrine or cytogenic gland. In cyclic fashion it secretes female sex hormones directly into the bloodstream and therefore is an endocrine gland.

Structure

KEY WORDS: **surface epithelium, cortex, ovarian follicle, tunica albuginea, medulla**

Each ovary is covered by a mesothelium that is continuous with that of the mesovarium. As this membrane extends over the surface of the ovary, the squamous cells become cuboidal and form the **surface epithelium** of the ovary. Sections of the ovary show an outer cortex and an inner medulla, but the boundary between the two regions is indistinct. The stroma of the **cortex** consists of a compact feltwork of fine collagenous fibers and numerous fibroblasts.

Elastic fibers are associated mainly with blood vessels and otherwise are rare in the cortex. Scattered throughout the cortical tissue are the **ovarian follicles,** whose size varies with their stage of development. Immediately below the surface epithelium, the connective tissue of the cortex is less cellular and more compact, forming a dense layer called the **tunica albuginea** (Fig. 18-1).

The stroma of the **medulla** consists of a loose connective tissue that is less cellular than that of the cortex and contains many elastic fibers and smooth muscle cells. Numerous large, tortuous blood vessels, lymphatics, and nerves also are present in the medulla.

Ovarian Follicles

KEY WORDS: **cortex, ovum, epithelial cell**

The follicles are located in the **cortex** of the ovary, deep to tunica albuginea, and consist of an immature **ovum** surrounded by one or more layers of **epithelial cells.** At birth, the human ovary contains 300,000 to 400,000 ova embedded in the cortical stroma, but only a few reach maturity and are ovulated. During the menstrual cycle, several follicles begin to grow and develop, but only one attains full maturity; the rest degenerate. The size of the follicle and the thickness of the epithelial envelope vary with the stage of development. During their growth, the follicles undergo a sequence of changes in which primordial, primary, secondary, and mature follicles can be distinguished.

Primordial Follicles

KEY WORDS: **unilaminar, follicular cell, primary oocyte, diplotene**

In the mature ovary, follicles are present in all stages of development. Most, especially in young females, are primordial or **unilaminar** follicles, which are found in the peripheral cortex, just beneath tunica albuginea (Fig. 18-2). These follicles consist of an immature ovum surrounded by a single layer of flattened **follicular cells** that rest on a basement membrane. At this stage, the ovum is a **primary oocyte,** suspended in the **diplotene** stage of the meiotic prophase. The primary oocyte is a large cell with a prominent vesicular nucleus and nucleolus. The Golgi

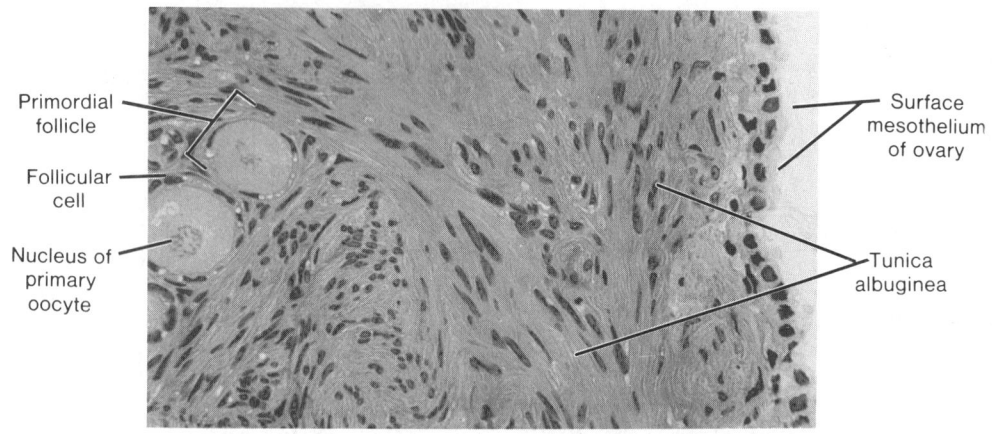

Fig. 18-1. A region of ovarian cortex illustrating the surface mesothelium and the tunica albuginea (LM, × 250).

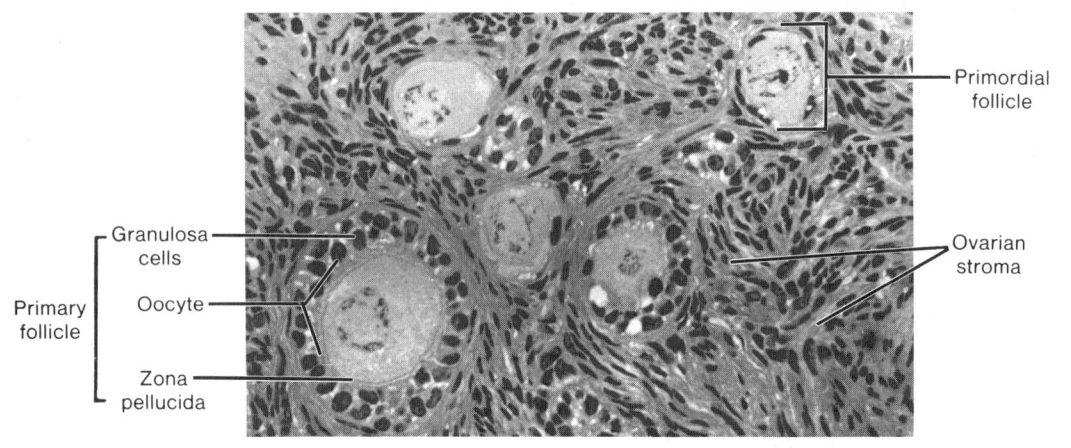

Fig. 18-2. Follicles within the ovarian cortex (LM, × 250).

apparatus is well developed, and spherical mitochondria tend to concentrate in the region of the centrosome. As the primordial follicle develops into a primary follicle, changes occur in the ovum, follicular cells, and adjacent connective tissue of the cortex.

Primary Follicles

KEY WORDS: zona pellucida, granulosa cell, stratum granulosum, theca folliculi, theca interna, theca externa

During growth of a primary follicle, the oocyte increases in size, and the Golgi complex, originally a single organelle

next to the nucleus, becomes a multiple structure scattered throughout the cytoplasm. Free ribosomes increase in number, mitochondria become dispersed, and granular endoplasmic reticulum, although not prominent, becomes more extensive. A few lipid droplets and lipochrome pigments appear in the cytoplasm, and yolk granules accumulate; these, however, are smaller and less numerous than in nonprimate species. As the oocyte grows, a clear refractile membrane, the **zona pellucida,** develops between the oocyte and adjacent follicular cells (Fig. 18-3). The zona pellucida consists of glycoproteins secreted primarily by the oocytes, although a lesser amount may be produced by

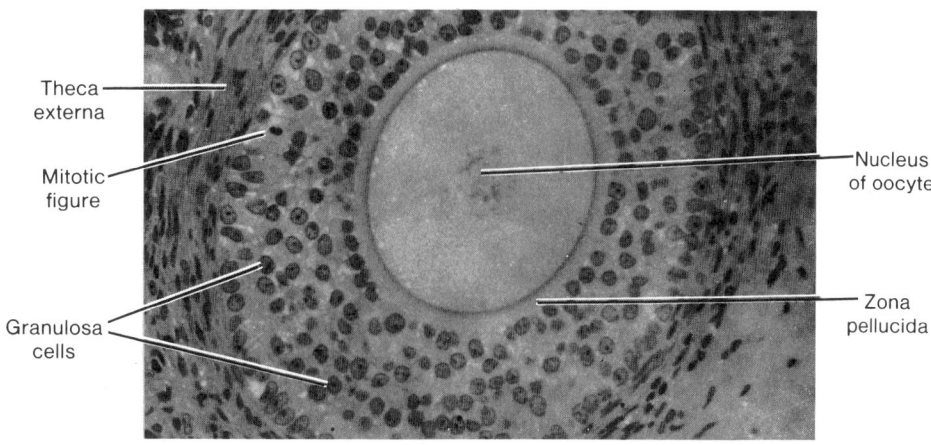

Fig. 18-3. Components of a primary ovarian follicle (LM, × 250).

surrounding granulosa cells. Microvilli extend from the oocyte into the zona pellucida.

As the ovum grows, the flattened follicular cells of the primordial follicles become cuboidal or columnar in shape and proliferate to form several layers. These cells now are referred to as **granulosa cells,** and the layer of stratified epithelium that they form is called **stratum granulosum.** Mitochondria, free ribosomes, and endoplasmic reticulum increase and the Golgi element becomes prominent. Irregular, slender cytoplasmic processes extend from the granulosa cells and penetrate the zona pellucida to make contact with processes from the oocyte. The processes of the oocyte and granulosa cells are united by gap junctions.

While these changes occur in the oocyte and follicular cells, the adjacent stroma becomes organized into a sheath, the **theca folliculi,** which surrounds the growing primary follicle. The theca folliculi is separated from the stratum granulosum by a distinct basement membrane. The cells differentiate into an inner secretory layer, the **theca interna,** and an outer fibrous layer, the **theca externa,** but the boundary between the layers is somewhat indistinct. The theca externa merges imperceptibly into the surrounding stroma. Many small blood vessels penetrate the theca externa to provide a rich vascular network to the theca interna. The stratum granulosum remains avascular until after ovulation.

Secondary Follicles

KEY WORDS: liquor folliculi, follicular antrum, antral follicle, cumulus oophorus, corona radiata

As the follicular cells continue to proliferate, the growing follicle assumes an ovoid shape and gradually sinks deeper into the cortex. When the stratum granulosum has become 8 to 12 layers thick, fluid-filled spaces appear between the granulosa cells. The fluid, the **liquor folliculi,** increases in amount and the spaces fuse to form a single cavity called the **follicular antrum.** Liquor folliculi contain growth factors, steroids, gonadotrophic hormone, and other substances several times their concentration in blood plasma. The follicle is now a secondary or **antral follicle,** and the oocyte has reached its full size and undergoes no further growth. The follicle, however, continues to increase in size, due in part to continued accumulation of liquor folliculi. The ovum is eccentrically placed in the follicle, surrounded by a mass of granulosa cells that projects into the fluid-filled antrum, forming the **cumulus oophorus** (Fig. 18-4). The cells of the cumulus are continuous with those lining the antral cavity. Granulosa cells that surround the oocyte form the **corona radiata** and are anchored to the zona pellucida by cytoplasmic processes.

Mature Follicles and Ovulation

KEY WORDS: endocrine gland, aromatase enzyme complex, secondary oocyte, ovulation, stigma (macula pellucida)

In the human female, a follicle requires 10 to 14 days to reach the stage of maturity. At maximum size, the follicle occupies the thickness of the cortex and bulges from the surface of the ovary. Fluid spaces appear between granulosa cells of the cumulus oophorus, and the connection

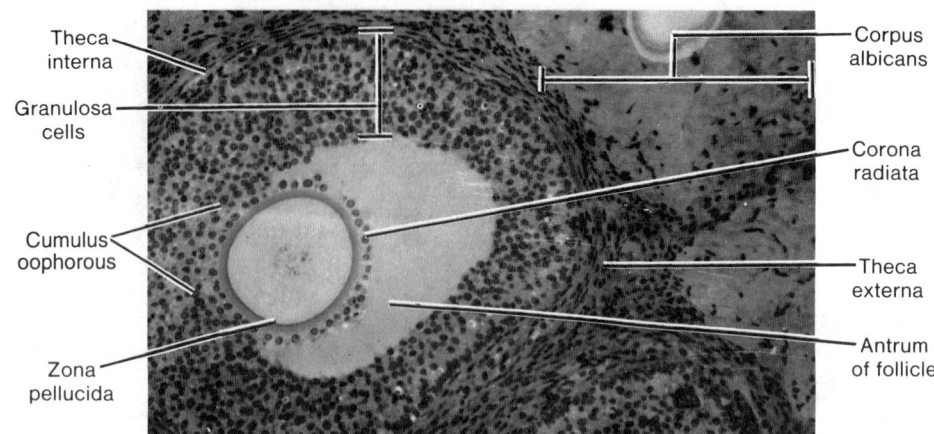

Theca interna
Granulosa cells
Cumulus oophorous
Zona pellucida
Corpus albicans
Corona radiata
Theca externa
Antrum of follicle

Fig. 18-4. Components of a secondary ovarian follicle (LM, × 100).

between the ovum and stratum granulosum is weakened. The theca folliculi has attained its greatest development, and the theca interna assumes the cytologic characteristics of a steroid-secreting **endocrine gland.** Androstenedione and testosterone are important androgens secreted by the theca interna that serve as substrates for estrogen biosynthesis by the granulosa cells. An **aromatase enzyme complex** within the granulosa cells converts the substrates to estradiol, a form of estrogen. Estradiol then diffuses back across the limiting basement membrane of the stratum granulosum to enter the circulation via capillaries within the theca interna. Just before ovulation, the oocyte completes the first meiotic division and gives off the first polar body. Thus at ovulation a **secondary oocyte** is liberated.

Rupture of the mature follicle and liberation of the ovum constitute **ovulation,** which normally occurs at the middle of the menstrual cycle. Immediately before ovulation, further expansion of the follicle occurs due to increased secretion of liquor folliculi. Where the follicle bulges from the ovary, its wall becomes thinner, and a small, avascular, translucent area appears. This is the **stigma,** or **macula pellucida.** The tunica albuginea thins out, and the surface epithelium of the ovary becomes discontinuous in this area. A collagenase produced by granulosa cells adjacent to the tunica albuginea appears to be responsible for the breakdown of collagen fibers at the site of the stigma. The stigma protrudes as a small blister and ruptures, and the ovum, with its adherent corona radiata, is extruded along with follicular fluid.

Corpus Luteum

KEY WORDS: estrogen, progesterone, granulosa lutein cell, theca lutein cell, luteinization, corpus albicans, relaxin

Following ovulation, the follicle is transformed into a temporary endocrine gland, the corpus luteum, which elaborates both **estrogens** and **progesterone.** The walls of the follicle collapse, and stratum granulosum is thrown into folds. Bleeding from capillaries in the theca interna may result in a blood clot in the center of the corpus luteum (Fig. 18-5). Granulosa cells increase greatly in size, take on a polyhedral shape, and transform into pale-staining **granulosa lutein cells.** Lipid accumulates in the cytoplasm of the cells, smooth endoplasmic reticulum becomes abundant, and mitochondria show tubular cristae. Cells of the theca interna also enlarge and become epithelioid in character to form **theca lutein cells** (Fig. 18-6). The lutein cells derived from the theca interna are somewhat smaller than granulosa lutein cells. The process of formation of granulosa and theca lutein cells is called **luteinization.**

With the depolymerization of the basement membrane between the theca interna and the granulosa cells, capillaries from the theca interna invade the lutein tissue to form a complex vascular network throughout the corpus luteum. Connective tissue from the theca interna also penetrates the mass of lutein cells and forms a delicate network about them. The fully formed corpus luteum secretes estrogens and progesterone; secretion of progesterone increases

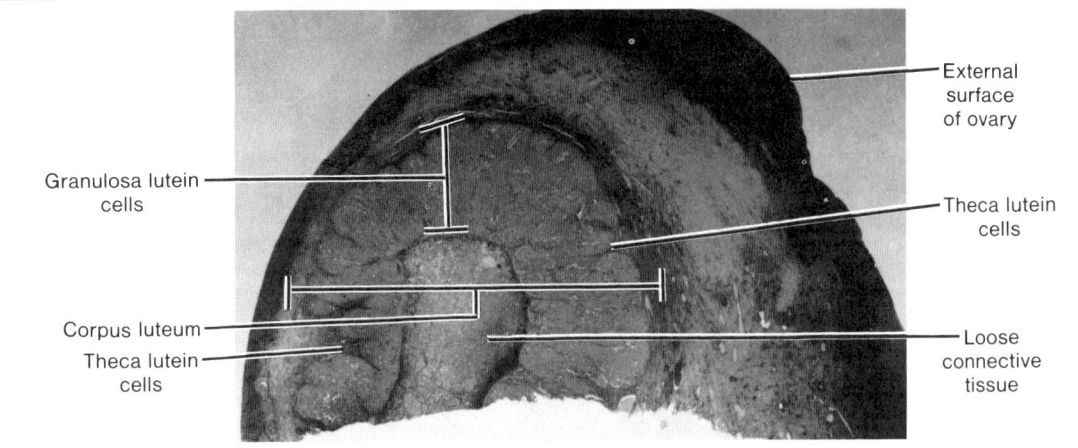

Fig. 18-5. A human corpus luteum (LM, × 25).

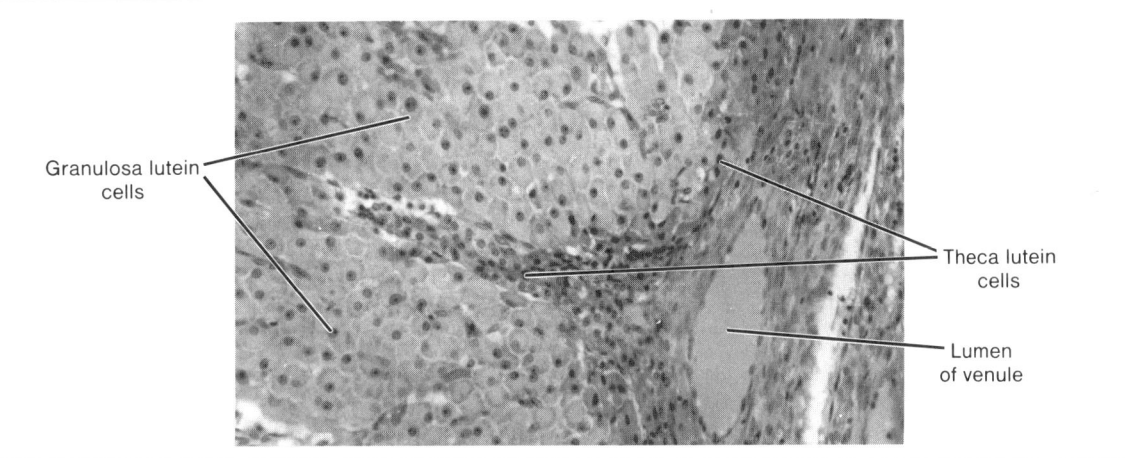

Fig. 18-6. Granulosa lutein and theca lutein cells of a human corpus luteum (LM, × 100).

rapidly during luteinization, and high levels are maintained until the corpus luteum involutes.

If the ovum is not fertilized, the corpus luteum persists for about 14 days and then undergoes involution. The cells decrease in size, accumulate much lipid, and degenerate. Hyaline material accumulates between the lutein cells, the connective tissue cells become pyknotic, and the corpus luteum gradually is replaced by an irregular white scar, the **corpus albicans** (Fig. 18-7). Over the following months, the corpus albicans itself disappears. If fertilization occurs,

the corpus luteum enlarges further and persists for about the first 6 months of pregnancy and then gradually declines. The corpora lutein cells of pregnancy secrete a polypeptide known as **relaxin.** Relaxin is thought to suppress uterine contractile activity during pregnancy and late in pregnancy enhances softening of the cervix and relaxation of the pelvic ligaments, resulting in a slight separation of the pubic symphysis prior to birth. Following delivery, involution of the corpus luteum is accelerated, resulting in the formation of a corpus albicans.

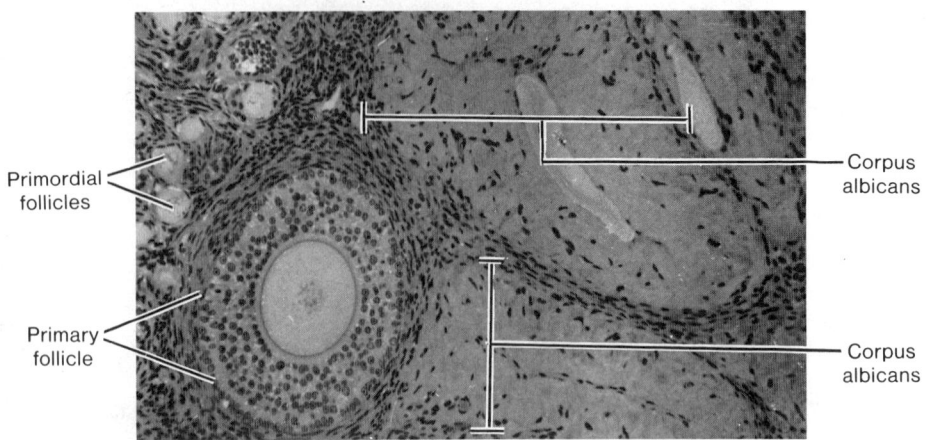

Fig. 18-7. A region of ovarian cortex showing follicles as well as two corpus albicans (LM, × 100).

Atresia of Follicles

KEY WORDS: **atresia, glassy membrane**

During the early part of each menstrual cycle, several primordial follicles begin to grow, but usually only one attains full development and is ovulated. The remainder undergo a degenerative process called **atresia,** which may occur at any stage in the development of a follicle. In atresia of a primary follicle, the ovum shrinks, degenerates, and undergoes cytolysis; the follicular cells show similar degenerative changes. The follicle is resorbed, and the small space that is left is rapidly filled by connective tissue from the stroma. Similar degenerative changes occur in larger follicles, but the zona pellucida may persist for a time after dissolution of the oocyte and follicular cells. Macrophages invade the atretic follicle and engulf the degenerating material, including fragments of the zona pellucida.

Cells of the theca interna persist longer than those of the stratum granulosum but also show degenerative changes. The theca cells increase in size, lipid droplets appear in the cytoplasm, and the cells assume an epithelioid character, similar to that of lutein cells. The cells assume a cordlike arrangement, the cords being separated by connective tissue fibers and capillaries. The basal lamina between granulosa cells and the theca interna frequently becomes thicker and forms a hyalized, corrugated layer, the **glassy membrane.** This structure is characteristic of growing follicles that have undergone atresia and aids in distinguishing a large atretic follicle from a corpus luteum. Other differ-

ences include degenerative changes in the granulosa cells and the presence of fragments of zona pellucida at the center of a follicle without an associated oocyte. Ultimately, the degenerated remains of the follicle are removed and a scar resembling a small corpus albicans is left. This, too, eventually disappears into the stroma of the ovary.

In some mammals, especially rodents, clusters of epithelioid cells are scattered in the stroma of the cortex. These interstitial cells contain small lipid droplets and bear a marked resemblance to luteal cells. It is thought that these interstitial cells arise from the theca interna of follicles that are undergoing atresia. In the human, interstitial cells are most abundant during the first year of life, the period during which atretic follicles are most numerous. In the adult, they are present only in widely scattered, small groups. Their exact role in ovarian physiology is unknown; in humans they elaborate androgens.

Other large epithelioid cells, the hilus cells, are found associated with vascular spaces and unmyelinated nerves in the hilus of the ovary. The cells appear similar to the interstitial cells of the testis and contain lipid, lipochrome pigments, and cholesterol esters. Hilus cells are most commonly found during pregnancy and at menopause, but their function is unknown. Tumors arising from these cells have a masculinizing effect.

The Ovarian Cycle

KEY WORDS: **follicle-stimulating hormone, luteinizing hormone, inhibin**

Maturation of ovarian follicles, their endocrine functions, and the phenomenon of ovulation are regulated by follicle-stimulating hormone (FSH) and luteinizing hormone (LH). These are gonadotropic hormones elaborated by the anterior pituitary. **Follicle-stimulating hormone** is responsible for maturation of follicles and stimulates the aromatase enzyme complex within granulosa cells to produce estrogens. Initial development of follicles is self-regulated and does not require FSH, but the hormone is essential for their maturation. Together with FSH, **luteinizing hormone** induces ripening of the mature follicle and ovulation. Alone, LH converts the mature follicle to a corpus luteum and induces it to secrete estrogen and progesterone. The cyclic nature of follicle formation and ovulation is the result of reciprocal interaction between pituitary gonadotropins and ovarian hormones.

As production of estrogens by the granulosa cells increases, release of FSH from the pituitary is inhibited, and the level of FSH falls below that needed for maturation of new follicles. However, the rising level of estrogen stimulates release of LH from the pituitary, resulting in ovulation, formation of a corpus luteum, and secretion of progesterone and estrogen by this body. Increasing levels of progesterone now inhibit release of LH from the pituitary, and as the level of LH declines, the corpus luteum no longer is maintained. With decline of the corpus luteum, estrogen levels diminish, the pituitary no longer is inhibited from secreting FSH, and a new cycle of follicle formation is initiated. The ovarian steroid hormones have only minor effects on the pituitary directly; their primary action is mediated through neurons in the hypothalamus. Both FSH and ovarian androgens stimulate granulosa cells to synthesize and release a glycoprotein called **inhibin.** Inhibin preferentially suppresses the synthesis and secretion of FSH by cells in the anterior pituitary. The ovary also secretes a protein called *follicular regulatory protein,* which is thought to be important in establishing the dominance of a single follicle over the several recruited at the beginning of the ovarian cycle. How this protein retards the growth of the other follicles is unknown.

Oviducts

The oviducts (uterine tubes) are a pair of muscular tubes that extend from the ovary to the uterus along the upper margin of the broad ligament. One end of the tube is closely related to the ovary and, at this end, is open to the peritoneal cavity. The other end communicates with the lumen of the uterus. The oviduct delivers the ovum, released at ovulation, to the uterine cavity and provides an environment for fertilization and initial segmentation of the fertilized ovum.

Structure

KEY WORDS: infundibulum, fimbria, ampulla, isthmus, intramural (interstitial) part

The oviduct is divided into several segments. The ovarian end, the **infundibulum,** is funnel-shaped, and its margin is drawn out in numerous tapering processes called **fimbria.** The infundibulum opens into the tortuous, thin-walled **ampulla** of the tube, which makes up slightly more than half the length of the oviduct. The ampulla is continuous with the **isthmus,** a narrower, cordlike portion that makes up about the medial one-third. The **intramural (interstitial) part** is the continuation of the tube where it passes through the uterine wall. As in other hollow viscera, the wall of the oviduct consists of several layers: an external serosa, an intermediate muscularis, and an internal mucosa.

Mucosa

KEY WORDS: plica, columnar epithelial cell, lamina propria, ciliated cell, nonciliated cell, cyclic change

The mucosa of the oviduct presents a series of longitudinal folds called **plicae.** In the ampulla the plicae have secondary and even tertiary folds to create a complex labyrinth of epithelial-lined spaces. In the isthmus the plicae are shorter with little branching, while in the intramural part the plicae form only low ridges. Throughout the tube, the plicae consist of a single layer of **columnar epithelial cells** resting on an incomplete basement membrane and a **lamina propria** of richly cellular connective tissue that contains a network of reticular fibers and fibroblasts. The epithelium decreases in height from ampulla to uterus and contains ciliated and nonciliated cells (Fig. 18-8). The **ciliated cells** are most numerous on the surface of the fimbria and progressively decrease in number through the ampulla, isthmus, and intramural portions. The **nonciliated cells** appear to be secretory and may help establish an environment that is suitable for the survival and fertilization of the ovum and for maintenance of the zygote. A third cell type, an undifferentiated cell with a darkly staining nucleus, may represent a precursor of the secretory cells or an exhausted secretory cell.

The epithelium shows **cyclic changes** associated with ovarian cycles. During the follicular phase, ciliated cells increase in height, reaching their maximum at about the time of ovulation, and there is evidence of increased preparations in the secretory cells. In the luteal phase, ciliated cells decrease in height and lose their cilia, and there is

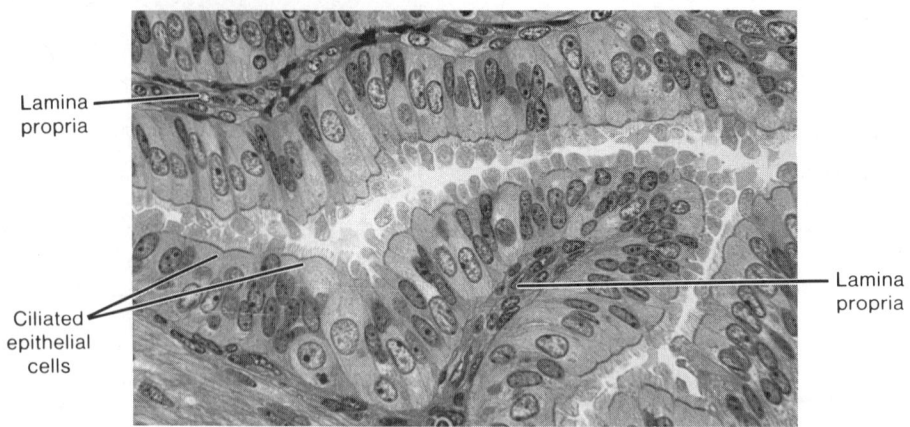

Lamina propria

Lamina propria

Ciliated epithelial cells

Fig. 18-8. The ciliated simple columnar epithelium lining the ampullary region of the oviduct (LM, × 250).

augmented secretory activity by the secretory cells. Loss of cilia is greatest in the fimbria and least in the isthmus. The cilia are responsive to steroid hormones: Estrogen appears to be responsible for the appearance and maintenance of cilia, and progesterone increases the rate at which they beat. Other ciliated cells of the body show no such hormone responsiveness.

Muscular Coat and Serosa

KEY WORDS: no submucosa, inner circular muscle, outer longitudinal muscle

The mucous membrane rests directly on the muscle coat, and there is **no submucosa.** The muscle coat consists of two layers of smooth muscle, but the layers are not sharply defined. The **inner layer is circular** or closely spiralled; the **outer layer of longitudinal muscle** is the thinner. The muscularis increases in thickness toward the uterus due to the increased depth of the inner layer. Externally, the oviduct is covered by a serosa that represents the peritoneal covering of the organ.

Uterus

The human uterus is a single, hollow, pear-shaped organ with a thick muscular wall; it lies in the pelvic cavity between the bladder and rectum. The nonpregnant uterus varies in size depending on the individual but generally is about 7 cm in length, 3 to 5 cm at its widest (upper) part, and 2.5 to 3.0 cm thick. It is slightly flattened dorsoven-

trally, and the cavity corresponds to the overall shape of the organ. The uterus receives the fertilized ovum and nourishes the embryo and fetus throughout its development until birth.

Structure

KEY WORDS: body, cervix, cervical canal, endometrium, myometrium, perimetrium

Several regions of the uterus can be distinguished. The bulk of the organ consists of the **body,** which comprises the upper expanded portion. The dome-shaped part of the body between the junctions with the oviduct constitutes the *fundus.* Below, the uterus narrows and becomes more cylindrical in shape: This region forms the **cervix,** part of which protrudes into the vagina. The **cervical canal** passes through the cervix from the uterine cavity and communicates with the vagina at the external os.

The wall of the uterus is made up of several layers that have specific names: The internal lining or mucosa is called the **endometrium;** the middle muscular layer forms the **myometrium;** and the external layer is referred to as the **perimetrium.** The perimetrium is the serosal or peritoneal layer that covers the body of the uterus and supravaginal part of the cervix posteriorly, and the body of the uterus anteriorly.

Myometrium

KEY WORDS: smooth muscle, internal layer, middle layer, stratum vasculare, outer layer

The bulk of the uterine wall consists of the myometrium, which forms a thick coat about 15 to 20 mm in depth. The myometrium consists of bundles of **smooth muscle** fibers separated by thin strands of·connective tissue that contain fibroblasts, collagenous and reticular fibers, mast cells, and macrophages. The muscle forms several layers that are not sharply defined because of the intermingling of fibers from one layer to another. Generally, however, internal, middle, and outer layers can be distinguished. The **internal layer** is thin and consists of longitudinal and circular fibers. The **middle layer** is the thickest and shows no regularity in the arrangement of the muscle fibers, which run longitudinally, obliquely, circularly, and transversely. This layer also contains many large blood vessels and has been called the **stratum vasculare.** The **outer layer** of muscle consists mainly of longitudinal fibers, some of which extend into the broad ligament, oviducts, and ovarian ligaments. Elastic fibers are prominent in the outer layer but are not present in the inner layer of the myometrium except around blood vessels.

In the nonpregnant uterus, the smooth muscle cells are 30 to 50 μm long, but during pregnancy, they hypertrophy to reach lengths of 500 to 600 μm or greater. New smooth muscle cells are produced in the pregnant uterus from undifferentiated cells and possibly from division of mature cells also. The connective tissue of the myometrium also increases in amount during pregnancy. In spite of a total increase in muscle mass, the layers are thinned during pregnancy as the uterus becomes distended. After delivery, the muscle cells rapidly decrease in size, but the uterus does not regain its original, nonpregnant dimensions.

The myometrium normally undergoes intermittent contractions that, however, are not intense enough to be perceived. The intensity may increase during menstruation to result in cramps. The contractions are diminished during pregnancy, possibly in response to the hormone relaxin. At parturition, strong contractions of the uterine musculature occur, causing the fetus to be expelled. Uterine contractions increase after administration of oxytocin, a hormone produced by the neurohypophysis, and in response to prostaglandins. A rise in the level of prostaglandins occurs just prior to delivery.

Endometrium

KEY WORDS: endometrial stroma, simple columnar epithelium, uterine gland, stratum basale, stratum functionale, radial artery, straight artery, spiral artery, lacuna

The endometrium is a complex mucous membrane that, in the human female, undergoes cyclic changes in structure and function in response to the ovarian cycle. The cyclic activity begins at puberty and continues until menopause. In the body of the uterus, the endometrium consists of a thick lamina propria (**endometrial stroma**) and a covering epithelium. The stroma resembles mesenchymal tissue and consists of loosely arranged stellate cells with large, round or ovoid nuclei supported by a network of fine connective tissue in which lymphocytes, granular leukocytes, and macrophages are scattered. There is no submucosa, and the stroma lies directly on the myometrium, to which it is firmly attached.

The stroma is covered by a **simple columnar epithelium** that contains ciliated cells and nonciliated secretory cells. The epithelium dips into the stroma to form numerous **uterine glands** that extend deeply into the stroma (Fig. 18-9), occasionally penetrating into the myometrium. Most are simple tubular glands, but some may branch near the muscle. There are fewer ciliated cells in the glands than in the covering epithelium. A basement membrane underlies the glandular and surface epithelia.

The endometrium can be divided into a stratum basale (basal layer) and a stratum functionale (functional layer), which differ in their structure, function, and blood supply. The **stratum basale** is the narrower, more cellular, and more fibrous layer and lies directly on the myometrium. Occasionally, small pockets of stratum basale may extend into the myometrium, between muscle cells. This layer undergoes few changes during the menstrual cycle and is not shed at menstruation but serves as the source from which the functional layer is restored.

The **stratum functionale** extends to the lumen of the uterus and is the part of the endometrium in which cyclic changes occur and which is shed during menstruation. The stratum functionale sometimes is divided into the *compacta,* a narrow superficial zone, and the *spongiosa,* a broader zone that forms the bulk of the functionalis.

The blood supply of the endometrium (Fig. 18-10) is unique and plays an important role in the events of menstruation. Branches of the uterine artery penetrate the myometrium to its middle layer, where they furnish arcuate arteries that run circumferentially in the myometrium. One set of branches from these arteries supplies the superficial layers of the myometrium, while other branches, the **radial arteries,** pass inward to supply the endometrium. At the junction of myometrium and endometrium, the radial branches provide a dual circulation to the endometrium. **Straight arteries** supply the stratum basale, while the stratum functionale is supplied by highly coiled **spiral arteries** (Fig. 18-11). As the latter pass through the functional layer, they provide terminal arterioles, which unite with a complex network of capillaries and thin-walled, dilated

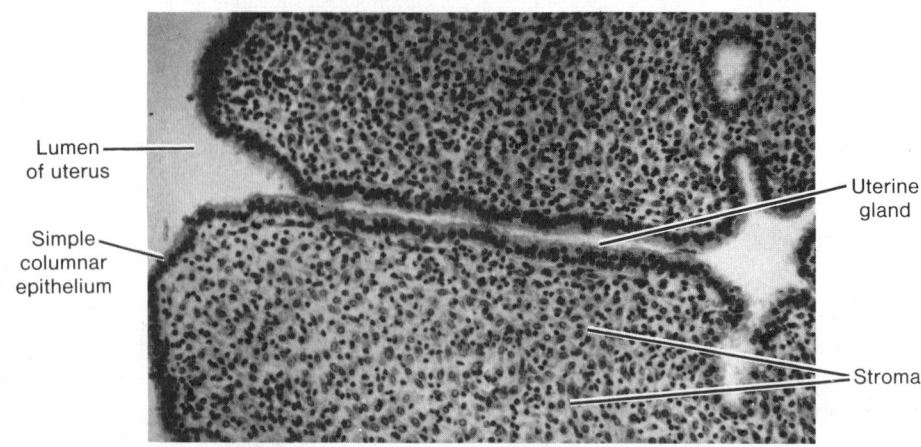

Lumen of uterus

Simple columnar epithelium

Uterine gland

Stroma

Fig. 18-9. A section of human endometrium showing the lining epithelium, a uterine gland, and numerous stromal cells (LM, × 100).

vascular structures, the **lacunae.** The venous system forms an irregular network of venules and veins with irregular sinusoidal enlargements and then drains into a plexus at the junction of myometrium and endometrium. During menstrual cycles, the spiral arteries constrict periodically, subjecting the functional layer to intermittent periods of anoxia. The distal portions of the arterial supply in the functionalis undergo degeneration and regeneration with each menstrual cycle, whereas the straight arteries of the basal layer show no such changes.

Cyclic Changes in the Endometrium

KEY WORDS: proliferative phase, maturation of follicle, secretory phase, corpus luteum, ischemic (premenstrual) phase, menstrual phase

During a normal menstrual cycle, the endometrium undergoes a continuous sequence of changes in which four phases can be described. The phases correlate with the functional activities of the ovaries and constitute the proliferative, secretory, ischemic (premenstrual), and menstrual phases.

The **proliferative phase** begins at the end of the menstrual flow and extends to about the middle of the cycle. This stage is characterized by rapid regeneration and repair of the endometrium. Epithelial cells in the remnants of the glands in the stratum basale proliferate and migrate over the raw surface of the mucosa; stromal cells also proliferate. In the early part of this period, the endometrium is

of limited thickness, and its glands are sparse and fairly straight and have small lumina. The epithelium of the glands and surface is simple cuboidal to low columnar, and mitoses are present in the glandular lining. As the proliferative phase advances, the endometrial glands increase in number and length and become more tortuous and more closely spaced, and the lumina widen. Toward the end of the proliferative phase, glycogen accumulates in the basal region of the glandular epithelium. The surface and glandular epithelia now are tall columnar with fewer ciliated and more secretory cells. The secretory cells have large numbers of small mitochondria, but the endoplasmic reticulum and Golgi complex are poorly developed. Microvilli are present on the free borders of the cells. The nuclei are round or oval and contain finely granular chromatin with one or more nucleoli. The spiral arteries lengthen but are lightly coiled and do not extend into the superficial third of the endometrium.

The proliferative phase corresponds to the **maturation of the ovarian follicle** up to the time of ovulation. Estrogen, secreted by the developing follicles, stimulates growth of the endometrium; some growth may continue for a day or two after ovulation.

During the **secretory phase,** the endometrium continues to increase in thickness as a result of hypertrophy of stromal and glandular cells, stromal edema, and increased vascularity. The glands lengthen and become irregularly coiled and convoluted and show wide, irregular lumina. The epithelial cells become larger, and their free surfaces show

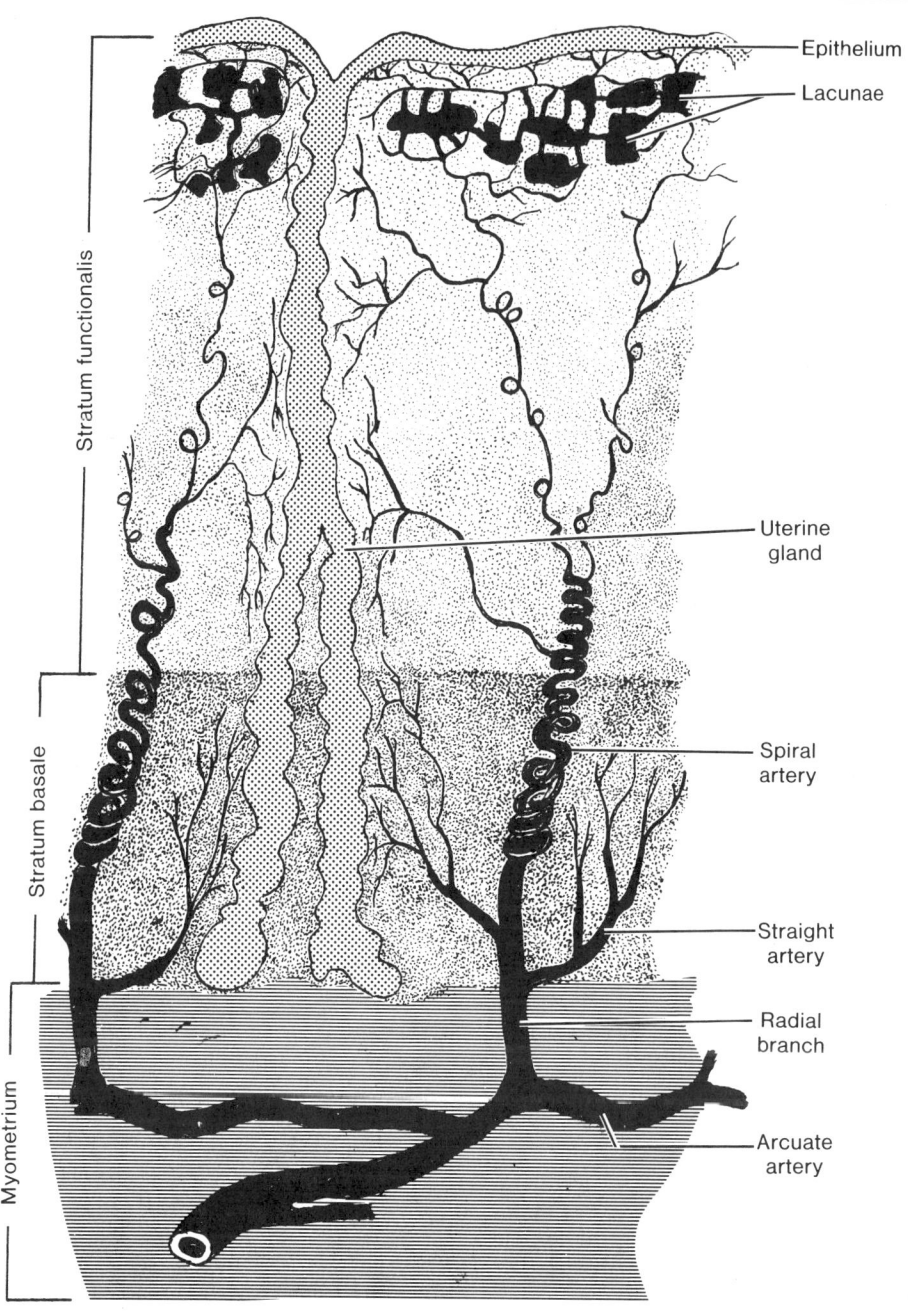

Fig. 18-10. Arrangement of blood vessels in the endometrium.

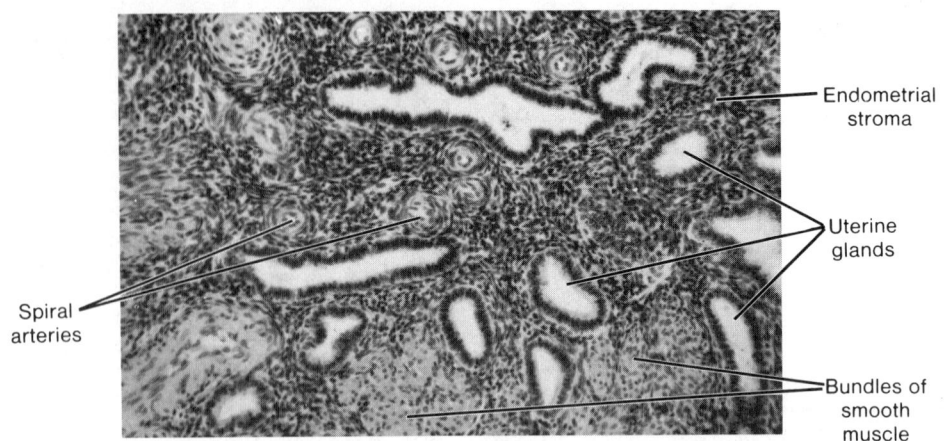

Fig. 18-11. The stratum basale of human endometrium. Note the spiral arteries (LM, × 100).

many long microvilli. Mitochondria are large, there is a rich endoplasmic reticulum, and the Golgi apparatus is prominent. Nuclei are enlarged and contain a distinct nucleolus. Glycogen and mucoid materials rapidly increase, first in the bases of the glandular cells, then in the apical portions of the cells, and then into the lumina of the glands. Elongation and coiling of the spiral arteries continue, and the vessels extend into the superficial part of the endometrium.

The secretory phase is associated with development of a **corpus luteum** and is maintained as long as the corpus luteum remains functional. Progesterone secreted by the corpus luteum is responsible for the secretory changes in the endometrium.

The **ischemic (premenstrual) phase** is characterized by intermittent constriction of the spiral arteries, resulting in vascular stasis and reduced blood flow to the functional layer. The outer zone of the endometrium is subjected to anoxia for hours at a time, resulting in breakdown of the stratum functionale. The stroma becomes increasingly edematous and infiltrated by leukocytes.

In the **menstrual phase,** the functional layer becomes necrotic and is shed. The spiral arteries also become necrotic, and blood is lost from the arteries and veins. Small lakes of blood form and coalesce, and overlying patches of mucosa are detached, leaving a denuded stromal surface. Sloughing of the endometrium continues until the entire functional layer has been discarded. Blood oozes from veins torn during the shedding of the endometrial tissue. The menstrual discharge consists of arterial and venous blood,

autolyzed and degenerated epithelial cells, and glandular secretions. The straight arteries of stratum basale do not constrict during menstruation, thus preserving the basal layer to provide for restoration of the endometrium during the following new proliferative stage.

The onset of the menstrual cycle coincides with the beginning involution of a corpus luteum. The hormonal relationships of the ovary and uterine mucosa are shown in Figure 18-12.

Cervix

KEY WORDS: **dense connective tissue, endocervix, nonkeratinized stratified squamous epithelium, secretory change**

The wall of the cervix differs considerably from that of the body of the uterus. Little smooth muscle is present, and the cervical wall consists mainly of **dense connective tissue** and elastic fibers. In the part that protrudes into the vaginal canal, smooth muscle is lacking. The cervical canal is lined by a mucosa, the **endocervix,** which forms complex, branching folds. The epithelial lining consists of tall, mucus-secreting columnar cells along with a few ciliated columnar cells. Numerous large, branched glands are present that are lined by mucus-secreting columnar cells similar to those of the lining epithelium. The cervical canal usually is filled with mucus. Occasionally the glands become occluded and filled with secretion, forming nabothian cysts.

The portion of the cervix that protrudes into the vaginal

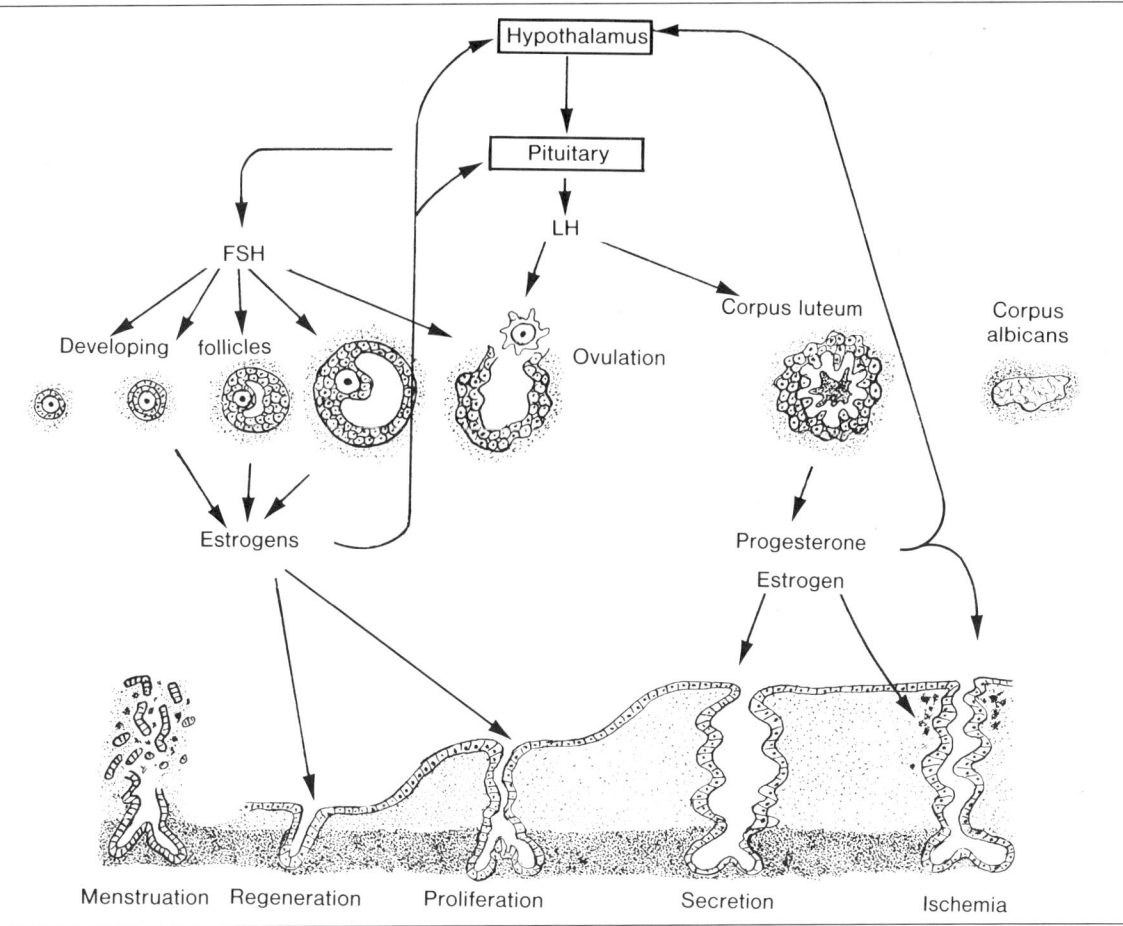

Fig. 18-12. Hormonal relationships of the ovary and uterine mucosa (FSH = follicle-stimulating hormone; LH = luteinizing hormone).

canal, the exocervix, is covered by a **nonkeratinized stratified squamous epithelium** whose cells contain much glycogen. The transition from the columnar epithelium of the cervical canal is fairly abrupt and usually occurs just inside the external os. The cervical mucosa does not take part in the cyclic changes of the body of the uterus and is not shed. The glandular elements do show **changes in secretory activity,** however. At midcycle there is a copious secretion of a thin alkaline fluid, probably the result of increased stimulation by estrogen. After ovulation and establishment of a corpus luteum, the amount of secretion decreases and the mucus becomes thicker and more viscous.

Vagina

The vagina is the lowermost portion of the female reproductive tract and is a muscular tube that joins the uterus to the exterior of the body. It represents the copulatory organ of the female. Ordinarily the lumen is collapsed and the anterior and posterior walls make contact.

Structure

KEY WORDS: mucosa, rugae, nonkeratinized stratified squamous epithelium, lamina propria, muscularis, adventitia

The vaginal wall consists of a mucosa, a muscularis, and an adventitia. The **mucosa** is thrown into folds **(rugae)** and consists of a thick surface layer of **nonkeratinized stratified squamous epithelium** overlying a lamina propria (Figs. 18-13 and 18-14). The epithelial cells contain abundant glycogen, especially at midcycle. The **lamina propria** consists of a fairly dense connective tissue that becomes more loosely arranged near the muscle coat. Immediately below the epithelial lining, elastic fibers form a dense network. Diffuse and nodular lymphatic tissues are found occasionally, and many lymphocytes, along with granular leukocytes, invade the epithelium. The vagina has no glands, and the epithelium is kept moist by secretions from the cervix.

The **muscularis** consists of bundles of smooth muscle fibers that are arranged circularly in the inner layer and longitudinally in the outer part. The longitudinal fibers become continuous with similarly oriented fibers in the myometrium.

The **adventitia** is a thin outer layer of connective tissue

Fig. 18-13. Surface cells lining the mucosa of the vagina (SEM, × 1,000).

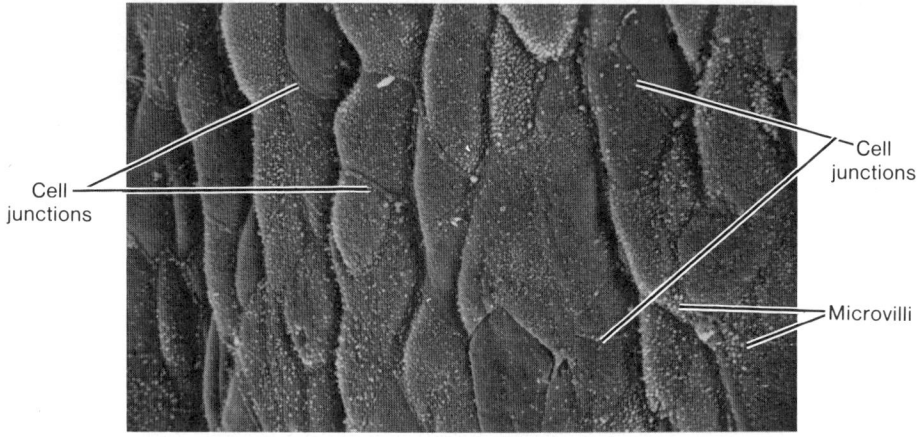

Fig. 18-14. The mucosal lining of the vagina. Note the surface cells and compare with Fig. 18-13 (LM, × 250).

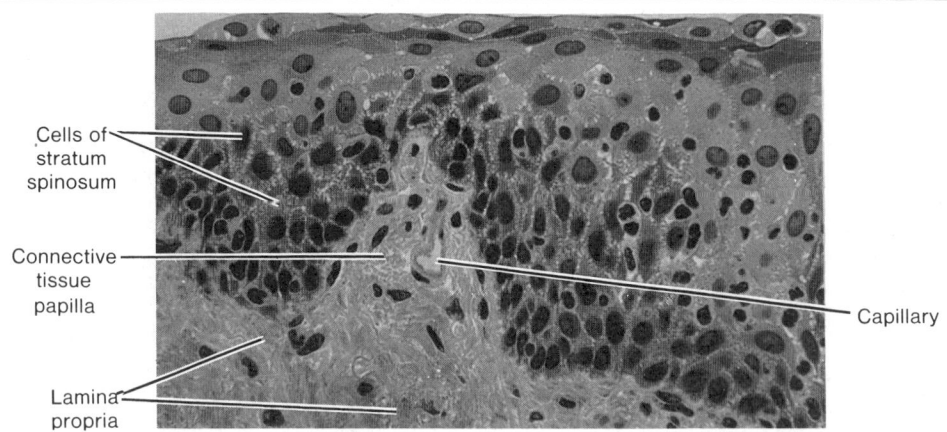

rich in elastic fibers. It merges imperceptibly into the surrounding loose connective tissue around other organs.

Cytology of the Cervical Canal

KEY WORDS: basal cell, parabasal cell, intermediate squamous cell, superficial squamous cell

The stratified squamous epithelium that lines the vagina and covers the exocervix is arranged in layers similar to those of the epidermis. Deepest is the basal layer (stratum germinativum), followed by an intermediate (spinous) layer and a superficial layer (stratum corneum) from which cells are shed. Desquamated cervical cells can be recovered from vaginal secretions, or cells can be obtained more directly by gentle abrasion of the exocervix. The cells can be studied in smear preparations, and examination of Papanicolaou's ("Pap") smears has become a standard procedure for early detection of cervical cancer.

In the normal adult female, four types of cells usually are recognized cytologically, loosely corresponding to the layers of the cervicovaginal epithelium. These include basal (lower basal) cells, parabasal (outer basal) cells, intermediate (precornified) squamous cells, and superficial (cornified) squamous cells. Germinal cells from the basal layer are exceedingly rare in smears from normal adults, and the basal and parabasal cells recognized cytologically are derived from different levels of the intermediate layer. They do represent less mature cells, and their presence indicates that immature cells are at higher than normal levels in the epithelium. These cells are associated with marked deficiency of estrogens and are common before puberty and during menopause.

Basal cells arise from the lower levels of the transitional zone and are rounded or oval cells, about 4 to 5 times the size of a granular leukocyte. The central nucleus is deeply stained, but a pattern of fine chromatin granules and dense patches can be made out. In some cells a Barr body is distinctly visible. The moderate amount of basophilic cytoplasm stains deeply and evenly. **Parabasal cells** arise higher up in the transitional zone and also are round or oval cells, but they are larger than basal cells with a more abundant cytoplasm that is less basophilic and often shows a somewhat "blotchy" pattern. The central nucleus remains about the same size but may be more dense than that of a basal cell. **Intermediate squamous cells** vary in size, but all appear as thin, polygonal plates with abundant transparent cytoplasm. The nucleus is smaller, vesicular, and centrally placed. The cytoplasm stains somewhat variably and may be lightly basophilic or show some degree of eosinophilia. **Superficial squamous cells** represent dead surface cells. They are large, with voluminous eosinophilic cytoplasm that is thin and transparent with sharply defined borders. Because of the thinness, the cytoplasm often is curled, wrinkled, and folded. The nucleus is very small — about one-half to one-third that of an intermediate squamous cell — and is densely stained and pyknotic. The superficial and intermediate squamous cells are the largest cells seen in a routine preparation and range from 40 to 50 μm in diameter. Occasionally, anucleate squamous cells may be found. They have a somewhat shrivelled appearance, and the site of the nucleus is suggested by a pale central zone. They represent cells that are more completely keratinized.

Cells that originate from the endocervix also may be present and often occur in small sheets or strips; their appearance depends on the orientation. From end on, the cells appear as groups or nests of small polyhedral or round cells with sharp cell boundaries and relatively large central nuclei. In profile, the cells show their columnar shape with the nuclei close to one pole. Strips of these cells give a "picket fence" appearance.

External Genitalia

The external genitalia of the female consist of the labia, clitoris, and vestibular glands.

Labia

KEY WORDS: labia majora, homologue of the scrotum, labia minora

The **labia majora** consist of folds of skin covering an abundance of adipose tissue. In the adult, the outer surface is covered by coarse hair with many sweat and sebaceous glands. The inner surface is smooth and hairless and also contains sweat and sebaceous glands. The labia majora are the **homologue of the scrotum** in the male. The **labia minora** consist of a core of highly vascular, loose connective tissue covered by stratified squamous epithelium that is deeply indented by connective tissue papillae. The deeper layers of epithelium are pigmented. Both surfaces of the labia minora are devoid of hair, but large sebaceous glands are present.

Clitoris

KEY WORDS: erectile, corresponds to dorsal penis, corpora cavernosa, glans clitoridis, no corpus spongiosum

The clitoris is an **erectile** body that **corresponds to the dorsal penis.** It consists of two **corpora cavernosa** enclosed in a layer of fibrous connective tissue and separated by an incomplete septum. The free end of the clitoris terminates in a small, rounded tubercle, the **glans clitoridis,** which consists of spongy erectile tissue. The clitoris is covered by a thin layer of stratified squamous epithelium with high papillae associated with many specialized nerve endings. The clitoris has **no corpus spongiosum** and therefore does not contain the urethra.

Vestibular Glands

KEY WORDS: stratified squamous epithelium, mucous cell, major vestibular gland, tubuloalveolar gland

The vestibule is the cleft between the labia minora and in it are the vaginal and urethral openings. It is lined by **stratified squamous epithelium** and contains numerous small vestibular glands concentrated about the openings of the vagina and urethra. The glands contain **mucous cells** and resemble the urethral glands of the male. A pair of larger glands, the greater or **major vestibular glands,** are present in the lateral walls of the vestibule. They are **tubuloalveolar glands** that secrete a mucoid lubricating fluid. The major glands correspond to the bulbourethral glands of the male.

Pregnancy

Pregnancy involves implantation of a blastocyst into a prepared uterine endometrium and subsequent formation of a placenta to nourish and maintain the developing embryo. Prior to implantation, fertilization of the ovum and cleavage of the resulting zygote occur in the oviduct.

Fertilization

KEY WORDS: female pronucleus, male pronucleus

Before an ovum can be fertilized, it must undergo maturational changes, chief of which is reduction of the chromosome complement to the haploid number. The oocyte passes through the early stages of the first meiotic division during fetal life, and it is only just before ovulation that the division is completed and the first polar body is given off. The resulting secondary oocyte immediately enters the second meiotic division, which, however, proceeds only to metaphase and is not completed until fertilization occurs.

At the time of ovulation, the oviduct shows active movements that bring the infundibulum and fimbria close to the ovary. Cilia on the surface of the fimbria sweep the ovum into the ampulla of the oviduct where fertilization, if it is to occur, takes place. The human ovum probably remains fertilizable for about 1 day, after which it degenerates if fertilization does not occur.

Of the millions of sperm deposited in the female tract, only one penetrates the ovum. There is no evidence for chemotactic attraction, and random movement brings sperm and ovum together. The successful sperm pierces the corona radiata and zona pellucida, possibly by lysis of the membrane by enzymes in the sperm acrosome, and the entire spermatozoon is engulfed by the cytoplasm of the ovum. Electron micrographs suggest that the plasma membranes of the sperm and ovum fuse, that of the spermatozoon being left at the surface of the ovum. Penetration is followed by release of electron-dense cortical granules that underlie the plasmalemma of the ovum and by immediate changes in the permeability of the zona pellucida, which thereafter excludes entry by competing sperm.

The ovum now completes the second maturation division and extrudes the second polar body. The remaining chromosomes (23) reconstitute and form the **female pronucleus.** The nucleus (head) of the spermatozoon swells and forms the **male pronucleus,** and the body and tail are resorbed. The two pronuclei move to the center of the cell, and two centrioles, supplied by the anterior centriole of the spermatozoon, appear. The chromatin of each pronucleus resolves into a set of chromosomes that align themselves on a spindle to undergo a normal mitotic division of the first cleavage. Each cell resulting from this division receives a full diploid set of chromosomes.

Cleavage

KEY WORDS: blastomere, morula, blastocele, blastocyst, inner cell mass, trophoblast

The zygote undergoes a series of rapid divisions called *cleavage* that result in a large number of cells, the **blastomeres.** The cells are bounded by the zona pellucida, and a mulberry-like body, the **morula,** is formed. Cleavage is a fractionating process: No new cytoplasm is formed, and at each division the cells become smaller until a normal, predetermined, cytoplasmic-nuclear ratio is reached. Thus the total size of the morula is not increased. Cleavage occurs as the morula slowly is moved along the oviduct by the waves of peristaltic contractions in the muscle coat. When the morula reaches the uterine cavity, it is at the 12- to 16-cell stage. At about this time, fluid penetrates the zona pellucida and diffuses between the cells of the morula. The fluid increases in amount, the intercellular spaces become confluent, and a single cavity, the **blastocele,** is formed. The morula has now become a **blastocyst** that forms a hollow sphere containing, at one pole, a mass of cells called the **inner cell mass** that will form the embryo proper. The capsule-like wall of the blastocyst consists of a single layer of cells, the **trophoblast.** After reaching a critical mass, the blastocyst breaks through the surrounding zona pellucida and remains free in the uterine cavity for about a day; then it attaches to the endometrium, which is in the secretory phase.

Implantation

KEY WORDS: secretory phase, syncytial trophoblast, cytotrophoblast, lacunae

At the time of implantation, the endometrium is in the **secretory phase** and, having been under the influence of progesterone from the corpus luteum for several days, has reached its greatest thickness and development. Implantation is initiated by close approximation of the trophoblast to the microvilli and surface projections of the uterine epithelial cells. At the points of contact, the trophoblast shows patches of coated vesicles and numerous lysosomes. The microvilli shorten and disappear, and the trophoblast extends finger-like processes between the uterine epithelial cells, and the two layers become closely locked by numerous tight junctions that develop between them. The uterine epithelial cells degenerate and are engulfed by the trophoblast, the cellular debris appearing in phagosomes within the trophoblast cytoplasm. Where it is fixed to the endothelium, the trophoblast proliferates to form a cellular mass between the blastocyst and maternal tissues. No cell boundaries can be made out in this cell mass, which is called the **syncytial trophoblast.** The syncytium continues to erode the endometrium at the point of contact, creating a ragged cavity into which the blastocyst sinks, gradually becoming more deeply embedded until it entirely lies within the endometrial stroma. The surface defect in the endometrium is closed temporarily by a fibrin plug. Later, proliferation of surrounding cells restores the surface continuity of the endometrial lining.

As the blastocyst sinks into the endometrium, the syncytial trophoblast rapidly increases in thickness at the original site of attachment and progressively extends to cover the remainder of the blastocyst. When completely embedded, the entire wall of the blastocyst consists of a thick outer syncytial trophoblast and an inner **cytotrophoblast,** which is composed of a single layer of cells with well-defined boundaries. The cytotrophoblast shows active mitosis and contributes cells to the syncytial trophoblast, where they fuse with and become part of that layer. The syncytial trophoblast continues to erode the uterine tissues,

opening up the walls of maternal blood vessels. Spaces appear in the syncytial trophoblast and these **lacunae** expand, become confluent, and form a labyrinth of spaces. Many of the spaces contain blood extravasated from eroded maternal blood vessels; this blood supplies nourishment for the embryo and represents the first step in the development of uteroplacental circulation.

Placenta

KEY WORDS: primary villus, chorion, secondary villus, chorionic plate, chorion frondosum, chorion laeve, tertiary placental villus, decidua, human chorionic gonadotropin (hCG)

As the lacunae enlarge, intervening strands of trophoblast form the **primary villi** that cover the entire periphery of the blastocyst. Each villus consists of a core of cytotrophoblast covered by a layer of syncytial trophoblast. Trophoblastic cells at the tips of the villi apply themselves to the endometrium and form a lining for the cavity in which the blastocyst lies. When the embryonic germ layers have been established, mesoderm grows out from the embryo as the **chorion** and forms a lining for the trophoblast that surrounds the blastocyst. Mesoderm extends into the primary villi to form a core of connective tissue and convert the villi to **secondary villi.** The deeply embedded portion of the chorion constitutes the **chorionic plate** from which numerous villi project to form the **chorion frondosum.** Villi on that part of the chorion facing the uterine cavity grow more slowly and are less numerous; ultimately, these villi disappear, leaving a smooth surface that forms the **chorion laeve.** Blood vessels develop in the mesenchymal cores of the secondary villi and soon make connection with the fetal circulation. With vascularization, the secondary villi become the definitive or **tertiary placental villi.**

At parturition, all but the deepest layers of the endometrium are shed; thus the superficial part of the endometrium is called the **decidua.** A feature of the stroma is the alteration of its cells to form enlarged, decidual cells that contain much glycogen. According to the relationship with the implantation site, three areas of the decidua are recognized. The *decidua capsularis* is the part that lies over the surface of the blastocyst, while the *decidua basalis* underlies the implantation site and forms the maternal component of the placenta. The endometrium lining the remainder of the pregnant uterus is the *decidua parietalis.* As the embryo grows, the decidua capsularis becomes increasingly attenuated. Eventually, the decidua capsularis makes contact with decidua parietalis on the opposite surface of the uterus, and the uterine cavity is obliterated.

The mature placenta consists of maternal and fetal components. The maternal part is decidua basalis; the fetal portion consists of the chorionic plate and the villi arising from it. Maternal blood circulates through the intervillous spaces and bathes the villi, of which there are two types. Some pass from the chorionic plate to decidua basalis as anchoring villi from which secondary and tertiary branches float in the intervillous spaces as the floating villi. Both types of villi consist of a core of loose connective tissue in which lie fetal capillaries. Covering each villus is an inner layer of cytotrophoblast cells, which have large nuclei and lightly basophilic cytoplasm containing considerable glycogen. External to the cytotrophoblast is a layer of syncytial trophoblast of variable thickness. The cells of the cytotrophoblast decrease in number in the latter half of pregnancy, and only a very few are present at term. The syncytial trophoblast also thins out to form a narrow layer.

The structure of the placenta is shown diagrammatically in Figure 18-15.

The placenta transfers oxygen and nutrients from the maternal to the fetal circulation and waste products from the fetal to the maternal circulation. Although maternal and fetal circulations are closely apposed, they remain separated by the syncytial trophoblast (and early in pregnancy by the cytotrophoblast also), a basement membrane, the

connective tissue of the villi, and the wall of the fetal blood vessels (Fig. 18-16). Transport of material between fetal and maternal blood appears to be regulated by the syncytial trophoblast. Being without cell boundaries and intercellular spaces, any materials passing into or out of the fetal blood must pass through the cytoplasm of the syncytium. Electrolytes, steroids, fatty acids, oxygen, and carbon dioxide traverse the syncytial trophoblast by passive diffusion. Other substances, such as glucose, cross this barrier by carrier molecules. In addition, the syncytial trophoblast plasmalemma contains receptors for macromolecules such as transferrin, insulin, and immunoglobulins that are taken in by receptor-mediated endocytosis and cross the epithelial barrier in transport vesicles. The cytotrophoblast appears to serve as the source of cells for the syncytial trophoblast.

The placenta also is a multipotential endocrine organ essential for monitoring pregnancy. The trophoblast secretes a glycoprotein, **human chorionic gonadotrophin (hCG),** that can be detected by the end of the first week after fertilization. It increases in amount to the fifth week of pregnancy and maintains and stimulates the corpus luteum to secrete estrogens and progesterone during early pregnancy. The placenta also secretes estrogens and progesterone that aid in monitoring the uterine environment for continued fetal development. Hormones, synthesized by the placenta, are released into the maternal blood by exocytosis from the syncytial trophoblast.

Fig. 18-15. Diagram of the structure of the placenta.

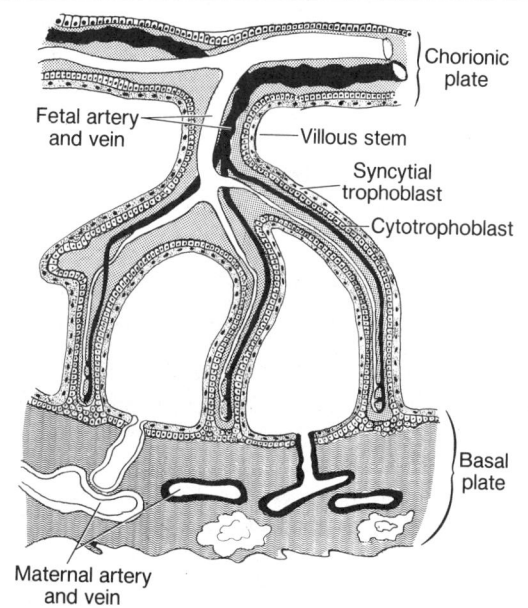

Mammary Glands

Mammary glands are present in both sexes but in males remain rudimentary throughout life. In the female, the size, shape, and structure vary with age and functional status. Prior to puberty, the female breasts are undeveloped but enlarge rapidly at puberty due mainly to accumulation of adipose tissue. They reach their greatest development at late pregnancy and lactation. In the adult female, the mammary glands are variably hemispherical and conical in shape, each surmounted by a cylindrical projection, the nipple. Surrounding the nipple is a slightly raised, circular area of pigmented skin, the areola.

The mammary gland is a compound tubuloalveolar gland and is considered to be of cutaneous origin, representing a modified apocrine sweat gland.

Structure of the Resting Gland

KEY WORDS: duct, simple cuboidal epithelium, lactiferous sinus, stratified cuboidal epithelium, stratified squamous epithelium, alveolus, myoepithelial cell

Each mammary gland consists of 15 to 20 individual pyramid-shaped lobes that radiate from a nipple. The lobes are separated by septa of dense connective tissue. Each lobe represents an individual gland and contains glandular tissue drained by its own ductal components, embedded in the intralobular connective tissue. The proportions of glandular and nonglandular tissue vary with functional status.

In the resting gland, the principal glandular elements are the **ducts,** which are grouped together to form lobules. The smallest branches of the ductal system are lined by **simple cuboidal epithelium,** but the lining increases in height as the ducts unite and pass toward the nipple. Just beneath the areola, the ducts expand to form the **lactiferous sinuses,** which are lined by a two-layered **stratified cuboidal epithelium.** As the duct ascends through the nipple, it becomes lined by **stratified squamous epithelium.**

Secretory units consist of clusters of **alveoli** surrounding a small duct. The alveolar wall consists of simple cuboidal epithelium resting on a basement membrane. Between the epithelial cells and the basement membrane are highly branched **myoepithelial cells** whose long, slender processes embrace the alveolus in a basket-like network. Alveoli are not prominent in the resting gland, and there is some question as to whether they are present at all. When present, they usually occur as small, budlike extensions of the terminal ducts.

The intralobular connective tissue around the ductules and alveoli is loosely arranged and cellular, whereas that surrounding the larger ducts and lobes (interlobular connective tissue) is variably dense and contains much adipose tissue.

Nipple and Areola

KEY WORDS: keratinizing stratified squamous epithelium, lactiferous duct, smooth muscle, areolar gland

The nipple is covered by **keratinizing stratified squamous epithelium** continuous with that of the skin overlying the breast. The dermis projects deeply into the epithelium, forming unusually tall dermal papillae. The skin of the nipple is pigmented and contains many sebaceous glands but is devoid of hair and sweat glands. The nipple is traversed by many **lactiferous ducts,** each of which drains a lobe and empties onto the tip of the nipple. The outer parts of the ducts also are lined by keratinizing stratified squamous epithelium. The dense collagenous connective tissue of the nipple contains bundles of elastic fibers, and much **smooth muscle** is present. The smooth muscle fibers are arranged circularly and radially and, on contraction, produce erection of the nipple.

The areola is the pigmented area of skin that encircles the base of the nipple. During pregnancy, the areola becomes larger and more deeply pigmented. It contains sweat, sebaceous, and **areolar glands.** The latter appear to be intermediate in structure between apocrine sweat glands and true mammary glands.

Fig. 18-16. Human placenta at the sixth month of gestation. Note that only the syncytial trophoblast remains covering placental villi (LM, × 250).

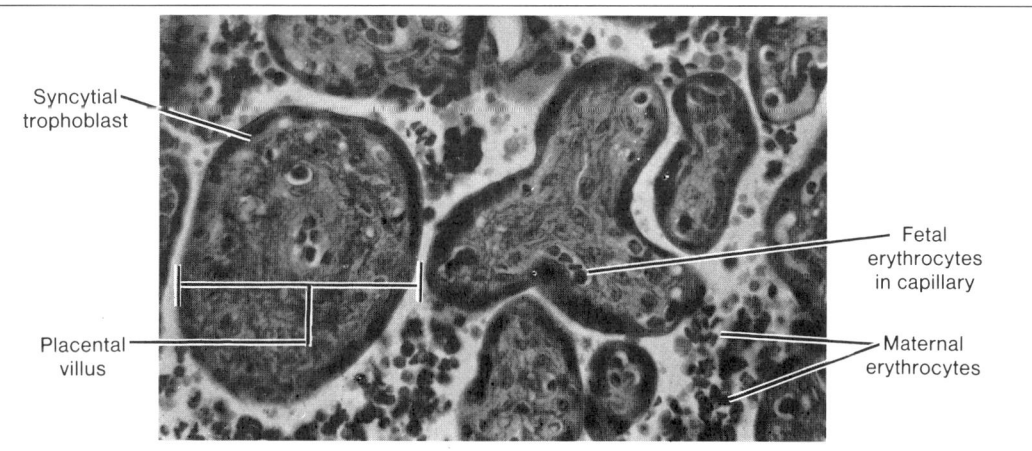

Syncytial trophoblast

Fetal erythrocytes in capillary

Placental villus

Maternal erythrocytes

Pregnancy and Lactation

KEY WORDS: proliferation of glandular tissue, colostrum, apocrine secretion

During pregnancy, mammary glands undergo extensive development in preparation for their role in lactation. During the first half of pregnancy, the terminal portions of the ductal system grow rapidly, branch, and develop terminal buds that expand to become alveoli. **Proliferation of glandular tissue** takes place at the expense of the fat and stromal connective tissue, which decrease in amount. The connective tissue becomes infiltrated with lymphocytes, plasma cells, and granular leukocytes. During late pregnancy, proliferation of glandular tissue subsides, but the alveoli expand and there is some formation of secretory materials. The first secretion, **colostrum,** is thin and watery. It is poor in lipid but contains a considerable amount of antibodies that provide passive immunity to the newborn.

True milk secretion begins a few days after parturition, but not all the breast tissue is functioning at the same time. In some areas, alveoli are distended with milk, the epithelial lining is flattened, and the lumen is distended; in other areas, the alveoli are resting and are lined by tall columnar epithelial cells (Fig. 18-17). Secreting cells have abundant granular endoplasmic reticulum, moderate numbers of relatively large mitochondria, and supranuclear Golgi complexes. Milk proteins are elaborated by the granular endoplasmic reticulum and in association with the Golgi body form membrane-bound vesicles. These are carried to the apex of the cell, where the contents are released by exocytosis. Lipid arises as cytoplasmic droplets that coalesce to form large spherical globules. The droplets and globules reach the apex of the cell and protrude into the lumen. Ultimately, they pinch off, surrounded by a thin film of cytoplasm and the detached portion of the plasmalemma. This method of release is a form of **apocrine secretion,** but only minute amounts of cytoplasm are lost. Immunoglobulins in milk are synthesized by plasma cells in the connective tissue surrounding the alveoli of the mammary glands. The secreted IgA is taken up by receptor-mediated endocytosis along the basolateral plasmalemma of mammary gland epithelial cells and transported in small vesicles to the cell apex, where it is discharged into the lumen by exocytosis. Passage of milk from alveoli into and along the ducts is due to contraction of the myoepithelial cells.

Average milk production by a woman breast-feeding a single infant is about 1200 ml/day. Milk consists of about 88 percent water, 6.5 percent carbohydrate, 3.3 percent lipid, and 1.5 percent protein. Lactose is the primary carbohydrate and casein the primary protein. In addition, immunoglobulins (IgE and IgA), electrolytes (Na^+, KI^+, Cl^-), minerals (Ca^{2+}, Fe^{2+}, Mg^{2+}), and other substances occur in milk. Women in the fourth and fifth months of lactation may secrete as much as 0.5 g of immunoglobulin per day in her milk. Such immunoglobulins are important in the re-

Fig. 18-17. A portion of a lobule from a lactating human mammary gland. Note the abundant alveoli (LM, × 100).

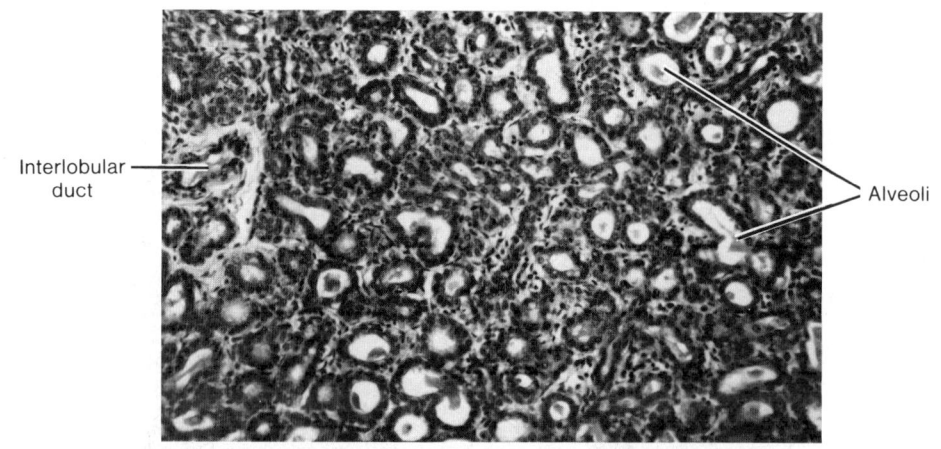

Interlobular duct

Alveoli

sistance of enteric infections and provide the infant with considerable passive immunity.

With weaning, lactation soon ceases, and the glandular tissue returns to its resting state. However, regression is never complete, and not all the alveoli completely disappear.

Hormonal Control

KEY WORDS: estrogen, progesterone, somatotropin, prolactin, adrenal corticoid, neurohormonal reflex, oxytocin

Before puberty, growth of the mammary gland parallels general body growth, but as the ovaries become functional, the mammary tissue comes under the influence of estrogen and progesterone. While some structural changes can be observed, the cyclic response of the breast is minor. During pregnancy, the glands are continuously stimulated by **estrogen** and **progesterone** from the corpus luteum and placenta. Generally, growth of the ductal system depends on estrogen, but for alveolar development, both progesterone and estrogen are required. To attain the full development in pregnancy, other hormones — **somatotropin, prolactin,** and **adrenal corticoids** — appear to be necessary.

At the end of pregnancy (birth), the levels of circulating estrogen and progesterone fall abruptly. In the absence of their inhibition, there is an increased output of prolactin by the hypophysis. Prolactin is a powerful lactogenic stimulus, and full lactation is established in a few days after birth. Maintenance of lactation requires continuous secretion of prolactin, which results from a **neurohormonal reflex** established by suckling. Periodic suckling also causes release of **oxytocin** from the neurohypophysis, which stimulates contraction of myoepithelial cells, resulting in release of milk from alveoli into and along the ducts.

Review Table 18-1. Key Features of the Female Reproductive System

	Ovum	Cells of Wall	Other
Structures Contained Within the Ovary			
Primordial follicle	Dominates follicle; nucleus large, central, and vesicular; prominent nucleolus	Unilaminar layer of flattened follicular cells	
Primary follicle	Large with central vesicular nucleus; zone pellucida may be visible	Two or more layers of granulosa cells; cuboidal in shape	Surrounding stroma organized into theca interna and externa
Secondary follicle	Ovum with large vesicular nucleus; zona pellucida prominent	Stratum granulosum 8 to 12 layers thick; large spaces between cells; formation of antrum	Well defined theca interna and externa
Mature follicle	Ovum eccentric in large follicle located near surface of ovary	Cells organized into cumulus oophorus and corona radiata; very large antrum	Well defined theca interna and externa; cells of theca interna plump
Atretic follicle	May or may not be present — shows signs of degeneration	Show degenerative changes; macrophages usually present; connective tissue infiltration	Cells of theca interna show degenerative changes; appearance of glassy membrane
Corpus luteum	Absent	Prominent granulosa lutein cells; smaller theca lutein cells	
Corpus albicans	Absent	Thin, dark staining nuclei of fibroblasts	Large bundles of thin collagen fibers; hyaline material

Review Table 18-1 (continued)

	Epithelial Lining	Support	Muscle
The Female Reproductive Tract			
Oviduct	Simple columnar; ciliated and nonciliated cells	Lamina propria, cellular, thin collagen, and reticular fibers	Well-defined inner circular smooth muscle; outer layer of scattered longitudinal fibers
Uterus	Simple columnar with groups of ciliated cells; extends into lamina propria to form tubular glands	Lamina propria = endometrial stroma, similar to mesenchyme; richly cellular, fine reticular fibers, few collagen fibers. *Note:* The oviduct and uterus are the only hollow organs with this type of lamina propria	Myometrium; thick coat of smooth muscle in 3 intermingling layers
Cervix	1. Tall columnar; some ciliated, some mucus-secreting cells 2. Vaginal part covered by nonkeratinized stratified squamous	Dense collagenous and elastic connective tissue	Smooth muscle much reduced in amount, lacking in vaginal part
Vagina	Stratified squamous nonkeratinized epithelium	Lamina propria of dense collagenous connective tissue. Dense elastic network below epithelial layer	Inner circular smooth muscle, outer longitudinal smooth muscle

Functional Review: Female Reproductive System

The ovary is a cytogenic gland releasing ova and also acts as a cyclic endocrine gland. During its growth in the follicles, the oocyte is nourished by blood vessels of the ovary via the theca interna. When a corpus luteum is formed after ovulation, estrogens and progesterone are produced and are responsible for development of the uterine mucosa prepared for reception of the blastocyst. The periodic nature of hormone production by the ovary establishes the menstrual cycle during which, in the absence of pregnancy, the uterine mucosa is shed. If pregnancy occurs, the corpus luteum persists and its hormonal activity maintains the endometrium in a prepared state.

Female reproductive function is regulated primarily by positive and negative feedback loops on neurons in the hypothalamic region of the brain and on cells of the anterior pituitary. The hypothalamic neurons release a single factor (gonadotropin-releasing factor) that regulates the release of both FSH and LH from cells of the anterior pituitary. Follicle-stimulating hormone influences the growth of late primary and secondary follicles and promotes the formation of estrogens. Luteinizing hormone stimulates ovulation and promotes corpus luteum formation. Estrogen suppresses FSH secretion by acting on both cells of the anterior pituitary and neurons within the hypothalamus. Inhibin secreted by follicular granulosa cells suppresses secretion of FSH directly by acting on cells of the anterior pituitary. Progesterone inhibits LH release; therefore, the corpus luteum lacks the stimulus essential for its maintenance and soon degenerates.

The oviduct is the site of fertilization of an ovum and also transports the zygote to the uterus as the result of muscular and ciliary actions. Conditions within the oviduct sustain the zygote as it undergoes cleavage during passage through the tube. Nutrition for the zygote is provided by material stored in the cytoplasm of the ovum.

On entering the uterine cavity, the blastocyst lies in secretions produced by the endometrium. The secretion is rich in glycogen, polysaccharides, and lipids, providing an excellent "culture medium" for the dividing cells of the blastocyst. The uterine endometrium, to which the blastocyst attaches, provides for the sustenance of the embryo throughout its development. A rich food supply for the implanting blastocyst comes from the secretions of the uterine glands and from products of the uterine stroma as the blastocyst burrows into the endometrium. These products are absorbed by the syncytial trophoblast and diffuse to the developing embryo. As the trophoblast continues to erode into the endometrium, blood-filled spaces form, and the blastocyst is bathed by pools of maternal blood that supply the embryo with nourishment. When a placenta has been established, the nutritional, respiratory, and excretory needs of the embryo are met by this fetal-maternal membrane. The placenta also has protective functions, preventing passage of particulate matter to the embryo. It also has endocrine functions, elaborating gonadotropins, estrogens, and progesterone. Gonadotropins are produced very early and maintain the corpus luteum during early pregnancy.

The basal layer of the endometrium is not shed at parturition or at menstruation and provides for the restoration of the uterine mucosa after these events have occurred. The myometrium, by its muscular contractions, expels the fetus at birth.

Changes in the secretory activities of cervical glands during the menstrual cycle may have some significance in fertility. During most of the cycle, the glands produce a thick, viscous mucus that appears to inhibit passage of sperm. The thin, less viscid mucus elaborated at midcycle appears to favor the migration of sperm.

The mammary glands provide nourishment for the newborn, which is delivered in an immature and dependent state. The first secretion, colostrum, has a high content of antibodies and may give passive immunity to the suckling young. Passage of milk from the alveoli to and along ducts is accomplished by contraction of myoepithelial cells under the influence of the hormone oxytocin.

Review Questions and Answers

Questions

93. What is the appearance of the ovarian follicle at the time of ovulation? Describe the ovum immediately after ovulation.
94. Describe the endometrium during the proliferative and secretory phases and relate the events to the ovarian cycle.
95. What parts of the placenta are derived from fetal components and what parts from maternal components?
96. Outline the nutritional supply to a newly developing organism from fertilization to birth.
97. What changes occur in the mammary gland as the result of pregnancy? How is lactation maintained ?

Answers

93. At the time of ovulation, the ovum is contained within a mature or graafian follicle, which occupies the thickness of the cortex and bulges from the surface of the ovary. The ovum becomes loosened from its attachment to the cumulus oophorus as fluid-filled spaces appear between the granulosa cells of the cumulus.

The cells of the theca interna have assumed the characteristics of an endocrine gland and secret androgens, which are converted to estrogens by aromatase enzymes of the granulosa cells. At the point where the follicle bulges from the ovary, its wall thins and the stigma, a small avascular, translucent area, appears and protrudes as a small blister.

With rupture of the stigma, ovulation occurs. The ovum has completed the first meiotic division and so a secondary oocyte is released, complete with its corona radiata and zona pellucida.

94. During the proliferative stage, the epithelial cells of the glandular remnants in the stratum basale proliferate rapidly and migrate to cover the denuded mucosal surface. The stromal cells also proliferate and the endometrium increases in thickness. These events are stimulated by estrogens elaborated by the developing follicles. In the secretory phase, the glandular elements elongate, become tortuous and coiled, and begin to secrete. Stromal edema and an increase in the vascularity of the endometrium also occur. The secretory phase is associated with the production of large amounts of progesterone by the corpus luteum.

95. The fetal components of the mature placenta consist of the chorionic plate (established from mesoderm of the embryo) and its villi. The maternal component consists of the decidua basalis, the superficial part of the endometrium that lies beneath the implantation site.

96. In the oviduct, the nutrition for the dividing zygote is derived from the cytoplasm of the ovum. Within the uterine cavity and prior to implantation, the blastocyst lies within the secretions of the endometrium, which are rich in glycogen, polysaccharides, and lipids. As the blastocyst implants in the endometrium, it encounters a rich food supply in the glandular secretions and products of the breakdown of the uterine stroma. Later in the process of implantation, the blastocyst becomes surrounded by blood-filled spaces in the endometrium that supply the nutritional needs. When the placenta has been fully established, the nutritional needs of the embryo are met by diffusion of materials between the closely related maternal and fetal circulations.

97. During the first half of pregnancy, the terminal portions of the ducts branch and develop terminal buds that will become the secretory alveoli. In later pregnancy the glandular portions increase in size and some slight secretion occurs. Growth of the ducts depends on stimulation by estrogens; development of the alveoli requires progesterone, estrogen, and other hormones. Lactation is maintained by a neurohormonal reflex established by suckling; prolactin and oxytocin are released by the pituitary, resulting in milk secretion and release.

19 Endocrine Glands

Endocrine glands lack ducts, and their secretory products (hormones) generally pass into the blood or lymphatic circulation. However, hormones also may be secreted directly into the intercellular space (paracrine secretion) to elicit a local effect on adjacent cells. In some instances cells may secrete a chemical messenger that acts on its own receptors and is self-regulatory (autocrine secretion). Hormones that are secreted into the vascular system eventually enter the tissue fluids and, depending on the specificity of the hormone, can alter the activities of just one organ or influence several. The organs that respond to a specific hormone are the target organs of that hormone. Hormones may be biogenic amines, steroids, polypeptides, proteins, or glycoproteins, depending on their gland of origin.

In some glands, such as the liver, the epithelial cells have exocrine and endocrine functions. Hepatocytes fulfill the definition of endocrine cells because they release glucose, proteins, and other substances directly into the hepatic sinusoids, but the same hepatocytes also secrete bile into adjacent canaliculi (the beginnings of the liver duct system) and thus are exocrine cells also. The pancreas, kidney, testis, and ovary are mixtures of endocrine and exocrine components in which separate or isolated groups of cells secrete directly into the vasculature and form the endocrine portion. The rest of the gland makes up the exocrine portion that usually is drained by ducts. The placenta is a unique, temporary endocrine gland that persists only for about 9 months in the pregnant human female. The brain and heart also have aspects of endocrine organs.

What are termed *classic endocrine glands* consist of discrete masses of cells with a relatively simple organization into clumps, cords, or plates supported by a delicate vascular connective tissue. Endocrine glands have a rich vascular supply, and the secreting cells have direct access to capillaries. The classic endocrine system consists, in part, of pineal, parathyroid, thyroid, adrenal, and pituitary glands.

A vast system of unicellular endocrine cells (glands) also exists, and collectively these cells form a diffuse endocrine system. It is represented by scattered unicellular glands that often lie at some distance from one another, separated by cells of a different type. These cells are present through much of the gastrointestinal tract, in the ducts of the major digestive glands, and in the conducting airways of the lungs. When the cells of this system are considered collectively, they form a mass larger than many of the individual organs of the classic endocrine system.

Pineal Gland

The human pineal gland is a somewhat flattened body 5 to 8 mm long and 3 to 5 mm wide. It is attached to the brain (roof of the diencephalon) by a short stalk that contains some nerve fibers and their supportive elements.

Structure

KEY WORDS: pinealocyte, glial cell, corpora arenacea (brain sand), melatonin, serotonin

The pineal is covered by a thin connective tissue capsule that is continuous with the surrounding meningeal (pial) tissue. Richly vascularized and innervated septa extend into the parenchyma of the gland and subdivide it into poorly defined lobules (Fig. 19-1). Two types of cells, pinealocytes and glial cells, are present and form clumps and cords.

Pinealocytes have relatively large lobulated nuclei. Long cytoplasmic processes often radiate from the cell body to form club-shaped endings near adjacent perivascular spaces or pinealocytes. Small, membrane-bound vesicles, some with electron-dense cores, are present in the swollen ends of the processes. Synaptic ribbons may be present also. Nearer the cell body, the cytoplasm contains abundant, fairly large mitochondria, numerous free ribosomes, profiles of smooth endoplasmic reticulum, lipid droplets, lipochrome pigment, lysosomes, and large numbers of microtubules and intermediate filaments. Numerous gap junctions link groups of pinealocytes electrically and metabolically. Although light and dark forms have been described, it is unknown whether each forms a distinct cell type or whether they represent differences in the activity of a single cell type.

Glial cells form an interwoven network within and around the parenchymal cords and clumps of the pineal gland. They are fewer in number than pinealocytes, and their

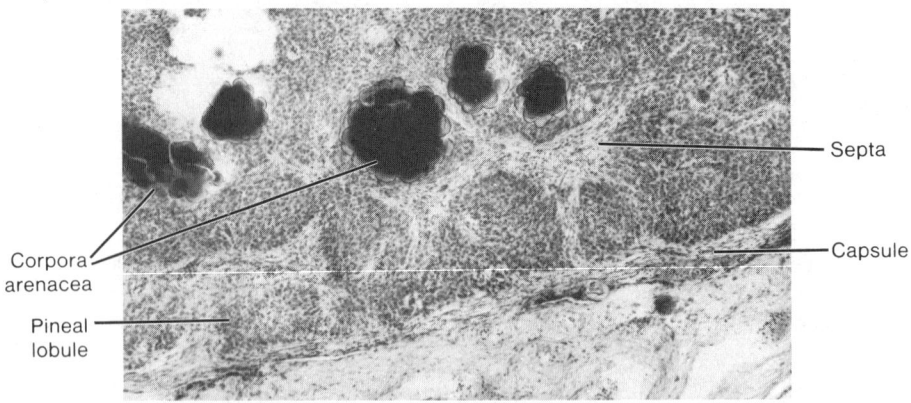

Fig. 19-1. A region of human pineal showing corpora arenacea and lobules (LM, × 400).

nuclei are smaller and stain more deeply. Glial cells show long cytoplasmic processes and often are regarded as a form of astrocyte. The cytoplasm contains many fine filaments, 5 to 6 nm in diameter, which may form large bundles.

Corpora arenacea (brain sand) are concretions found in the pineal gland of humans. These irregularly shaped structures occur in the capsule and substance of the gland. They consist mainly of calcium carbonates and phosphates within an organic matrix. In humans they increase with age, but their significance is unknown.

The pineal gland has an extensive blood supply and is provided with many postganglionic sympathetic nerve fibers from neurons in the superior cervical ganglia.

The pineal gland produces **melatonin,** an indolamine. In mammals such as the rat, pinealocytes synthesize **serotonin,** a precursor of melatonin that varies cyclicly (circadian and seasonal) in amount and changes in relation to the amount of light. Serotonin levels fall and melatonin levels rise during darkness. These circadian rhythms are controlled by the release of norepinephrine from the sympathetic fibers that enter the pineal from the superior cervical ganglia, which in turn is controlled by the light perceived by the retina. In some rodents and birds the pineal, through secretion of melatonin, plays an active role in the reproductive cycles, which are influenced by the length of daylight (photoperiod). The pineal in these species acts as a neuroendocrine organ, influencing the gonads in response to light. The function of melatonin in humans, however, is unknown. Pineal-specific peptides such as arginine vasotocin are thought to influence or mediate hypophyseal function, but their exact role is uncertain.

Parathyroid Glands

The parathyroid glands are small, brownish, oval bodies 4 to 8 mm long and 2 to 5 mm wide. There usually are four in humans, but as many as six or more may lie within the capsule of the thyroid gland or may be embedded in the substance of the middle third of the thyroid. In about 8 percent of the human population, additional parathyroid tissue is present in the thymus, the association being due to the common developmental origin of these structures from the third pharyngeal pouch.

Structure

KEY WORDS: chief (principal) cell, oxyphil cell, parathyroid hormone

Each parathyroid gland is surrounded by a thin connective tissue capsule and is subdivided by delicate trabeculae that extend from the capsule. Blood vessels, lymphatics, and nerves course through the trabeculae to enter the substance of the parathyroid. In older persons fat cells often are abundant and may form about 50 percent of the gland. The parenchyma consists of closely packed groups or cords of epithelial cells supported by a delicate framework of reticular fibers that contains nerve fibers and a rich capillary network.

The parenchyma contains two types of cells. **Chief (principal) cells** are the more numerous and measure 8 to 10 μm in diameter. They have round, centrally placed, vesicular nuclei, and the cytoplasm contains the usual organelles as well as large accumulations of glycogen and lipid droplets. In addition to small, dense, membrane-bound granules, lipofuscin granules often are present also.

Oxyphil cells form only a small part of the cell population and may occur singly or in groups (Fig. 19-2). They are not present in significant numbers until puberty and then increase with age. Oxyphil cells are larger than chief cells, and their cytoplasm stains intensely with eosin and is packed with large, elongated mitochondria that show numerous cristae. Between the mitochondria are small accumulations of glycogen and occasional profiles of granular endoplasmic reticulum. The small nuclei stain deeply and often appear pyknotic.

Numerous cells that are intermediate in appearance between chief and oxyphil cells also have been described. Chief cells may be the primary parenchymal element of the parathyroid gland, and other cell types (such as the oxyphil and intermediate cells) may represent a modification or a stage of development of the chief cell.

The parathyroid secretes **parathyroid hormone** (PTH), a polypeptide hormone that regulates the calcium level of the blood. Nearly half the blood calcium is bound; the remainder is present as free ions. It is the concentration of calcium ion that governs the secretion of parathyroid hormone. If the concentration drops below normal, the hormone is secreted and acts directly on osteoclasts and osteocytes to mobilize calcium ion from bone and on the renal tubules to promote absorption of calcium ion and inhibit absorption of phosphate ion from the glomerular filtrate. Parathyroid hormone also promotes synthesis and release of 1,25-vitamin D_3 by the proximal convoluted tubules of the kidney. This form of vitamin D_3 increases calcium absorption in the small intestine. Increase in the concentration of calcium ion in the blood results in a decrease in the amount of PTH released. Blood levels of calcium are kept from exceeding the optimum by a second calcium-regulating hormone, calcitonin, produced by the parafollicular cells of the thyroid gland.

Parathyroid hormone is essential for life. Complete removal of the parathyroid glands results in a precipitous drop in blood calcium followed by tetany and death.

Thyroid Gland

The thyroid gland normally weighs between 25 and 40 g in humans and is located in the anterior of the neck, below the cricoid cartilage. It consists of two lateral lobes and an isthmus that forms a narrow bridge across the trachea between the two lobes.

Structure

KEY WORDS: follicle, colloid, principal (follicular) cell, colloid resorption droplet, thyroglobulin, triiodothyronine, tetraiodothyronine (thyroxin), binding protein, parafollicular cell (light cell, C cell), calcitonin (thyrocalcitonin)

The parenchyma of the thyroid gland is enclosed in a connective tissue capsule and organized into spherical structures, the **follicles,** which vary considerably in diameter and contain a gelatinous material called **colloid** (Fig. 19-3). The walls of the follicles consist of a simple epithelium

Fig. 19-2. Human parathyroid exhibiting both chief and oxyphil cells (LM, × 100).

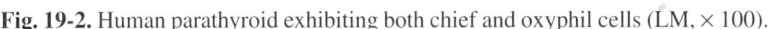

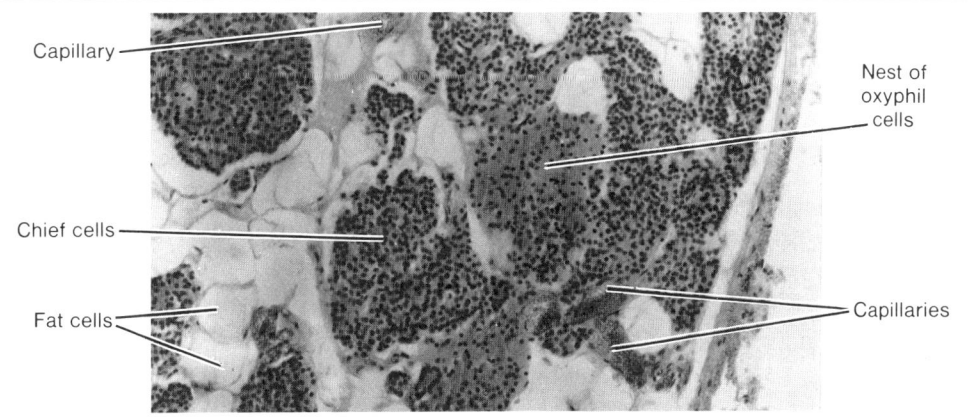

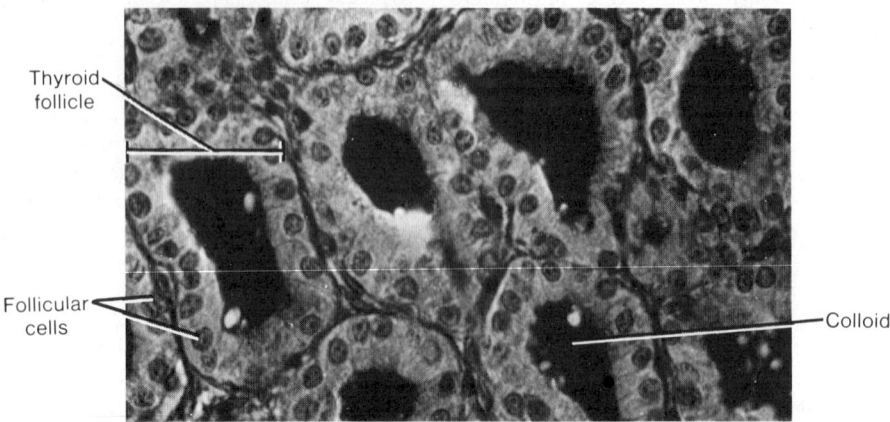

Fig. 19-3. Human thyroid. Note the colloid in each follicle (LM, × 100).

that rests on a thin basal lamina, about 50 nm thick. Each follicle is supported by a delicate reticular network that contains a vast capillary plexus, numerous nerve fibers, and blindly ending lymphatic vessels.

The follicular epithelium consists mainly of **principal (follicular) cells,** which usually are squamous to cuboidal in shape; depending on the functional status of the thyroid, the cells may become columnar. The nuclei are spherical, contain one or more nucleoli, and have a central position. The lateral cell membranes are united at the apex by junctional complexes, and the luminal surfaces bear short microvilli. The basal plasmalemma is smooth, with no infoldings. Mitochondria are evenly distributed throughout the cytoplasm and vary in number according to the activity of the cell. Active cells take on a cuboidal shape and show many profiles of granular endoplasmic reticulum. Inactive cells are squamous, and granular endoplasmic reticulum is sparse. Golgi complexes usually have a supranuclear location in active cells, and the apical cytoplasm contains numerous small vesicles, lysosomes, and multivesicular bodies. **Colloid resorption droplets** also are found in the apical cytoplasm (Figs. 19-4 and 19-5).

The thyroid is a unique endocrine gland in that the secretory product, colloid, is stored extracellularly in the lumen of the follicle. Colloid consists of mucoproteins, proteolytic enzymes, and a glycoprotein called **thyroglobulin,** the primary storage form of thyroid hormone. Synthesis of thyroglobulin occurs in the principal cells along the same basic intracellular pathway as glycoprotein in cells elsewhere in the body. Amino acids are synthesized into polypeptides in the granular endoplasmic reticulum and then carried in transport vesicles to the Golgi complex,

where the carbohydrate moiety is conjugated to the protein. From the Golgi complex, the glycoprotein (noniodinated thyroglobulin) is transported to the apical surface in small vesicles, from which it is discharged by exocytosis into the follicular lumen and stored as part of the colloid.

While thyroglobulin is being synthesized, thyroperoxidase is assembled in the granular endoplasmic reticulum and then passes through the Golgi complex and is released by small vesicles at the apical surface of the cells. Follicular cells have a unique ability to take up iodine from the blood and concentrate it. The iodine subsequently is oxidized to iodide by intracellular peroxidase and used in the iodination of tyrosine groups in thyroglobulin. Formation of monoiodotyrosine and diiodotyrosine is thought to occur in the follicle, immediately adjacent to the microvillus border of the follicular cells. When one molecule of monoiodotyrosine is linked to one of diiodotyrosine, a molecule of **triiodothyronine** is formed. Coupling of two molecules of diiodotyrosine results in the formation of **tetraiodothyronine (thyroxin).** The thyronines make up a small part of the thyroglobulin complex but represent the only constituents with hormonal activity. Thyroglobulin and the thyronines are stored in the lumen of the follicle until needed.

When thyroid hormone is required, droplets of colloid are sequestered from the follicular lumen by endocytosis and enter the apical cytoplasm of the follicular cells. Lysosomes coalesce with the vacuoles and hydrolyze the contained thyroglobulin, liberating mono-, di-, tri-, and tetraiodothyronine into the cytoplasmic matrix. The mono- and diiodotyrosines are deiodinated, and the iodine is reused by the cell. Thyroxin and triiodothyronine mole-

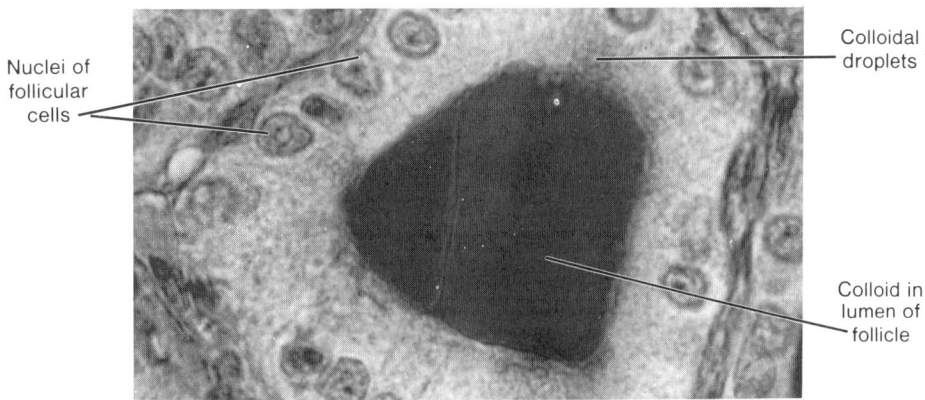

Nuclei of follicular cells

Colloidal droplets

Colloid in lumen of follicle

Fig. 19-4. Follicular cells of a human thyroid follicle, some of which contain colloidal resorption droplets (LM, × 400).

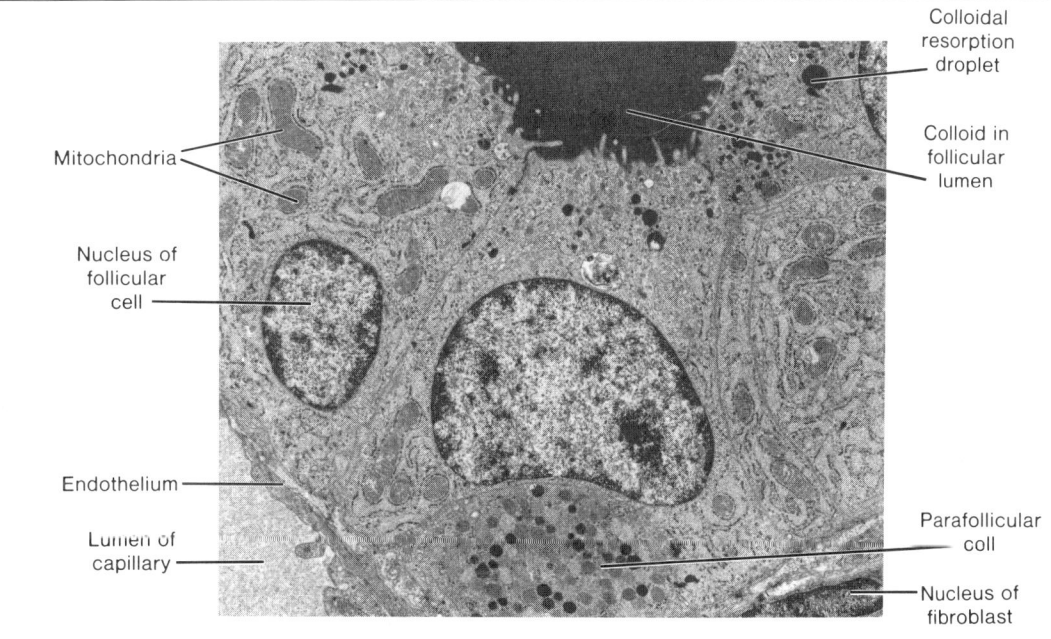

Colloidal resorption droplet

Mitochondria

Colloid in follicular lumen

Nucleus of follicular cell

Endothelium

Lumen of capillary

Parafollicular coll

Nucleus of fibroblast

Fig. 19-5. Follicular cells and a portion of a parafollicular cell in a thyroid follicle (LM, × 3,500).

cules are the thyroid hormone and are released at the base of the cell into blood and lymphatic capillaries. Thyroxin is transported in the blood plasma complexed to a binding protein. Triiodothyronine, which hormonally is the more potent of the two but is not as abundant, is not as firmly bound to the protein.

Synthesis of thyroid hormone is shown in Figure 19-6.

Binding protein is important in regulating the amount of free hormone that circulates in the blood plasma. Thyroxin and triiodothyronine are released from the binding protein only in response to metabolic needs, thus preventing rapid fluctuations in the levels of these hormones. As

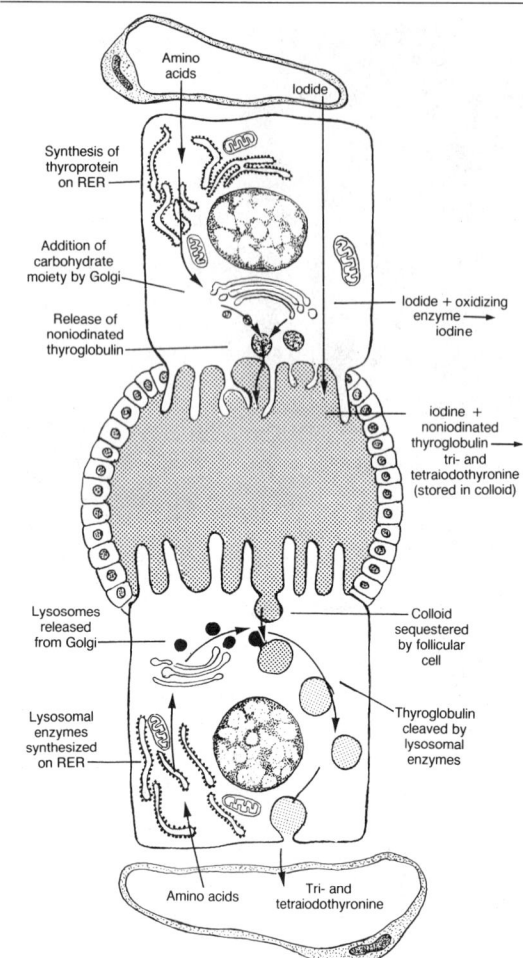

Fig. 19-6. Synthesis of thyroid hormone.

in the follicular epithelium appear to be sandwiched between follicular cells and lie immediately adjacent to the basal lamina; parafollicular cells never directly border on the lumen of the follicle. Between follicles, the cells may occur singly or in small groups. In electron micrographs, the parafollicular cells show numerous moderately dense, membrane-bound secretory granules that measure 10 to 50 nm in diameter. The cytoplasm also contains occasional profiles of granular endoplasmic reticulum, scattered mitochondria, and poorly developed Golgi complexes.

Parafollicular cells secrete **calcitonin (thyrocalcitonin),** another polypeptide hormone that regulates blood calcium levels. Calcitonin lowers blood calcium by acting on osteocytes and osteoclasts to suppress resorption of calcium from bone and its release into the blood. Thus calcitonin has an effect opposite that of PTH, serves to control the action of PTH, and helps regulate the upper levels of calcium concentration in the blood.

Adrenal Gland

The adrenal glands in humans are a pair of flattened, triangular structures with a combined weight of 1 to 2 g. One adrenal is situated at the upper pole of each kidney. The adrenal is a complex organ consisting of a cortex and medulla, each differing in function, structure, and embryonic origin.

Adrenal Cortex

KEY WORDS: zona glomerulosa, zona fasciculata, zona reticularis, mineralocorticoid, glucocorticoid

The adrenal gland is surrounded by a thick capsule of dense irregular connective tissue that contains scattered elastic fibers. The capsule contains a rich plexus of blood vessels — mainly small arteries — and numerous nerve fibers. Some blood vessels and nerves enter the substance of the gland in the trabeculae that extend inward from the capsule and then leave the trabeculae to enter the cortex.

The parenchyma of the adrenal cortex consists of continuous cords of secretory cells that extend from the capsule to the medulla, separated by blood sinusoids. The cortex is subdivided into three layers according to the arrangement of the cells within the cords. These layers consist of an outer zona glomerulosa, a middle zona fasciculata, and an inner zona reticularis (Fig. 19-7). The cytologic changes from one zone to the next are gradual.

Zona glomerulosa forms a narrow band just beneath the capsule. The columnar cells are arranged into ovoid

they are released and elicit their effects, the hormones are replaced by other thyroxin and triiodothyronine molecules from the thyroid. The activity of the thyroid is regulated by a thyroid-stimulating hormone (TSH) secreted by the anterior lobe of the pituitary.

Thyroid hormone has general effects on the metabolic rate of most tissues, and among its functions are influencing carbohydrate metabolism, the rate of intestinal absorption, heart rate, body growth, and mental activities.

The thyroid also contains a smaller number of cells variously called **parafollicular, light,** or **C cells,** which are present in the follicular epithelium and in the delicate connective tissue between follicles (see Fig. 19-5). The C cells

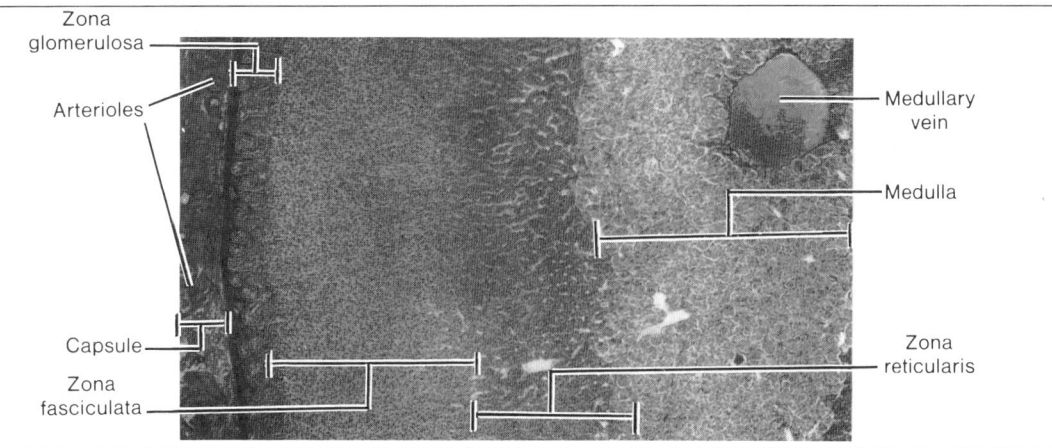

Fig. 19-7. Human adrenal showing the three zones of the cortex as well as the medulla (LM, × 40).

groups or arcades and have centrally placed, spherical nuclei. In electron micrographs, the cells show a well-developed smooth endoplasmic reticulum and numerous mitochondria that are evenly distributed throughout the cytoplasm. Occasional lipid droplets and scattered profiles of granular endoplasmic reticulum also are present.

Zona fasciculata forms the widest zone of the cortex and consists of long cords that usually are one or two cells thick. The cords run parallel to one another, separated by sinusoids that are lined by an attenuated endothelium. The cells of zona fasciculata are larger than those of the other two zones and have a polyhedral shape, and their spherical nuclei are centrally placed; binucleate cells are common. The cytoplasm contains numerous rounded mitochondria with tubular cristae, abundant smooth endoplasmic reticulum, and well-developed granular endoplasmic reticulum. Many lipid droplets also are present and contain neutral fat, fatty acids, and fatty acyl esters of cholesterol; these represent stored precursors for the synthesis of the steroid hormones secreted by the zona fasciculata. Short microvilli often are present on the plasmalemma adjacent to sinusoids.

Zona reticularis forms the innermost zone and is made up of a network of irregular, anastomosing cords that also are separated by sinusoids. In general, the cells of this zone resemble those of zona fasciculata except that they are smaller, the cytoplasm contains fewer fat droplets, the nuclei stain more deeply, and lipofuscin granules are prominent. Light and dark forms of the cells have been described, but their significance is not known.

The adrenal cortex synthesizes and secretes various steroid hormones, which generally can be placed in two broad categories: mineralocorticoids that control water and electrolyte balance and glucocorticoids that affect carbohydrate metabolism. Both are derived from cholesterol. Enzymes for the synthesis of the hormones are located mainly in the smooth endoplasmic reticulum and mitochondria of the adrenal cortical cells. The pathway involves transfer of precursor molecules and intermediate products back and forth between the two organelles. Adjacent lipid droplets provide the substrates needed for the biosynthetic processes.

The **mineralocorticoids** (aldosterone and deoxycorticosterone) are secreted primarily by zona glomerulosa. Aldosterone acts on the distal tubules of the kidney to increase the rate of sodium absorption from the glomerular filtrate and, at the same time, to increase potassium excretion. It also lowers the concentration of sodium in the secretions of sweat and salivary glands and intestinal mucosa. Aldosterone secretion is controlled primarily by the renin-angiotensin system, which is sensitive to changes in blood pressure and to the concentration of sodium and potassium in the blood plasma. The exact mechanism is unknown, but it has been suggested that the juxtaglomerular apparatus of the kidney receives signals created by decreased arterial renal blood pressure (which reduces the degree of stretch on the juxtaglomerular cells) or by a decrease in the amount of sodium detected by the macula densa. In response to the signal, juxtaglomerular cells release renin, which, together with a converting enzyme, ultimately transforms angiotensinogen to angiotensin II. The latter stimulates the zona glomerulosa to secrete aldosterone. Angiotensin II also is a vasoconstrictor and directly increases blood pressure; aldosterone increases blood pressure by increasing the sodium ion concentration. Both agents influence juxtaglomerular cells and the release of renin. Adrenocorticotropic hormone from the pituitary may have a minor role in the activity of the zona glomerulosa.

Glucocorticoids (cortisol, cortisone, and corticosterone) are secreted by zona fasciculata and zona reticularis. The hormones act mainly on the metabolism of fats, proteins, and carbohydrates, resulting in an increase in blood glucose and amino acid levels and in the movement of lipid into and out of fat cells. Cortisol also suppresses the inflammatory response and some allergic reactions. A weak androgen called *dehydroepiandrosterone* also is secreted by cells of the zona fasciculata and zona reticularis, although cells of the reticularis are said to be the more active in androgen secretion.

Secretion by zona fasciculata and zona reticularis is controlled by adrenocorticotropic hormone (ACTH), a polypeptide hormone secreted by the anterior pituitary. Adrenocorticotropic hormone stimulates the synthesis and release of steroids, increases blood flow through the cortex, and promotes growth of the two inner zones of the adrenal cortex. The adrenal cortex is essential for life.

Adrenal Medulla

KEY WORDS: chromaffin cell, catecholamine, epinephrine, norepinephrine

The adrenal medulla is composed of large round or polyhedral cells arranged in clumps or short cords (see Fig. 19-7). These are the **chromaffin cells,** so named because they show numerous brown granules when treated with chromium salts (the chromaffin reaction). The parenchyma is supported by a framework of reticular fibers, and this stroma contains numerous capillaries, veins, and nerve fibers. Sympathetic ganglion cells are present also and may occur singly or in small groups. Chromaffin cells secrete **catecholamines.** By special histochemical means, two types of chromaffin cells have been identified, one containing **epinephrine** and the other **norepinephrine.**

Ultrastructurally, chromaffin cells are characterized by numerous electron-dense granules, 100 to 300 nm in diameter, that are limited by a membrane. The granules from cells that secrete norepinephrine have intensely electron-dense cores, whereas cells that elaborate epinephrine have homogeneous, less dense secretory granules. Both cell types show profiles of granular endoplasmic reticulum, scattered mitochondria, and well-developed Golgi complexes that lie close to the nucleus.

Although not essential for life, hormones of the adrenal medulla help the organism meet stressful situations. Epinephrine increases cardiac output, elevates blood glucose, and increases the basal metabolic rate. Norepinephrine acts primarily to elevate and maintain blood pressure by causing vasoconstriction in the peripheral segments of the arterial system. Secretion of both hormones is controlled by the sympathetic nervous system.

Blood Supply

The rich plexus of small arteries in the capsule of the adrenal provides cortical arteries that enter the parenchyma of the adrenal to empty into the vast network of cortical sinusoids that surrounds the cords of epithelial cells. The attenuated endothelium of the sinusoids has numerous fenestrations and is supported only by a thin basal lamina and a delicate network of reticular fibers. The sinusoids pass through all three layers of the cortex and near the corticomedullary junction begin to merge to form large collecting veins. These drain the medulla and finally empty into a single, large suprarenal vein that drains the entire gland. The cortex has no separate venous drainage. Some arteries enter the cortex in the trabeculae and provide a direct arterial supply to the medulla. Hence the medulla receives blood from cortical sinusoids and medullary arteries. The capillaries that surround the medullary cords also are lined by a thin, fenestrated endothelium.

In addition to their systemic effects, adrenocorticosteroids (especially glucocorticoids) may affect the adrenal medulla, and the steroid-rich blood from the cortex may influence the secretion of epinephrine or norepinephrine by the chromaffin cells. Some experimental evidence indicates that glucocorticoids induce an enzyme that converts norepinephrine to epinephrine. In humans, epinephrine is the major catecholamine secreted.

Paraganglia

KEY WORDS: chief cell, norepinephrine, supporting cell

Clusters of chromaffin cells enclosed in thick capsules of dense irregular connective tissue normally occur outside the adrenal medulla associated with the sympathetic nervous system. These form the paraganglia, which are highly vascularized structures that contain two types of cells:

chief and supporting cells. **Chief cells** contain numerous membrane-bound, electron-dense granules that are similar to those of chromaffin cells of the adrenal medulla. They are believed to secrete **norepinephrine.** The chief cells are surrounded, in part or completely, by elongated **supporting cells** that lack secretory granules. Some paraganglia, such as aortic chromaffin bodies, are paired and quite large. Paraganglia can be important clinically if they secrete abnormally high amounts of catecholamine.

Hypophysis

The hypophysis or pituitary is a complex endocrine gland located at the base of the brain, lying in the sella turcica, a small depression in the sphenoid bone. It is attached to the hypothalamic region of the brain by a narrow stalk and has vascular and neural connections with the brain. The gland weighs about 0.5 g in the adult but is slightly heavier in females and in multiparous women may exceed 1 g in weight. Despite its small size, the pituitary gland produces several hormones that directly affect other endocrine glands and tissues.

Organization

KEY WORDS: adenohypophysis, neurohypophysis, pars distalis (anterior lobe), pars intermedia, pars tuberalis, pars nervosa, infundibular stalk, median eminence

The hypophysis consists of two major parts, an epithelial component called the **adenohypophysis** and a nervous component that forms the **neurohypophysis.** The adeno-hypophysis is further subdivided into the **pars distalis (anterior lobe)** that lies anterior to the residual lumen of Rathke's pouch; **pars intermedia,** which forms a thin partition behind the residual lumen; and **pars tuberalis,** which is an extension of pars distalis surrounding the neural stalk (Fig. 19-8).

The neurohypophysis also contains three portions. The major part is **pars nervosa,** which lies just posterior to pars intermedia and is continuous with the **infundibular stalk** and the **median eminence** (Fig. 19-9). Pars intermedia of the adenohypophysis and pars nervosa of the neurohypophysis often are regarded collectively as forming a posterior lobe. The hypophysis is surrounded by a thick connective tissue capsule.

Vascular Supply

KEY WORDS: superior, middle, and inferior hypophyseal arteries; common capillary bed

The pituitary is supplied by three pairs of arteries: the **superior, middle,** and **inferior hypophyseal arteries,** all of which supply directly to the neurohypophysis. The middle hypophyseal artery previously was believed to traverse the parenchyma of the adenohypophysis, but other evidence has shown that it too passes directly to the neurohypophysis. Thus the adenohypophysis has no direct arterial blood supply but is linked to the neurohypophysis by a **common capillary bed** that extends throughout all three regions of the hypophysis and connects with the capillary bed of the hypothalamus. Large-bore, thin-walled veins (portal veins) also have been described and play a role in forming a link between the capillary plexus associated with

Fig. 19-8. A sagittal section of human hypophysis (LM, × 20).

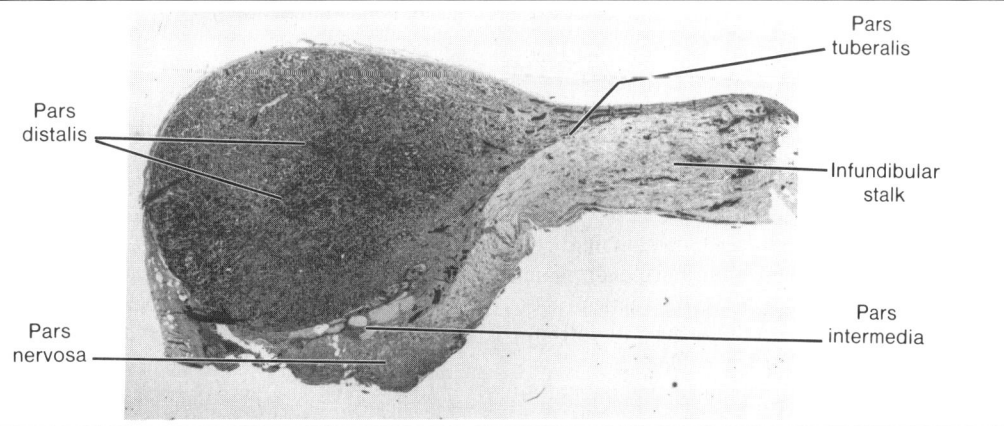

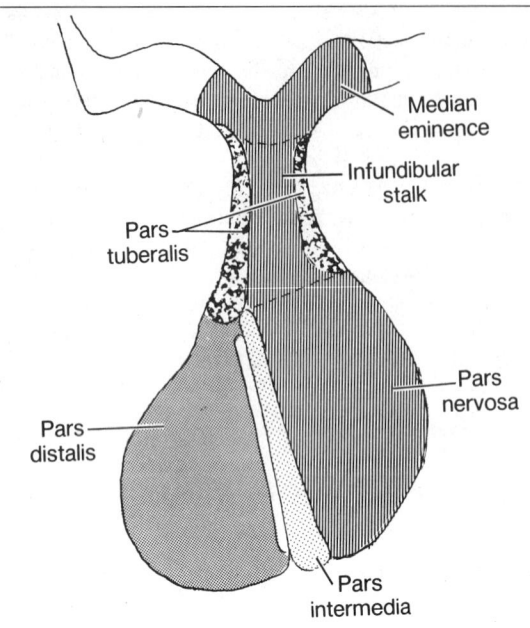

Fig. 19-9. Subdivisions of the hypophysis and their relationships.

the median eminence and infundibular stalk and that of the pars distalis.

Venous drainage occurs through confluent pituitary veins that carry blood from the adenohypophysis, pars intermedia, and neurohypophysis through a common trunk to the venous systemic circulation. Only a few lateral hypophyseal veins extend directly from the adenohypophysis to the cavernous sinus. The only direct drainage of the neurohypophysis is by the neurohypophyseal limbs of the confluent veins located at the lower end of the neurohypophysis. Pars intermedia is relatively avascular and few capillaries traverse it. The most prominent vessels of pars intermedia are confluent pituitary veins.

Pars Distalis (Anterior Lobe)

KEY WORDS: **chromophilic cell, acidophilic cell, somatotrope, somatotropin (growth hormone), mammotrope, prolactin, basophilic cell, corticotropes, adrenocorticotropic hormone, thyrotrope, thyrotropin (thyroid-stimulating hormone), gonadotrope, follicle-stimulating hormone, luteinizing hormone, interstitial cell–stimulating hormone, chromophobe**

Pars distalis makes up about 75 percent of the hypophysis and primarily consists of epithelial cells arranged in clumps or irregular cords (Fig. 19-10). The parenchyma and surrounding sinusoids are supported by a delicate network of reticular fibers. The parenchymal elements mostly consist of chromophobic and chromphilic cells, distinguished by whether or not they take up stain. **Chromophilic cells** are further subdivided into acidophilic and basophilic cells according to the staining properties of their secretory granules. All three cell types — chromophobes, acidophils, and basophils — are present in pars distalis.

Acidophilic cells are large, round or ovoid cells, 14 to 19 μm in diameter, with well-developed juxtanuclear Golgi bodies. The secretory granules stain with eosin in standard preparations. Two types of acidophils are present: somatotropes and mammotropes. The cytoplasm of **somatotropes** contains a well-developed granular endoplasmic reticulum and numerous spherical, electron-dense secretory granules that measure 350 to 400 nm in diameter. Somatotropes produce **somatotropin (growth hormone),** a peptide hormone that is important in controlling body growth and acts primarily on epiphyseal cartilage. The second type of acidophil, the **mammotrope,** tends to be more scattered within the cellular cords. Active cells contain granular endoplasmic reticulum, Golgi complexes, and scattered lysosomes. The cytoplasm also contains irregular granules that measure 550 to 615 nm in diameter. Mammotropes secrete **prolactin,** which promotes mammary gland development and lactation. The concentration of prolactin rises progressively during pregnancy but is kept from showing its lactogenic effect by the high concentration of progesterone and estrogen. The titer of these hormones decreases after birth, and the lactogenic properties of prolactin are expressed.

Basophilic cells of pars distalis consist of three types of cells: corticotropes, thyrotropes, and gonadotropes. **Corticotropes** generally are round or ovoid and larger than most acidophilic cells. They are most numerous in the central anterior portion of pars distalis. In humans, the cells contain numerous secretory granules, 350 to 400 nm in diameter, that are similar in appearance to the granules of somatotropes. The cytoplasm often contains lipid droplets and small filaments ranging from 6 to 8 nm in diameter. Corticotropes synthesize a glycoprotein prohormone, protropin, that is then cleaved to form ACTH, beta-lipotropic hormone (LPH), melanocyte-stimulating hormone (alpha-MSH), and beta-endorphin. The different hormones produced by acidophils and basophils in the adenohypophysis also are produced as prohormones, but how these are cleaved to form the active hormone is not understood.

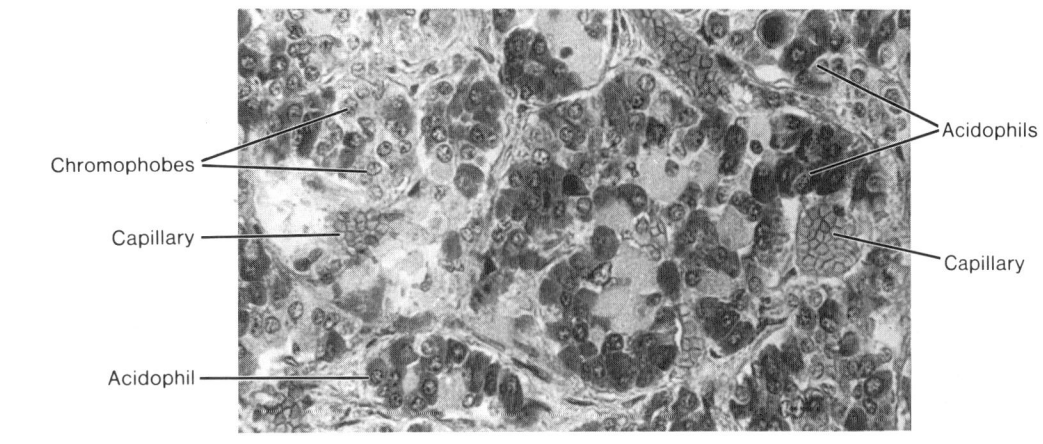

Fig. 19-10. Epithelial cells of human pars distalis (LM, × 250).

Adrenocorticotropic hormone stimulates the zona fasciculata and zona reticularis of the adrenal to produce glucocorticoids.

The **thyrotrope** type of basophil often is angular in shape and may form small groups deep in the parenchymal cords, at some distance from the sinusoids. At 130 to 150 nm in diameter, their secretory granules are the smallest of any granules in the parenchymal cells of the hypophysis. Thyrotropes secrete a glycoprotein, **thyrotropin (thyroid-stimulating hormone),** which stimulates the thyroid to produce and release thyroid hormone.

Gonadotropes are the third type of basophil cells in the pars distalis. They are generally smaller than corticotropic basophils, often show only a scattering of granules, and usually are round in shape. They often lie immediately adjacent to sinusoids. The cytoplasm shows profiles of granular endoplasmic reticulum, prominent Golgi complexes, and scattered lysosomes. The few spherical secretory granules are about 150 nm in diameter. This type of basophil secretes two glycoprotein hormones. **Follicle-stimulating hormone** (FSH) stimulates growth of ovarian follicles in the female and, in the male, stimulates synthesis of androgen-binding protein by Sertoli cells in the seminiferous epithelium, thereby activating and promoting spermatogenesis. **Luteinizing hormone** (LH) is essential for ovulation, for stimulating secretion of estrogen by ovarian follicles, and for promoting luteinization of follicles following development induced by FSH. The male equivalent of luteinizing hormone, **interstitial cell–stimulating hormone** (ICSH),

is secreted by this type of basophil also. It stimulates production of testosterone by the interstitial cells of the testis. Testosterone is essential for sperm maturation and for development and maintenance of secondary sex organs and characteristics.

The remaining general cell type, the **chromophobes,** usually are small and are confined to the interior of the parenchymal cords. Lacking granules, they fail to color with routine stains, hence their name. Their existence as a separate type has been debated, and they may represent a reservoir of cells from which chromophilic cells originate. This accords with the hypothesis that cells of the pars distalis cycle, first accumulating secretory granules and then releasing them.

Occasionally, cystlike follicles, lined by nonsecretory (follicular) cells and filled with a colloid type of material, may be seen in the pars distalis. Microvilli and occasional tufts of cilia extend from the apices of the cells into the follicular lumen. The follicular cells are unusual in that they also have long basal processes that extend between adjacent secretory cells. Similar cells called *folliculostellate cells* are not organized into follicles and extend long branching processes between secretory cells. The processes of folliculostellate cells, which form a network, are united by gap junctions. Their cytoplasm contains intermediate filaments (gliofibrillary acid protein) and suggests a function similar to that of glial cells in the central nervous system, i.e., providing a supporting framework and monitoring a microenvironment for the secretory cells.

Pars Tuberalis

Pars tuberalis forms a sleeve, up to 60 μm thick, around the infundibular stalk. It is thickest anteriorly and may be incomplete posteriorly. Pars tuberalis is characterized by longitudinally arranged cords of parenchymal cells separated by sinusoids. The parenchyma of pars tuberalis is continuous with that of pars distalis and consists of acidophils, basophils, and undifferentiated cells. The latter usually are columnar, and their cytoplasm shows numerous small granules and large amounts of glycogen. Nests of squamous cells may be present. Pars tuberalis has no known distinct hormonal function.

Pars Intermedia

KEY WORDS: **chromophobe, basophil, melanocyte-stimulating hormone**

Pars intermedia is rudimentary in humans, in whom it forms only about 2 percent of the hypophysis. It consists of **chromophobe** and **basophil cells,** the latter often encroaching into pars nervosa. Colloid-filled cysts lined by chromophobic or basophilic cells also may be present, and remnants of Rathke's cleft may persist as fluid-filled cysts lined by ciliated columnar epithelium (Fig. 19-11). Pars intermedia in some species is responsible for the secretion of **melanocyte-stimulating hormone** (MSH), a polypeptide hormone. Its role in mammals is unknown, but it may

be involved in melanocyte production. Only the precursor molecule for MSH occurs in humans. MSH contains amino acid sequences that are identical to portions of the ACTH molecule, and the occurrence of these structurally similar regions is believed to account for the melanocyte-stimulating effect of ACTH.

Control of the Adenohypophysis

KEY WORDS: **hypothalamus, releasing hormone, inhibitory hormone, somatostatin**

Secretions from target endocrine glands that are under control of the adenohypophysis influence cells in the adenohypophysis through regulatory neurons in the median eminence of the **hypothalamus.** These neurons produce **releasing hormones,** small peptides that enter the capillaries surrounding the infundibular stalk and then are transported to the capillaries of pars distalis, where they stimulate specific cells to release hormones. Releasing factors are known for somatotropin, thyrotropin, corticotropin, and gonadotropin, and their amino acid sequences have been determined. **Inhibitory hormones** such as **somatostatin** (a peptide of 14 amino acids) also are secreted by hypothalamic neurons. Somatostatin inhibits secretion of growth and thyrotropic hormones. Inhibitory hormones affecting the remaining cell types are said to be produced by hypothalamic neurons, but their biochemical structure and nature are unknown.

Fig. 19-11. Cysts of human pars intermedia (LM, × 100).

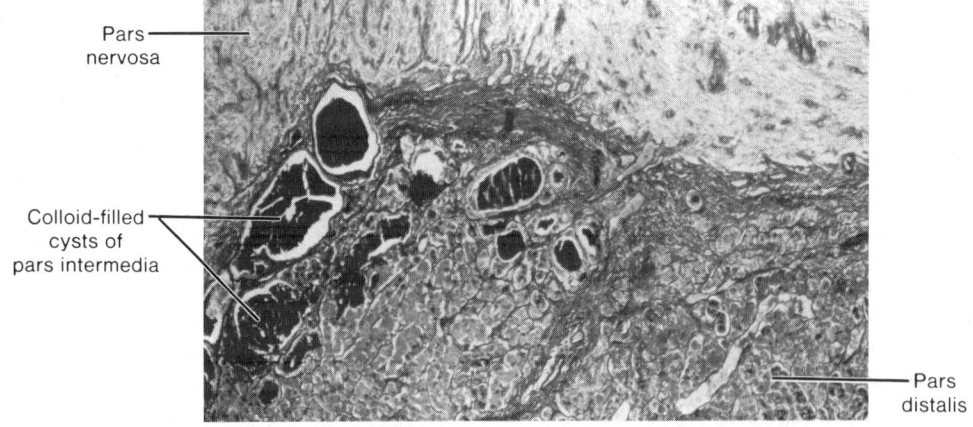

Pars nervosa

Colloid-filled cysts of pars intermedia

Pars distalis

Neurohypophysis

KEY WORDS: **hypothalamohypophyseal tract, Herring bodies, pituicytes, neurophysin, oxytocin, antidiuretic hormone**

Macroscopically, the neurohypophysis consists of the median eminence, infundibular stalk, and infundibular process (pars nervosa). The greater part of the neurohypophysis consists of unmyelinated nerve fibers of the **hypothalamohypophyseal tract,** which originates mainly from neurons in the supraoptic and paraventricular nuclei. The tract receives additional nerve fibers from other hypothalamic regions. The fibers end blindly in the pars nervosa, close to a rich capillary plexus. The axons comprising the hypothalamohypophyseal tract commonly contain spherical masses of secretory material that vary considerably in size (Fig. 19-12). These are the **Herring bodies,** which consist of large accumulations of dense secretory granules that vary from 120 to 200 nm in diameter. Hormones are synthesized in the perikarya of neurons in the supraoptic and paraventricular nuclei. They then pass down the axons to be stored in the nerve terminals that form the pars nervosa. When released, the hormones cross a thin, fenestrated endothelium to enter the capillaries in the pars nervosa and thence into hypophyseal veins to finally enter the systemic circulation.

Scattered among the nerve fibers are cells called **pituicytes,** which vary in size and shape and may contain pigment granules. Pituicytes are considered to be the equivalent of neuroglial cells of the central nervous system, but whether they have only a supportive function or actively participate in the secretory processes of adjacent nerve terminals is not known.

Two cyclic polypeptide hormones, oxytocin and antidiuretic hormone (vasopressin), are stored and released from the pars nervosa. During axonal transport, each hormone is bound to a carrier protein called **neurophysin,** and it is the neurophysin-hormone complex that forms the major part of the Herring bodies. **Oxytocin** is released in lactating mammals by means of a neuronal reflex that is initiated during suckling and transmitted to the hypothalamus through the cerebral cortex. Oxytocin causes contraction of myoepithelial cells around the alveoli of the mammary gland and helps to express milk into the ductal system. Large exogenous doses of oxytocin enhance the contraction of uterine muscles during parturition, but it is not certain that this is a normal, physiologic function. **Antidiuretic hormone** (ADH) acts primarily on the collecting tubules of the kidney to increase their permeability to water, thus promoting its resorption from the glomerular filtrate. The perikarya of neurons in the supraoptic and paraventricular nuclei are thought to act as osmoreceptors and to secrete in response to an increase in the osmolarity of the body fluids. Pharmacologic doses of ADH cause contraction of vascular smooth muscle and elevate blood pressure. This hormone is sometimes called vasopressin.

Fig. 19-12. Herring body in the infundibular stem of a human neurohypophysis (LM, × 250).

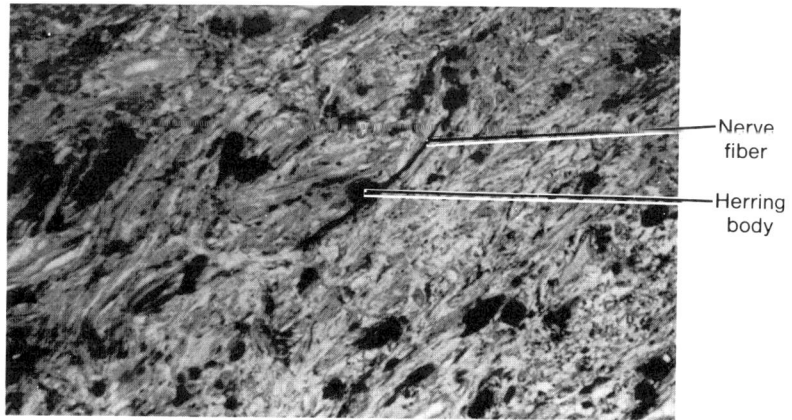

Nerve fiber

Herring body

Review Table 19-1. Key Features of Classic Endocrine Glands

	Arrangement of Cells	Cell Types	Other Features
Neurohypophysis	Vary in size; arranged around unmyelinated nerve fibers	Pituicytes	Herring bodies
Adenohypophysis	Irregular cords and plates	Basophils, acidophils, chromophobes	Numerous capillaries between cords
Parathyroids	Irregular cords and plates	Chief cells (majority), oxyphils; occur singly or in nests	Abundant fat in interlobular connective tissue
Thyroid	Follicles	Follicular cells, parafollicular cells; single (adjacent to follicles) or in small interfollicular nests	Colloid in follicular lumen
Adrenal cortex			
Zona glomerulosa	Ovoid groups or arcades	Columnar cells, occasional lipid droplets	Near capsule
Zona fasciculata	Long cords	Polyhedral cells, numerous lipid droplets	Cords of cells separated by sinusoids
Zona reticularis	Irregular, anastomosing cords	Polyhedral cells, few lipid droplets	Cords of cells separated by sinusoids
Adrenal medulla	Irregular clumps or short cords	Large, pale-staining chromaffin cells	Ganglion cells, "sympathetic neurons"; sinusoids
Pineal	Poorly defined lobules, clumps, and cords	Pinealocytes, glial cells	Corpora arenacea

Functional Review: Endocrine Glands

The parenchymal cells of the endocrine glands synthesize hormones that may regulate specific tissues or have more general, systemic effects. Endocrine glands control and coordinate many of the physiologic activities of the body and often act in concert with the nervous system. Neurons often perform endocrine functions by secreting peptides directly into the bloodstream. Antidiuretic hormone and oxytocin are synthesized by neurons in the supraoptic and paraventricular nuclei of the hypothalamus, travel down their respective axons, and are released into the vasculature of the pars nervosa. From here they enter the hypophyseal veins and make their way into the general circulation. Antidiuretic hormone promotes absorption of water from the glomerular filtrate, thus conserving body water and concentrating the urine. The stimulus for secretion of ADH is an increase in the osmolality of the blood plasma. The cell bodies of neurons in the paraventricular and supraoptic nuclei are thought to act as osmoreceptors and to secrete ADH in response to this stimulus, as well as to stimuli from other pressure-sensitive regions of the vascular system. Oxytocin, which is released in response to the suckling reflex, stimulates contraction of myoepithelial cells and helps to express milk from the secretory units into the ductal system of the mammary gland.

Other hypothalamic neurons act on mediator cells, the gonadotropes of the adenohypophysis, and through them control follicular growth, ovulation, and hormone production in the female and spermatogenesis and hormone production in the male. Several hypothalamic neurons, acting through releasing hormones, influence somatotropes and thyrotropes of the pars distalis to control the levels of growth hormone and thyroid hormone, which control body growth and metabolism, respectively.

Adrenocorticotropic hormone elaborated by corticotropes of the pars distalis stimulates the zona fasciculata and zona reticularis of the adrenal cortex to secrete and release glucocorticoids. Like other epithelial cells in the pars distalis, corticotropes are controlled by inhibitory and releasing hormones secreted by hypothalamic neurons.

Unlike most of the classic endocrine glands, the chromaffin cells of the adrenal medulla are under the direct control of the sympathetic nervous system. When stimulated to secrete, the chromaffin cells release two active catecholamines — epinephrine and norepinephrine — which, although not essential for life, are important in meeting stressful "fight or flight" situations by increasing cardiac output, metabolic rate, and blood pressure.

Other classic endocrine glands are not under direct nervous stimulation or control by neural hormones. The activity of the parathyroid gland, for example, is influenced primarily by the concentration of calcium ion in the circulating blood. If the concentration falls below a specific level, PTH is secreted and acts on the following: osteocytes and osteoclasts to promote release of calcium ion from bone, kidney tubules to increase calcium absorption from the glomerular filtrate, and proximal convoluted tubules to release 1,25-vitamin D_3, which increases calcium absorption in the small intestine. Blood calcium is kept from rising above optimum levels by the hormone calcitonin, which is released by the parafollicular cells of the thyroid gland. Calcitonin counters the action of PTH by suppressing the resorptive activities of osteoclasts and osteocytes. Thus regulation of the concentration of calcium ion in the blood is controlled by two endocrine glands that act in opposition to one another.

Parathyroid hormone is the major hormone regulating the excretion of phosphate and retention of calcium ion by the kidney, but it also controls the synthesis of 1,25-dihydroxyvitamin D_3 by cells in the proximal tubules of the kidney. 1,25-Dihydroxyvitamin D_3 is the hormonal form of vitamin D_3, which is now considered to be a prohormone. Like vitamin D_3, 1,25-dihydroxyvitamin D_3 is lipid-soluble, and its mechanism of action is similar to that of steroid hormones. In the proximal small intestine, it induces an increase in mucosal weight and the length of villi and microvilli and accelerates the rate of mucosal turnover. It also is involved in regulating calcium stores in the skeleton. Receptors for the hormone occur in a variety of tissues, including differentiating cells of the immune system.

The hypothalamus is the major neuroendocrine regulatory region of the brain and is restricted by the blood-brain barrier, whereas the neurohypophysis is not. This may explain why oxytocin, ADH, and the releasing and inhibitory hormones released by neurons in the hypothalamic region must travel considerable distances along axons before being released into capillaries in the region of the median eminence, infundibular stalk, and infundibular process, finally to enter the hypophyseal or systemic circulations.

Review Questions and Answers

Questions

98. Briefly discuss the relationship that exists (a) between the hypophysis and the adrenal and (b) between the adrenal and the kidney.
99. In what way is the thyroid unique among endocrine glands?
100. Describe the functional relationship between the thyroid and parathyroid glands with regard to blood calcium levels.
101. What is a common capillary bed? How is it important?
102. How is secretion by the pars distalis controlled?

Answers

98. (a) Secretory neurons in the hypothalamic region of the brain synthesize and secrete a small peptide known as corticotropin-releasing hormone. This hormone is synthesized in the perikaryon, transported down the axon, and released in the region of the infundibular stem. Here it passes into a rich capillary plexus stemming from the superior hypophyseal arteries and enters the portal veins of the hypophyseal portal system, which conduct it to the sinusoids of the pars distalis. The corticotropin-releasing hormones pass through the thin, fenestrated endothe-

lium of the sinusoids and enter the intercellular fluids of the pars distalis, where they elicit the secretion of adrenocorticotropic hormone (ACTH) from the corticotropes, a specific population of basophils. Adrenocorticotropic hormone passes from the hypophysis by way of the hypophyseal veins to enter the general systemic circulation. Ultimately, it reaches the zona fasciculata and zona reticularis of the adrenal cortex and stimulates these specific zones to synthesize and release glucocorticoids. (b) The zona glomerulosa of the adrenal cortex is functionally related to the juxtaglomerular apparatus of the kidney. It is thought that either the juxtaglomerular cells sense changes in renal arterial pressure and/or that cells of the macula densa sense differences in sodium ion concentration in the glomerular filtrate. Whatever the case, the juxtaglomerular cells release an enzyme called renin that, acting through the renin-angiotensin system, stimulates the zona glomerulosa to release mineral corticoids, principally aldosterone. Aldosterone has its primary activity on the distal convoluted tubules of the kidney, where it promotes sodium absorption and potassium excretion. Aldosterone also promotes sodium retention by the salivary glands, sweat glands, and intestinal mucosa.

99. The thyroid is unique among endocrine glands in its ability to store secretory products extracellulary in the lumen of a follicle. Individual follicular cells synthesize thyroglobulin, which is then transported to and temporarily stored in the lumen of the follicle in a complex of materials called colloid. When stimulated to secrete their endocrine product (thyroid hormone), the follicular cells sequester colloid from the lumen of the follicle by endocytosis. The resulting colloid droplets, which contain iodinated thyroglobulin, fuse with lysosomes whose hydrolytic enzymes act on the thyroglobulin to release triiodothyronine and tetraiodothyronine (thyroxine) into the cytoplasm. Other iodinated compounds released during this process are broken down intracellularly, and the iodine is reutilized by the follicular cells. Thyroxine and triiodothyronine are released at the base of the cell and enter surrounding capillaries and lymphatics. The thyroid also has a unique ability to concentrate iodine.

100. The parathyroid glands secrete parathyroid hormone in direct response to lowered calcium ion levels in the circulating blood plasma. Parathyroid hormone stimulates osteocytes and osteoclasts to resorb calcium from the bone matrix and also acts on the tubules of the kidney to promote calcium ion absorption from the glomerular filtrate. At the same time it decreases the retention of phosphate, which is lost in the urine. It is through these mechanisms that parathyroid hormone functions to elevate the level of blood calcium.

Parafollicular cells (C cells) of the thyroid gland, on the other hand, secrete a peptide hormone called calcitonin that inhibits bone resorption by osteocytes and osteoclasts, thereby preventing blood calcium levels from rising above optimal levels. Calcitonin counters the action of parathyroid hormone and keeps the circulating levels of blood calcium within normal limits.

101. A common capillary bed forms a vital link between the hypothalamic neurons of the brain and the epithelial cells of the pars distalis. It consists of two capillary plexuses, one in the area of the median eminence and infundibular stalk and one in the pars distalis. These are united to some extent by a series of thin-walled, large-bore veins called the portal veins. Neurohormones released from axons in the region of the infundibular stalk enter the surrounding capillary plexus, and via this vascular network enter the sinusoids that surround the epithelial cells of the pars distalis. In this way neurohormones (releasing and inhibiting hormones) released in the region of the infundibular stem have a direct vascular pathway to target cells in the pars distalis.

102. The cells of the pars distalis are controlled by releasing and inhibitory hormones that are secreted from the axons of the hypothalamic neurons. It is thought that there are chemically distinct releasing and inhibitory hormones for most hormones secreted by the pars distalis.

Appendix: Key Features in the Recognition of Structures with Similar Morphology

Appendix Table 1. Key Features in Identifying Cartilage, Bone, and Tendon

	Arrangement of Cells and Lacunae	Other Features
Hyaline cartilage	Glasslike matrix; lacunae randomly distributed, slitlike in appearance near perichondrium	Isogenous groups; territorial matrix
Elastic cartilage	Matrix more fibrous in appearance; elastic fibers (need to be stained selectively)	Large isogenous groups
Decalcified bone	Lacunae show organization within lamellae of osteons; haversian canals	Tide marks; bone marrow; canaliculi
Ground bone	Haversian systems; interstitial and circumferential lamellae	Canaliculi between lacunae
Fibrocartilage	Dense fibrous connective tissue a dominant feature with a small amount of ground substance; lacunae few in number and scattered between fibers	Round chondrocytes
Tendon	Fibroblast nuclei are densely stained and elongate; lie in parallel rows between fibers of collagen	Lacunae absent

Appendix Table 2. Key Features in the Recognition of Connective Tissue, Muscle, and Nerve

	Characteristics of Nuclei	Shape and Orientation of Cells and Fibers	Additional Features
Loose areolar connective tissue	Fibroblast nuclei stain either intensely or lightly, appear scattered within extracellular matrix	Fibroblasts and extracellular fibers are randomly oriented; extracellular fibers stain intensely.	Scattered fat cells, adipose tissue
Peripheral nerve	Numerous Schwann cell nuclei, stain lightly, may be stippled in appearance	Schwann cells (nuclei) parallel to nerve axons	Undulating or wavy appearance of discrete bundles; limited by perineurium; nodes of Ranvier and axons may be visible
Smooth muscle	Single, oval-shaped, light-staining	Small, spindle-shaped cells; single central nucleus; well-stained cytoplasm	Cells organized into sheets or layers; scant intervening connective tissue
Dense regular connective tissue	Dense-staining; elongate fibroblast nuclei organized in rows	Large, dense-staining fibers arranged in parallel; fibers and fibroblasts orderly arranged in same direction	
Skeletal muscle	Numerous light-staining peripheral nuclei within each cell	Cells arranged in same direction per fascicle; striations of cell visible in longitudinal section	Connective tissue visible around individual cells, organizes several cells into fascicles
Cardiac muscle	Usually a single, large, light-staining central nucleus; occasional cell (fiber) shows 2–3 nuclei	Large branching cells oriented in same direction as others in a specific layer; cross striations as for skeletal muscle	Intercalated discs

Appendix Table 3. Key Features in Identifying Major Glandular Structures

	Secretory Units	Ductal System	Other Features
Parotid	All acini (alveoli) and tubules are serous	Prominent intercalated ducts; intralobular duct system prominent	Numerous fat cells
Pancreas	Acini and tubules are serous; cells pyramidal	Centroacinar cells, prominent intralobular ducts	Islets of Langerhans
Lacrimal gland	All acini and tubules are serous; cells low columnar	Inconspicuous	Occasional fat cells; gland small in size
Submandibular gland	Most acini and tubules are serous; some mucous tubules capped with serous demilunes	Very prominent intralobular duct system, especially striated ducts; numerous profiles in each lobule	Occasional fat cells
Sublingual gland	Mucous and serous acini and tubules equal in abundance; mucous tubules capped by serous demilunes	Duct system not prominent	Occasional fat cells
Thyroid	Large follicles filled with gelatinous colloid	Absent	Nests of parafollicular cells between follicles
Lactating mammary gland	Large, expanded alveoli contain thin secretion	Ducts visible in interlobular connective tissue	Lobulated
Nonlactating mammary gland	Alveoli poorly developed	Intralobular and interlobular ducts visible	Lobulated, loose connective tissue and fat abundant
Parathyroid gland	Irregular cords of serous cells; individual or clumps of oxyphil cells	Absent	Abundant fat cells in gland and surrounding connective tissue
Adenohypophysis	Irregular cords or clumps of epithelial cells; basophils, acidophils, chromophobes	Absent	Abundant capillaries in surrounding connective tissue
Adrenal	Cortex: cells arranged into epithelial cords separated by sinusoids; three zones Medulla: large, light-staining cells	Absent	Cortex: dark-staining Medulla: light-staining
Ovary	Ovarian follicles in cortex	Absent	Surface covering mesothelium; cortex; medulla of connective tissue and blood vessels
Kidney	Parenchyma consists of tubules forming the nephrons; large scattered, renal corpuscles	Large, light-staining collecting ducts in medulla and medullary rays	Cortex: dark-staining tubules Medulla: light-staining tubules running in parallel
Testis	Seminiferous tubules, spermatogenic cells	Rete testis, efferent ductules	Tunica albuginea, septa, interstitial cells

Appendix Table 4. Key Features in Identifying Major Tubular Structures

Organ	Epithelial Lining	Muscle Coats	Special Features
Esophagus	Nonkeratinized stratified squamous	Prominent muscularis mucosae; muscularis externa of inner circular and outer longitudinal layers; skeletal (upper ¼), mixed smooth and skeletal (middle ¼), smooth (lower ½)	Glands present in submucosa; may occur in lamina propria at distal and proximal ends
Small intestine	Simple columnar, striated border; goblet cells	Muscularis mucosae and externa, inner circular, outer longitudinal smooth muscle	Villi; intestinal glands; glands in submucosa of duodenum; Peyer's patches in ileum
Colon	Simple columnar, striated border; more goblet cells	Outer longitudinal layer of muscularis externa thinner except for taeniae coli	No villi, smooth luminal surface, intestinal glands
Appendix	Simple columnar, goblet cells	Muscularis mucosae obscured by lymphocytic infiltration	No villi; much nodular and diffuse lymphatic tissue, small size
Trachea and main bronchi	Ciliated pseudostratified columnar; goblet cells	Thin muscularis mucosae	Cartilage rings
Aorta	Simple squamous (endothelium)	Tunica media dominant layer; smooth muscle and elastic laminae	Lumen usually circular in outline; vasa vasorum
Muscular artery	Endothelium	Prominent muscular tunica media	Internal and external elastic laminae
Vena cava and large veins	Endothelium	Tunica adventitia is main layer; collagen, elastic, and smooth muscle fibers	Lumen often irregular
Vagina	Nonkeratinized stratified squamous	No muscularis mucosae; external coat of inner circular and outer longitudinal smooth muscle	Prominent venous plexus in lamina propria; no glands
Oviduct	Ciliated simple columnar	External muscle: inner circular, outer longitudinal smooth muscle; cellular lamina propria	Ampulla: mucosa elaborately branched, lumen complex
Ureter	Transitional epithelium	Proximal ⅔, inner longitudinal, outer circular smooth muscle Distal ⅓, inner longitudinal, middle circular, outer longitudinal	Isthmus: mucosa less branched, lumen a single space Stellate lumen
Vas deferens	Pseudostratified columnar	Inner longitudinal, middle circular, outer longitudinal smooth muscle	Small, irregular lumen bounded by a thick muscle coat

Appendix Table 5. Key Features in Identifying Major Abrupt Junctions

	Changes in Epithelium	Other Features
Olfactory: respiratory junction of nasal cavity	Olfactory *to* ciliated pseudostratified columnar with goblet cells	Thick lamina propria containing nerves, serous glands, and venous sinuses *to* a thin, vascular lamina propria
Oropharynx: nasopharynx junction at edge of soft palate	Nonkeratinized stratified squamous *to* ciliated pseudostratified columnar with goblet cells	Central core of skeletal muscle, mucous/seromucous glands in connective tissue underlying epithelium
Integument: oral mucosal junction at vermillion border of lip	Typical thin skin with hair follicles, sebaceous glands, and sweat glands *to* a thick, nonkeratinized stratified squamous	Typical dermis of thin skin *to* a thick submucosa with tall connective tissue papillae extending into the oral epithelium; labial salivary glands; skeletal muscle core between the skin and oral mucosa
Esophageal: gastric junction	Thick, nonkeratinized stratified squamous *to* simple columnar	Branched tubular glands of esophagus *to* gastric pits and simple branched tubular cardiac glands of stomach
Gastrointestinal junction	Simple columnar *to* simple columnar with striated (microvillus) border and goblet cells	Gastric pits, pyloric glands *to* villi, intestinal glands, Brunner's glands in submucosa
Rectoanal junction	Simple columnar with striated border and goblet cells *to* nonkeratinized stratified squamous epithelium	Intestinal glands and muscularis mucosae of rectum *lost* at rectoanal junction
Vaginal-cervical junction	Nonkeratinized stratified squamous *to* tall, light-staining simple columnar	Coarse, vascular lamina propria devoid of glands, muscularis of smooth muscle *to* dense connective tissue wall containing cervical glands, nabothian cysts

Board-Style Final Exam

Multiple Choice Questions

Chapter 1. The Cell

Pick the one correct answer.

1. Organelles most notable for producing and degrading hydrogen peroxide are
   A. lysosomes.
   B. mitochondria.
   C. Golgi bodies.
   D. peroxisomes.
   E. genomes.

2. Dense granules of RNA typically occur, among other places, within the
   A. pars granulosa of cytoplasmic nucleoli.
   B. pars granulosa of membrane-bound nucleoli.
   C. pars fibrosa of nuclear nucleoli.
   D. pars granulosa of nuclear nucleoli.
   E. nucleolus-associated chromatin.

3. Euchromatin of a cell is
   A. active in controlling the cell's metabolic processes.
   B. concentrated within the cell's karyosomes.
   C. metabolically active during mitosis.
   D. more heavily stained by nuclear stains than is heterochromatin.
   E. typically complexed chemically with histones.

4. The nucleus is
   A. always found one per cell.
   B. an essential organelle present in all cells.
   C. enclosed within a nucleolar membrane.
   D. enclosed within inner, intermediate, and outer unit membranes.
   E. completely covered externally by the fibrous lamina.

5. Which set of items below includes only those which are either cytoplasmic inclusions or which are found within some kind of cytoplasmic inclusion?

   A. lipofuscin, glycogen, karyosome, lipid droplets
   B. melanin, lipid droplets, endoplasmic reticulum, glycogen
   C. hemosiderin, secretory granules, glycogen, lipofuscin
   D. endoplasmic reticulum, melanin, lipid droplets, Golgi bodies
   E. secretory granules, lipofuscin, melanosome, autosome

6. Which one of the characteristics below is *not* true for messenger RNA (mRNA) in typical living human cells?
   A. contains a message encoded in successive sets of three nucleotides called codons
   B. functionally the same as transfer RNA
   C. attaches to ribosomes in the cytoplasm
   D. formed within the nucleus
   E. formation occurs on a template of DNA

7. Fluid-phase micropinocytosis is a cellular process that
   A. employs pseudopodia.
   B. is directly visible with the aid of a light microscope.
   C. occurs with the formation of small fluid-filled vesicles.
   D. is narrowly selective or specific in the materials it takes in.
   E. characterizes cell breakdown.

8. Granular endoplasmic reticulum is most abundant in cell types that are involved in
   A. transport of calcium ions.
   B. producing steroid hormones.
   C. lipid metabolism.
   D. detoxification of foreign compounds.
   E. protein synthesis.

Chapter 2. Epithelium

Pick the one correct answer.

1. Which of the following statements is true of the basal lamina of mammalian epithelia?
   A. It is the same as an epithelium's basement membrane.
   B. It is a product of the underlying connective tissue.
   C. It normally constitutes a barrier to the diffusion of nutrients and electrolytes to basal epithelial cells.
   D. It is composed of laminin, a proteoglycan rich in heparan sulfate and collagen.
   E. It has a lamina densa directly adjacent to the basal surface of overlying epithelial cells.

2. Which one of the epithelial subtypes listed below typically is also simple squamous?
   A. mesothelium
   B. distended transitional
   C. germinal
   D. keratinized
   E. pseudostratified

3. In epithelia, a junctional complex
   A. is the same as a desmosome.
   B. consists of two adjacent sets of internal laminae.
   C. generally consists of three regions: zonula occludens, zonula adherens, and macula adherens.
   D. usually contains a hemidesmosome.
   E. is composed of the peripheral cytoskeletons of two or more adjacent cells.

4. Microvilli of epithelial cells
   A. are restricted in occurrence to the lining of the digestive system.
   B. are generally visible with light microscopy.
   C. develop into cilia.
   D. increase the absorptive capacity of the cells.
   E. are motile appendages, chiefly of the basal plasma membrane.

5. Myoepithelial cells
   A. are simple squamous epithelial cells that have contractile ability.
   B. consist of actin-containing contractile cells associated with some gland cells.
   C. represent striated muscle cells that lie adjacent to certain epithelial cells.
   D. generally lie outside the basement membrane of the adjacent epithelial cells.
   E. are incompletely differentiated mesenchymal cells.

Chapter 3. General Connective Tissue

Pick the one correct answer.

1. Name the material *not* formed by connective tissue.
   A. adipose tissue
   B. scar tissue
   C. stroma of a gland
   D. parenchyma of a gland
   E. elastic fibers

2. Mature general connective tissue consists *chiefly* of
   A. intercellular material that contains only ground substance.
   B. matrix that contains fibers and ground substance.
   C. connective tissue cells.
   D. tissue fluid (under normal conditions).
   E. mesenchyme.

3. In loose connective tissue, collagen fibers
   A. provide the greatest tensile strength.
   B. are composed of unit fibrils showing repeating transverse bands at 340-nm intervals.
   C. are composed of macromolecules of tropocollagen consisting of three globular proteins.
   D. contain alpha units that have uniform amino acid compositions and sequences in different parts of the body.
   E. can be synthesized only by fibroblasts.

4. Which *one* of the following is true of reticular fibers?
   A. concentration into bundles
   B. secreted by reticulocytes in the red bone marrow
   C. a major fibrous component of scar tissue
   D. well seen with light microscopy after routine histology/pathology staining procedures (e.g., hematoxylin-eosin, Masson's, etc.)
   E. with electron microscopy they are seen to have the same banding pattern as collagen fibers

5. Which *one* of the following is true of elastic fibers?
   A. are more variable in diameter than collagen fibers
   B. their microfibrils are composed of cross-banded lengths of elastin
   C. their elastin contains glycine and hydroxylysine
   D. their elastin contains glycine and proline
   E. their microfibrillar protein differs from elastin itself in containing desmosine and isodesmosine

6. Ground substance of general connective tissue
   A. typically has a large content of free water, the intercellular fluid.
   B. contains glycoproteins, glycosaminoglycans, and proteoglycans.
   C. does not contain collagen or tropocollagen.

D. owes its changeability in viscosity and permeability to its content of glycoproteins.

E. retards the spread of microorganisms, chiefly because of its contained network of long-chain carbohydrates.

Chapter 4. Special Connective Tissue: Cartilage, Bone, and Joints

Each question has one to five correct answers among the choices A through E.

1. Which of the following are characteristic of hyaline cartilage?
   A. matrix that contains ground substance and collagen fibers
   B. capable of interstitial growth
   C. incapable of appositional growth
   D. chondrocytes within lacunae
   E. where exposed to synovial fluid it usually is covered by a perichondrium

2. Elastic cartilages of the human body include those within the
   A. external ear (pinna).
   B. epiglottis.
   C. intervertebral discs.
   D. symphysis pubis.
   E. growing ends of long bones.

3. The periosteum of a diaphysis
   A. can proliferate bone, increasing shaft length.
   B. can proliferate bone, increasing shaft diameter.
   C. can proliferate bone in an epiphyseal ossification center.
   D. can proliferate bone on the outer surface of the bone marrow cavity.
   E. enables strong tendon attachments during bone growth and repair.

4. Osteocytes in compact bone
   A. lie in lacunae.
   B. are interconnected by canaliculi housing their cytoplasmic processes.
   C. have nexus junctions with their neighbors.
   D. show cytoplasmic constituents suggesting active protein synthesis.
   E. are metabolically inert.

5. Adult bone is
   A. rigid and, in its composition and microarchitecture, is fixed for life.
   B. preceded developmentally by a cartilage precursor or model, as in the flat bones of the skull.
   C. without internal blood vessel channels.

D. cancellous within the centers of the epiphyses of long bones.

E. subject to resorption by osteoclasts.

6. Specific "structural units" of compact bone are the
   A. interstitial lamellae.
   B. circumferential lamellae.
   C. osteons.
   D. haversian systems.
   E. Volkmann's canals.

7. Major chemical constituents of bone matrix include
   A. calcium phosphate.
   B. calcium carbonate.
   C. type I collagen.
   D. type III collagen.
   E. minerals in the form of hydroxyapatite crystals.

8. Joints or articulations that involve skeletal elements that have little if any movement in relation to each other include
   A. fibrous joints.
   B. sutures.
   C. synostoses.
   D. gomphoreses.
   E. diarthroses.

9. Synovial fluid within synovial joints is
   A. in direct contact with articular cartilage.
   B. elaborated by the synovial membrane.
   C. an important route for nutrients to the deeper chondrocytes of the neighboring articular cartilage(s).
   D. an ultrafiltrate of blood plasma, plus added mucin.
   E. very viscous, due especially to its content of highly polymerized hyaluronic acid.

Chapter 5. Special Connective Tissue: Blood

Each question has one to five correct answers among the choices A through E.

1. Polychromatophilic erythrocytes
   A. are also called reticulocytes.
   B. have typically lost their nucleus.
   C. have patches of bluish cytoplasm (after usual staining procedures) due to some remaining ribonucleoprotein material.
   D. have areas of acidophilic, hemoglobin-rich cytoplasm.
   E. in peripheral blood can provide an indication of the rate of erythrocyte proliferation.

2. Blood platelets of mammals
   A. are small but are the *most* numerous of the formed elements of blood.
   B. are released by megakaryocytes as cytoplasmic fragments.
   C. have a granulomere with alpha particles containing a wide variety of platelet-specific and platelet-nonspecific proteins.
   D. have a marginal bundle formed by a system of microtubules.
   E. are in many ways the functional equivalents of the thrombocytes of some other classes of vertebrates.

3. As a group, granular leukocytes in humans
   A. have a nonlobate nucleus.
   B. have a multilobate nucleus in some cell subtypes.
   C. have nuclear lobation that usually increases in distinctness and number with cellular development and aging.
   D. include blood monocytes having cytoplasmic granules.
   E. are normally formed in the adult by reticular cells in the spleen.

4. Human neutrophil granulocytes normally
   A. comprise 5 to 15 percent of the circulating leukocytes.
   B. comprise 20 to 45 percent of the circulating leukocytes.
   C. comprise 55 to 70 percent of the circulating leukocytes.
   D. contain a myeloperoxidase that complexes with peroxide to produce activated oxygen, having bactericidal activity.
   E. have specific granules (type B granules) that contain lysozyme, an enzyme complex that can act against components of bacterial cell walls.

5. Human eosinophil granulocytes normally
   A. comprise 10 to 13 percent of the circulating leukocytes.
   B. comprise 20 to 35 percent of the circulating leukocytes.
   C. have a decrease in numbers in relation to the alarm reaction and adrenocortical endocrine secretion.
   D. have large cytoplasmic granules containing a high content of myeloperoxidase.
   E. have large cytoplasmic granules containing a major basic protein that is responsible for their eosinophilia.

Chapter 6. Special Connective Tissue: Hemopoietic Tissue

Each question has one to five correct answers among the choices A through E.

1. Bone marrow in adult humans is a major organ for proliferation of cells maturing into
   A. lymphocytes.
   B. monocytes.
   C. plasma cells.
   D. red blood cells.
   E. granular leukocytes.

2. Adult human red bone marrow
   A. contains a network of reticular fibers.
   B. contains primarily regions of well-vascularized adipose tissue.
   C. is anatomically continuous with endosteum.
   D. lies within the Howship's lacunae.
   E. contains both fixed tissue cells and free cells of blood.

3. Primitive stem cells of adult human red bone marrow
   A. include pluripotent stem cells.
   B. include cells that when transfused can give rise to splenic colony-forming units.
   C. include restricted stem cells, destined to proliferate only one specific cell line.
   D. can be increased numerically by the appropriate administration of particular antimitotic drugs.
   E. include some that resemble lymphoid cells.

4. In human proerythroblasts
   A. hemoglobin synthesis has not begun.
   B. nuclear chromatin masses are large and dense appearing.
   C. mitotic divisions no longer occur.
   D. some cytoplasmic regions are eosinophilic.
   E. the cytoplasm is basophilic and relatively small in volume.

5. The loss of a nucleus from precursor cells of the erythrocytes
   A. gives rise to the normoblasts.
   B. may involve an active expulsion mediated by contractile protein.
   C. is preceded by nuclear pyknosis.
   D. is preceded by nuclear vesiculation.
   E. can be followed by loss of the expelled nuclei through the activity of phagocytes.

6. *The* fundamental stimulus for production of erythrocytes is
   A. erythropoietin.
   B. the kidney.

C. the spleen.

D. hypoxia.

E. hyperoxygenation.

7. *Myelocyte* is a cell category and maturational stage
   A. normally found in circulating blood.
   B. that precedes metamyelocyte in the sequence of stages leading to mature granular leukocytes.
   C. at which the differentiation of specific cytoplasmic granules can first be seen.
   D. at which mitotic activity normally still occurs.
   E. that contributes to erythropoiesis.

8. Lymphocytes from pluripotent stem cells in bone marrow emigrate and then they and/or certain of their daughter cells can be found in
   A. blood.
   B. spleen.
   C. lymph nodes.
   D. Peyer's patches.
   E. tonsils.

9. Cells commonly within or adjacent to lymphatic (lymph) nodules include
   A. reticular cells.
   B. small lymphocytes.
   C. macrophages.
   D. plasma cells.
   E. lymphoblasts.

10. Germinal centers of lymphatic nodules
    A. start in development during the third trimester of fetal life.
    B. are inhibited during later development by antigenic stimuli.
    C. are inhibited in their development in experimental laboratory mammals that are raised and maintained in a germ-free environment.
    D. are centers of lymphocyte production.
    E. are essential for antibody production.

Chapter 7. Muscle

Each question has one to five correct answers among the choices A through E.

1. A skeletal muscle fiber is
   A. a multinucleated cell.
   B. in direct contact with the endomysium.
   C. the smallest independent structural unit of the muscle.
   D. *not* changed in size by degree of use or disuse.
   E. composed of contractile units called sarcomeres.

2. Tissues and tissue components closely associated within a fascicle of skeletal muscle include

A. reticular fibers.

B. collagen fibers.

C. capillaries.

D. motor end-plates.

E. sarcolemmas.

3. Myofibrils of skeletal muscle fibers
   A. do not run the length of the fiber.
   B. are too small for resolution by the light microscope.
   C. have a cross-banding pattern identical to that of the whole fiber.
   D. individually (each one) lie *within* a membrane-encircled sarcotubule.
   E. have A-I band junctions at which a pair of dilated sarcotubules, the terminal cisternae, pass around the sarcoplasmic reticulum.

4. The triad of skeletal muscle is composed of
   A. three sarcomeres.
   B. two sarcosomes and one T-tubule.
   C. two terminal cisternae and one T-tubule.
   D. two T-tubules and one terminal cisterna.
   E. a set of three closely adjacent T-tubules.

5. In smooth muscle cells
   A. only very thin myofibrils have been found.
   B. there is no distinct sarcomere organized.
   C. strong contractions cannot be maintained for long without muscle fatigue.
   D. a structurally organized tubule system of the sarcoplasmic reticulum is generally present.
   E. cell boundaries inhibit the free intercellular passage of electrical impulses.

Chapter 8. Nervous Tissue

Each question has one to five correct answers among the choices A through E.

1. The physiologic activities of a neuron depend in part on the _____ within the perikaryon.
   A. chromatin
   B. melanosomes
   C. lipofuscin granules
   D. neurotubules
   E. Nissl substance

2. Neuronal cytoplasmic processes called dendrites
   A. do *not* contain Nissl substance.
   B. do *not* contain microtubules.
   C. often have spines or gemmules on their surface.
   D. have synapses that may sometimes number hundreds of thousands per cell.
   E. do *not* contain neurofilaments.

3. Axons near their parent neuron cell body generally differ from dendrites in
   A. being more numerous.
   B. conducting impulses away from the perikaryon.
   C. having more ribonucleoprotein bodies.
   D. being more slender and smoother.
   E. being more nearly constant in diameter.

4. Cytoplasmic transport within axoplasm typically can be
   A. antigrade and fast, transferring membranous organelles to the telodendria.
   B. antigrade and slow, transferring small nutrient molecules and calcium to the axon terminals.
   C. retrograde, transferring some proteins and small molecules to the telodendria.
   D. aided by neurotubules.
   E. retrograde, facilitating invasion of the central nervous system by some toxins and viruses.

5. Within the central nervous system, gray matter differs from white in containing per unit volume
   A. more neuronal cell bodies.
   B. fewer small blood vessels.
   C. more axons covered with thick myelin sheaths.
   D. more axons covered with thin myelin sheaths.
   E. fewer interneuronal synaptic contacts.

6. The neuroglia of the human central nervous system
   A. are less abundant than the neurons.
   B. account for less than one-fourth of brain weight.
   C. include Schwann cells.
   D. include cells that form myelin sheaths.
   E. include the microglia.

7. The ependymal cells of the adult human central nervous system
   A. typically form a simple cuboidal to columnar epithelium.
   B. line the spinal cord's central canal and the brain's ventricles.
   C. have long, thin cytoplasmic processes (extensions) basally.
   D. line the subarachnoid space.
   E. where they adjoin the tela choroidea, contribute to choroid plexuses.

8. Gray matter of the human spinal cord
   A. has two dorsal and two ventral horns.
   B. contains perikarya of multipolar sensory neurons, particularly in the ventral horns.
   C. contains dorsal horn motor neurons of very large size.
   D. includes the intermediolateral horns containing preganglionic sympathetic fibers.

E. is subdivided into anterior, lateral, and posterior funiculi.

9. Tracts in the adult human central nervous system
   A. collectively predominate in the contents of the gray matter.
   B. are composed chiefly of dendrites.
   C. contain neuroglia.
   D. contain many myelin sheaths.
   E. contain many perikarya of motor neurons.

10. Following transection of a peripheral nerve
    A. degenerative changes occur only in the peripheral axonal processes.
    B. the axon and its myelin sheath show wallerian degeneration distal to the transection.
    C. within a few hours mitochondria and neurofilaments become vesicular and begin to break up.
    D. Schwann's cells degenerate along with the adjacent myelin layers.
    E. myelin degenerates distally into droplets of simpler lipids that are phagocytized there by microglia.

Chapter 9. The Eye

Each question has one to five correct answers among the choices A through E.

1. The layers of the eye's cornea include
   A. Bowman's membrane, which lies between the inner surface of the substantia propria and the corneal epithelium.
   B. the stratified squamous corneal epithelium.
   C. the pseudostratified columnar corneal epithelium.
   D. the substantia propria or stroma, containing keratocytes.
   E. Descemet's membrane, lying between the outer surface of the substantia propria and the corneal epithelium.

2. The middle and vascular coat of the eyeball
   A. is called the uvea.
   B. includes the choroid.
   C. includes the ciliary body.
   D. includes the sclera.
   E. includes the iris.

3. The ciliary epithelium
   A. covers the inner surface of the ciliary body.
   B. elaborates vitreous humor that flows into the anterior chamber of the eye.
   C. elaborates aqueous humor, which is similar to blood plasma in its relative percentage concentrations of proteins and electrolytes.

    D. lines the ciliary processes from which zonule fibers extend, holding the lens in place.

    E. produces zonule fibers, mainly by its nonpigmented cell portion.

4. The neural retina

    A. is chiefly anchored within the eyeball by a cellular interdigitation with the pigment epithelium.

    B. posteriorly has an exactly central region containing the macula lutea.

    C. posteriorly has a small depression called the fovea centralis, where photoreceptor cells are most tightly clustered and most precisely organized.

    D. about 3 mm lateral to the macula lutea has a "blind spot."

    E. contains an optic disc about 3 mm medially (nasal side) from the macula lutea.

5. The photoreceptor cells of the human retina

    A. are represented by rod cells and cone cells.

    B. include rod cells whose outer segments each have hundreds of flattened membranous sacs or discs containing rhodopsin.

    C. include rod cells that after exposure to light have rhodopsin that changes from cis to trans and breaks down.

    D. include rod cells that with exposure to light undergo a hyperpolarization of the plasmalemma.

    E. include cone cells that, as a group, respond to light of relatively low intensity through effects on a single kind of visual pigment.

Chapter 10. The Ear

Each question has one to five correct answers among the choices A through E.

1. The tympanic cavity of the middle ear

    A. contains ossicles that convey vibrations from the tympanic membrane to the inner ear via the round window.

    B. contains ossicles that consist of fibrocartilages suspended by connective tissue ligaments.

    C. is normally filled with a serous fluid that circulates through middle and inner ear chambers.

    D. lacks muscles affecting movement of the ossicles.

    E. is connected to the nasopharynx via the eustachian tube.

2. The membranous labyrinth of the inner ear

    A. contains perilymph.

    B. contains endolymph that, like extracellular fluid, has contents of sodium and potassium ions that are low and high, respectively.

    C. contains endolymph produced by the stria vascularis of the cochlear duct.

    D. has a portion known as the vestibular labyrinth, important in containing the cochlea and its sensory apparatus for hearing.

    E. is tightly bound to the petrous part of the temporal bone.

3. In the inner ear, the sensory neuroepithelial region of a semicircular canal

    A. is a crista ampullaris.

    B. is a macula of the utricle.

    C. contains type I and type II sensory hair cells.

    D. contains a gelatinous cupula supporting crystalline otoliths.

    E. contains a gelatinous cupula in which the microvilli of the hair cells are embedded.

4. The modiolus of the adult human cochlea

    A. contains afferent multipolar neurons of the spiral ganglion.

    B. is composed of spongy bone.

    C. has a spiral lamina extending from it into the cochlear duct.

    D. has a spiral ligament extending from it to the basilar membrane.

    E. has a spiral lamina extending from it to the stria vascularis.

5. Cells of the organ of Corti

    A. transform vibrations of the basilar membrane into nerve impulses.

    B. are isolated from the vibrations transmitted along small blood vessels.

    C. lie on the vestibular membrane.

    D. lie within the scala vestibuli.

    E. include the inner and outer phalangeal cells that support hair cells.

Chapter 11. Cardiovascular and Lymph Vascular Systems

Each question has one to five correct answers among the choices A through E.

1. Muscle tissues of the heart wall

    A. are composed entirely of cardiac muscle.

    B. include large amounts of smooth muscle in the atrial endocardium.

    C. include part of the conduction system of the heart.

    D. are mainly concentrated in the myocardium.

    E. include the pectinate muscles spiraling in the walls of the ventricles.

2. Tissue structures of the epicardium include
   A. mesothelium.
   B. fat.
   C. fibroelastic connective tissue.
   D. visceral pericardium.
   E. parietal pericardium.
3. The so-called skeleton of the heart in adult humans
   A. is composed of a mixture of hyaline, elastic, and fibrous cartilages.
   B. is formed in part by the annuli fibrosi, composed of Purkinje fibers.
   C. includes the septum membranaceum.
   D. is composed chiefly of dense irregular connective tissue.
   E. is structurally supported by the atrioventricular bundle (of His).
4. Within a normal adult human heart, impulses for cardiac muscle contraction
   A. are transmitted directly from one cardiac muscle fiber to another.
   B. typically pass in this sequence: sinoatrial node to atrial cardiac muscle to atrioventricular node to atrioventricular bundle.
   C. typically pass in this sequence: Purkinje cells to AV node to atrial cardiac muscle fibers to sinoatrial node to atrioventricular bundle.
   D. depend on innervations from the autonomic nervous system.
   E. involve a "pacemaker" in the epicardium next to the right atrium.
5. The innermost tissue layer of blood vessels is
   A. the lamina propria.
   B. the tunica intima.
   C. the tunica adventitia.
   D. continuous with the endocardium of the heart.
   E. lined with endothelium.

Chapter 12. Lymphatic Organs

Each question has one to five correct answers among the choices A through E.
1. Human tonsils that
   A. are single and medial in position are the palatine and lingual.
   B. have the deepest crypts are the pharyngeal and lingual.
   C. filter lymph have the most extensively branching crypts.
   D. are functionally active proliferate lymphocytes.
   E. are functionally active have afferent lymph vessels.

2. Mammalian lymphatic organs with distinctive tissue capsules include
   A. nodules with independent fibroelastic capsules.
   B. the spleen, having a capsule containing much smooth muscle, depending on the animal group.
   C. the thymus, located in the dorsal mesogastrium.
   D. Peyer's patches within the intestinal lamina propria.
   E. the appendix.
3. Human lymph nodes have
   A. their afferent lymphatic vessels collected at the hilus.
   B. their efferent lymphatic vessels collected within the cortical rays.
   C. a cortex containing trabeculae, lymphatic nodules, and diffuse lymphatic tissue .
   D. a medulla containing medullary cords with numerous plasma cells, macrophages, and lymphocytes.
   E. B-lymphocytes that are concentrated typically in the nodular regions.
4. The human spleen serves
   A. to filter the lymph that passes through it from afferent to efferent lymph vessels.
   B. to filter blood that passes through it.
   C. to release lymphocytes, especially those produced in the red pulp.
   D. to release lymphocytes, especially those produced in the white pulp.
   E. to remove from the blood aged and defective erythrocytes whose remains are phagocytized.
5. The human thymus
   A. after puberty typically shows gradual involution, accompanied by replacement of lymphoid tissue by connective and adipose tissues.
   B. contains thymic corpuscles consisting of concentric layers of epithelial reticular cells.
   C. contains widespread macrophages that can be seen to have engulfed degenerating cells.
   D. attains its greatest relative size by the time of birth.
   E. contains reticular cells only in its medulla.
6. Lymphocytes of the human thymus
   A. appear to be functionally inert until after they are released and then pass through the spleen.
   B. include a great many that are produced there as T-cells and that are themselves and their daughter cells elsewhere involved in cell-mediated immune responses.
   C. after release have a relatively short life span.
   D. after release circulate and recirculate through the other lymphatic organs via lymph and blood flows.

E. are active in mitosis and give rise to daughter cells that will emigrate and engender immunologic responses to certain bacteria, viruses, and fungi.

7. Lymphocytes of the T-cell series
A. do not typically nor directly elaborate conventional antibodies.
B. can cooperate with B-cells to produce antibodies against foreign erythrocytes.
C. can be readily distinguished from those of the B-cell series when studied with routine staining procedures and light microscopy.
D. participate in the mechanism of rejection of tissue grafts of significantly different genetic source(s).
E. give rise to immunologic "memory cells" at widespread systemic sites.

8. Endocrine (hormonal) activity of the thymus is
A. suggested by the characteristic secretory granules of the T-cells.
B. known best in terms of peptide factors or fractions isolated and tested from thymic extracts.
C. known best in terms of steroidal factors or fractions extracted and tested from thymic tissue.
D. believed to include some regulatory and stimulatory actions on the thymus itself.
E. known in part in relation to thymosin and its restoration of T-cell deficiencies in thymectomized mice.

Chapter 13. Skin

Each question has one to five correct answers among the choices A through E.

1. Normal cells within the human epidermis include
A. stratified squamous epithelial cells.
B. keratinocytes.
C. Langerhans' cells.
D. endothelial cells.
E. melanocytes.

2. Human epidermis proliferates epithelial cells in the
A. stratum basale.
B. stratum lucidum.
C. stratum corneum.
D. stratum granulosum.
E. stratum germinativum.

3. Melanocytes in human skin
A. can be found in the dermis.
B. originate from cells derived from the neural crest cells of early development.
C. contain melanosomes.

D. when absent, constitute the usual basis for complete albinism.
E. when unable to synthesize melanin, constitute the usual basis for complete albinism.

4. "Appendages" of human skin include
A. lanugo.
B. sebaceous glands.
C. apocrine glands.
D. nails.
E. eccrine sweat glands.

Chapter 14. Respiratory System

Each question has one to five correct answers among the choices A through E.

1. Pseudostratified columnar epithelium in the human nasal cavity
A. contains, in part, goblet cells that have a serous secretion.
B. contains, in part, goblet cells that have a mucous secretion.
C. contains, in part, ciliated cells.
D. lines the vestibule.
E. includes the olfactory epithelium.

2. Olfactory cells of the olfactory epithelium in humans
A. are bipolar neurons.
B. are unipolar sensory cells.
C. can regenerate.
D. apically have modified nonmotile cilia, the olfactory hairs.
E. basally have thin dendritic processes that combine to form the fila olfactoria.

3. Intrapulmonary bronchi
A. are supported by C-shaped cartilages in their walls.
B. are not supported by cartilages in their walls.
C. are lined with simple columnar epithelium through which the ducts of compound mucous glands pass.
D. are lined with pseudostratified ciliated columnar epithelium with goblet cells.
E. have a layer of spiralling smooth muscle external to the submucosa.

4. Terminal bronchioles
A. are lined with ciliated cuboidal cells mixed with goblet cells.
B. are lined with a mixture of ciliated and nonciliated cells.
C. do not contain any visible mucus-secreting cells or glands.
D. support some respiratory alveoli from their walls.

E. distally connect directly with alveolar ducts.

5. Interalveolar septa of the human lung
    A. normally contain a rich capillary network.
    B. contain a supporting framework of elastic and collagenous fibers.
    C. lack perforations or pores.
    D. contain, as the thinnest blood-air barrier of the lung, in sequence: capillary endothelial cell, fused basal laminae of the latter and next epithelium, attenuated epithelium of the alveolar lining cell, and a thin film of fluid.
    E. contain, as the thinnest blood-air barrier of the lung, in sequence basal lamina of capillary endothelial cell, interstitial tissue, basal lamina of alveolar lining cell, attenuated epithelial cell of the alveolar wall, and a thin film of fluid.

Chapter 15. Digestive System

Each question has one to five correct answers among the choices A through E.

1. In which two layers below are ganglion cells of the autonomic nervous system most common?
    A. epithelium of the mucosal lining
    B. lamina propria
    C. submucosa
    D. between muscle layers of the muscularis externa
    E. adventitia
2. Unicellular glands of the alimentary tract's lining epithelium and a segment in which they are common:
    A. goblet cells, middle part of the esophagus.
    B. neutral mucus-secreting cells, fundic stomach.
    C. goblet cells, colon.
    D. Paneth cells, fundic stomach.
    E. goblet cells, ileum.
3. Submucosal glands of the alimentary tract and a tract segment in which they are common in humans:
    A. gastric glands, fundic stomach.
    B. Brunner's glands, duodenum.
    C. Brunner's glands, ileum.
    D. intestinal glands, colon.
    E. goblet cell glands, appendix.
4. The mucosa of the fundic stomach contains
    A. Peyer's patches.
    B. parietal cells.
    C. mucous neck cells.
    D. chief cells.
    E. gastric pits.
5. Villi of the alimentary tract occur in the
    A. large intestine.

B. appendix.
    C. stomach.
    D. jejunum.
    E. ileum.
6. The liver's blood vascular system includes
    A. branches of the portal vein from the kidney.
    B. branches of the portal vein from the gastrointestinal tract.
    C. hepatic sinusoids that receive blood from branches both of the portal vein and the hepatic artery.
    D. hepatic sinusoids that transmit blood to the branches of the hepatic vein.
    E. hepatic sinusoids that transmit blood to the central veins.
7. Hepatocytes
    A. occur in plates within liver lobules.
    B. commonly include some that are polyploid, as can be recognized from variation in nuclear size.
    C. have microvilli bordering bile canaliculi.
    D. have microvilli bordering bile ducts.
    E. produce and secret bile.
8. The exocrine pancreas is
    A. a compound tubuloacinar gland containing holocrine secretory cells.
    B. composed of secretory cells of several types and duct cells that transmit their secretions to the duodenum.
    C. dominated by zymogenic secretory cells with abundant granular endoplasmic reticulum.
    D. active in secreting amylase, which breaks down starch and glycogen to disaccharides.
    E. active in secreting several kinds of proteolytic enzymes.
9. In the development and growth of human teeth, formative cells and a product from each:
    A. mesenchyme cells, odontoblasts.
    B. ameloblasts, dentin.
    C. odontoblasts, dentinal fibers.
    D. mesenchyme cells, cementoblasts.
    E. ameloblasts, enamel.
10. Which are the correctly paired cell products and their cellular sources in the endocrine pancreas?
    A. secretin, S-cells
    B. cholecystokinin, I-cells
    C. glucagon, alpha cells
    D. insulin, delta cells
    E. pancreatic polypeptide, beta cells

Chapter 16. Urinary System

Each question has one to five correct answers among the choices A through E.

1. In the following pairs select those which show a direct proximal to distal sequence.
   A. urethra, urinary bladder
   B. renal sinus, minor calyx
   C. glomerular capsule, proximal convoluted tubule
   D. loop of Henle, medullary collecting duct
   E. papillary ducts, minor calyces

2. Contents of the juxtaglomerular apparatus include the
   A. macula densa.
   B. vascular pole of the glomerular capsule.
   C. efferent arteriole of the glomerulus.
   D. renin-producing cells.
   E. angiotensinogen-producing cells.

3. A simple cuboidal epithelium typically lines the lumen of
   A. a minor calyx.
   B. the thin part of Henle's loop.
   C. the proximal convoluted tubule.
   D. the urethra.
   E. the arched collecting duct.

4. Factors and one of their direct and cellular targets that contribute to the production of a hypertonic urine:
   A. antidiuretic hormone, some collecting ducts.
   B. renin, glomerular capsule cells.
   C. increased cellular permeability, thin part of Henle's loop.
   D. vascular countercurrent exchange system, medullary vasa recta.
   E. Vascular countercurrent exchange system, cortical vasa recta.

5. Surface specializations and the epithelial cells in which they occur in the urinary system include
   A. microplicae and large numbers of apical cytoplasmic vesicles, principal (light) cells of some collecting ducts.
   B. apically a central cilium surrounded by short microvilli, intercalated cells of some collecting ducts.
   C. apically a brush border consisting of closely packed cilia, cells of the proximal convoluted tubules.
   D. apically a brush border consisting of closely packed microvilli, cells of the proximal convoluted tubules.
   E. pores (fenestrations), glomerular capillaries.

Chapter 17. Meiosis and Male Reproductive Organs

Each question has one to five correct answers among the choices A through E.

1. Meiosis
   A. is characterized by two cell divisions, but only one includes replication of DNA.
   B. in its first cell division produces haploid cells.
   C. in its second cell division produces sister chromatids.
   D. in its first prophase is similar to that of mitosis.
   E. in its first prophase has tetrads formed, each representing homologous chromosomes.

2. Each testis
   A. functions in part as a holocrine gland.
   B. functions in part as an endocrine gland.
   C. is covered anteriorly and laterally with mesothelium.
   D. has a thick fibrous capsule that is part of the visceral layer of the tunica vaginalis.
   E. contains about 250 elongate compartments called lobuli testis.

3. Interstitial tissue of the testis contains
   A. fibroblasts.
   B. macrophages.
   C. testosterone synthesizing and secreting Leydig cells.
   D. mast cells.
   E. mesenchyme cells.

4. Synthesis of testosterone in certain testicular cells depends on
   A. enzymes of the smooth endoplasmic reticulum.
   B. mitochondrial enzymes.
   C. cytoplasmic crystals of Reinke.
   D. a gonadotropic hormone, ICSH, from the pars distalis of the anterior pituitary gland.
   E. sufficient growth of the beard.

5. In the seminiferous tubules there are
   A. contractile myoid cells.
   B. Sertoli cells that proliferate precursors of spermatozoa.
   C. spermatogonia that have a supportive function primarily.
   D. cells secreting testicular fluid.
   E. cells releasing androgen-binding protein.

6. The blood-testis barrier separates
   A. the adluminal compartment from the spermatogonia.

B. the adluminal compartment from the basal compartment.

C. the adluminal compartment from haploid cells differentiating into sperm cells.

D. the basal compartment of the seminiferous tubules from blood vessels of the interstitial tissue.

E. the basal compartment of the seminiferous tubules from small vascular channels within the peritubular membrane.

7. Lining epithelium of the tubuli recti is
A. stratified columnar.
B. simple squamous.
C. in direct contact with that of the efferent ductules.
D. in direct contact with that of the rete testis.
E. a pavement of undifferentiated primary spermatocytes.

8. The ductus deferens is
A. lined with stratified columnar epithelium.
B. lined with pseudostratified columnar epithelium.
C. functionally active in muscular propulsion of sperm during ejaculation.
D. connected directly to the prostatic urethra by its ampulla, which has a thicker muscularis.
E. swollen proximally, forming the colliculus seminalis.

9. Among the human male's accessory sex glands,
A. the seminal vesicles' secretory cells are active in protein synthesis.
B. the seminal vesicles are significant in storing the semen prior to ejaculation.
C. the seminal vesicles' secretion provides a source of energy for sperm.
D. the prostate contains 30 to 50 compound tubuloalveolar glands, whose numerous main ducts drain independently into the membranous urethra.
E. the prostate gland's lumina commonly contain concretions, especially around the time of puberty.

10. Mesenchymal tissue in which the testis cords develop later forms the
A. septuli testis.
B. tunica albuginea.
C. connective tissue of the mediastinum.
D. interstitial tissue.
E. cells of Leydig.

Chapter 18. Female Reproductive System

Each question has one to five correct answers among the choices A through E.

1. Human ovarian follicles typically lie
A. within capsules, each one of which is formed and surrounded by the tunica albuginea.
B. within the medullary stroma.
C. within the cortical stroma.
D. within a mesothelial epithelium that surrounds each follicle individually.
E. in grapelike clusters called uveal tracts.

2. Of the types or stages of human ovarial follicles,
A. primordial follicles reach their maximum number at a woman's age of 20 to 30 years.
B. primordial follicles reach their maximum number around the time of birth.
C. primordial follicles have a primary oocyte surrounded by a two- to three-celled layer of cuboidal follicle cells.
D. primary follicles contain a primary oocyte surrounded by simple squamous epithelium of granulosa cells.
E. secondary follicles have reached maximum follicle size.

3. Ovulation in the human female
A. in its timing centers around the middle of the menstrual cycle.
B. most commonly involves the release from the follicle of a primary oocyte.
C. most commonly involves follicular release of a secondary oocyte.
D. is immediately preceded by a marked increase in liquor folliculi.
E. involves a follicular rupture facilitated by collagenase produced by certain of the granulosa cells.

4. Ovarian follicular atresia occurs
A. chiefly at menopause.
B. chiefly with the formation of corpora lutea.
C. as the fate of a minority of the follicles growing during each menstrual cycle.
D. generally only in early stages of follicular development.
E. with a relatively thick glassy membrane forming from the basal lamina of the granulosa cell epithelium.

5. The distal to proximal sequence of named segments of the oviduct is
A. fimbria, ampulla, isthmus, infundibulum.
B. infundibulum, ampulla, isthmus, pars interstitialis.
C. isthmus, ampulla, infundibulum, uterine tube.
D. fimbria, infundibulum, isthmus, ampulla, pars interstitialis.
E. ampulla, infundibulum, isthmus, pars interstitialis.

6. In comparison with the body of the uterus, the adult cervix has
    A. a wall composed chiefly of dense connective tissue.
    B. a junctional zone between smooth and skeletal muscle layers in its muscularis.
    C. a copious secretion of thin alkaline fluid, especially during menses.
    D. a nonkeratinized stratified squamous epithelium throughout its length.
    E. a cyclic shedding of only the proximal part and superficial layer of its mucosa.

7. The vaginal wall of mature women typically has
    A. a keratinized stratified squamous epithelium.
    B. a mucosa containing plicae circulares.
    C. short branched tubular mucous glands in its lamina propria.
    D. a serosal outer covering over about one-half its length anteriorly.
    E. a muscularis of smooth muscle formed by inner circular and outer longitudinal layers.

8. Tissue components of the human clitoris include
    A. a corpus spongiosum of erectile tissue.
    B. a glans clitoridis containing spongy erectile tissue.
    C. a single corpus cavernosum containing dense irregular connective tissue.
    D. a covering of pseudostratified columnar epithelium.
    E. an abundance of high connective tissue papillae containing sensory nerve endings.

9. Fertilization in the human female
    A. takes place in the isthmus of the oviduct.
    B. takes place in the ampulla of the uterine tube.
    C. involves a single spermatozoon penetrating the ovum.
    D. involves the oocyte just prior to the second maturation division.
    E. is rapidly followed by reduction in permeability of the penetrated ovum's zona pellucida, which thereafter blocks entry by contending sperm.

10. A human blastocyst
    A. forms from a morula 5 to 10 days after fertilization.
    B. has an inner cell mass that will form the embryo proper.
    C. has a capsule-like wall, the trophoblast, consisting of a layer one cell thick.
    D. after being freed in the uterine cavity, will implant in the endometrium during the proliferative phase of the endometrial cycle.

E. during implantation extends fingerlike processes of its trophoblast between uterine epithelial cells; numerous tight junctions then form, locking them together.

Chapter 19. Endocrine Glands

Each question has one to five correct answers among the choices A through E.

1. Human parathyroid glands
    A. contain chief cells that constitute the most numerous of its endocrine cells.
    B. have oxyphil endocrine cells, notable after puberty, and then increasing with further aging.
    C. secrete parathyroid hormone, whose action on calcium metabolism is nearly opposite to that of calcitonin.
    D. secrete parathyroid hormone, which raises blood levels of calcium.
    E. secrete parathyroid hormone, which promotes synthesis and release of 1,25-vitamin D_3 by the kidney.

2. Thyroid follicles
    A. most commonly have a simple cuboidal to simple squamous epithelium formed of principal (follicular) cells.
    B. generally have follicular cells that are lower (squamous) in height when they are most active in synthesis and transport of thyroid hormone.
    C. surround a lumen containing colloid, representing primarily stored lipid materials.
    D. store halogen salts in their colloidal droplets.
    E. contain parafollicular cells that lie outside the principal cells and that secrete iodine salts.

3. Endocrine cells of the adrenal
    A. cortex are developmentally derived from neural crest cells.
    B. medulla secrete hormones that are catecholamines.
    C. cortex secrete hormones that are steroids.
    D. cortex are arranged in irregular parenchymal cords, between which chromaffin cells are common.
    E. medulla can regenerate after trauma or surgical removal.

4. The pituitary adenohypophysis
    A. embryologically develops from the infundibulum of the diencephalon.
    B. contains chromophilic and chromophobic endocrine cells.

C. contains somatotropes and mammotropes, both of which have cytoplasmic secretory granules that are acidophilic.
D. includes the pars intermedia and the pars tuberalis.
E. contains some basophilic endocrine cells; two of these that have basophilic cytoplasmic secretory granules are thyrotropes and gonadotropes.

5. The hypothalamohypophyseal tract
A. contains axons from the neurosecretory neuroendocrine cells of the supraoptic and paraventricular nuclei of the hypothalamus.
B. transports oxytocin and antidiuretic hormone to the neurohypophysis, where they are stored and released.
C. contains neuroglia.
D. contains abundant neuronal perikarya.
E. descends the pituitary stalk.

—————————————— *Answer Key* ——————————————

Chapter 1. The Cell

1. D
2. D
3. A
4. B
5. C
6. B
7. C
8. E

Chapter 2. Epithelium

1. D
2. A
3. C
4. D
5. B

Chapter 3. General Connective Tissue

1. D
2. B
3. A
4. E
5. D
6. B

Chapter 4. Special Connective Tissue: Cartilage, Bone, and Joints

1. A, B, D, E
2. A, B
3. B, E

4. A, B, C
5. D, E
6. C, D
7. A, B, C, E
8. A, B, C, D
9. A, B, C, D, E

Chapter 5. Special Connective Tissue: Blood

1. A, B, C, D, E
2. B, C, D, E
3. B, C
4. C, D, E
5. C, D, E

Chapter 6. Special Connective Tissue: Hemopoietic Tissue

1. A, B, C, D, E
2. A, C, E
3. A, B, C, D, E
4. D, E
5. B, C, E
6. D
7. B, C, D
8. A, B, C, D, E
9. A, B, C, D, E
10. C, D

Chapter 7. Muscle

1. A, B, C, E
2. A, B, C, D, E
3. C

4. C
5. B

Chapter 8. Nervous Tissue

1. A, D, E
2. C, D
3. B, D, E
4. A, D, E
5. A
6. D, E
7. A, B, C, E
8. A, D
9. C, D
10. B, C

Chapter 9. The Eye

1. B, D
2. A, B, C, E
3. A, D, E
4. B, C, E
5. A, B, C, D

Chapter 10. The Ear

1. E
2. C
3. A, C, E
4. B
5. A, B, E

Chapter 11. Cardiovascular and Lymph Vascular Systems

1. C, D
2. A, B, C, D
3. C, D
4. A, B, E
5. B, D, E

Chapter 12. Lymphatic Organs

1. D
2. B
3. C, D, E
4. B, D, E
5. A, B, C, D

6. A, B, D, E
7. A, B, D, E
8. B, D, E

Chapter 13. Skin

1. A, B, C, E
2. A, E
3. A, B, C, E
4. A, B, C, D, E

Chapter 14. Respiratory System

1. B, C, E
2. A, C, D
3. D
4. B, C
5. A, D

Chapter 15. Digestive System

1. C, D
2. B, C, E
3. B
4. B, C, D, E
5. D, E
6. B, C, D, E
7. A, B, C, E
8. C, D, E
9. A, C, D, E
10. C

Chapter 16. Urinary System

1. C, E
2. A, D
3. C, E
4. A, D
5. D, E

Chapter 17. Meiosis and Male Reproductive Organs

1. A, C, E
2. A, B, C, E
3. A, B, C, D, E
4. A, B, D
5. A, D, E

6. A, B
7. D
8. B, C
9. A, C
10. A, B, C, D, E

Chapter 18. Female Reproductive System

1. C
2. B
3. A, C, D, E
4. E
5. B

6. A
7. E
8. A, B, E
9. B, C, D, E
10. B, C, E

Chapter 19. Endocrine Glands

1. A, B, C, D, E
2. A
3. B, C
4. B, C, D, E
5. A, B, C, E

Index

Page numbers followed by the letter *f* refer to figures; page numbers followed by the letter *t* refer to tables.